AAPG Treatise of Petroleum Geology

The American Association of Petroleum Geologists
gratefully acknowledges and appreciates
the leadership and support of the
AAPG Foundation in the development of the
Treatise of Petroleum Geology

THE BUSINESS OF PETROLEUM EXPLORATION

EDITED BY
RICHARD
STEINMETZ

TREATISE OF PETROLEUM GEOLOGY
HANDBOOK OF PETROLEUM GEOLOGY

TREATISE EDITORS:
NORMAN H. FOSTER
AND
EDWARD A. BEAUMONT

Published by
The American Association of Petroleum Geologists
Tulsa, Oklahoma, U.S.A. 74101

Published August 1992

ISBN: 0-89181-601-1

Association Editor: Susan A. Longacre
Science Director: Gary D. Howell
Publications Manager: Cathleen P. Williams
Special Projects Editor: Anne H. Thomas
Production: Kathy A. and Dana M. Walker, Editorial Technologies, Renton, WA

CONTENTS

Preface to the *Handbook of Petroleum Geology*.......xi
—Norman H. Foster and Edward A. Beaumont

Acknowledgments.......xii

About the Editor.......xiii

PART I. OVERVIEW OF THE BUSINESS.......1

Chapter 1. Introduction.......3
—Richard Steinmetz

Chapter 2. Some Basic Business Concepts.......7
—Donald R. Jones

Chapter 3. History of Exploration Expenditures in the United States: 1956–1989.......13
—Robert E. Megill

Chapter 4. Exploration: A Misunderstood Business.......21
—John Lohrenz

PART II. NATURE OF THE BUSINESS.......27

Chapter 5. Dealing with Exploration Uncertainties.......29
—Ed Capen

Chapter 6. Estimating Prospect Sizes.......63
—Robert E. Megill

Chapter 7. Chance of Success and Its Use in Petroleum Exploration.......71
—Peter R. Rose

Chapter 8. Selecting and Assessing Plays.......87
—David A. White

Chapter 9. Risk Behavior in Petroleum Exploration.......95
—Peter R. Rose

PART III. ECONOMIC ASPECTS OF THE BUSINESS.......105

Chapter 10. Exploration Economics.......107
—Robert E. Megill

Chapter 11. Development Economics 117
—*Field Roebuck*

Chapter 12. Oil and Gas Property Evaluation 125
—*Gene B. Wiggins III*

Chapter 13. PC Software Considerations for Economic Evaluation of Investments in the Petroleum Industry 163
—*John M. Stermole*

Chapter 14. Databases: Basic Considerations Using Well History Data 175
—*Robert W. Meader*

PART IV. MANAGING THE BUSINESS 185

Chapter 15. Managing an Exploration Program for Success 187
—*Bill St. John*

Chapter 16. It All Begins with People 201
—*Marlan W. Downey*

Chapter 17. Building Creative Environments and Leading Creative People 205
—*Kenneth F. Wantland*

Chapter 18. Trend Analysis 215
—*Michael D. Smith and Donald R. Jones*

Chapter 19. Annual Budget: A Short-Term Action Plan for Exploration 237
—*Donald R. Jones*

Chapter 20. Funding Oil and Gas Ventures 257
—*Lee T. Billingsley*

Chapter 21. Putting a Deal Together 267
—*Albert H. Wadsworth, Jr.*

Chapter 22. Enhancement Techniques for Prospect Presentations 271
—*Robert F. Ehinger*

PART V. LEGAL, POLITICAL, ETHICAL, AND ENVIRONMENTAL ASPECTS OF THE BUSINESS 277

Chapter 23. Basic Oil and Gas Law in the United States 279
—*Samuel B. Katz*

Chapter 24. Types of International Petroleum Contracts: Their History and Development 297
—*Samuel B. Katz*

Chapter 25. Analysis of International Oil and Gas Contracts....................307
—Sidney Moran

Chapter 26. International Exploration by Independents............................319
—Robert G. Bertagne

Chapter 27. Ethics in the Business of Petroleum Exploration...................331
—Robert W. Spoelhof

Chapter 28. The Environmental Realities of Petroleum Exploration 345
—G. Rogge Marsh

Glossary 357
—James A. McCaleb

Index...379

TREATISE OF PETROLEUM GEOLOGY
ADVISORY BOARD

AMERICAN ASSOCIATION OF PETROLEUM GEOLOGISTS FOUNDATION
TREATISE OF PETROLEUM GEOLOGY FUND*

Major Corporate Contributors
($25,000 or more)

Amoco Production Company
BP Exploration Company Limited
Chevron Corporation
Exxon Company, U.S.A.
Mobil Oil Corporation
Oryx Energy Company
Pennzoil Exploration and Production Company
Shell Oil Company
Texaco Foundation
Union Pacific Foundation
Unocal Corporation

Other Corporate Contributors
($5,000 to $25,000)

ARCO Oil & Gas Company
Ashland Oil, Inc.
Cabot Oil & Gas Coporation
Canadian Hunter Exploration Ltd.
Conoco Inc.
Marathon Oil Company
The McGee Foundation, Inc.
Phillips Petroleum Company
Transco Energy Company
Union Texas Petroleum Corporation

Major Individual Contributors
($1,000 or more)

John J. Amoruso
Thornton E. Anderson
C. Hayden Atchison
Richard A. Baile
Richard R. Bloomer
A. S. Bonner, Jr.
David G. Campbell
Herbert G. Davis
George A. Donnelly, Jr.
Paul H. Dudley, Jr.
Lewis G. Fearing
Lawrence W. Funkhouser
James A. Gibbs
George R. Gibson
William E. Gipson
Mrs. Vito A. (Mary Jane) Gotautas
Robert D. Gunn
Merrill W. Haas
Cecil V. Hagen
Frank W. Harrison
William A. Heck
Roy M. Huffington
J. R. Jackson, Jr.
Harrison C. Jamison
Thomas N. Jordan, Jr.
Hugh M. Looney
Jack P. Martin
John W. Mason
George B. McBride
Dean A. McGee
John R. McMillan
Lee Wayne Moore
Grover E. Murray
Rudolf B. Siegert
Robert M. Sneider
Estate of Mrs. John (Elizabeth) Teagle
Jack C. Threet
Charles Weiner
Harry Westmoreland
James E. Wilson, Jr.
P. W. J. Wood

The Foundation also gratefully acknowledges the many who have supported this endeavor with additional contributions.

*Based on contributions received as of June 30, 1992.

PREFACE TO THE *HANDBOOK OF PETROLEUM GEOLOGY*

The Business of Petroleum Exploration is one of four books of the *Handbook of Petroleum Geology*, which is part of the *Treatise of Petroleum Geology*. The Treatise comprises three different publication sets: the *Reprint Series*, the *Atlas of Oil and Gas Fields*, and the *Handbook of Petroleum Geology*. The Treatise is AAPG's Diamond Jubilee project, commemorating AAPG's 75th Anniversary in 1991.

With input from an advisory board of more than 250 geologists and geophysicists from around the world, we designed this entire effort so that the set of publications will represent the cutting edge in petroleum exploration knowledge and application. The Reprint Series provides useful literature from various geological, geophysical, and engineering publications. In some cases, reprinted articles are from obscure sources. The Atlas is a collection of detailed field studies that illustrate the myriad ways oil and gas are trapped. It is also a guide to the petroleum geology of the basins where these fields are found. Field studies like those published in the Atlas are the documentation of petroleum geology. They form the basis for all of our assumptions regarding petroleum geology. From the standpoint of the explorationist who is building and selling prospects, details from field studies can be stored as memories and used to build stronger prospects and convince doubters of the validity of any unique features of a prospect—nothing is more convincing than a close analogy. The third part of the Treatise, the *Handbook of Petroleum Geology*, is a professional explorationist's guide to the methodology and technology used to find fields similar to those described in the Atlas.

The Handbook set is divided into four volumes, each of which addresses one of the four steps of oil and gas prospecting—evaluation of source rocks and migration, evaluation of reservoir quality and properties, evaluation of trapping conditions, and evaluation of economic opportunity. Accordingly, the four volumes of the Handbook are *Source and Migration Processes and Evaluation Techniques*; *Reservoirs*; *Traps*; and *The Business of Petroleum Exploration*. These publications should be kept close at hand so that when a question arises, an answer can be found quickly and easily.

The *Business of Petroleum Exploration* contains discussions of most of the economic considerations made when exploring for, producing, or purchasing petroleum reserves. This is the first book published by AAPG that encompasses so many aspects of the business of finding and producing petroleum. This book is unique because it goes beyond the usual geological aspects to cover most of the business considerations made in petroleum exploration. Frequently, geologists have not been involved in making business decisions, yet they need to be part of the decision-making process if exploration is to remain a cornerstone of the petroleum industry. One of the aims of this book is to provide readers with a foundation of practical business knowledge.

We thank Richard Steinmetz for planning, compiling, and editing this significant volume. When one donates the time and energy to compile a book such as this one, he does it because he wants to give something back to his profession. The design of this book is the product of many years of the editor's hard-won experience. If you ever have the opportunity to meet Richard Steinmetz, please thank him. We also want to thank all of the authors who so enthusiastically contributed to *The Business of Petroleum Exploration*.

Edward A. Beaumont and Norman H. Foster
Editors of the *Treatise of Petroleum Geology*

Acknowledgments *AAPG members and all readers of this Handbook owe a hearty thanks to the contributing authors. They spent a great deal of time on their chapters and have much to offer. I particularly thank Ted Beaumont, Norm Foster, Jim McCaleb, Bob Megill, Field Roebuck, Pete Rose, and Don Harris for their ideas and assistance. I thank those at AAPG headquarters who worked to make this book a reality—especially Anne Thomas, Cathleen Williams, and Linda Farrar. I would like to thank Kathy Walker of Editorial Technologies for her copyediting, production, and layout of the book. Finally, I give heartfelt thanks to my wife, Janet, who was patiently supportive during the four years I worked on this Handbook.*

—Richard Steinmetz

About the Editor

Richard Steinmetz was born in Cuxhaven, Germany, in 1932. He received his A.B. degree (with honors) in geology from Princeton University, his M.S. in mineralogy from Pennsylvania State University, and his Ph.D. in geology from Northwestern University in Evanston, Illinois. His early work in geology included several summers of field work in Wyoming, petrography of uranium-bearing sandstones, design of sampling programs, and statistical analysis of cross-bedding.

From 1956 to 1958, Richard was employed as a subsurface and wellsite geologist for Plymouth Oil Company in Midland, Texas. From 1961 to 1967, he worked for Pan American Petroleum, predecessor to Amoco Production Company. He spent the first three years in the Fort Worth Division office doing stratigraphic and reservoir studies of the Woodbine Group in the East Texas basin and the remaining three years at the Tulsa Research Center working on sandstone petrology and sedimentation. From 1967 to 1971, Dr. Steinmetz was Assistant Professor of Geology at Texas Christian University, where he taught sedimentation, sedimentary petrology, stratigraphy, geostatistics, physical and historical geology, and oceanography.

In 1971, he rejoined Amoco Production Company at their Tulsa Research Center. While there, from 1971 to 1977, he worked on research projects in the Uinta basin in Utah, on the Athabasca tar sands of Canada, and on rock-fluid properties in very deep basins. He also taught numerous company sedimentation and technical seminars domestically and overseas.

From 1977 to 1980, Richard was Project Geologist in Amoco's Denver office, overseeing exploration projects in the Williston basin, northern Wyoming, and Montana. From 1980 to 1984, he was Regional Consulting Geologist and then Regional Geologist in Amoco's New Orleans office, which was responsible for exploration in the offshore Gulf of Mexico, southeastern United States, and offshore east coast. These years included the peak of the boom in the United States, and Denver and New Orleans were two of the hot spots.

From 1984 to 1987, Richard was Assistant Manager of Geology and then worked in exploration planning at Amoco's headquarters in Chicago. From 1987 to the present, he has been a Senior Consulting Geologist in Amoco's Houston office, working on competitor analysis in Worldwide New Ventures and most recently in the Exploration Analysis group.

Dr. Steinmetz is a member of AAPG, and he served as Elected Editor and Executive Committee member from 1983 to 1985. He is also a member of SPE and AAAS and is a Certified Petroleum Geologist. He is a member of the Houston and New Orleans Geological Societies and the Rocky Mountain Association of Geologists.

PART I

OVERVIEW OF THE BUSINESS

The four chapters in this part serve as a starting point and background for the subject of this Handbook, *The Business of Petroleum Exploration*. They include an introduction, a presentation of the basic concepts, the history of exploration expenditures in the United States, and an explanation of why exploration is a misunderstood business. The introductory chapter covers the purpose, background, and utility of this Handbook. Chapter 2 on basic concepts focuses on the business side of exploration, setting forth terms and measures that are used in many other chapters. The next chapter on exploration expenditures gives some historical perspective on the U.S. exploration business. Chapter 4 is the author's personal essay on why exploration is a unique business. A synopsis of each of the four chapters follows.

Chapter 1, "Introduction," by Richard Steinmetz, gives an overview and perspective of the Handbook rather than a review of the subject itself. The author states that the primary purpose of the Handbook is to present various aspects of the business side of petroleum exploration. Its basic assumption is that all businesses have a common goal—to make a profit—which means making money. In these times of economic and political uncertainty, this Handbook should prove informative and helpful to a wide range of people: from anyone seeking to make a profit in the exploration business, to explorationists working as consultants, to university students majoring in geoscience, petroleum engineering, or business.

Chapter 2, "Basic Concepts," by Donald R. Jones, starts by stating that any economic enterprise must have consistent, unbiased, and quantitative measurements of its performance in order to manage its ongoing business efficiently and profitably. The author then briefly discusses accounting versus cash flow measures, the time value of money, and discounted cash flow rate of return. He next tabulates various types of success measures for the short term (quarterly to annually), intermediate term (2 to 5 years), and long term (projected over the life of a reserve). He finishes with some rules of thumb for "quick and dirty" guidelines for investment decisions on exploration prospects.

Chapter 3, "History of Exploration Expenditures in the United States: 1956 through 1989," by Robert E. Megill, is a brief summary of the subject data published annually by API. The author presents these data in tables and time-series graphs. He discusses the value to individual companies, small and large, of knowing these historical exploration expenditures in order to (1) observe key trends in exploration, (2) make a scoping estimate of industry-wide economic results of exploration, and (3) gauge their own expenditure levels.

Chapter 4, "Exploration: A Misunderstood Business," by John Lohrenz, has a philosophical message. The author argues that exploration is a unique business because of the peculiar qualities of the inventories (hydrocarbon reserves in the ground) that explorationists discover. It is impossible to conduct precise audits or controlled experiments based on countable, tangible product units (barrels of oil). This is indicated by the large variety of terms used to describe resources and reserves: potential, proved, developed, in-place, economically recoverable, revised, and so forth. It is also impossible to track how much more inventory (oil) was withdrawn due to a specific investment or to ascertain when and how a given unit of inventory became withdrawable.

Chapter 1

Introduction

Richard Steinmetz

Amoco Production Company
Houston, Texas, U.S.A.

It has always been my rule in business to make everything count.
—J. D. Rockefeller

PURPOSES OF HANDBOOK

- To present the business side of petroleum exploration.
- To describe how the business is conducted.
- To present methods for dealing with the unique risks inherent to the business.
- To expand the knowledge of explorationists and petroleum engineers by covering a wide variety of topics relevant to the business of petroleum exploration.
- To help those professionals be more successful explorers and effective producers by summarizing subjects not always taught in school or on the job.
- To improve the economics and quality of oil and gas prospects worked up by explorationists, thereby leading to a greater percentage of successful wells.
- To focus explorationists' thinking on economically attractive deals and prospects.
- To show explorationists the importance of going beyond science and technology to the essential requirement of all investors or companies—making an adequate return on the money invested.
- To have explorationists consider the value of their information and ask if more detailed information is really worth the additional cost, or will it negate the economics of the play or prospect?

AUDIENCE FOR HANDBOOK

- Anyone seeking to make a profit in finding and producing oil and gas.
- AAPG members, petroleum geologists, petroleum engineers, geophysicists, landmen, negotiators, attorneys, and MBAs.
- Employees of large companies, who tend to be specialized, and employees of small companies.
- Independents and consultants.
- Oil and gas operators and landowners.
- Anyone trying to sell a deal, play, or prospect.
- Managers with large and small companies.
- U.S. government agencies such as DOE, USGS, MMS, BLM, BIA, and FERC.
- Foreign government agencies and oil companies.
- Investors, bankers, and lenders.
- University geoscience and mineral economics departments and business schools.
- University faculty and students and recent graduates working for oil companies.
- Public, university, and departmental libraries.

BASIC THEME

This Handbook is a first for the American Association of Petroleum Geologists, a book of original contributions emphasizing the business and business management side of petroleum exploration. Its basic theme is that all businesses have a common goal—to make a profit that ensures survival. That means making money. A simple equation expresses the success of any business:

$$\text{Profit} = \text{Volume} \times (\text{Price} - \text{cost} - \text{taxes})$$

Volume of hydrocarbons is difficult to quanitify. Upon discovery , volume is estimated on the basis of drilling and test results of one to three wells, plus spatial results from seismic and subsurface data. After confirmation and infill wells are drilled and more production and pressure data become available, petroleum engineers can then calculate reserves with greater accuracy. In the petroleum industry, *price* is beyond company or individual control, and prediction of future oil prices has a large element of the unknown. *Cost*, including overhead, can be controlled to a significant extent by companies and individuals. *Taxes* are not discussed in detail in this Handbook. They are a complex subject that would require its own set of handbooks. Taxes are largely beyond the control of any one company or individual.

In summary, the previous equation indicates that making a profit involves several difficult factors and largely hinges on controlling costs. Explorationists and petroleum engineers make their contribution to the business in two ways: (1) increasing or maintaining volume and (2) reducing or controlling costs.

There are many ways to define and measure "making money" in petroleum exploration. They are discussed by the various contributors to this Handbook. We should remember what Marlan Downey is quoted as saying (Barrett, 1991, p. 23): "My own view is that it is easy to find oil. It's hell to make money."

UTILITY OF HANDBOOK

The oil exploration business is dynamic and ever changing, along with the economy, global events, and politics. At present, the survival of many companies and governmental entities, seems doubtful. Yet, successful explorationists, always on the go, always learning, do survive and prosper. As A. M. Skov (1991, p. 955) has said:

> The successful managers of tomorrow will be the young men and women of today who have experience in a wide range of issues. They must understand the social, political, legal, environmental, as well as the engineering and economic, implications of any decision.

This Handbook is for those people and should be a part of every explorationist's library. It is a practical guide directly related to the day-to-day work of explorationists. It is a comprehensive compilation of practical information and economic performance measures. It includes many real-world examples and problem-solving techniques.

The Handbook emphasizes economic over technical success. It details what an individual needs to know to be an economically successful oil or gas finder. That includes deciding what to do, making successful evaluations, selling prospects and getting funding, and making a profit.

Most chapters cover the what, why, how, and now what. Some topics are covered in more than one chapter. This seeming redundancy is beneficial because the reader will learn about the topic and/or measure in different applications and from different points of view. Each chapter represents a synopsis or overview of the subject. They are not intended as extensive treatises on each subject. References cited at the end of each chapter can be read for more details.

Many ideas and procedures are presented in this Handbook. They are the result of years of experience of each contributor. They represent ideas that the authors have found to work. They do not necessarily represent their companies' methodologies or philosophies. Neither should they be taken to be complete in detail nor complete legally. Other knowledgeable professionals could have different approaches and techniques that are equally valid for solving the same problems.

Because the oil and gas business must adjust to present recessionary economic times, more and more staffs of traditional companies are going to be made up of explorationists employed as consultants or contract workers. This Handbook should also prove useful to these people.

Because the business is dynamic and broad in scope, this Handbook cannot be the final word, nor does it cover all possible topics. No specific coverage, for example, is provided of exploration strategy, reevaluating old plays, the land function, leasing strategies, environmental liability laws, effective report writing, gas contracts, postdrilling appraisal, or abandonment costs. Other pertinent handbooks are listed at the end of this chapter.

HANDBOOK AUTHORS

A highly knowledgeable and experienced group of authors have contributed to this Handbook. They have a wide variety of backgrounds and have learned the business in the school of hard knocks. They have generously donated their time and talents, and worked at their own expense, to make this Handbook a reality. A brief biography of each author is given at the end of each chapter.

References Cited

Barrett, A., 1991, Out of Alaska: Financial World, Aug. 20, p. 18-23.

Skov, A. M., 1991, On the road again: Journal of Petroleum Technology, v. 43, n. 8, p. 955 and 966.

Other Relevant Handbooks

Bradley, H. B., ed., 1987, Petroleum engineering handbook: Richardson, TX, Society of Petroleum Engineers, 1800 pp.

Harbaugh, J. W., J. H. Doveton, and J. C. Davis, 1977, Probability methods in oil exploration: New York, John Wiley and Sons, 269 pp.

Hyne, N. J., 1991, Dictionary of petroleum exploration, drilling and production: Tulsa, OK, Penn Well Books, 750 pp.

McCray, A. W., 1975, Petroleum evaluation and economic decisions: Englewood Cliffs, NJ, Prentice-Hall, 544 pp.

Moody, G. B., ed., 1961, Petroleum exploration handbook: New York, McGraw-Hill, 840 pp.

RICHARD STEINMETZ, Senior Consulting Geologist for Amoco Production Co. in Houston, received an A.B. degree in Geology from Princeton University in 1954, an M.S. in Mineralogy from Pennsylvania State University in 1956, and a Ph.D. in Geology from Northwestern University in 1961. He worked as a subsurface and well site geologist for Plymouth Oil Co. in Midland, Texas, from 1956 to 1958. He worked for Pan American Petroleum (predecessor to Amoco Production Co.) in Fort Worth and at the Tulsa Research Center from 1961 to 1967, and spent the next 4 years as Assistant Professor of Geology at Texas Christian University in Fort Worth. Since 1971 Dr. Steinmetz has worked for Amoco Production in Tulsa, Denver, New Orleans, Chicago, and Houston in research, management, and staff positions with various domestic and international responsibilities. He is a Certified Petroleum Geologist and a member of AAPG (Elected Editor and Executive Committee 1983–1985), SPE, AAAS, the Houston and New Orleans Geological Societies, and the Rocky Mountain Association of Geologists.

Chapter 2

Some Basic Business Concepts

Donald R. Jones
Consultant
Houston, Texas, U.S.A.

Annual income twenty pounds, annual expenditure nineteen nineteen six, result happiness. Annual income twenty pounds, annual expenditure twenty pounds ought and six, result misery.
—Mr. Micawber in *David Copperfield*

INTRODUCTION

Any economic enterprise, whether it is a one-person shop or a giant multinational corporation, must have consistent, unbiased, and quantitative measurements of its success (or lack of success). Without such measurements, it is impossible to manage the ongoing business efficiently. Recent history has shown that, even for industries in state-owned, centrally planned economies, the measurement of success must relate to making an economic profit or else the industry, and ultimately the entire economy, will fail spectacularly.

In short, success must be measured by the KISS (keep it simple, stupid) principle that *income must exceed outgo.* Profit (success) is the excess of income over outgo, and the larger it is, the more successful an enterprise is. This definition may seem unbearably patronizing to the reader, but most discussions of oil and gas economics obscure this fundamental fact in a fog of business school jargon. All the various econometric, accounting, taxation, and engineering methods of measuring economic success are simply variations of the time span being analyzed and the timing of recognition of the various income and expense components.

Let's expand the KISS principle to a definition of success (profit or loss) for an enterprise of any size engaged in oil and gas exploration and development (including acquisition of producing properties):

Profit = Income + Drawdown of cash reserves – Outgo

Income includes sales revenue, interest received, borrowed funds received, increased capitalization, and asset sales. Outgo includes operating expenses, taxes, interest expense, debt repayment, and asset acquisition costs. Note that this is not an accounting definition nor a tax definition nor even a business school definition. It is a *survival* definition. If this measure is negative at any time, the enterprise can continue to exist only at the forbearance of its creditors.

ACCOUNTING MEASURES

Traditional accounting reporting methods are required by federal and state law in the United States for all economic entities. Even privately held businesses must use traditional accounting methods (as modified by the special provisions of the tax laws) to report their income taxes. The traditional reports deal separately with capital items (assets and liabilities) and current operations (profit and loss). Assets are valued at cost less a gradual (usually) reduction for DD&A (depreciation, depletion and amortization) to represent their wearing out and eventual need for replacement. Current profit and loss calculations include these charges.

Thus, the current operating cash profit of any enterprise is understated substantially by the "paper charges" (that is, no money is actually paid out to anyone) for depreciation, depletion, amortization, and special write-downs of assets. Also, the assets are not reported at their actual value, but at cost less accumulated DD&A. For an oil and gas company, whose principal asset is reserves, it is obvious that the company's value is not what it cost to obtain those reserves but rather how many reserves the company

Table 1. Accounting Concept

Year	Investment	Revenue	Depreciation	Operating Expenses	Profit
1	$5,000	$2,500	$1,000	$500	$1,000
2		2,500	1,000	500	1,000
3		2,500	1,000	500	1,000
4		2,500	1,000	500	1,000
5		2,500	1,000	500	1,000
Totals	$5,000	$12,500	$5,000	$2,500	$5,000

Table 2. Cash Flow Concept

Year	Investment	Revenue	Depreciation	Operating Expenses	Net Cash Flow
0	$5,000				$–5,000
1		$2,500		$500	2,000
2		2,500		500	2,000
3		2,500		500	2,000
4		2,500		500	2,000
5		2,500		500	2,000
Totals	$5,000	$12,500		$2,500	$5,000

has and how much they will bring in future profits. So the traditional balance sheet type of calculation is not much help in evaluating an oil and gas company or a specific project proposal, nor is the traditional profit and loss statement.

Why, then, do we bother to make accounting reports? First, because they are required by law, and second, and just as important, because they are used by bankers, stock analysts, and investors as measures of success to make their decisions concerning our companies. These are people whose good opinion we both value and need. It is fashionable to deride management as short sighted and not being able to see beyond the next quarter's earnings. But a management that ignores, for very long, the perception of its company's success as reported in its accounting statements will find their own positions, and often the company's position as well, in serious peril of loss or takeover.

CASH FLOW MEASURES

The practical evaluation problems presented by traditional accounting methods have led to the general use, by the operating and management segments of the oil and gas industry, of *cash flow* measurements instead of accounting measurements. The primary differences between cash flow and accounting are that, for cash flow, the following are true:

1. Only cash items are included, that is, real transactions where money either comes in or goes out. *There are no DD&A type charges.*
2. All transactions are recognized in their *full cash amount* at the time the cash changes hands. For example, the entire cost of a well is charged as an expense when it is completed and *paid for.* The income from a well is recognized when the oil is produced *and sold.*
3. Transactions are considered on a "money-forward" basis, that is, past expenditures are considered as "sunk costs" and past income is treated similarly. This is especially important in on-going investment decisions. How much money we have already spent on drilling a well is irrelevant (in a truly objective decision process) to the decision to attempt a completion. (An exception to this rule is the tax treatment of past expenditures in after-tax analyses.)

Cash flow analysis is used to look back and calculate the success of a completed project or to estimate the success of a proposed project. It is simply a "spreadsheet" listing of all the expense and income items of a project in their appropriate time periods, along with a summation of costs, expenses income, and net cash flow over each time period and the total project.

Tables 1 and 2 (from Megill, 1988) show a simple example of the difference between the accounting and cash flow approaches for a single investment of $5,000.00, which will earn a revenue of $2,500.00 yearly for 5 years while incurring $500.00 of yearly expenses. Note that on the cash flow basis, payout (the time when we profit enough to get our investment back) occurs at the real-world (cash) point of 2-1/2 years on the cash flow analysis. (Year "0" is a convention used in cash

Table 3. Discounted Cash Flow Concept

Year	Investment	Revenue	Operating Expense	Net Cash Flow	10% Discounted Net Cash Flow	20% Discounted Net Cash Flow
0	$5,000			$–5,000	$–5,000	$–5,000
1		$2,500	$500	2,000	1,818	1,667
2		2,500	500	2,000	1,653	1,389
3		2,500	500	2,000	1,503	1,157
4		2,500	500	2,000	1,366	965
5		2,500	500	2,000	1,242	804
Total	$5,000	$12,500	$2,500	$5,000	$2,582	$982

flow to denote time zero, that is, the beginning of the project and the date to discount back to.)

TIME VALUE OF MONEY

In the oil and gas industry, many projects have long lifetimes—as long as 30 to 40 years. The need to analyze such long-lived projects leads us to *discounted cash flow* methods. In essence, these methods account for the fact that a dollar received at some time in the future is not worth as much as a dollar received right now. This is true because a dollar received today can be invested and will grow to be worth more in the future.

For example, how much would you pay for the right to receive a dollar one year from today? Obviously, something less than a dollar, but how much less? If you can earn money on investments at the rate of, say, 10% a year, then the logical answer to the question is the amount of money you would have to invest today at 10% per year so as to receive $1.00 one year from now. Rounding off, that number is $0.91, that is, if you deposit $0.91 today at a simple 10% annual interest rate, in one year you will have

$$0.91 + 0.10(0.91) = 0.91 + 0.09 = 1.00.$$

In practice, various interest rates and various ways of compounding the interest are used. The $0.91 is called the *present worth*, the *present value*, or the *net present value* of the $1.00. Its application in a cash flow analysis is called a discounted cash flow. So, without going into the details of the interest calculations at this point, let's add another column to the cash flow analysis. Table 3 is an example of a discounted cash flow analysis. It tells us, for example, that at a 10% earning rate, the value of the $2,000 of net cash flow we will receive in the fifth year is only $1,242. Also, if we make an investment of $5,000 today, the total net present value resulting from that investment (at a 10% discount rate) is only $2,582. Note that this is considerably less than the project's total undiscounted net cash flow of $5,000.

This lesser valuation results only from the time value of money; we must make the entire investment at once, but the income is spread out over five years. Note that this concept *has nothing to do with inflation;* it assumes no change in the buying power of the future dollars.

If our company has a higher earnings rate, say, 20%, then the effect of discounting is even greater, that is, future income is worth less (see the 20% discounted cash flow column in Table 3). A negative discounted cash flow will occur at some high enough discount rate. This means that the project does not generate enough income, relative to the investment, to earn that interest rate.

DISCOUNTED CASH FLOW RATE OF RETURN

For every cash flow stream, there is an interest rate that will discount the cash flow stream to exactly zero, that is, the positive discounted net cash flows will exactly equal the negative discounted cash flow from the investment. This interest rate, by definition, is the *discounted cash flow rate of return (DCFROR).* We find it by trial and error.

Table 4 is a continuation of Table 3 with two added columns showing 30% and 28.6% discounted cash flow. Note that the total for 30% discounted cash flow is negative, that is, 30% is too high a discount rate. By interpolation between 20% (too low) and 30% (too high), we find the DCFROR of 28.6%. This is the earning rate of the project as measured by the DCFROR method. Variations of this method exist, but the basic concept is the same. For this project as it stands, we say it will "earn" a rate of return of 28.6%.

Suppose, however, that this project is being offered for sale by a promoter who wants a cash up-front fee of $982. The total cost to the buyer at time zero is $5,982, and the net cash flow at time zero is $–5,982. The DCFROR is no longer 28.6% since that would result in a total discounted net cash flow of $–982. Instead, the DCFROR is now 20% since this would result in a total discounted net cash flow of $0.

Another way to understand the amount of DCFROR,

Table 4. Discounted Cash Flow Rate of Return Concept

Year	Net Cash Flow	20% Discounted Net Cash Flow	30% Discounted Net Cash Flow	28.6% Discounted Net Cash Flow
0	$-5,000	$-5,000	$-5,000	$-5,000
1	2,000	1,667	1,538	1,555
2	2,000	1,389	1,183	1,208
3	2,000	1,157	910	939
4	2,000	965	700	730
5	2,000	804	539	568
Totals	$5,000	$982	$-130	$0

Table 5. Confirmation of DCFROR Concept

Year	BOY[a] Investment	Net Cash Income	Earnings @ 28.6% of BOY[a] Investment	Investment Recovery	End of Year Investment
0					$5,000
1	$5,000	$2,000	$1,433	$567	4,433
2	4,433	2,000	1,270	730	3,703
3	3,703	2,000	1,061	939	2,764
4	2,764	2,000	792	1,208	1,556
5	1,556	2,000	444	1,556	0
Totals		$10,000	$5,000	5,000	

[a]Beginning of year.

which is particularly helpful to those trained to think in accounting terms, is that the DCFROR is the interest rate earned on the remaining unamortized investment. Table 5 (from Megill, 1988) demonstrates this point. At the beginning of the first year, the unamortized investment is $5,000; 28.6% of $5,000 is $1,433. Of the $2,000 cash flow for the year, $1,433 is assigned to earnings and $567 to investment recovery. For the second year, the unamortized investment is $5,000 – $567 = $4,433, the earnings are $4,433 x 28.6% = $1,270, and the investment recovery is $2,000 – $1,270 = $730. Note that by using this scheme, the total investment recovery is $5,000, exactly enough to recover the $5,000 investment. The DCFROR (28.6%) is the only interest rate that will exactly recover the investment.

MEASURES OF SUCCESS

In addition to being categorized as accounting (financial or "book" basis) methods or cash flow methods, success measurements have been devised that relate to reserves and/or production/sales, measures that incorporate risk adjustments, and combinations of these methods. In addition, the various success measures can be categorized according to the time span for the calculations. Short-term measures are quarterly or annual, intermediate-term measures are 3 to 5 years, and so on. The following outline lists some of the potential ways to measure success, that is, either to estimate the value of a project or company or to calculate its earnings.

I. Short-term measures (quarterly or annual)
   A. Financial ("book") basis
      1. Net income and net income per share
      2. Return on capital invested and return on shareholders equity
      3. Net assets and net assets per share
      4. Current share price
      5. Yield (dividend rate per share)
      6. P/E ratio
   B. Cash flow basis
      1. Cash flow profit and cash flow profit per share
      2. Cash flow profit per unit of production or sales (unit = equivalent bbl or equivalent MCF)
   C. Present value basis
      1. SEC P.V. of proved reserves (and per share)
      2. "Company" P.V. of proved reserves (and per share) (using "company" price and cost projections and discount rate)
   D. Combined financial and present value basis
      1. Net assets (other than oil and gas proper-

ties) less debt plus SEC P.V. of proved reserves (and same per share)
    2. Net assets (other than oil and gas properties) less debt plus "company" P.V. of proved reserves (and same per share)
  E. Reserves and/or production/sales basis
    1. Proved reserve change
    2. Production/sales rate
    3. Reserves/production ratio ("reserves life index")

II. Intermediate term measures (2 to 5 years)
  A. Financial ("book") basis
    1. Net income growth rate (and per share)
    2. Average return on capital employed and average return on shareholders equity
    3. Growth rates for ROIs above
    4. Share price growth rate
    5. Net assets growth rate
    6. Yield growth rate
    7. P/E ratio changes
  B. Cash flow basis
    1. Average annual cash flow (and per share)
    2. Annual cash flow growth rate (and per share)
    3. Average annual cash flow per unit of production/sales
    4. Growth rate for annual cash flow per unit of production/sales
  C. Present value basis
    1. Growth rate for SEC P.V. of proved reserves (and per share growth rate)
    2. Growth rate for "company" P.V. of proved reserves (and per share growth rate)
  D. Combined financial and present value basis
    1. Growth rate for net non–oil and gas assets less debt plus SEC P.V. value of proved reserves (and per share growth rate)
    2. Growth rate for net non–oil and gas assets less debt plus "company" P.V. of proved reserves (and per share growth rate)
  E. Reserves and/or production/sales basis
    1. Proved reserves growth rate (and per share growth rate)
    2. Production rate growth
    3. Reserves/production ratio changes

III. Long-term measures (measured at one time point but projected over lifetime of reserve estimates)
  A. *Pro forma* projections of financial statements
  B. Present value of proved reserves plus risk-adjusted P.V. of probable and possible reserves from exploration, development, secondary, and so on
  C. Present value from item B above but including interest expense
  D. DCFROR for Proved reserves only and "dry hole costs" for all other ventures
  E. DCFROR for item D above but including interest expense
  F. DCFROR for proved reserves plus risk-adjusted DCFROR for probable and possible reserves
  G. DCFROR for item F above but including interest expense
  H. Risk-adjusted DCFROR vs. "cost of capital" or vs. ROR from alternate "low risk" investments
  I. Investment efficiency ("portfolio analysis") including risk adjustments

SOME RULES OF THUMB

As the previous outline demonstrated, the ways to measure success are limited only by imagination. There is no one "best" measure to evaluate all proposals, nor are there any absolute rules for ranking investment opportunities and selecting the best ones. The chapters that follow will provide the reader with an in-depth understanding of the economics side of the oil and gas business. He or she can then use the insight provided by various economic measures to make intelligent investment decisions.

In the meantime, we provide some rules of thumb for "quick and dirty" guidelines for investment decisions on exploration prospects (modified from Peter Rose, pers. comm.):

1. **40/1 Hurdle Rate.** If the undiscounted gross value of total production from the accumulation is not at least 40 x the cost to discover it (land and exploration drilling), forget it. As suggested by Hardin (1958), this was originally 50/1 for new field wildcats.
2. **Rate of Return.** If the discounted rate of return is not greater than certain established values, depending on perceived risk, forget it. Recommended values are

   High Risk = 30–40%

   Medium Risk = 20–30%

   Low Risk = 15–25%

   Problems: This method uses ROR as a risking tool. This is mixing apples with oranges, but some very capable prospect screeners who use it consistently find it to be very useful. Also, experience suggests that explorationists perceive risk poorly, that is, "low-risk" projects turn out to fail more often than we predict.
3. **Pay-Out.** If the project will not pay out in less

than x years, turn it down. For individual wells, this is about 2-1/2 years. For entire new field projects, it is about 4 or 5 years, depending on size.
Problem: This indicates nothing about the volumes or profits generated after pay-out. This is a better tool for smaller ventures, particularly where pipelines and other needed facilities are already in existence. Also, discounted pay-out may be a more realistic measure.

4. **Cost of Finding (COF) Rule.** Present industry costs of finding (1989) are about $5/bbl for oil and $0.75/MCF for gas. If the project's finding costs are higher, reject the project (must be risk adjusted to be valid). Industry cost of finding includes all dry hole costs, so divide the prospect's COF by chance of success to risk adjust and make a valid comparison.
Problem: Conventional "COF" found in financial analyses also includes development costs.
5. **Net Revenue Interest (NRI) Rule.** Any wildcat deal promoted so that the net revenue interest is less than 75% is looked at very skeptically in most quarters. One key factor here is whether the promoter is also recovering his or her geological and geophysical (G&G) cost and lease expenses. Strive for more than 80% NRI.
6. **$1,000,000 Profit per BCFG Rule.** For any gas prospect, if the perceived venture as described will not return $1 million profit for each billion cubic feet of producible gas, don't do it. This includes development drilling costs. Note that this is a fairly conservative measure.
7. **First Well Pays Rule.** The first well should earn enough to pay for land, seismic surveys, and drilling. Otherwise, the deal may be too tough or deliverability too "skinny." The idea here is that development wells should carry no additional burden.
8. **Three Offsets Rule.** Don't take any new field wildcat venture in which at least three offset development wells are not clearly feasible if the wildcat well proves to be successful.
9. **Double Dipping Rule.** Turn down any submittal that requires both an "acceptance bonus" and a carried interest.
10. **Dry Hole Carrying Capacity Rule.** A risk-related exploration hurdle is that each NFW prospect must have a present value for the cash flow after time zero which is greater than

Cost through completed discovery well
+ Probability failure
x (Cost through abandonment of prospect)
÷ Probability of success.

Note that this makes successful wildcats pay for themselves and carry other dry holes (Newendorp, 1975).

References Cited

Hardin, G. C., 1958, Economic factors in the geological appraisal of wildcat prospects: Gulf Coast Association of Geological Societies Transactions, v.8, p. 14-19.

Megill, R. E., 1988, The business side of geology: A collection of articles reprinted from the AAPG Explorer, 1984–1986, Tulsa, OK, AAPG, p. 7-9 to 7-11.

Newendorp, P. D., 1975, Decision analysis for petroleum exploration, Chapter 2: Tulsa, OK, Petroleum Publishing Co., p. 42-45.

DONALD R. JONES, an independent consultant in petroleum evaluation in Houston, received a B.A. degree in Geology from UCLA in 1953 and an M.S. in Operations Research from the University of Houston in 1970. After serving 2 years in the U.S. Air Force as an intelligence officer, he was employed in 1955 by Marathon Oil Co. as an exploration geologist in Midland, Texas, and subsequently in Corpus Christi. In 1961 he became a log analyst at Marathon's Research Center in Littleton, Colorado. Mr. Jones was employed in 1966 by Offshore Operators Inc. (a predecessor of TransOcean Oil Inc.) as an evaluation engineer in Houston and later served as a reservoir engineer and Manager of Reservoir Engineering. In 1980 he became Vice-President of Exploration and Production of American Exploration Co. in Houston, and from 1987 to 1988 was Vice-President of Engineering for Oxoco Inc. He is a member of AAPG, SPE, SPWLA, and the Houston Geological Society.

Chapter 3

History of Exploration Expenditures in the United States: 1956–1989

Robert E. Megill
Consultant
Kingwood, Texas, U.S.A.

Money speaks sense in a language all nations understand.
—Aphra Behn, 1680

INTRODUCTION

In setting up a predrilling evaluation, explorers go to numerous sources to get vital cost information. Past drilling records help to estimate the cost of drilling an exploratory well. Various departments in a company aid in estimating taxes, operating costs, development well schedules, and producing rates. It is often useful to look at expenditures in their entirety. Then one can ascertain relationships and information about the industry as a whole, as well as the developing trends in expenditures.

Records of exploratory expenditures for the United States have been kept by the American Petroleum Institute (API) and the Bureau of the Census for most of the past 50 years. The current source of data for exploration expenditures is the *Survey on Oil and Gas Expenditures* published annually by API. It not only contains a record of exploration expenditures but also lists expenditures in development drilling as well as producing costs for existing fields. An estimate of net revenue (total revenue minus royalty payments) is included in the annual survey.

The cost categories assigned to exploration have changed little since the first record in 1944. Until 1966, all dry holes were assigned to exploration; successful wildcats were included with development wells. From 1966 on, only exploratory wells, successful and dry, have been included in exploration expenditures.

An example of the summary sheet of expenditures for the U.S. petroleum industry is shown in Table 1, which is for the year 1989. Not only are expenditures shown for the total United States but also for the lower 48 onshore, the lower 48 offshore, and Alaska. For the total United States, the percentage change from the previous year is listed for each category. In 1989, the petroleum industry spent $7.5 billion in exploration, $13.9 billion in development, and $23.7 billion in production of existing fields. A total of $45.1 billion was spent in the three categories. Total revenue for the petroleum industry at the wellhead was $66 billion; thus, the $45 billion spent was a reinvestment of 68% of the money received in 1988.

When exploration expenditures are grouped, they include (among others) the following general categories:

- Exploratory drilling
- Acquiring undeveloped acreage
- Land, leasing, and scouting
- Geological and geophysical (G & G)
- Test hole contributions
- Direct overhead, general and administrative (G & A) overhead, and other miscellaneous

In addition to following these categories by broad geographic areas, the annual survey also shows exploration, development and producing expenditures for the 19 largest companies and for the balance of the petroleum industry.

Table 1. 1989 Estimated Expenditures for Exploration, Development, and Production by Geographic Area in the United States (in Millions of Dollars)

	Total United States	Percentage Change	Lower 48 Onshore	Lower 48 Offshore	Total Alaska
EXPLORATION EXPENDITURES					
Total drilling and equipping expenditures	$2,596	–15.0%	$1,478	$1,059	$59
Total completed well costs	2,395	–20.3	1,414	965	16
Oil wells	242	–24.1	191	45	6
Gas wells	553	–5.0	400	153	0
Dry holes	1,600	–23.9	823	767	10
Work-in-progress	201	NM	64	94	43
Acquiring undeveloped acreage	1,759	–42.4	896	795	68
Land, leasing, and scouting	237	2.0	205	27	5
Geological and geophysical	955	–25.2	527	372	56
Lease rentals	263	–1.7	167	68	28
Test hole contributions	15	108.6	1	1	13
Direct and G & A overhead and other	1,677	–10.4	NA	NA	NA
Total exploration expenditures	$7,502	–23.1%	NA	NA	NA
DEVELOPMENT EXPENDITURES					
Total drilling and equipping expenditures	$8,279	3.5%	$6,045	$1,818	$416
Total completed well costs	7,259	–3.7	5,593	1,309	357
Oil wells	2,542	–19.4	1,844	399	299
Gas wells	3,489	15.6	2,944	543	2
Dry holes	1,228	–10.1	805	367	56
Work-in-progress	1,020	NM	452	509	59
Lease equipment	2,776	2.6	1,231	1,412	133
Fluid injection and recovery programs	920	–12.1	723	36	161
Direct and G & A overhead and other	1,875	–23.2	NA	NA	NA
Total development expenditures	$13,850	–2.4%	NA	NA	NA
PRODUCTION EXPENDITURES					
Total direct operating	$14,904	–1.5%	$11,203	$2,594	$1,107
Operating and maintenance	11,597	–5.3	9,029	1,834	734
Well workovers	1,643	9.0	1,029	484	130
Other direct expenses	1,664	21.0	1,145	276	243
Total indirect operating	8,838	7.5	NA	NA	NA
Windfall profits tax[a]	–20	NM	–11	–17	8
Other taxes (excluding income taxes)	4,278	2.8	2,850	85	1,343
G & A overhead and other indirect expenditures	4,580	9.5	NA	NA	NA
Total production expenditures	$23,742	1.7%	NA	NA	NA
TOTAL EXPENDITURES	$45,094	–4.7%	NA	NA	NA

[a]Windfall profits tax includes refunds due to overpayment in prior years.

NA—not available; NM—not meaningful.

Source: American Petroleum Institute, 1989.

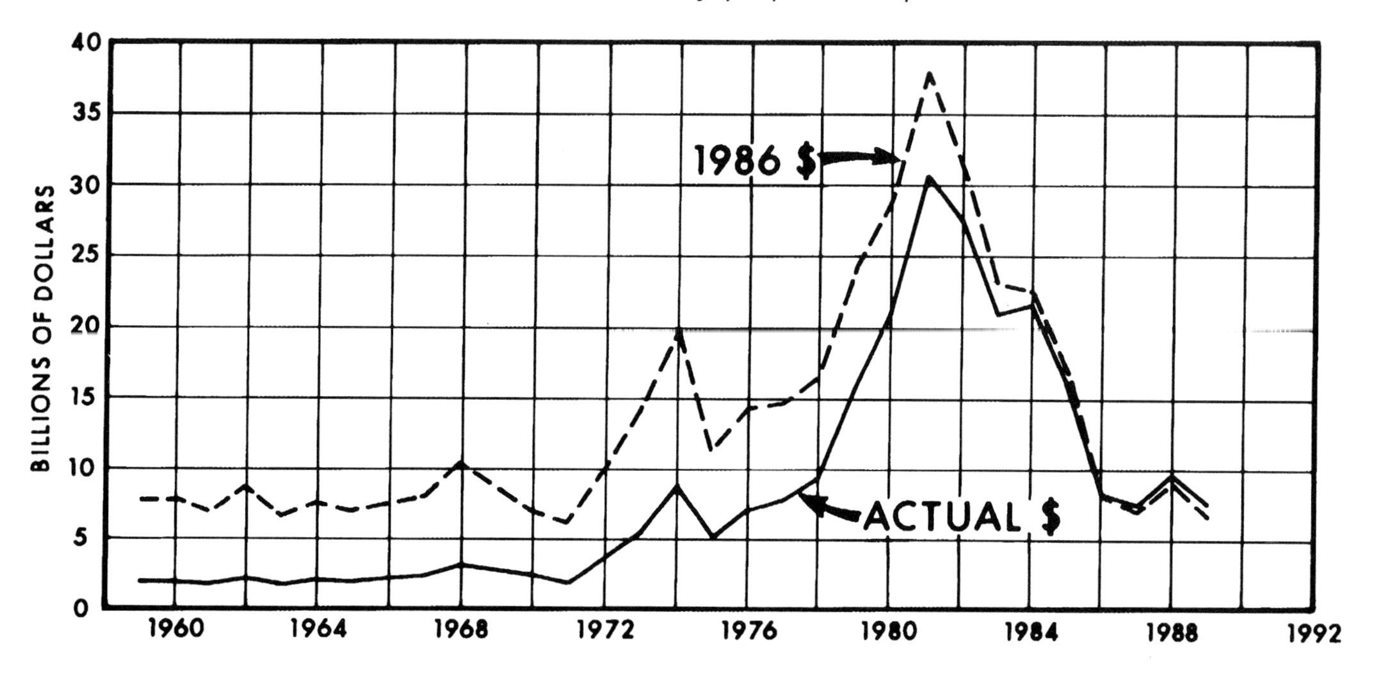

Figure 1. Exploration expenditures in the United States.

HISTORY OF EXPLORATION EXPENDITURES IN THE UNITED STATES

Table 2 shows the exploration expenditures by categories for the United States from 1956 through 1989. Data for 1957 and 1958 were not compiled by API, but a continuous record exists from 1959 through 1989. Expenditure data for the years 1973 through 1982 were compiled by the Bureau of the Census. The API resumed compilation in 1983; in the fall of each year it releases expenditure data for the prior year. From 1956 through 1989, the petroleum industry spent about $267 billion searching for new oil and gas fields.

The data have not always been on an exactly comparable basis, as was mentioned earlier. Looking at Table 2, one can see the marked changes in overhead in both 1966 and 1970, indicating a change in definition of the category to include more expenditures. The total expenditures, however, represent the most reasonable estimate of the amount of money spent in the search for new fields.

The dollar amounts shown are as of the year spent; they have not been corrected to a common dollar base. Such a correction is shown graphically in Figure 1. The dashed line is the correction to 1986 dollars (CPI), and the solid line is the "as spent" expenditure. On a constant dollar basis, the petroleum industry is now spending annually about what it spent in the 1960s. The peaks in 1974 and 1981 were associated with high crude oil and gas prices as well as the anticipation of higher future prices. Table 3 shows the calculations for the correction of expenditures to constant dollars. It also shows net revenue, the percentage exploration expenditures of net revenue, and overhead as a percentage of total exploration expenditures.

Not until 1970 were the figures for overhead expenses given on a consistent basis. In Figure 2, overhead is shown as a percentage of exploration expenditures for the period 1970 through 1989. Until the mid-1980s, overhead averaged less than 10% of total expenditures. The peak of 24% in 1986–1987 is associated with markedly reduced total expenditures. Will the overhead level return to the 10% range in the future? Or has the industry changed its methods of operation so as to increase permanently the nonfield expenditures in exploration (for example, 3-D seismic efforts)? In the next few years, the answers to these questions will be known.

VALUE OF KNOWING TOTAL EXPENDITURES

Knowing total expenditures for the industry allows a view of the industry's optimism or pessimism, an observation of trends, and a look at expenditure levels. These expenditures form the basis for estimating the economic results of the search for new fields. Expenditures combined with the economic output of discoveries allow an estimate of the economic yardsticks associated with the search for new oil and gas fields. By combining producing rates with prices, development costs, taxes, and operating expenses, a

Table 2. Exploration Expenditures by Year in United States (in Millions of Actual Dollars)

Year	Exploratory Drilling[a]	Lease Purchases	Land, Leasing, and Scouting	Geology and Geophysics	Lease Rentals and TWC[b]	Overhead, Including G & A	Total
1956	909	561	—	360	—	287	2,117
57	—	—	—	—	—	—	—
58	—	—	—	—	—	—	—
59	821	554	-	320	193	124	2,012
1960	774	626	104	277	193	71	2,045
61	774	428	115	280	189	65	1,851
62	847	815	108	299	197	58	2,324
63	790	376	117	300	193	69	1,845
64	854	570	100	336	177	72	2,109
1965	849	438	102	355	166	61	1,971
66	775	577	70	378	208	128	2,136
67	793	829	86	392	174	122	2,396
68	836	1,578	82	373	213	136	3,218
69	944	1,137	93	387	167	168	2,896
1970	815	714	98	349	168	332	2,476
71	775	642	100	361	167	348	2,393
72	910	1,722	105	372	177	386	3,672
73	944	3,552	76	399	165	310	5,446
74	1,580	5,774	91	568	187	459	8,659
1975	2,124	1,615	113	653	222	583	5,310
76	2,467	3,024	126	692	229	646	7,184
77	3,230	2,587	164	856	279	727	7,843
78	4,137	2,912	164	1,046	310	883	9,452
79	5,619	7,036	224	1,218	335	1,185	15,617
1980	8,679	7,899	275	1,871	457	1,650	20,831
81	13,546	11,188	376	2,823	634	2,107	30,674
82	12,936	8,163	458	2,998	690	2,613	27,858
83	7,045	8,052	317	2,540	635	2,491	21,080
84	7,363	7,807	379	2,619	611	2,719	21,498
1985	6,297	4,040	381	2,392	557	2,732	16,399
86	3,048	1,335	301	1,244	508	2,032	8,468
87	2,373	1,685	228	1,194	293	1,826	7,599
88	3,054	3,053	232	1,276	274	1,871	9,760
89	2,596	1,759	237	955	278	1,677	7,502

[a]All dry holes until 1966.

[b]Test well contributions added in 1966.

Data are from API, Annual Survey of Oil and Gas Expenditures, for the years 1956–1972 and 1983–1989, and the Bureau of the Census, Annual Survey of Oil and Gas, for the years 1973–1982 (see reference list).

Table 3. Exploration Expenditures in the United States Corrected to Constant Dollars Net Revenue and Overhead

Year	Inflator CPI 1986 = 1.00	Exploration Expenditures (million $)		Net[a] Wellhead Revenue (billion $)	Percentage	
		Actual Dollars	1986 Dollars		Exploration Expenditures of Net Revenue	Overhead of Total Expenditures
1956	4.176	2,117	8,841	7.3	29	—
57	4.043	—	—	—	—	—
58	3.931		—	—	—	—
59	3.903	2,012	7,853	7.9	26	
1960	3.838	2,045	7,845	8.1	25	—
61	3.799	1,851	7,032	8.4	22	—
62	3.762	2,324	8,742	8.7	27	—
63	3.712	1,845	6,847	9.1	20	—
64	3.655	2,109	7,708	9.2	23	—
1965	3.606	1,971	7,107	9.4	21	—
66	3.506	2,136	7,489	10.0	21	—
67	3.401	2,396	8,149	10.8	22	—
68	3.264	3,218	10,504	11.4	28	—
69	3.095	2,896	8,963	12.2	24	—
1970	2.928	2,476	7,250	13.1	19	13.4
71	2.805	2,393	6,712	13.8	17	14.5
72	2.717	3,672	9,977	13.9	27	10.5
73	2.559	5,446	13,936	15.7	34	5.7
74	2.304	8,659	19,950	24.5	36	5.3
1975	2.112	5,310	11,215	27.9	19	11.0
76	1.996	7,184	14,339	29.4	24	9.0
77	1.875	7,843	14,706	34.1	23	9.3
78	1.742	9,452	16,465	38.4	25	9.3
79	1.565	15,617	24,441	48.4	32	7.6
1980	1.378	20,831	28,705	74.9	28	7.9
81	1.250	30,674	38,343	108.5	28	6.9
82	1.136	27,858	31,647	102.6	27	9.4
83	1.101	21,080	23,209	111.0	19	11.8
84	1.056	21,498	22,702	115.9	19	12.6
1985	1.020	16,399	16,727	106.8	15	16.7
86	1.000	8,468	8,468	61.2	14	24.0
87	0.965	7,599	7,330	66.7	11	24.0
88	0.976	9,760	9,526	59.9	16	19.2
89	0.916	7,502	6,872	66.1	11	22.4

[a]Gross revenue minus royalty.

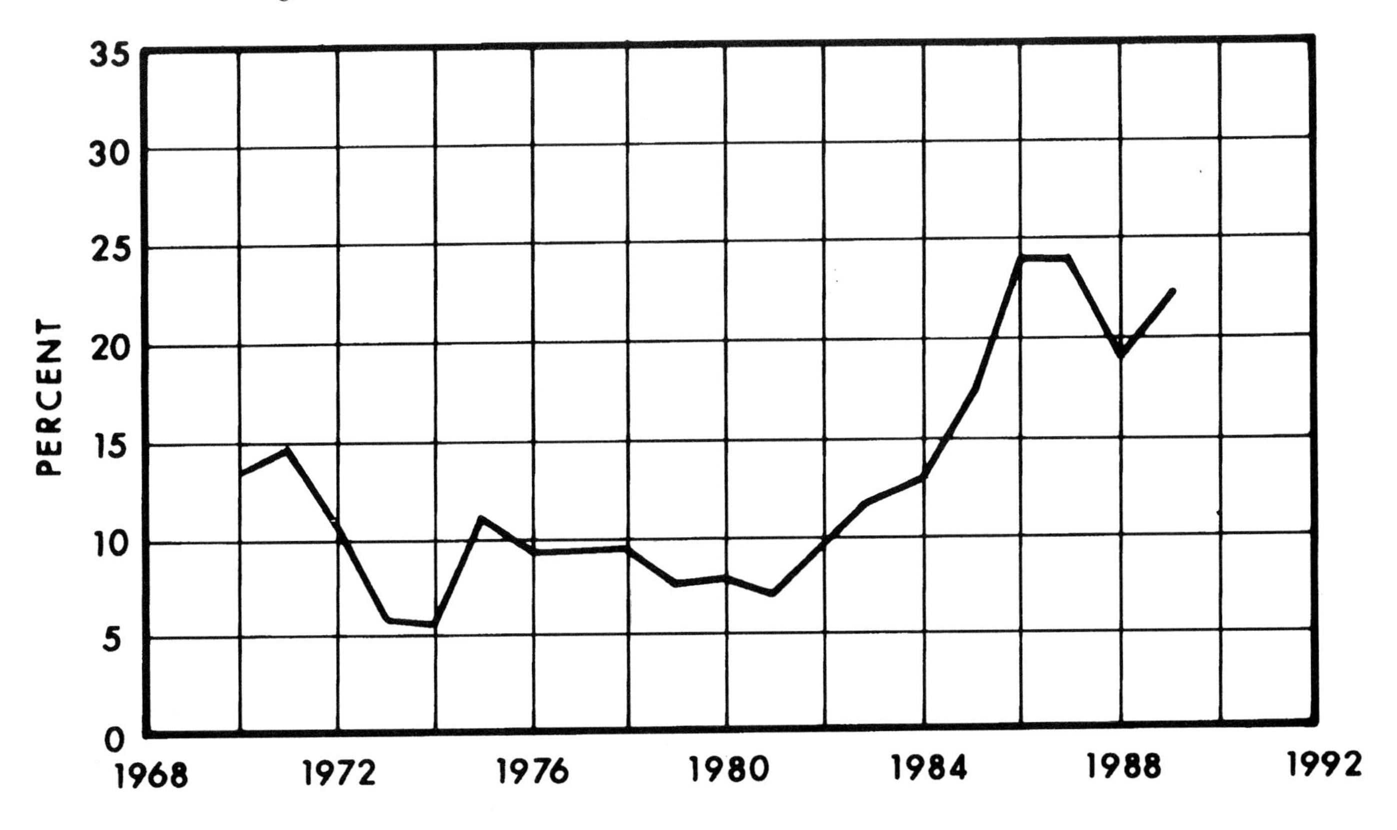

Figure 2. Overhead as a percentage of exploration expenditures in the United States.

cash flow stream from discoveries can be calculated and the economic health of exploration in the United States can be determined. In addition, because expenditure data are shown for the lower 48 onshore, the lower 48 offshore, and Alaska, economic yardsticks can be determined for each of these areas.

Because exploration expenditures are the first step in making an estimate of economic results of exploration, they are vital information for both explorers and analysts. They allow the observation of key trends in exploration. Figure 3 shows an example of one of these trends. It shows the percentage exploration expenditures are of total net revenue at the wellhead. For much of history, that percentage remained between 20 and 25%; however, in the mid-1980s, it reached new lows. The low points occurred in 1987 and 1989 when only 11% of net revenue was spent in the search for new oil and gas fields.

Similar expenditure data can be assembled for other parts of the world. In some countries where exploration has only begun in recent decades, government agencies have maintained detailed records of petroleum expenditures. Such information provides valuable input for the explorer or analyst wanting to view the monetary magnitude of an industry's search for new fields.

Finally, the separation by API of the expenditure data for the 19 largest companies provides a comparison of an individual company's U.S. expenditure pattern with an appropriate reference base.

References Cited

American Petroleum Institute, 1989, Annual survey of oil and gas expenditures: Washington, D.C., API, 17 pp.

Bureau of the Census, 1982, Annual survey of oil and gas: Bureau of the Census, Washington, D.C., MA-13K(82)-1.

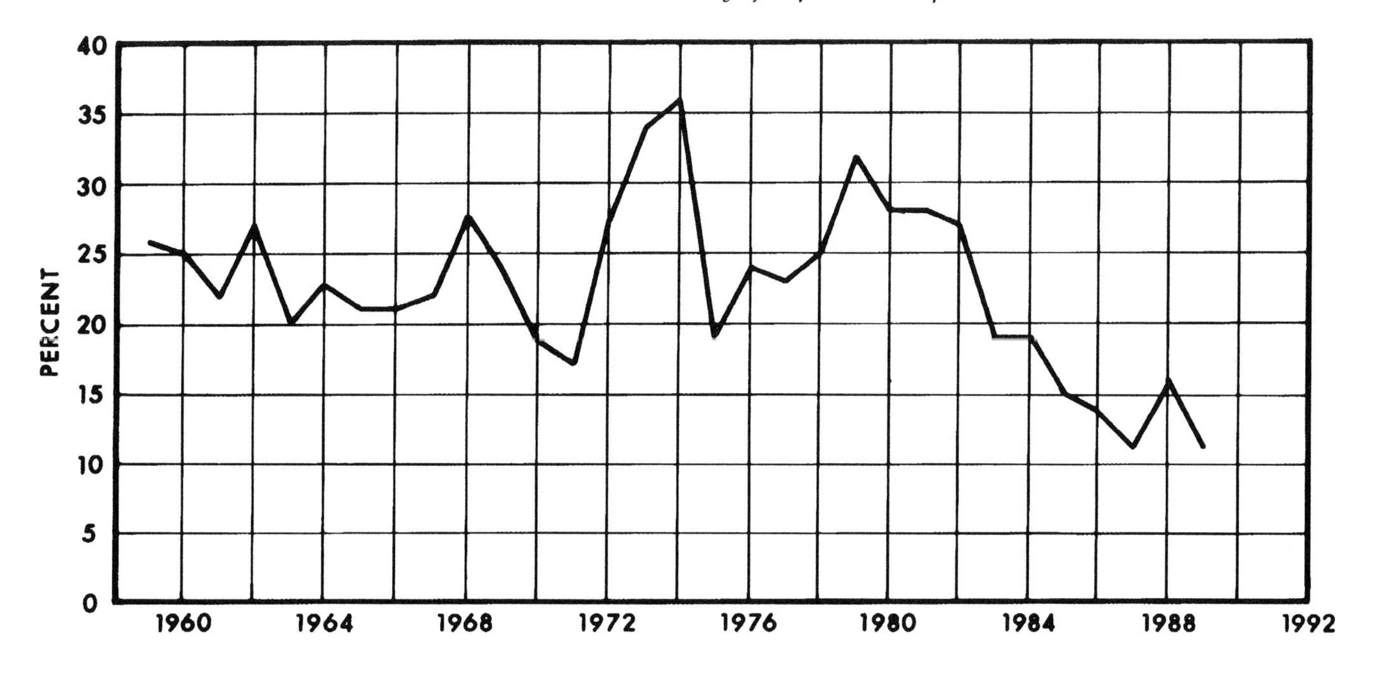

Figure 3. Percent exploration expenditures of total net revenue at the wellhead in the United States.

ROBERT E. MEGILL, a consulting geologist in Kingwood, Texas, received a B.S. in Geological Engineering from the University of Tulsa. He worked for Exxon, including predecessor companies Carter and Humble, from 1941 to 1984. He served as staff geologist, petroleum economist, reserve geologist, Head of Reserve Economics, Head of Exploration Economics, Manager of Planning, and Coordinator of Planning and Coordinator of Economic Evaluation. Mr. Megill has published numerous technical articles and is widely known for his books and courses on Introduction to Risk Analysis and Introduction to Exploration Economics. He is a member of AAPG and for the last 7 years has written a monthly column in The Explorer on "The Business of Petroleum Exploration."

Chapter 4

Exploration: A Misunderstood Business

John Lohrenz

Louisiana Tech University
Department of Petroleum Engineering and Geosciences
Ruston, Louisiana, U.S.A.

. . . when one transfers x barrels of oil from an inventory perceived to hold X barrels of undiscovered oil, there is no requirement that the remaining undiscovered oil inventory is (X – x) barrels.

—John Lohrenz

INTRODUCTION

Oil and gas exploration is a unique kind of business. Exploration, like all product businesses, survives by providing products from an inventory to an accepting market. The uniqueness of exploration arises out of the peculiar quality of the inventories exploration creates.

Businesses providing a vast and ever-changing panoply of products to markets are a focus of several disciplines' energetic study and analysis. Management science, operations research, economics, finance, and, of course, business administration are all disciplines that have trained generations of practitioners and continue to do so. Maintenance of inventories of products is a central problem. What sales from inventory should be sought at what price? What are proper inventory levels given the costs of both adding to and maintaining inventory?

The product inventory problem is robust, pertinent, and meaningful, and it merits the voluminous and protracted attention received from keen business practitioners. Prototypical business practitioners, be they trained by years of business hurly-burly, or sophisticated MBAs with arrays of mathematical algorithms and computers, are not normally prepared, however, to recognize the unique nature of exploration's inventories. Put together such a business practitioner with an explorationist and misunderstandings, hidden and open, are inevitable and predictably rife.

The first purpose here is to articulate the inherited inventory handling paradigms of business practitioners in relation to exploration's inventories. To do so, standard pedagogy in business administration is used and a case study of an exploration venture is presented. A second purpose is to show the burdens that the misunderstandings create. The result is not just business plans that go awry, but public policies that have effects opposite from those intended.

THE CASE STUDY NOT STUDIED: EXPLORATION

The case study is a powerful instructional tool in business education. The salient operant rules, constraints, assumptions, and uncertainties are stated as quantitatively as possible for the business at hand. Results sought are not just decisions and strategies to undertake, but an understanding of the interrelationships that affect outcomes.

A case study of an exploration venture described to suit a generic product inventory problem follows.

Definitions of the Products and Markets:

- The products are oil and gas. For simplicity, oil can be considered as the single product.
- All of any oil withdrawn from the inventory can be absorbed by the market at the market price.
- Assume an effective unit market price of, say, \$25 per barrel. The unit market price has fluctuated in the past and will likely continue to do so. One can select from among "experts" predicting future prices and price volatility.
- The unit market price is independent of the amount withdrawn from the inventory.

Acquiring an Inventory and Its Properties:

- Inventories can be acquired by paying a fee.
- The size of the fee is uncertain, but it can be estimated for any particular inventory considered.
- While some choices may exist regarding sites where inventories can be acquired, once acquired, inventories are site specific and cannot be moved except to the market.

❖

So far, a business practitioner studying this case would not only dispute any uniqueness to exploration, but find exploration a most simple version of an inventory problem. Common complications do not exist. For example, the unit price and quantity of sales are not related, which eliminates the cumbersome task of determining products' price–supply functions. Practitioners accustomed to dealing with complicated optimum siting of inventories would find exploration's site-specific inventories unusually simple.

❖

- An overall statistical correlation appears to exist between the size of the fee and inventory acquired; however, the correlation is far from unimodal. A huge inventory purchased with a small fee is not uncommon. Likewise, a huge fee can be used to purchase an empty inventory.
- Any particular fee to acquire an inventory may actually acquire none at all.
- If an inventory is acquired with a fee, the precise quantity will never be known and cannot be counted. The quantity of inventory acquired can be estimated, but only with huge uncertainties that may or may not decrease with time.
- Subsequently, the quantity of the inventory can be repeatedly estimated, but substantial uncertainties will be omnipresent.
- Further, only some fraction of the total of any inventory acquired can ever be withdrawn for sale to the market.
- One can seek to increase the fraction of inventory that can be withdrawn by more investment. How the fraction varies with investment can be estimated; however, uncertainties are ineradicable. Further, no audit could ever determine which particular investment caused any increase in withdrawable inventory.
- One can also seek to withdraw from inventory to the market at higher rates. That, too, requires additional investment. Faster withdrawal rates can also affect the fraction, either higher or lower. Such effects can be estimated, but with inherent uncertainties.

❖

Consider now how the credulity of an ordinary business practitioner with usual product inventory expertise has been assaulted. Every grocer can, at any time, determine the exact quantity of a particular cereal in inventory—by counting, if by no other way. Further, the grocer can determine when and how each cereal box entered the inventory and can track its egress to a customer. At the appropriate time, the grocer can acquire another inventory of x units of the cereal. The purchase price is determinable with acquisition. Delivery of the x units purchased is presumed.

Suppose a grocer, ordering 17.7 million units of cereal, would have to do business as follows:

1. Pay a fee, the full amount to be determined subsequently. The fee secures certain rights and pays for activities that *may* attain the objective of acquiring 17.7 million units of marketable cereal.
2. For the fee, whatever the amount, 8.6% of the time an inventory will be acquired; 91.4% of the time no inventory will result, but the fee will not be returned.
3. If an inventory is acquired, the *weighted average* quantity of withdrawals that can be made will be the 17.7 million units sought. In contrast, 50% of the time, the quantity that can be withdrawn will be less than 0.33 million units.

The numbers used here were drawn from an actual study of exploration history through 1982 in the East Texas geological province (Lohrenz, 1984). The name of the product has only been changed from oil to cereal. A vender of cereals would, no doubt, recoil from any opportunity to acquire an inventory in this manner.

❖

The Economic Outcomes

- Clearly, one has choices, fraught with uncertainties, regarding (1) the rate of withdrawals from an existing inventory and (2) the fraction of initially acquired inventory that can be made withdrawable. One can make the choices to maximize economic outcomes deemed desirable.
- Four different economic outcomes can occur when a fee is paid with the hope of acquiring an inventory:
    1. No inventory is acquired; the total fee is lost.
    2. An inventory is acquired, but its size is deemed so small that additional investment to bring the inventory to market is not justified. The total fee is lost.
    3. An inventory is acquired and deemed large enough to justify additional investment to bring the inventory to market, but not large enough to compensate for the fee. There is a net loss, but less than the total fee.
    4. An inventory is acquired and deemed to justify additional investment to bring the inventory to market and is large enough to more than compensate for the fee.

The first three outcomes are economic failures; only the last outcome is an economic success. Thus, to be successful in ventures with this inventory, the aggregated gains from the last outcome must more than compensate for the aggregated losses of the first three.

- Finally, for each and every increment of every expenditure (fee or investment) made to bring the inventory to market, one can obtain technical estimates of any changes in the ultimately withdrawable inventory or rates of withdrawal that will or did result. Even for the best estimates, substantial uncertainties will remain. Yet these estimates are the only information that will become available. *One can never run a controlled experiment or audit that counts precisely how much more inventory was withdrawn due to a specific investment. One can never track or examine a unit of inventory and ascertain when and how the unit became withdrawable—or why it did not.*

Exploration as a General Inventory Problem Case Study

The case study specifications show how exploration's inventories confound the powerful and elegant paradigms of keen business practitioners for general product inventory problems. What, then, precisely is it that makes exploration and its inventories unique?

Let's explore an answer that is often heard: the risks and uncertainties of exploration are inherently and uniquely large. The answer is not illogical nor wholly wrong. However, *every* investor in inventories for sale to the market prudently considers risks and uncertainties. No purely riskless investments exist. Business practitioners have reasonably argued that the huge risks and uncertainties do not *ipso facto* make exploration or any other business unique. A case for the uniqueness of exploration based only on huge risks and uncertainties is not complete. Practitioners who already know about and accept risks and uncertainties with impacts ranging from huge to miniscule will not be convinced that exploration is unique.

A better answer emerges by considering the following methodologies routinely useful throughout product inventory businesses:

- A consumer product marketer conducts controlled tests showing how an investment in a new package or display can increase market purchases.
- A botanist seeking lower cost acquisition of feed grain inventories conducts controlled experiments showing how investment in a new variety can yield specific increases to marketable inventory.
- A chemist seeking more desirable reaction products runs reproducible experiments examining discernible variables, say, pressure, temperature, and reactant concentration, to determine the best reactant conditions and compositions.
- An auditor, perhaps engaged to check a retailer's product inventory, counts every product by type at two different times and then checks whether the difference between the two times equates with inventory additions and sales.

None of these methodologies can guarantee a perfect answer, but they can be replicated. Replication yields an answer that can be statistically bounded. Replication incurs costs, of course. Usually, more replication and costs provide an answer within tighter statistical bounds.

However, this is not the case in exploration and the earth sciences, for which the powerful methodologies of general business practitioners do not suffice. One cannot conduct reproducible experiments on an undiscovered or even discovered oil and gas reservoir. One cannot "ask" a barrel of oil which incremental investment caused it to issue forth to the market from its *in situ* lair. A census taker can ask individual members of a population their birth date and site and, subject to an error rate that can be checked, expect reliable answers. In contrast, no earth scientist can query a barrel of oil or MCF of gas to determine its origins, discovery date, or reasons why the oil or gas is available for query. A census taker, given sufficient budget and energy, can count every member of an assigned population. But no earth scientist can ever actually count *in situ* barrels of oil or MCFs of gas, withdrawable or not.

The impossibility of conducting controlled experiments and audits based on countable, tangible product units for inventories is the feature that makes the exploration business unique. The assertion has a corollary I have argued in detail elsewhere (Lohrenz, 1988). Usually, the inherent presumption is that one can presume that investing more in information leads to answers converging, at least statistically, to some "truth." The presumption is not a given in exploration and throughout the earth sciences, however. Investments in information may, and frequently do, lead to "answers" that diverge, not converge.

IGNORANCE OF EXPLORATION'S INVENTORY: SOME OUTCOMES NOT SOUGHT

The first rule for a business that markets products is that "you must know the territory." If business practice had a set of unchangeable commandments engraved into stone, surely that rule would be in the set. One can only wonder why there is no parallel rule to wit: "you must also know the inventory." Every salesperson, from the lowest drummer in the boondocks to the highest executive on the top floor of headquarters, must know about the territories and quantities *plus the qualities of the inventories* involved. The rule intoning knowledge of inventories should have suasion equal to that for territories.

In essence, the previous section described how general business training and practice inures practitioners to knowledge of exploration's inventories. There has been no dearth of technically sound efforts to communicate the unique qualities of exploration's inventories. Adjectives in the earth science literature to describe a particular kind of oil and/or gas reserve inventory are plentiful. A partial list follows:

- Proved, probable, possible, marginal, potential, inferred, indigenous
- Demonstrated, estimated, speculative
- In-place, recoverable
- Discovered, undiscovered, demonstrated
- Primary, secondary, tertiary, quaternary
- Risked, unrisked
- Economic, subeconomic, noneconomic
- Searchable, undeveloped, developable
- Producible, declining, geriatric

Each of these adjectives was coined in an attempt to describe the qualities of an *in situ* inventory of oil that cannot be counted or audited like an inventory of oil in a tank battery. All of these adjectives are attempts to communicate an exploration inventory's unique qualities.

Exceptions to the pattern of misunderstood exploration inventories do exist. Brown (1989) recently wrote that "the analogy between petroleum reserves and inventories . . . is especially useful as a *peculiar sort of semifinished inventory*" (italic added for emphasis). Brown, an economist who serves here as a business practitioner of superior perceptiveness, distinguished between undeveloped and developed (presumably "finished") proved reserves.

Brown is a rare and enlightened exception, however. Even the title of an article, "Exploration for Petroleum and the Inventory of Proven Reserves" (Flam and Moxnes, 1987), which states a concern that Norway's "exploratory effort occurs at appropriate rates and times," shows the usual misunderstanding. Rose and Rosenthal (1989) studied Federal Outer Continental Shelf (OCS) policies that would lease tracts for exploration only when reserves exceed some "minimum field size desirable to lease today." The long history of misunderstandings of exploration's inventories has hardly been marred and has certainly not been reversed.

One particular misunderstanding is, no doubt, the most pervasive. When x barrels of oil are removed from a tank that contained X barrels, one can expect the remaining tank contents to be $(X - x)$ barrels. However, when one transfers x barrels of oil from an inventory perceived to hold X barrels of undiscovered oil, there is no requirement that the remaining undiscovered oil inventory is $(X - x)$ barrels. In fact, exploration and other activities, if successful, show a pattern of additions to inventories from which withdrawals are taken. Brown (1989) found it necessary to note this "odd" quality and wrote, "Thus, another peculiarity of reserves as semifinished inventory: establishing productive capacity (development) often creates new semifinished inventory." Brown recognized that $(X - x)$ algebra did not apply and that this was a peculiar quality of his inventory.

Brown's work appears in a 1989 issue of *The Energy Journal*, the flagship scholarly journal of the International Association for Energy Economics. Brown and his intended audience are leading edge scholars in business practice treating oil and gas production. Thus, his preachments about the "peculiarities" of oil and gas inventories were only for the cognoscente.

What are the outcomes of making the mistake of using $(X - x)$ algebra on exploration's inventories? One outcome can be both profound and insidious. The mistake is at the foundation of the methodology that seeks (a) to estimate current and future inventories of oil and gas exploration, (b) to estimate future demands for oil and gas, (c) to then predict the extent of depletion of the inventories through time, and finally, (d) to promote policies consistent with that predicted depletion. This methodology has been and is routinely applied to exploration's inventories in both private and government sectors. The outcome is profound in that sought-for quantitative charts of alternatives to the future are presented. If not fatally flawed, the outcome would indeed be an analyst's *opus magnus*.

$(X - x)$ algebra in exploration, however, contains an insidious shift of focus. This algebra implies that the business decision criterion is optimal management of the inventory from the time when $X = X$ to $X = 0$, akin to a physician supervising the demise of a terminal patient. The inventories X cannot be counted or audited, only estimated. $(X - x)$ algebra charts a course on a map that can and does change at whim. Ever-changing maps are the bane of navigators, but ever-changing estimates of X are signatures of the activity of exploration's inventories.

$(X - x)$ algebra underlies the rationale of the judge who, in 1974, set X equal to (an estimated) 19 billion bbl of recoverable oil to come from the Atlantic OCS. At last count, over $3 billion dollars in bonuses and 46 wells later, x is still precisely zero and it is fair to surmise that X is an even more uncertain estimate than before. The refineries that the judge approved to process the 19 billion bbl have (mercifully for investors) not been built.

The Energy Policy Project of the Ford Foundation (1974) gathered eminent scholars to analyze oil and gas planning options. Their report found the U.S. Geological Survey's Atlantic OCS estimates flawed, to wit:

> Big fluctuations in recent Geological Survey estimates of the oil resource potential . . . —which changed from 114 billion barrel in 1972 to '8 to 16' billion barrels in 1974—illustrate the magnitude of the government's ignorance about the resources.

The tone of the business practitioner who penned those words is clear. Official functionaries had been sent to determine the size of an uncountable inventory. The functionaries brought back different estimates, which the practitioner wanted to use as a count. Observing that the functionaries were not providing stable counts, the practitioner believed the reason must be the functionaries' ignorance. Yet here, the practitioner's own ignorance of the qualities of exploration's inventories was revealed. The practitioner made the mistake of using $(X - x)$ algebra and was distraught that the functionaries would not (and could not) supply an X to use in their aberrant algebra.

The outcomes of ignorance of the qualities of exploration's inventories, of using $(X - x)$ algebra, are predictably and demonstrably undesirable. Projections crumble, predictions become nonsense, and technical discourse erodes. The ignorance does not simply give rise to inconsequential bickering among scholars and experts. The ignorance burdens basic plans and fundamental policy making throughout all levels of industry and government.

CONCLUSIONS

The syntax of an ordinary product inventory problem has been used here to describe oil and gas exploration. The qualities of exploration's inventories are unique and not ordinary. Withdrawals from exploration inventories do not necessarily deplete inventories. Unlike normal inventories, exploration's inventories cannot ever be counted or audited; only estimates can be adduced. That is why exploration is a unique business and not amenable to the paradigms used elsewhere in business practice.

When the unique nature of exploration's inventories are not recognized and the general business practice paradigms are applied, the outcomes carry a malady that cannot be cured. Projections disintegrate, subsequent events show that predictions were foolish, and technical arguments bloom with inanities. Decision making crucial to the future of an organization or nation becomes riddled with the lack of confidence that results when great plans repeatedly flounder.

Those who know the qualities of exploration's inventories, who know how ludicrous $(X - x)$ algebra is, may think even a minute ray of light should obliterate any last vestige of the burdening ignorance. That ignorance is identified here as the presumption that estimates of exploration's inventories can be used as counts like a grocer counts cans of corn and peas on shelves. The source of the ignorance is not the inability to comprehend complex esoterica requiring sophisticated preparation to understand. The source is blindness to the profoundly simple.

Yet the paradigms for ordinary product inventory problems are so powerful and widely effective that business practitioners tend to be inured to recognition of the unique qualities of exploration's inventories. One cannot be sanguine that business practitioners will soon cease using $(X - x)$ algebra. Perhaps all that has been done here is to articulate for explorationists why their business is misunderstood.

As the misunderstandings persist, explorationists will continue to confront practitioners intoning that "it is possible to explain the search for oil in the USA in terms of a combination of economic behavioral functions, lag schemes, and linking identities" (Drollas, 1986). Explorationists will still opt to explain their inventories in terms of equations, pictures, and so on depicting rocks and fluids never seen.

Exploration's inventories, uncountable like a retailer's stock in house, originate as ideas warehoused in cerebella of assorted explorationists. Audit counts with weights and measures in those warehouses are likewise impossible.

References Cited

Brown, K. C., 1989, Reserves and reserve production ratios in imperfect markets: The Energy Journal, v. 10, n. 2 (April), p. 177-186.

Drollas, L. P., 1986, The search for oil in the U.S.A., an econometric approach: Energy Economics, v. 8, n. 3 (July), p. 155-164.

Flam, S. D., and E. Moxnes, 1987, Exploration for petroleum and the inventory of proven reserves: Energy Economics, v. 9, n. 1 (Jan.), p. 190-194.

Energy Policy Project of the Ford Foundation, 1974, A time to choose: America's energy future: Cambridge, MA, Ballinger Publishing Company, p. 279-280.

Lohrenz, J., 1984, Histories of oil and gas exploration outcomes in thirty-two onshore geological provinces: Report prepared for and released by Gulf Oil Company, Sept. [Available from the author.]

Lohrenz, J., 1988, Net values of our information: Journal of Petroleum Technology, v. 40, p. 499-503.

Rose, M., and D. Rosenthal, 1989, The timing of oil and gas leasing on the Outer Continental Shelf: Theory and policies: The Energy Journal, v. 10, n. 2 (April), p. 109-131.

JOHN LOHRENZ is Associate Professor and Coordinator of Research in the Department of Petroleum Engineering and Geosciences at Louisiana Tech University in Ruston. He assumed that position in 1989 after years of service with oil industry companies and the Conservation Division (predecessor to the MMS) of the U.S. Geological Survey. While with the Geological Survey, he originated the widely used LPR data bases about revenues from Federal government mineral leases. He has published extensively on oil and gas economics. Dr. Lohrenz received a B.S. degree from Kansas State University, an M.S. from the University of Oklahoma, and a Ph.D. from the University of Kansas, all in Chemical Engineering. He joined SPE in 1960 and became a Distinguished Member in 1986. He also has served as Review Chairman for the Journal of Petroleum Technology.

PART II

NATURE OF THE BUSINESS

The five chapters in this part set forth some fundamentals of petroleum exploration. They cover exploration uncertainties, estimating prospect sizes, chance of success, selecting and assessing plays, and risk behavior. Each chapter can be read separately, but the chapters are arranged in progressive order. First, any explorationist or investor deals with uncertainty and needs to understand it (Chapter 5). Then, before drilling a prospect, its reserve size must be estimated (Chapter 6) and its chance of success determined (Chapter 7). Exploration prospects and related producing fields can be grouped into "plays", and assessment of the relative merits of different plays is usually the next step in the exploration business (Chapter 8). Lastly, an appreciation of risk (and risk aversion) is important to any serious explorationist or investor (Chapter 9). A synopsis of each chapter follows.

Chapter 5, "Dealing with Exploration Uncertainties," by Ed Capen, is extensive and very readable and includes some novel approaches developed by the author. He uses mathematical statistics and graphical procedures to define and analyze the hierarchy of uncertainties in petroleum exploration. He presents procedures for determining the distribution of possible oil reserve outcomes from single horizon prospects to multiprospect portfolios. The author concentrates on the pragmatic application of probability and statistics to exploration information by using real-world examples. A "roadmap" for uncertainty calculations is included.

Chapter 6, "Estimating Prospect Sizes," by Robert E. Megill, deals with how to estimate the size of potential prospect reserves. It covers the importance of using the full range of possible outcomes, rather than a single number, for the expected reserve size of a prospect. The author recommends that an effective way to demonstrate this is to compare predrill reserve estimates with postdrill results for a series of exploratory wells.

Chapter 7, "Chance of Success and Its Use in Petroleum Exploration," by Peter R. Rose, includes the following major topics: (1) geological chance factors and probability of success, (2) chance of commercial success, (3) subjective probability estimates in exploration, and (4) probability of discovery and prospect inventories. The author recommends that, by using the procedures presented here, explorationists can consistently assess the chances of success. Notable is Figure 1, which portrays different definitions of "success" in relation to probabilities of occurrence and general reserve levels.

Chapter 8, "Selecting and Assessing Plays," by David A. White, presents a prospect grading system that directly postulates success ratio and number of future fields from the geology of the target play itself. Figure 2, which relates prospect size distributions and the mean and maximum sizes for the largest prospect within each distribution, is a major contribution. The author concludes that the strengths of this assessment process lie in systematically documenting assumptions and in having multiple opportunities that will average out to success in the long term.

Chapter 9, "Risk Behavior in Petroleum Exploration," by Peter R. Rose, uses the term *risk* in the sense of *risk aversion,* which weighs the magnitude of investment against (1) availability of funds, (2) potential gain, (3) potential loss, and (4) probabilities of each outcome. The author states that risk aversion causes explorationists to make inconsistent investment decisions and that it costs exploration companies real money in lost opportunities, bad choices, and wasted investment dollars. At the end of the chapter, he presents several ways to improve risk decisions.

Chapter 5

Dealing with Exploration Uncertainties

Ed Capen

Consultant
Dallas, Texas, U.S.A.

If you come to a fork in the road, take it.
—Yogi Berra

PURPOSE

Exploration for oil and gas should fulfill the most adventurous in their quest for excitement and surprise. Unfortunately, the investment community does not thrive on that kind of nourishment—especially the surprises. So how do we take a surprise-prone business and structure recommendations so that both management and investors gain enough comfort to bolster their confidence? At the same time, how do explorationists build credibility when the nature of their job is, most always, to be wrong?

This chapter tries to cover that tall order. We'll touch on the magnitude of the uncertainty (which is far greater than in most other businesses), the effects of not knowing target sizes very well, how to build uncertainty into analyses naturally, how to tie reserves and chance estimates to economics, and how to look at the portfolio effect of an exploration program. With no apologies, we'll be using a different language for some readers—the language of *uncertainty*, which means probability and statistics. These tools will allow us to combine largely subjective exploration information with the more analytical data from the engineering and economic side.

Forget pie in the sky; we'll be concentrating on the pragmatic use of this material because, as a business man or woman in exploration, you have little leisure for chasing theoretical zephyrs. The sole purpose of this chapter is to help you make more money!

A FABLE

Malcom has been asked by his management to do a risk assessment on his prospect before they will consider a well. He drags out his maps, musters his intuition, and begins work. No one tells him how to go about the task. He looks at two approaches.

He can do a historical study of the trend and argue that his prospect looks like the average of those discovery sizes over the last 5 years. He decides that a few hours along that path might be instructive, so he heads for the library. What he finds disturbs him. "Small" describes the fields already discovered, so small, in fact, that he cannot figure out why anyone would have drilled for them or even completed them. Being thoughtful, he assumes that they must have been unlucky and, like himself, thought ahead of time that they had bigger targets. But, he reasons, since his maps show something much larger than the 2 million bbls of oil (MMBO) that is the historical average for the last 5 years, he will profit by staying away from these data.

Malcom tells himself that the historical approach would not be a responsible way of doing risk assessment because it does not properly honor the data that correspond to his unique prospect. In fact, he's never quite understood the statistical approach because it seems to disregard, even belittle, the explorationist's feelings about his prospect. (Note that Malcom is already in trouble.)

He next decides to take his own data and think about the possible boundaries on key parameters such as productive area, net pay, and recovery in barrels per acre-foot. Looking at the contours on his map, he figures his best guess would be about 600 acres with maybe ±100 acres either way. He thinks he'll see about 30 ft of net sand with a range of 20 to 40 ft. Based on the seismic reflection, he thinks he's got a little better porosity than others have found so he goes with 400 ± 50 bbls/acre-ft. He multiplies the low sides, the middle ones, and the high sides together to get reserves uncertainty:

Low:	500 acres x 20 ft x 350 bbl/acre-ft = 3.5 MMBO
Medium:	600 acres x 30 ft x 400 bbl/acre-ft = 7.2 MMBO
High:	700 acres x 40 ft x 450 bbl/acre-ft = 12.6 MMBO

Malcom then looks at the chance his well will strike hydrocarbons. He notices in the data where about 25% of the new field wildcats in the trend have been hitting. Because his target surpasses what the industry has found, he decides on a 35% chance of discovery.

So Malcom now strolls into the management meeting and tells them he has a prospect of about 7 MMBO with an uncertainty range of 4 to 13 MMBO and a 35% chance of hitting. After some probing of the geological setting, management wants to know if the prospect will support itself and three dry prospects. Malcom says he doesn't know. Management wants to know the chance that the well would make money, including G & G and overhead. Malcom doesn't know. Management wants to know how the company will look at the end of the year if it participates in 30 prospects like Malcom's. Malcom says he doesn't know. One manager remembers that most companies got big because of one major discovery, so he wants to know the chance of finding 100 MMBO on Malcom's prospect. Malcom said he doesn't know, but he guesses zero since the largest field he considered was a little less than 13 MMBOE. Management asks how many uneconomic fields were discovered on this trend in the last year or so. Malcom says, "Quite a few, but my prospect looks so much better than the average drilled by the industry that I didn't really look at the small ones."

One manager shows Malcom Figure 1, saying that she obtained it from another company working the area. She asks Malcom why the other company pictured the trend so differently. Malcom does not understand the picture, or at least not the reserves numbers around a portion of the circumference. He laughs and suggests that the competitor must be stupid to go around completing all those uneconomic wells at less than 1.5 MMBO. The manager, trying to be patient, replies that she doesn't think the competitor wanted to complete these wells either, but if you drill and find something small, it's better to recover a portion of your costs than none at all. She reminds Malcom that the most frequent field size is one well producing about 25,000 bbls. "How, Malcom, do you account for that possibility in the numbers you've shared with us?"

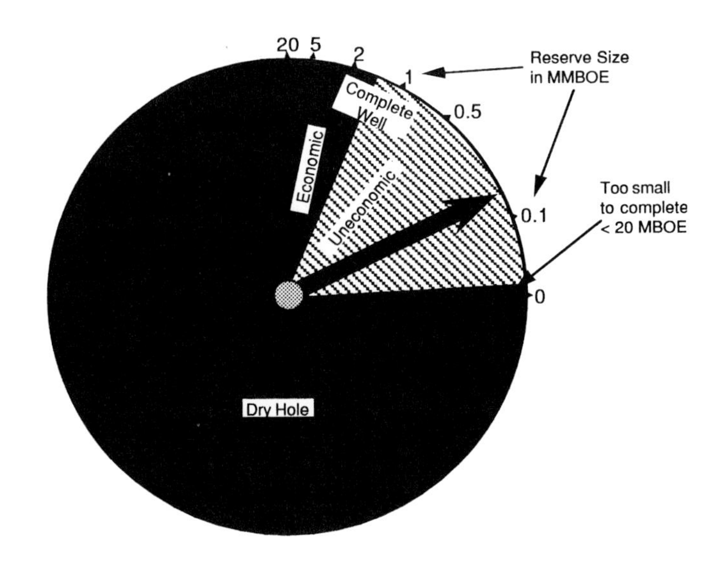

Figure 1. Exploration wheel.

Let's not prolong Malcom's misery any longer. You can see that he had not prepared for the kind of intelligent questions asked by enlightened and knowledgeable managers.

REAL EXAMPLES OF UNCERTAINTY

Look back at the spinner chart in Figure 1 to see what useful information jumps out at you. (Notice that this presentation mode does not require the observer to have a degree in math or statistics to understand the thrust of the diagram.) Clearly, spinning that arrow only once gives a very low chance of making an adequate return, much lower than captured in the well success ratio of 25%. Decision makers need to consider such thinking, particularly if the well uses a significant part of the budget and gambler's ruin becomes a possibility.

The 3 to 13 MMBO range given by Malcom covers a very small part of the wheel, less than 5%. The spinner wheel suggests that Malcom's procedure led him to a biased result. He did not allow for as wide a range of possibilities as nature will likely present him, and in addition, he probably felt motivated to choose his potential field sizes from the high end of the range without accounting for all the small accumulations that he might find. As it turns out, the spinner wheel accurately mirrors the outcome of a single new field wildcat well in the East Texas basin. The numbers on the upper right quadrant were actually placed there with care based on historical data showing discovery chances and sizes.

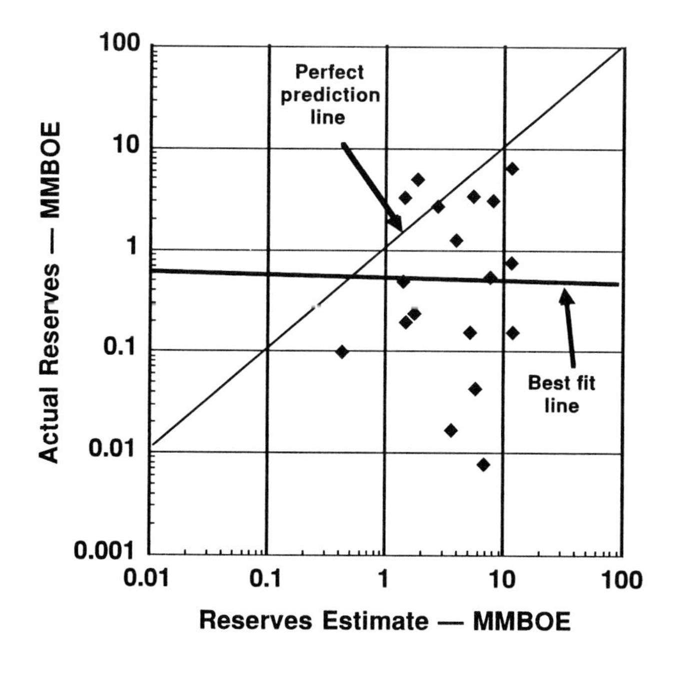

Figure 2. Estimates versus actuals for onshore lower 48, Company X.

Most explorationists dislike all their hard work being characterized as equivalent to spinning a pointer in a random fashion. You might like to see the evidence for treating the problem that way. It all stems from the rather large uncertainties that exist in the oil patch.

Don't take my word for large uncertainties; ask for evidence. Look at Figure 2 and notice the lack of correlation between reserves estimates (including potential) and the actuals established after discovery. (Even the actuals are just estimates, but far more accurate than the original predrilling estimate.) If you snicker at this picture and claim it could not happen to you, then you should keep careful records at your company to confirm how well you do. Remember, the examples you will see come from real companies, making real estimates and investing real money.

The exploration business has arranged itself so that it does not need conventional estimating accuracy. Successful companies either got lucky or exposed themselves to sufficient opportunities that only being right on average was necessary. In fact, any driller who waits until his uncertainties have vanished will find that others willing to drill with less comfort have already tested his prospect. Hence, large differences exist between estimates and actuals.

The eye does not always make intelligent judgments about what it sees when it comes to uncertainties, so we dutifully resort to the tools of mathematical statistics. Since this chapter does not propose to teach that subject, the writer will make some statements and hope those readers unfamiliar with the techniques will either accept them or go to any good statistics text for help.

The question I'd most like to ask about such data is this: what's the chance, given all the scatter, that the estimates and the actuals actually correlate as explorationists so desperately hope? Statistics does not answer that question, but it does get close. If the best fitting line through the data runs horizontally, then we would unhappily conclude that the estimates in no way relate to the actuals, that is, knowing one would not help to know the other at all.

Consider the equation of a straight line:

$$\text{Actual} = a + (b)(\text{Estimate})$$

How do the actual discovery sizes relate to our predrilling estimates? If the relationship were perfect, as depicted by the solid line in Figure 2, then $a = 0$ and $b = 1$. The slope, b, of the line is 1 for a 45° line.

$$\text{Actual} = 0 + (1)(\text{Estimate})$$

It turns out that the slope of the best fitting line through the data is about 0, far less than the slope of 1 that describes a 45° line. We did the line fit (called regression) using logarithms consistent with the log-log scale of the plot:

$$\ln \text{actual} = -0.63 - (0.045)(\ln \text{estimate})$$

Most statistical software for personal computers calculates t and F statistics to help answer this question: Is the line's slope different enough from 0 for me to accept that the estimates and actuals are correlated? This chapter is long enough without going into procedures that standard statistics texts handle adequately. Statistics, as a field of mathematics, attempts to extract truth from data—even highly scattered data and even if the "truth" is not particularly welcome. Perhaps this chapter will generate some interest in looking at this marriage of the exploration and statistics disciplines for the purpose of making better decisions.

Let's use the four graphs in Figure 3 to illustrate the concepts for those who need a refresher. For each set of X,Y data, we try to establish whether we can find a significant relationship between the variables. The word *significant* has a special meaning when used in this context—the data must pass a hypothesis test with some stated probability. In graph A, the X's and Y's are so scattered that no line fits better than a horizontal one. That is, knowing one variable gives no help in estimating another. Graph B shows that rare occurrence in analysis in which the X's and Y's correlate almost perfectly and knowing one variable allows you to estimate the other with very high confidence. Much real data looks more like graphs C and D. In C, we have the strong hint of a relationship because the slope of the best line is not horizontal and the points lie relatively close to the line. In graph D, we have the same best fit line, but with more scatter. The data in both C and D yield exactly the same regression equation, but with different confidences. In comparing reserve estimates and actuals, we'll be looking mostly at data resembling A or D.

For Company X's data (Figure 2), all the hard work that went into the reserves size evaluations resulted in estimates completely uncorrelated with actual discovery size. We have no knowledge about whether the company involved made money on this part of their exploration effort. You can see the difficulty in knowing what to pay for leases given this unsettling result.

Let's ask a few other questions of this data. First, what evidence do we have of bias—that tendency to overestimate or underestimate reserves? One reasonable approach would be to look at the ratio of estimates to actuals, or E/A. For an average E/A value, you'd probably like to have 1, that is, the estimates as close as possible to the actuals. What if, for example, you experienced the following two estimates leading to associated actuals:

E	A	E/A
10	0.1	100
0.1	10	0.01

The average of these two ratios is 100.01/2 = 50.005. If you want an average of 1 to show evidence of good work, you didn't get it. Yet the estimates show no bias. Your accuracy is not very good, but you will likely miss on the high side about as often as the low and by the same amount. We need to find a method such that estimating behavior similar to the example gets rewarded as unbiased.

You might notice that simply adding up the estimates and dividing by the sum (Σ) of the actuals yields

$$\sum \text{Est.} / \sum \text{Act.} = 10.1/10.1 = 1$$

So maybe we need this kind of ratio to judge bias. It has its problems, unfortunately. What if you experienced the following?

E	A	% Error
1000	1100	10
1	0.1	900
5	0.25	1900
3	1	200

$$\sum \text{Est.} / \sum \text{Act.} = 1009/1101.35 = 0.92$$

In this business, a value of 0.92 would be great, which means an estimating bias of less than 10%. Yet we had one error to the low side of 10% and three errors to the high side of 900%, 1900%, and 200%. Swamped by one large result, this performance measure just tells how well you do on the big ones. Since a company can waste a lot of money by missing size estimates on smaller accumulations, we ought to work until we find a measure that works correctly for a whole range of sizes.

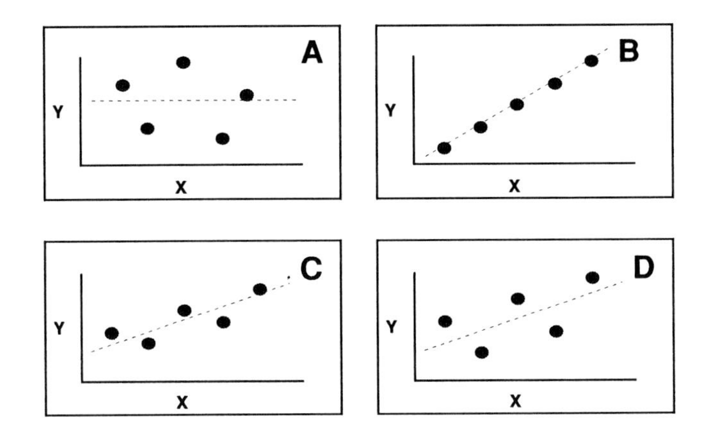

Figure 3. Regression principles.

We propose doing it by looking at the natural logarithm (ln) of the ratio in our first example:

E	A	E/A	ln(E/A)
10	0.1	100	4.6
0.1	10	0.01	–4.6

$$\text{Bias} = (1/n)\sum \ln(\text{E/A}) = (4.6 - 4.6)/2 = 0 \qquad (1)$$

Remember that if the logarithm of a number is 0, then the number must be 1. We want this log ratio to be 0 to correspond with E/A = 1. The variable n is the number of estimates in the sum, that is, the number we divide by to get the average. How does this formula work for the second example?

E	A	% Error	E/A	Ln(E/A)
1000	1100	10	0.91	–0.095
1	0.1	900	10	2.3
5	0.25	1900	20	3.0
3	1	200	3	1.1

$$(1/n)\sum \ln(\text{E/A}) = (1/4)(-0.095 + 2.3 + 3.0 + 1.1) = 1.57$$

The value 1.57 does not look much like 0, which is consistent with just looking at the data and noticing the big errors. Since the 1.57 is a logarithm, we exponentiate to get the estimating bias:

$$e^{1.57} = 4.8$$

This result means we have a *factor* of five bias. Errors in exploration tend to be so large that we can't make sense of percentage errors such as the 1900% and 900% that appeared in the previous tables, so we resort to factors. If you guess something weighs 100 lbs and it turns out to weigh 50 lbs, you missed by a factor of two. Similarly if you guessed 100 lbs and the object turned out to be 200 lbs, you still erred by a factor of two.

In addition to bias, it is helpful to get some idea of average error factor. You could have very high errors and still be unbiased. You could be off to the high side about as often as to the low side. You can tolerate this kind of error if you know about it. Bias, however, can cause economic hardship. Don't fall for the old saw that holds optimistic bias to be acceptable "because explorationists have to be upbeat about their prospects." Being upbeat is one thing; kidding yourself and your investor is quite another.

Average error factor, again, makes use of the logarithms of errors. Say E/A = 10 for one estimate. Then E/A = 0.1 for another. You were off by a factor of 10 both times, just in different directions. Do the same calculation by averaging the absolute values (terms enclosed by the vertical bars, | |) of the log ratios, that is, make all signs positive before averaging.

$$\text{Average error factor} = \text{AEF} = e^{(1/n)\Sigma|\ln(E/A)|} \qquad (2)$$

For the two estimate examples in the preceding paragraph, we would have

$$\begin{aligned}\text{AEF} &= e^{(1/2)(|\ln(10)|+|\ln(0.1)|}\\ &= e^{(1/2)(2.3+2.3)} = e^{4.6/2} = e^{2.3} = 10\end{aligned}$$

See, no mystery. We know 10 is the right answer, and you should have comfort that this method gets us there. Now let's do the calculation for the four-estimate example a few paragraphs back.

E	A	% Error	E/A	|ln(E/A)|
1000	1100	10	0.91	0.095
1	0.1	900	10	2.3
5	0.25	1900	20	3.0
3	1	200	3	1.1

$$\begin{aligned}\text{AEF} &= e^{(1/n)\Sigma|\ln(E/A)|} = e^{(1/4)(0.095+2.3+3.0+1.1)}\\ &= e^{1.62} = 5.1\end{aligned}$$

Notice that the only difference between bias and error factor is that, in bias, we want the low-side errors to cancel the high-side ones, thus we don't take absolute values. For error factor, we want to see all the error whether high side or low side so we disregard the sign of the error to get our average. As it turns out, there's not much difference between bias and AEF in this example—4.8 versus 5.1.

A note on bias. Measuring bias leads to some tricky math far beyond the scope of this chapter. Fractions such as E/A or worse, A/E, cause gross problems because of the nonlinear influences of the denominator. To keep from having to introduce some strange ratio correction factor (Capen and Clapp, 1974), we've recommended a bias measurement on the estimate of mean log reserves rather than on mean reserves. Since the mean of the logarithm translates into the median or 50/50 reserves point, we actually measure bias on the median rather than on the mean. You will not lose much by accepting this compromise.

Why did we just go through all this? To get you familiar with the concepts of uncertainty and how to think about variability in data. With simple examples you should see the reasoning behind some of the ways to look at data. Now back to Company X. We want to measure the bias in those data using the logarithm of ratios as in the recent example. For Company X (you have no tabulated data for this calculation, but you can estimate if from the charts.):

$$\begin{aligned}\text{Bias} &= e^{(1/n)\Sigma\ln(E/A)} &&= e^{1.94} = 7\\ \text{AEF} &= e^{(1/n)\Sigma|\ln(E/A)|} &&= e^{2.23} = 9.3\end{aligned}$$

For this exploration effort, Company X experienced an average estimating error on reserves of about a factor of 10 and a bias of seven times too high. What kind of response can you expect from such calculations? Sometimes utter rejection. Usually the company's top rationalizer will appear and, with Figure 2 in hand, will proceed to explain away every point that's not reasonably close to the 45° line. ". . . The lowest point on the chart was caused by a young explorationist who failed to interpret an obvious fault. We should not count that piece of data because we don't normally make that kind of error around here. The point plotted at 12 MBO actual I remember well. It should count as a dry hole instead. Someone on the well goofed in reading the tests. Also, the 500 MBO estimate with the 100 MBO actual has to go. That well came about due to a log correlation bust by another inexperienced explorationist. We have a much better training program now. . . ." And so on. Readers who tend to fall for such lines might also wish to check out a bridge for sale in Brooklyn.

To dissuade you from thinking that such data are rare, look at Figure 4. In working with any such chart, I propose going through four steps. (1) Look at the graph and observe the scatter. What would be the consequence of such errors? (2) Ask whether the best fit line has a slope sufficiently greater than zero, given the scatter of the data, for us to believe that the actuals are really related to the estimates. Perform the statistical "t" test. (See Chapter Appendix on scientific method and hypothesis testing.) (3) Look for possible bias and test its significance. (4) Check for the average

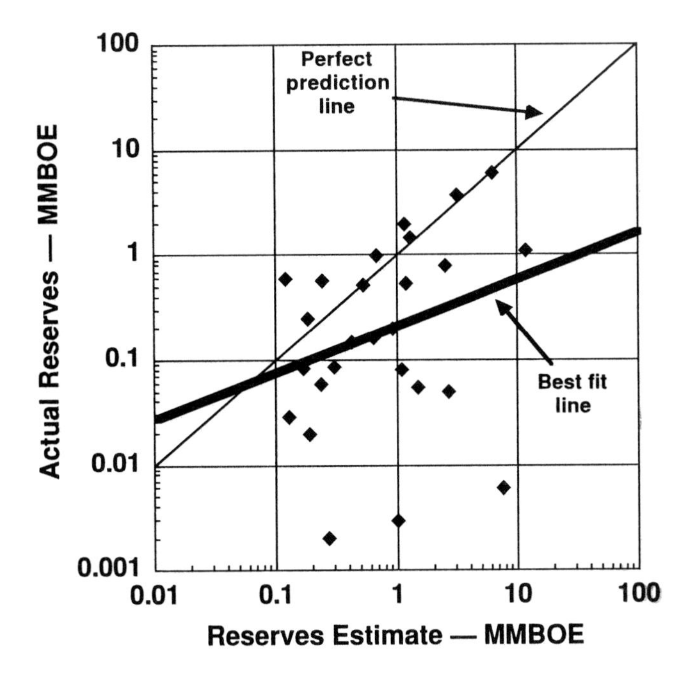

Figure 4. Estimates versus actuals for onshore lower 48, Company Y.

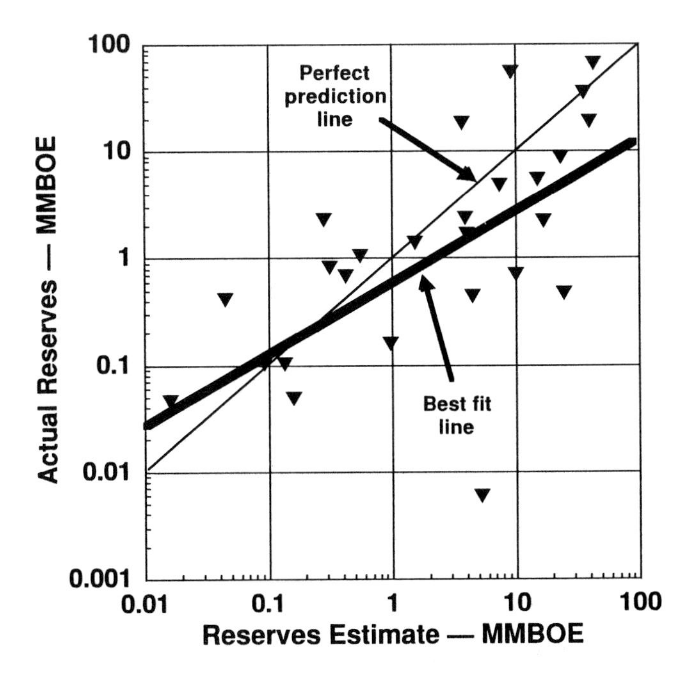

Figure 5. Estimates versus actuals for onshore lower 48, Company Z.

error factor along with its range of uncertainty. Now let's illustrate these steps using Figure 4.

1. Do you find the scatter "acceptable"?
2. $\ln A = -1.64 + 0.46 \ln E$

We find too much scatter to warrant our claiming correlation between the actuals and the estimates (that is, failed the statistical *t* test).

3. Bias = $e^{(1/n)\Sigma \ln(E/A)} = 4.5$

The "*t*" test shows that we must reject the hypothesis claiming no bias.

4. AEF = $e^{(1/n)\Sigma |\ln(E/A)|} = 6.1$

Look at Company Z in Figure 5. Again, here is a company working the onshore lower 48, but with some surprisingly different results. This company had good people as did the first two, but it did not spend nearly as much on technology and had many more joint ventures. Again, let's check out the four steps.

1. The scatter looks very wide and discomforting.
2. $\ln A = -0.26 + 0.65 \ln E$

The slope coefficient of 0.65, along with the scatter, is sufficiently larger than zero to claim a *significant* correlation between the actuals and the estimates. Explorationists did well.

3. Bias = 1.7

The tendency to estimate too high by about 70% is much better than we found in the other two charts, but still fails to support the hypothesis of no bias.

4. AEF = 4.6

How many explorationists think they experience an average error factor of almost five? Again, if you think you do better, prove that fact to yourself by making plots like these.

Now we'll look at some data for Company W from overseas (Figure 6), involving estimates of original oil in place (OOIP) rather than reserves. The scatter does not appear so extreme, but one would probably expect that from international data. In most non-U.S. areas, no one develops the kinds of fields routinely produced in the United States. So we would not expect to see the 100,000 bbl and 1 MMBO fields. We don't really know what size estimates accompanied the "too small to keep" accumulations. The lesson here is that for these performance measures, do not mix data from areas having greatly different economic cutoffs, such as Alaska and Kansas.

Also, since these data are OOIP instead of reserves, we would expect smaller variation because the error of getting from OOIP to reserves will not appear in these data.

2. $\ln A = 1.77 + 0.49 \ln E$

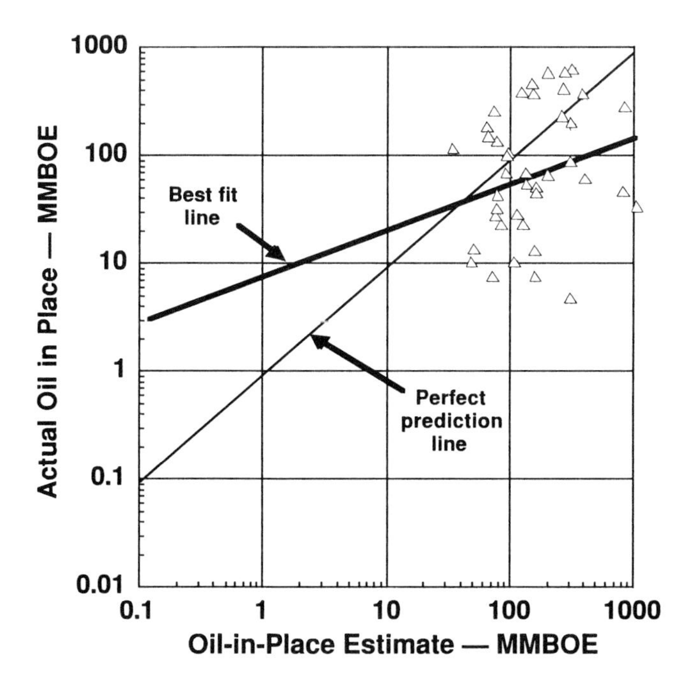

Figure 6. Estimates versus actuals for international data OOIP, Company W.

The 0.49 slope coefficient barely measures significant with our "*t*" test. However, if we had reserves estimates instead of OOIP, the increase in scatter would probably tip the evidence in favor of claiming no significant correlation between estimates and actuals.

3. Bias = 2.1

Here's another measurement helped considerably by the relatively high cutoff for commerciality. Even so, I think this data represents better estimating than two of the three we have already looked at. We cannot distinguish Company Z's 1.7 from Company W's 2.1, so I would argue that they are the same.

Be very careful how you interpret bias in reserves estimates where competitive bidding determines the owner. The winner's curse bias (winner tends to overestimate value) will appear as explorationists' overoptimism, when in fact the two kinds of biases are unrelated (Capen et al., 1971). There are so many variables that a rule of thumb is difficult to give, but we estimate that the "curse" provided about a factor of two bias in the Gulf of Mexico. That means if your explorationists do a great estimating job, you would still expect to find about half as many reserves as you thought when you made the purchases.

4. AEF = 3.5

This statistic also gets help from the economic truncations on the reserves distribution. The same

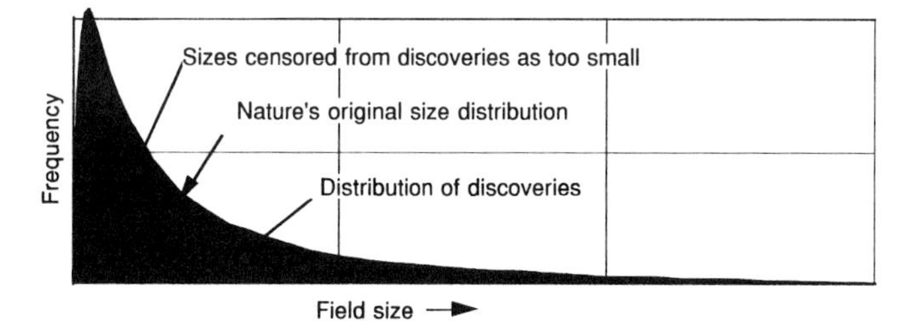

Figure 7. Nature's size distribution compared to the distribution of discoveries.

explorationists working U.S. onshore would probably exhibit an AEF of 5.

Although not presented here, I have examined data from the Gulf of Mexico, where the industry also experiences an economic censoring of the low-end data. We see results very similar to the international data from Company W.

How Can Industry Survive Such Errors?

You might ask how all this can be true—that if explorationists can come no closer than these charts suggest, how do they remain in business? Let's make some educated guesses. A company with an estimating bias suffers a penalty by making a lower return than it calculated; it probably does not lose money on average. And fortunately, no one has to pay development costs for oil and gas that does not exist. Furthermore, a company's lease payments may not always tie that closely to its value estimate, that is, explorers get a bargain now and then. Finally, the competition may have the same problem.

DISTRIBUTION PICTURES

Because the lognormal form of uncertainty makes a useful tool for analysis, we need to make sure everyone understands the basic idea of lognormal. Figure 7 shows pictures of two different lognormals. We have good reason to believe that nature stores her mineral reserves lognormally, as depicted by the light gray, highly skewed distribution shown in Figure 7. We know that multiplication of variables leads to lognormal distributions, so we need only look for nature's multiplications to see how oil and gas fields get that way. The volume of a sand-based reservoir would be proportional to the product of the following:

- drainage area, in itself a product of a length and width;
- duration of deposition;
- volume of precipitation;

- severity of wind and rain storms influencing erosion;
- the elevation difference between the highest and lowest points in the drainage basin; and
- subsidence rate of the area of sedimentation.

For reserves, multiply sand volume by source exposure time, porosity, and the pressure differential driving the hydrocarbons. We have ample multiplications to create a highly skewed distribution that looks enough like a lognormal to let that model serve us.

We know that we do not discover most of the very small accumulations that nature provides in such abundance, although we certainly "find" them often enough. That is, we often run into very small, uneconomic pockets of hydrocarbons. The smaller the accumulation, the less likely we will hit it with our drills, meaning that for every such encounter, thousands will escape detection. Thus, we have the dark gray distribution in Figure 7 showing industry discoveries in some basin. The lower end is censored, meaning a less abrupt cutoff than would occur with a truncation. The difference between the discovered distribution and nature's illustrates that the basin has much remaining oil and gas, some of it in very large fields. I made no attempt to "tilt" the discovery distribution toward the larger fields because of the earlier charts suggesting that explorationists have a difficult time picking the big ones ahead of time, but certainly some tendency in that direction must occur, particularly during the earlier stages of basin exploration.

The significance of this will become more apparent later as we get into using lognormal analysis. By some mathematical luck, being able to use the lognormal distribution for variables such as reserves, net pay, and recovery allows us to solve a series of equations to get all kinds of useful answers about how much risk and uncertainty surrounds a particular prospect, play, or drilling program. That means we can code the solutions onto simple spreadsheet programs for personal computers or even programmable calculators. No more messy and time consuming Monte Carlo runs. No more worry about how many trials to run. And perhaps no more misleading triangular distributions "because they're easy to understand."

Once we accept the lognormal model, we can see what these uncertainties might look like on an actual prospect. As the caption in Figure 8 states, the distributions have a mean of 100; only the variance or spread differs among them. As expected, these curves each have an area of 1.0 beneath them, where the 1.0 equals the probability of all possible events. The probability of getting less than 100 would be the area under the curve between zero at the left and 100 on the right. Looking at the curve with variance = 1, the area from 0 to 100 is about 0.75. Therefore, the chance of having something smaller than 100 is 75%. In this graph, the rectangles

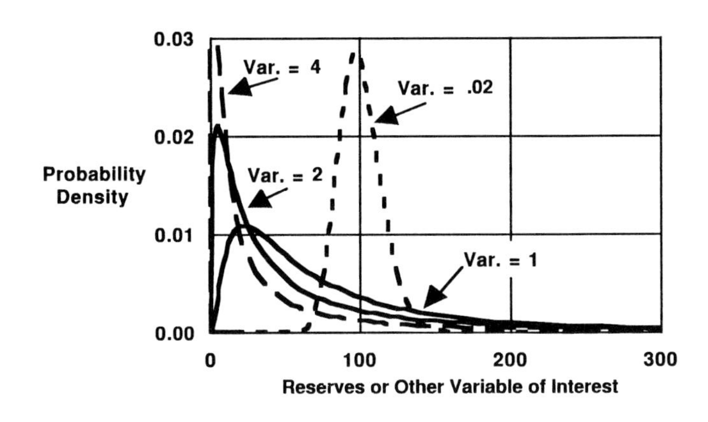

Figure 8. Lognormal densities, with mean = 100 and four different variances.

that make up the grid each have an area of 1.0 (100 length and 0.01 height).

The nearly symmetrical curve of variance = 0.02 describes the uncertainty one might have on a well-defined producing field—about ±20%. If you examine that curve, you will find most of the area under the curve between 80 and 120. Notice that the peak (we call the highest point on the distribution curve is the *mode*, or most likely point), however, is not at 100 but slightly lower. Lognormals have four parameters you may wish to understand. Any two of the four uniquely determine the distribution:

Mode:	That value of the variable that has the highest chance of occurring.
Median:	The point at which there is an equal likelihood of being above or below.
Mean:	The probability weighted average of everything that can happen.
Variance:	A measure of the spread of the distribution.

The first three parameters always occur in this order, with the mode being the smallest and the mean the largest. Sometimes people are tempted to use either a most likely or a median for exploration decisions. Do not do this! The most likely field size in the United States is about 10,000 bbls, the average of many one-well fields. You can't afford to drill for a "most likely" size.

The idea of lognormal may seem intimidating at first, but it needn't be if we examine it step by step. First, consider the normal distribution, the familiar bell shaped curve in Figure 9. These particular curves have a mean of 0, and one has variance = 1 while the other has variance = 3. More variance signifies more spread and a bigger chance of getting an x of greater than 2, for example. Now look at the area under the dotted line to the right of 2 on the x axis and compare it to the area

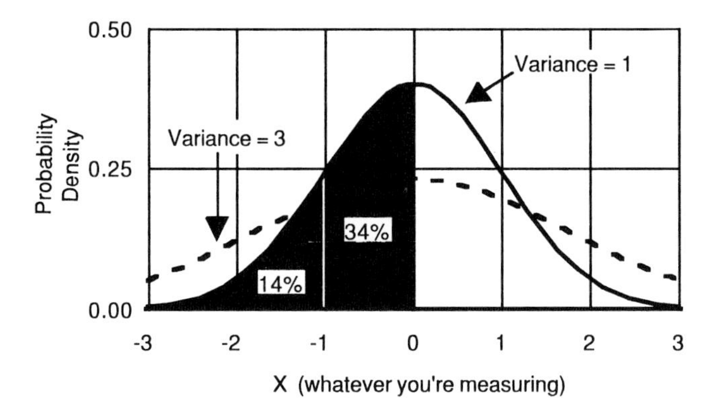

Figure 9. Probability density for normal distributions, with mean = 0, variance = 1, 3.

under the solid line to the right of 2. The eye tells you there is about four times as much area under the dotted curve, meaning four times the probability. Remember, both curves have a total area of 1. As it turns out, this estimate is about right. The solid curve has about 2.3% of its area beyond 2, while the dotted curve has 12.5% of its area there.

Two parameters, mean and variance, define a normal distribution. (The concept of variance sometimes confuses people, so if you want to get into this subject, you should to consult a good textbook.) The unit of variance is the square of the variable of interest. For example, if you plot the distribution of heights of American males, it will center around 69 in. with a variance of 5.5 in. squared. Who can visualize *inches squared* as a parameter in height measures? To make sense of such a number, we take the square root and call it the *standard deviation*, which in this case is about 2.3 in.

For the *solid* line in Figure 9, variance = standard deviation = 1, and the x axis shows how standard deviation units relate to the normal curve itself. Going 1 standard deviation unit away from the mean to the left creates an area equal to about 34% of the total area under the curve. Between the first and second standard deviation units, we have an area of about 14% of the total. That leaves about 2% to the left of –2 since half the area must be to the left of the mean for this symmetrical distribution.

If you wanted to plot some of these distributions, you would use what mathematical statisticians call a *probability density*, named this because it shows the chance density for all values of the variable of interest. For the normal curve,

$$f_N(x) = \left(1 / \sigma\sqrt{2\pi}\right)e^{-(x-\mu)^2/2\sigma^2}$$

$$= \text{Height of curve} \qquad (3)$$

where μ is the mean and σ^2 is the variance.

If you wish to plot lognormals, use the following density function:

$$f_{\ln}(y) = \left(1 / y\sigma\sqrt{2\pi}\right)e^{-(\ln y-\mu)^2/2\sigma^2} \qquad (4)$$

The variable y denotes our quantity of interest, say reserves or pay thickness. You'll notice the similarity between this formula and that of the normal density. Instead of an x in the exponent, we have ln y (the natural logarithm of y). That correctly suggests that the logarithm of y follows a normal distribution. The y appears in the denominator of the first term where no x appears for the normal distribution. Also, μ is the mean of the natural logarithm (ln y) and σ^2 is the variance of the natural logarithm of y.

CUMULATIVE LOGNORMALS AND PLOTTING DATA

You can put the formulas just discussed onto spread sheets and make all kinds of plots to see how they work. But while these pictures help us understand what's happening, they fall short of letting us do much quantitative work. So we must introduce another kind of plot that is very useful to the explorationist: the *cumulative probability plot*.

Remember the previous eyeball estimates we made of what portion of the area under the curves lay beyond a certain value on the x axis? An explorationist needs to deal with exactly these kinds of questions. What's the chance of having more than 10 MMBO given a productive well? What chance should one assign to getting noncommercial amounts of hydrocarbon given that the exploratory well flows some oil or gas? What chance do you have of hitting the "mother lode," say something above a billion barrels? What if the size of discovery dictates the development strategy? You'd have to know that prior to drilling the exploratory well because the economics of that well must include the outcome of the success, that is, the development leg.

If you recall our heights of American males and "area under curve" discussions earlier, and then remember that integral calculus finds areas under curves, it becomes apparent how a mathematician would solve our problem. For example, the *chance* of a randomly chosen adult American male being over 78 in. tall is

$$\int_0^{78} f_N(x)dx$$

where the integration takes place from 78 in. to infinity. If you had to sit down with integral tables or a computer every time you wanted to answer one of the questions just posed, you see that such questions would go largely unanswered.

Fortunately, graphics comes to our rescue. We can use a piece of lognormal graph paper and a straight edge to solve the problem. On this kind of paper, we construct the scales nonlinearly such that a straight line describes a lognormal distribution. In Figure 10, both lines go through 10 at the 50% point so each has a median of 10. On the solid line, "A" indicates a 10% chance of less than 8, "B" a 35% chance of less than 9, and "C" a 90% chance of less than 12. The dashed line shows more variance, thus, it's a steeper line. On it we read "D," a 40% chance of less than 9, and "E," a 60% chance of less than 12. By subtracting, we get a 20% chance of falling between 9 and 12. Once you have the line, you can get any number you need to aid your analysis of exploration uncertainty because this is the line that completely describes that uncertainty.

You need to learn how to travel this road both ways. Sometimes you start with the line and get from it the mean and variance. Sometimes you start with mean and variance and need to construct the line. Earlier we showed cross plots of reserves estimates versus actuals, that is, data. From those kind of data, we can calculate the mean and variance needed to construct our line.

$$\text{Mean of } x = \mu = \sum p_i x_i \qquad (5)$$

which is the sum of all values of x times the individual probabilities of those values. However, we usually do not have all the values; we just have a sample of n values. So we can never calculate the mean. Instead, we calculate the *sample average*. All the values in our sample, if it is truly a random sample, should have the same chance, $1/n$. So we have

$$\text{Sample average} = \bar{x} = \left(\sum x\right)/n \qquad (6)$$

where n is the number in the sample, and

$$\overline{\ln y} = \left(\sum \ln y\right)/n$$

We'll use this sample average as if it were μ. We have a similar problem with variance in that, with data, we can only get a sample variance, which we denote s^2 and use for σ^2. These substitutions sometimes introduce small errors, but not enough to cause us any problems.

$$s^2 = \left[\sum (\ln y)^2 - \left(\sum \ln y\right)^2/n\right]/(n-1) \qquad (7)$$

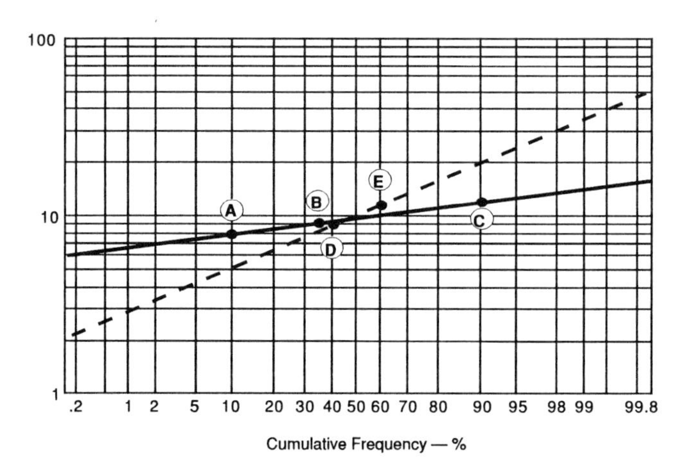

Figure 10. Different variances.

and

$$s = \sqrt{s^2}$$

Let's illustrate these principles with some average porosity data from Canadian oil fields. The numbers come from an Alberta government publication starting on page 1 and selecting every fifth field. Since we have 36 data points here, they represent 180 fields. Then we take the data and sort them from low to high, as shown in Table 1. Typical of lognormal data, we see many low porosities and very few high porosities. Let's plot the data and see if we feel comfortable with our choice of lognormal. For each porosity, we must get its natural logarithm (Table 1, column 3). Finally, we need to see where that porosity will plot on a cumulative frequency graph, as previously discussed. The question we ask of each piece of data is, what percentage of the points in the sample have values less than or equal to the point in question? Look at point number 36 in Table 1. It's clear that 100% (36/36) of the points in the sample equal 0.27 or less. Just as clear at the other end is that 0.028 (1/36) of the points equals 3% or less porosity. So it looks like we would use the simple formula of

$$\text{Cumulative frequency} = (\text{Point number})/(\text{Total number of points}) \qquad (8)$$

Actually, we employ a slightly different formula to make the plotted points spread more evenly across the graph. We add 1 to the denominator. This device leaves as much probability space above the highest point as we have below the lowest point. That trick gives us some intuitive comfort since we know the highest point in the sample may be surpassed if we gather enough data.

$$\text{Cumulative frequency} = (\text{Point number})/(1 + \text{Total number of points}) \qquad (9)$$

Table 1. Canadian Field Porosities

Number	Porosity	Ln Porosity	Cumulative Frequency
1	0.03	−3.51	2.7%
2	0.04	−3.22	5.4%
3	0.05	−3.00	—
4	0.05	−3.00	10.8%
5	0.06	−2.81	—
6	0.06	−2.81	—
7	0.06	−2.81	—
8	0.06	−2.81	21.6%
9	0.07	−2.66	—
10	0.07	−2.66	—
11	0.07	−2.66	—
12	0.07	−2.66	—
13	0.07	−2.66	35.1%
14	0.09	−2.41	—
15	0.09	−2.41	40.5%
16	0.10	−2.30	—
17	0.10	−2.30	—
18	0.10	−2.30	—
19	0.10	−2.30	51.4%
20	0.12	−2.12	54.1%
21	0.13	−2.04	56.8%
22	0.14	−1.97	—
23	0.14	−1.97	62.2%
24	0.15	−1.90	—
25	0.15	−1.90	—
26	0.15	−1.90	70.3%
27	0.18	−1.71	—
28	0.18	−1.71	75.7%
29	0.19	−1.68	—
30	0.19	−1.66	—
31	0.19	−1.66	—
32	0.19	−1.66	—
33	0.19	−1.66	89.2%
34	0.20	−1.61	91.9%
35	0.22	−1.51	94.6%
36	0.27	−1.31	97.3%

Where several points have the same value, we plot only one point to save clutter. See the results in Figure 11.

GOODNESS OF FIT

In working with probability distributions, we often want to know whether the data really come from a given distribution. For example, explorationists say they can understand why pay thickness and productive area might be lognormal because of nature's multiplicative process at work. But they cannot fathom why productivity (barrels per acre-foot) would be lognormal. I have not worked that out yet either; I only observe that the data tend to be skewed so that we find much rock with low porosity and very little with high porosities. That property makes the lognormal hypothesis a reasonable one to work with.

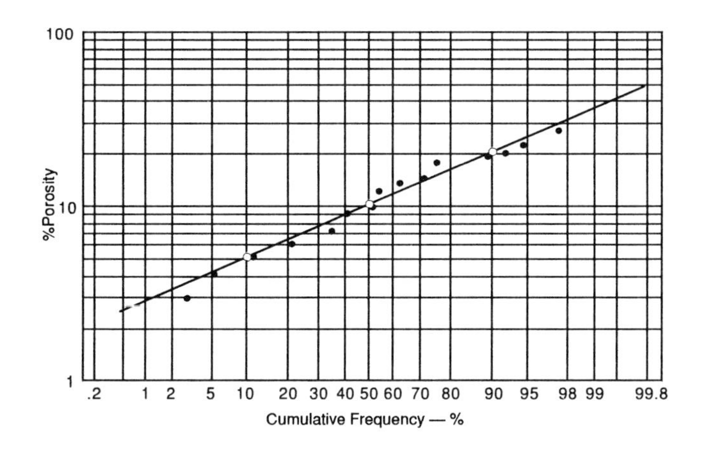

Figure 11. Porosities from Canadian producing zones.

But since we have some real data here, let's introduce a test. The straight line in Figure 11 represents a theoretical lognormal distribution with the same mean and variance as the sample data on porosities. Remember that the calculation of mean and variance does not require an assumption of the form of the distribution (triangular, binomial, normal, lognormal, etc.), so in making that calculation, we do not bias our answer either for or against lognormal. We're just saying we have some data from an unknown distribution but with known mean and variance.

Next we build a hypothesis and see if the data can knock it down. Our hypothesis is that the data in question come from a lognormal distribution with these defining parameters:

$$s^2 = \left[\sum(\ln y)^2 - \left(\sum \ln y\right)^2/\mathrm{n}\right]/(\mathrm{n}-1) = 0.3$$

$$s = \sqrt{s^2} = 0.548$$

$$\overline{\ln y} = \left(\sum \ln y\right)/n = 2.348$$

Then we construct a lognormal line consistent with these values. Recall that the median or 50% point goes with the average of ln y. To find out what porosity that logarithm corresponds to, just exponentiate, as follows:

$$\text{Median porosity} = e^{2.348} = 10.5\%$$

We need at least one more point to plot our line, but we'll get two just as an aid to checking the arithmetic. Let's plot the 90% point and the 10% point. Remember that we calculated s, the sample standard deviation, to get a measure of the spread of the distribution. For a normal or bell-shaped distribution, one standard deviation above the median plots at about the 84% point on the probability graph. Because of symmetry,

Table 2. Williston Basin Discovery Sizes

Size Class	Upper Limit MMBOE	Frequency	Cumulative Frequency	Cum. Probability Kol.–Smir. Test	Cum. Probability Plot
1	0.006	6	6	0.030	0.030
2	0.012	6	12	0.060	0.060
3	0.024	7	19	0.095	0.095
4	0.047	6	25	0.126	0.125
5	0.095	8	33	0.166	0.165
6	0.19	14	47	0.236	0.235
7	0.38	18	65	0.327	0.325
8	0.76	33	98	0.492	0.490
9	1.52	37	135	0.678	0.675
10	3.04	23	158	0.794	0.790
11	6.07	22	180	0.905	0.900
12	12.1	11	191	0.960	0.955
13	24.3	5	196	0.985	0.980
14	48.6	2	198	0.995	0.990
15	97.2	1	199	1.000	0.995
16	194.	0	199	1.000	0.995
17	388.	0	199	1.000	0.995
18	777.	0	199	1.000	0.995
19	1550.	0	199	1.000	0.995
20	BIG!	0	199	1.000	0.995

one standard deviation below the median plots at about the 16% point. Two standard deviations beyond the median plot at the 2.3% and 97.7% points, respectively. You can look these numbers up in any statistics book.

All we need to do is to see how many standard deviations out from the median lies the 90% point on the high side and the 10% point on the low side. The three-digit approximation is 1.28. Thus,

$$\text{10\% point} = e^{\mu - 1.28\sigma} \quad (10)$$

$$\text{50\% point} = e^{\mu} \quad (11)$$

$$\text{90\% point} = e^{\mu + 1.28\sigma} \quad (12)$$

We substitute our estimates for μ and σ to get:

$$\text{10\% point} = e^{2.348 - 1.28 \times 0.548} = 5.2$$

$$\text{50\% point} = e^{2.348} = 10.5$$

$$\text{90\% point} = e^{2.348 + 1.28 \times 0.548} = 21.1$$

If you look at Figure 11, you'll see three small, white circles plotted at these three coordinates with a straight line through them.

First do an "eyeball" test. Do you think the data points (black dots) in Figure 11 fall reasonably close to the straight line? Looks pretty good to me. But we need a more formal test than that. Look for the point that lies farthest from the lognormal line. If the farthest point lies close to the line, we should accept that as evidence that the data does indeed come from a lognormal distribution. The question is, how close is close? Kolmogorov and Smirnov (see Massey, 1951) examined that very question and came up with some tables based on probabilities. Using 36 data points, they found that 20% of the time, data sampled from true lognormal distributions would have its most errant point more than 19 percentage points away from the line. If you look at the graph of the porosity data, you'll see that the point farthest from the line lies only 10 percentage points away. So it looks like true lognormal data would resemble our data most of the time, giving us comfort in our assumption of lognormal data. Does that prove absolutely that we have lognormal data? No, but the evidence allows us to proceed with the assumption of lognormal without fear of significant error. To read more on this goodness of fit test, see Hoel (1970) or another more recent statistics textbook.

PROSPECT UNCERTAINTIES VERSUS BASIN VARIANCE

Keep in mind that you will deal with two different size distributions. One comes from real data or a data analog and tells you about the range of historical discoveries in your basin, trend, or play. The other comes from prospect-specific data and describes uncertainty about a specific prospect.

For example, consider Williston basin discoveries from 1971 through 1980. The numbers in Table 2 come from data compiled by Energy and Environmental Analysis (EEA) of Arlington, VA, and are appreciated reserves (Root, 1982), meaning they include a judgment about the ultimate reserves that will be associated with the individual fields rather than just being reported as proved in the early 1980s when EEA gathered the numbers.

Table 2 has two cumulative probability columns, one using the $n + 1$ divisor for plotting and the other using n with the Kolmogorov–Smirnov test. This test expects the n divisor. With 199 data points, it would hardly matter which divisor one used, but we often do not have the luxury of so many points. If you try to use the $n + 1$ divisor and attempt a Kolmogorov–Smirnov test, you could be off by as much as $100/(n + 1)$ percentage points, that is, you would have plotted a point that far away from where the test expects it.

Figure 12 shows the Williston basin data. For this example, we took a straight edge and drew the line that came closest to the most points—an eyeball fit. That trick saves our having to compute the mean, variance, and two points from which to plot the line. In a later example, we'll show that the eye sometimes causes problems when asked to do this task. The nonlinear scale makes it tough. The point plotted at 0.38 MMBOE, and 32% looks about 6 percentage points off the straight line lognormal. That is within the Kolmogorov–Smirnov limit of 9 percentage points for accepting lognormal. Be careful not to let the eye deceive you. The nonlinear probability scale makes it look like the points at the bottom of the graph get farther and farther away from the lognormal line. Measured in percentage points, however, none lie as far as that point plotted at 0.38 MMBOE.

Let's calculate variance from our eyeball lognormal line. Remember that the 90% point is 1.28 standard deviations out from the 50% point, μ.

$$\sigma = (\ln 90\%\ \text{point} - \ln 50\%\ \text{point})/1.28 \qquad (13)$$
$$\sigma = (\ln 8 - \ln 0.7)/1.28$$
$$= [2.080 - (-0.357)]/1.28 = 1.90$$
$$\sigma^2 = 3.6$$

With μ and σ^2, we can now calculate a mean:

$$M = e^{\mu + \sigma^2/2} \qquad (14)$$
$$M = e^{-0.357 + 3.6/2} = e^{1.44} = 4.2\ \text{MMBOE}$$

If you locate that number on the graph in Figure 12, you'll see that it lies just to the right of the 80% point. That means less than 20% of the fields discovered over the 10-year period fell above the average discovery. The result comes about from the standard characteristics of a lognormal distribution with high variance rather than being an "oddity."

How could you check this graphical method? Well, we could just average all the fields in the sample and see if we get about 4 MMBOE. But we don't have all the fields. We *do* have fields in 15 different size categories. Accepting the mid-point of each size category as the size to use for each field in the category, we calculate an average size of 2.8 MMBOE. Is 2.8 MMBO close enough to 4.2 MMBO to believe that both methods get about the same answer? That depends on your tolerance for ambiguity. It turns out that you can believe they are different if you can accept a 15% chance of being wrong.

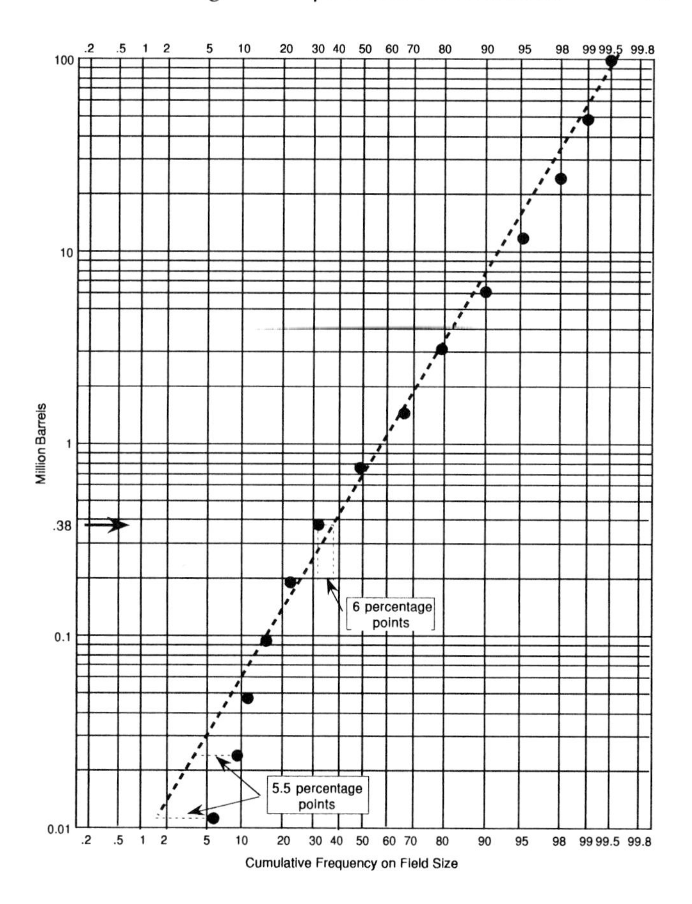

Figure 12. Williston basin discoveries 1971–1980.

One would think that the range of field sizes in a basin would have to be larger than the uncertainty about any one prospect. We can use some of the company data on estimates and actuals to test this hypothesis. Look again at the chart of Company X in Figure 2. We want to calculate the variance of the logarithm of those estimates about their respective actuals. If we calculate

$$\text{Var}\,[\ln(E/A)] = \text{Var}[\ln E - \ln A] \approx \text{Var}\,[\ln E]$$

by assuming no uncertainty about the actuals, then the variance of the ratio equals the variance of the estimates, E. (At least prior to field abandonment, "actuals" are just superior estimates. Calculations [not given here] show that the variance of the actuals is about 0.1. Thus, whether or not we put it in affects our answer very little.)

For Company X: Var[ln E] ≈ 4.9
For Company Y: Var[ln E] ≈ 4.5
For Company Z: Var[ln E] ≈ 4.7

These three variances represent uncertainty about a single prospect. Hypothesis: Prospect uncertainty (variance) exceeds basin variance. Answer: True. (We use a chi-square test on variances, but the details of that test are not given here.) It looks like the uncertainty of the estimate actually exceeds the range of field sizes. This probably happens because the estimates never include those small producers that industry completes for a few thousand barrels because they pay out the completion costs. The E/A ratios for reserves estimates can attain 100 or more. Interestingly, this phenomenon suggests that one would have better estimates if he just used the basin average for all prospects!

Be careful if you use estimating variance as a performance measure for your exploration staff. The uncertainty may be more a function of where one works than the quality of his or her skills. Observe in Figures 2, 3, 5, and 6 how the estimating variance goes down drastically if we only include those discoveries over 1 MMBOE. So it happens that those working offshore or in frontier areas have a built-in advantage since the industry requires large discoveries for commerciality.

A word of caution before we try to make too much of these previous calculations. The Williston basin from which our field variance comes is not heavily represented in the data from Companies X, Y, and Z. Preferably, we would want to compare a company's estimating variance against the field size distribution in its areas of interest.

Most of us have a tough time relating this log variance to magnitude of uncertainty. As an aid, look at Figure 13. The vertical scale shows what is called *estimating error*. For a variance of 4.7, we find an error of 17. That means that median estimates lie within a factor of 17 of actuals about 80% of the time. For comparison, producing properties with variances about 0.1 have an error factor of 1.6.

Does it make sense that explorationists should have such large prospect uncertainties? Probably. First, most of us are culturally deprived about uncertainty. Bosses pay us to tell them how much we know rather than how little, so we don't concentrate much on calibrating uncertainty. Many decision makers in the oil business cannot stand uncertainty and ambiguity, the hallmarks of the exploration business. Such attitudes force explorationists to come up with single "likely" maps or interpretations. Working explorationists know that they can interpret data on a prospect in many different and legitimate ways. They can construct a number of maps from these interpretations, some pessimistic and some optimistic. Ideally, prospect uncertainty must encompass several of these maps.

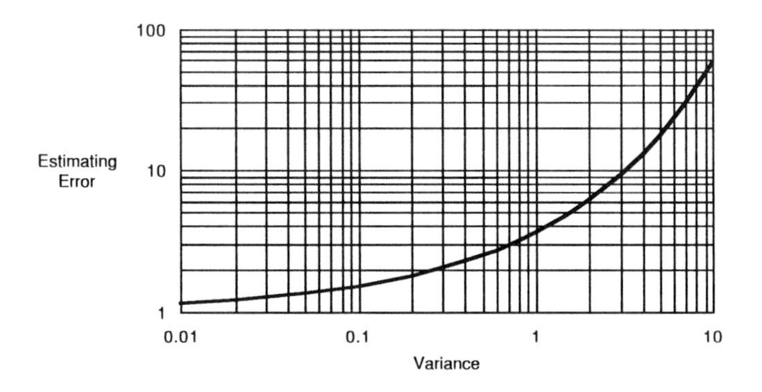

Figure 13. Errors and variance.

A SINGLE HORIZON PROSPECT

Using your maps and some log probability graph paper, you can analyze your uncertainty. You'll want probability ranges on your three most critical variables: net pay thickness, productive area, and recovery in barrels per acre-foot.

Net Pay

Ideally you should start with some data on net pay in your trend from which you can construct a cumulative frequency distribution on lognormal graph paper. If you have no reason to believe your prospect differs from those in the data base, you can simply use that data. Pick the 10% point and the 90% point, connect the two points, and you're finished with net pay. If you haven't enough data or have no data, or you believe the data does not represent your prospect, you'll have to apply judgment. If you think your seismic lines tell you something about pay thickness, you can use that information—being careful to remember seismic limitations. It may indicate gross rather than net pay, or you may err on velocity estimates. Leave yourself some room; admit uncertainty.

Let's say that when you have finished your deliberations, you have a picture that looks like Figure 14. You feel you have a 10% chance of getting less than 10 ft and a 10% chance of getting more than 30 ft. You notice that the line crosses 50% at about 18 ft. You believe in equal chances of getting less than or more than 18 feet of net pay given the prospect produces. Remember, these numbers apply not to an individual well net pay but to the average over the prospect.

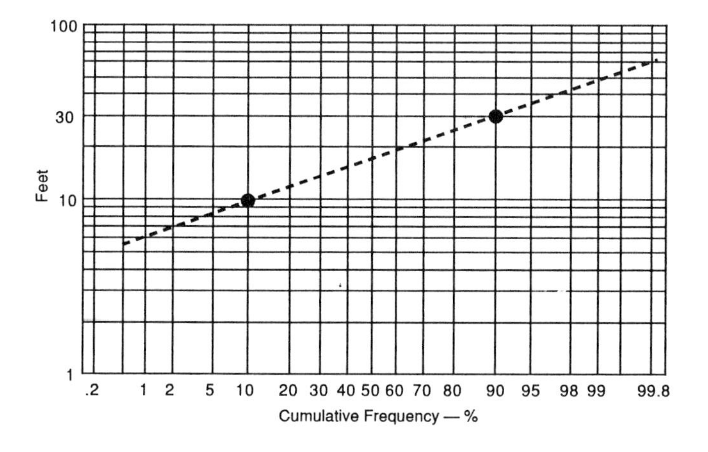

Figure 14. Net pay thickness.

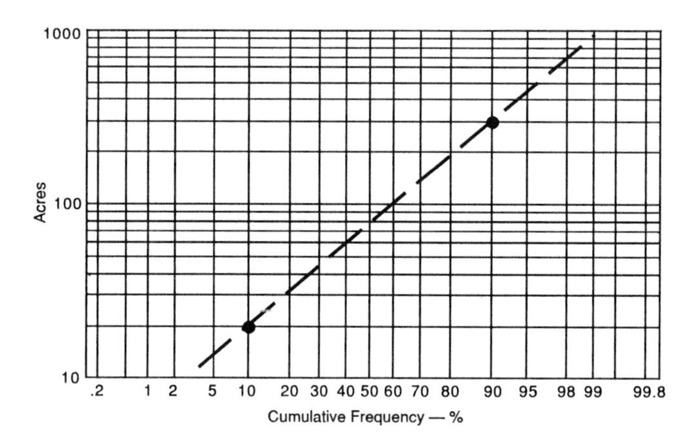

Figure 15. Productive area.

Productive Area

The same arguments apply as before for pay thickness—substitutes for good historical data are hard to find. In a trend, most explorationists see the size variation of prospects as differences in productive area and therefore would not want to use historical averages. After all, they have specific data leading to a unique map. I cannot quarrel with that logic, but events do not seem to work that way. The maps often do not turn out very close to reality so that, if you tie yourself to a range of productive areas depicted by your favorite map, you'll either greatly underestimate the potential for downside disaster or you'll severely underestimate the upside potential, that is, you will not give yourself enough chance for some extraordinary discovery.

When you first start trying to estimate these ranges, you may feel terribly alone. It will seem as if there's no help anywhere. Persevere. It takes experience and, ultimately, data. You need to start making short-cut productive area estimates for all the wells you can possibly get information on, including those drilled by your competitors. Keep records on your estimates, making sure to check on the actuals as soon as you can. After a time, you'll begin to see some patterns develop. For example, you may discover that only one-third of your mean estimates fall between the 10% and 90% points of your ranges. By definition, you expect 80% of your estimates in that 80% range. Having only 35% in an 80% range means you need to widen your range considerably even if you feel uncomfortable doing so.

For example, say your distribution looks like Figure 15. Your 80% range goes from 20 to 300 acres. This exercise attempts to describe what nature provides rather than what we want her to provide. Later we'll use truncation corrections to allow for the fact that small accumulations are not commercial.

Finally, you must prepare a similar chart for a recovery variable in bbls/acre-ft or MCF/acre-ft. Say your trend has fields that are mostly between 150 and 300 bbls/acre-ft, as in Figure 16. We know that multiplying net pay times productive area times recovery gives reserves. But we want to multiply the distributions to get the uncertainty in reserves. Many people instinctively believe that if you want to get the 10% point for reserves, you simply multiply the 10% points for area, net pay, and recovery. Not so. But it turns out that those who like that kind of fix can get one that works well. If you multiply the 23% points together, you get the 10% point, and if you multiply the 77% points together, you get the 90% point.

Figure 17 shows exactly that. We've plotted the three input variables as well as reserves so you can see how they fit together. It is important to note that this trick works for multiplying three lognormal variables with equal variances. We have three variables that we can reasonably assume to be lognormal, although not of equal variance. The area variable exhibits far more uncertainty than the other two. I've done a number of sensitivities to see what kind of error results from not satisfying that last condition, and I find the error to be about 3%. In our business, that's a direct hit! It would take you 16,000 Monte Carlo trials to do as well and you would wait much longer for your results. Table 3 shows the conversion. So the 10% point for reserves becomes 84 MBO, and the 90% point becomes 1.06 MMBO.

Now if your curiosity is up, you'll notice that multiplying 23% points yielded a still lower cumulative, namely, 10%. But multiplying 77% points together gives a higher cumulative, that is, 90%. Is it possible that somewhere in between we could find a cumulative for the components of reserves that multiplies to get its own cumulative? Yes, the 50% point. You'll notice that the reserves line does indeed cross at about the 300 MBOE mark, consistent with the 317 MBO value obtained previously.

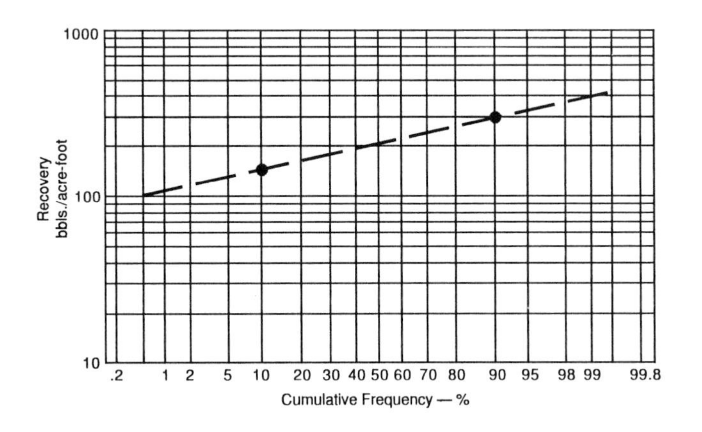

Figure 16. Recovery.

To get the mean reserves from that line, we use an approximation called *Swanson's rule*. Take the 10% point on the reserves curve and let it "represent" the bottom 30% of the distribution. Then take the 50% point and let it represent the middle 40% of the distribution, and finally, take the 90% point and let it represent the upper 30% of the distribution. Then take the weighted average of those three values, weighted by the portion of the whole distribution each represents.

Approximation of mean
= 0.3(10% point) + 0.4(50% point) + 0.3(90% point)
= 0.3(84000) + 0.4(317000) + 0.3(1060000)
= 448 MBO (15)

The quality of this approximation degrades as the variance grows because the 90% point becomes less representative of the upper region of the distribution (Figure 18). As you know from reading this chapter so far, you needn't be concerned about unsystematic calculation errors of less than 50% because exploration's inherent uncertainty proves to be far greater. Swanson's rule, however, introduces a systematic error or bias. Its users will tend to find themselves estimating too low prospect after prospect.

For people with personal computers or even hand-held calculators, life gets easier. Although a surprise to many people, the common Monte Carlo solution to this problem is pure overkill. Because the lognormal distribution fits the data, we can write down easily solved formulas for the answers we seek. First, recognize that the multiplication of variables is equivalent to adding their logarithms and then exponentiating. Thus, if reserves (R) equals

$$R = (\text{Pay})(\text{Area})(\text{Recovery})$$

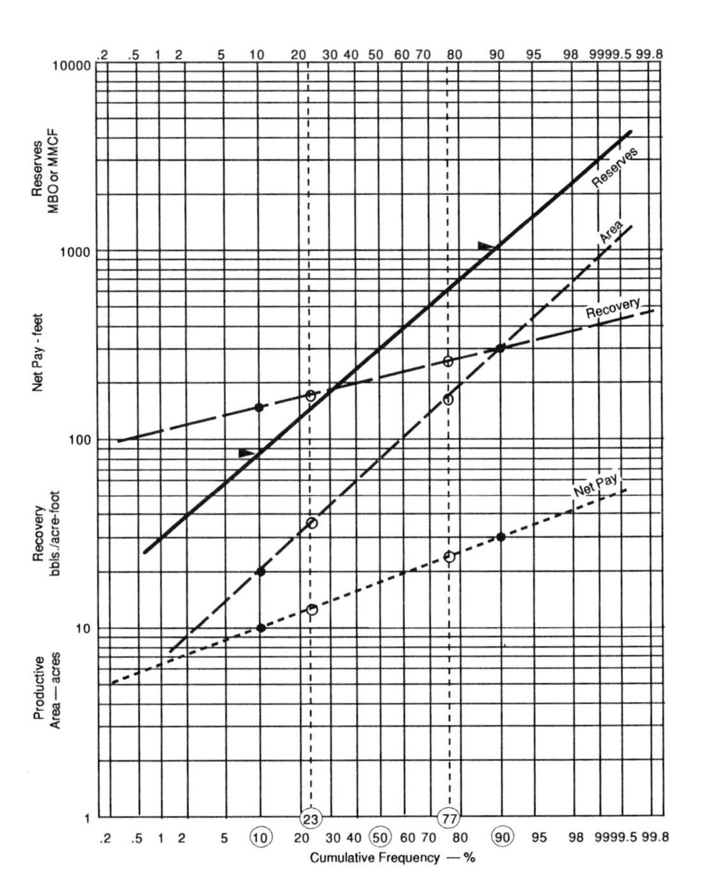

Figure 17. Reserves and components.

then

$$\ln(R) = \ln(\text{Pay}) + \ln(\text{Area}) + \ln(\text{Recovery})$$

and

$$R = e^{[\ln(\text{Pay}) + \ln(\text{Area}) + \ln(\text{Recovery})]}$$

So we can get to mean reserves either from the logarithms or by multiplying the variables pay, area, and recovery.

Introducing some simple notation here called *expected value* will make our calculations easier. (Don't try to reason this out; it's an arbitrary definition.)

Mean of variable X
= Expected value of variable X = E(X)

Thus, E(R), is the *mean* or *expected reserves*. From the rules of mathematical statistics, we know that, in general,

$$E(X + Y) = E(X) + E(Y) = \text{Sum of means always} \quad (16)$$

$$E(XY) = E(X)E(Y) = \text{Product of means if } X \text{ and } Y \text{ are independent} \quad (17)$$

$$E(X/Y) = E(X)E(1/Y) \text{ if } X \text{ and } Y \text{ are independent} \quad (18)$$
$$\neq E(X)/E(Y) \text{ (Beware of this common error)}$$

Table 3. Graphical Solution to Obtaining 10% an Reserves

Cumulative Probability	23%	77%	50%
Net Pay (ft)[a]	13	24	18
Area (acres)[a]	36	170	80
Recovery (bbl/acre-ft)[a]	180	260	220
Reserves (bbl)	84,240	1,060,800	316,800

[a]Read from graph.

$$\mathrm{Var}(X+Y) = \mathrm{Var}(X) + \mathrm{Var}(Y) = \text{Sum of variances if } X \text{ and } Y \text{ are independent} \quad (19)$$

This tell us that

$$E(\ln R) = E[\ln(\text{Pay})] + E[\ln(\text{Area})] + E[\ln(\text{Recovery})] \text{ always, and}$$
$$E(R) = E(\text{Pay})E(\text{Area})E(\text{Recovery}) \text{ usually since the variables are not very dependent.}$$

We are almost done. You need to be able to move quickly between exploration variables, such as reserves and pay thickness, and their logarithms. The following four equations allow you to do that, using reserves as the example variable:

$$M = e^{\mu + \sigma^2/2}, \text{ usually measured from data} \quad (14)$$
$$V = M^2(e^{\sigma^2} - 1), \text{ usually calculated from the formula} \quad (20)$$
$$\mu = \ln M - \sigma^2/2, \text{ usually calculated from data or this formula} \quad (21)$$
$$\sigma^2 = \ln(V/M^2 + 1), \text{ usually measured from data} \quad (22)$$

where

M = mean reserves, area, etc.
V = variance of reserves, area, etc.
μ = mean of natural log reserves, area, etc.
σ^2 = variance of natural log of reserves, net pay, etc.

It is advantageous to work in a hybrid system of M and σ^2. We want to talk about mean feet, mean acres, and mean barrels. But we want our measure of uncertainty (the variance) to be independent of the mean. An uncertainty of three ought to always have the same meaning no matter what the size of the mean M. If you look at the equation for V, you'll see that it has the square of the mean as a term, hardly the kind of number that is independent of the mean!

Now back to the example on the lognormal graph paper. Let's construct another table of values and cumulative probabilities, this time analytically (Table 4). The column under μ comes from taking the logarithm

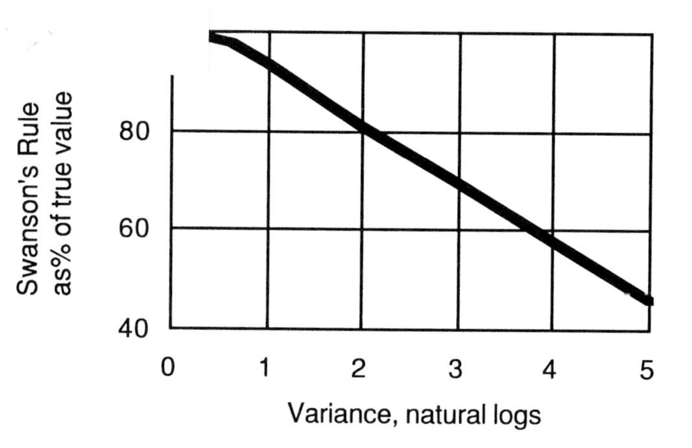

Figure 18. Accuracy of Swanson's rule.

of the 50% points. To get σ^2 is just a little trickier. Earlier we pointed out that the 90% point lies about 1.28 standard deviation units away from μ. Likewise, the 10% point falls 1.28 standard deviation units from μ in the other direction. Let's do a little arithmetic and see how that works. In general,

$$\mu = \ln 10\% \text{ point} + 0.5(\ln 90\% \text{ point} - \ln 10\% \text{ point}) = (\ln 90\% \text{ point} + \ln 10\% \text{ point})/2 \quad (23)$$
$$50\% \text{ point} = e^{\mu} \quad (11)$$
$$\sigma^2 = [(\ln 90\% \text{ point} - \ln 10\% \text{ point})/(2 \bullet 1.28)]^2 \quad (24)$$
$$M = e^{\mu + \sigma^2/2} \quad (14)$$
$$10\% \text{ point} = e^{\mu - 1.28\sigma} \quad (10)$$
$$90\% \text{ point} = e^{\mu + 1.28\sigma} \quad (12)$$

So we apply these equations to the net pay:

$$\mu = (\ln 30 + \ln 10)/2 = 2.85$$
$$50\% \text{ point} = e^{\mu} = 17.3$$
$$\sigma^2 = [(\ln 30 - \ln 10)/(2 \bullet 1.28)]^2 = 0.185$$
$$M = e^{2.85 + 0.185/2} = 19.0$$

Then to the area:

$$\mu = (\ln 300 + \ln 20)/2 = 4.35$$
$$50\% \text{ point} = e^{\mu} = 77.5$$
$$\sigma^2 = [(\ln 300 - \ln 20)/(2 \bullet 1.28)]^2 = 1.12$$
$$M = e^{4.35 + 1.12/2} = 136$$

And finally to the recovery:

$$\mu = (\ln 350 + \ln 150)/2 = 5.43$$
$$50\% \text{ point} = e^{\mu} = 229$$
$$\sigma^2 = [(\ln 350 - \ln 150)/(2 \bullet 1.28)]^2 = 0.11$$
$$M = e^{5.43 + 0.11/2} = 241$$

Table 4. Analytical Solution to Obtaining 10% and 90% Reserves and Mean

Cumulative Probability	10%	90%	50%	μ	σ^2	Mean
Net pay (ft)	10	30	17.3	2.85	0.185	19.0
Area (acres)	20	300	77.5	4.35	1.12	136
Recovery (bbl/acre-ft)	150	300	229	5.43	0.11	241
Reserves (MBO)	66.8	1400	306	12.63	1.42	622

Those of you with sharp eyes will notice slightly different 10%, 50%, and 90% points in Table 4 compared to Table 3. In Table 3, the numbers came from a graph. In Table 4, they were calculated from the 10% and 90% points. This is not an issue to be concerned about, however. Very little about exploration has high enough precision to be influenced by minor errors in reading graphs. Even so, once set up on a computer, the calculations will take place much faster than reading graphs.

It should now be clear that since

$$\ln(\text{Res.}) = \ln(\text{Pay}) + \ln(\text{Area}) + \ln(\text{Rec.})$$

then

$$\mu_{\text{Res.}} = \mu_{\text{Pay}} + \mu_{\text{Area}} + \mu_{\text{Rec.}}$$
$$= 2.85 + 4.35 + 5.43 = 12.6 \qquad (16a)$$

and

$$\sigma^2_{\text{Res.}} = \sigma^2_{\text{Pay}} + \sigma^2_{\text{Area}} + \sigma^2_{\text{Rec.}}$$
$$= 0.185 + 1.12 + 0.11 = 1.42 \qquad (19a)$$

$$\sigma_{\text{Res.}} = 1.19$$
$$M = e^{12.63+1.42/2} = 622{,}000$$
$$10\%\text{ point} = e^{12.63-1.52} = 66{,}800$$
$$90\%\text{ point} = e^{12.63+1.52} = 1{,}400{,}000$$
$$50\%\text{ point} = e^{12.63} = 306{,}000$$

You can also get mean reserves by multiplying the means of area, pay, and recovery:

$$\text{Mean} = (136\text{ acres})(19\text{ ft pay})(241\text{ bbls/acre-ft})$$
$$= 623\text{ MBO}$$

The difference between 622 and 623 MBO can be attributed to round-off error. The mean arrived at analytically exceeds by almost 40% the 448 MBO value we got from reading graphs and applying Swanson's rule. Since we're multiplying, even small errors in reading graphs can grow large.

Although this analytical solution may seem long and involved, that's only because we introduced it in small steps. If you work your way back through all this, you'll find that you can use personal computer spread sheet programs to get the answers you seek quickly since the answers you want all come from the 10% and 90% points you first obtained. In fact, you don't have to resort to graphs at all unless you wish to. I would recommend using graphs, however, because they force you to think about what you're doing. Once you put down 10% and 90% points, you have locked in a 50% point. Does it fit with your intuition? Does it fit the data? If not, you may need to adjust your 10% and/or 90% points.

If you do not like the idea of using 10% and 90% points, pick the ones you do like. Some people feel better using 25% and 75% points. You'll have an equal chance of being inside or outside that range, and that gives your intuition a chance to make a judgment. This choice for a range would cause a change in the formulas shown here from 1.28 to 0.674, a number you can find in any table of the normal distribution.

MULTIPLE HORIZONS

Analytical solutions to the multiple horizon problem also work. Figure 19 shows what I call a *hybrid binomial* lognormal distribution because it has elements of each. First, the binomial rationale is that the well either flows hydrocarbons (or will flow with proper stimulation) or it does not. The sum of the chances for these two events, of course, equals 1.0. Given that hydrocarbons flow, then we are on the lognormal portion of the distribution defined by some mean and variance (or perhaps the 10% and 90% points). Although hydrocarbons flow, we may still have a dry hole if the flow does not warrant setting a completion string, a possibility illustrated by the textured portion in the left-hand tail of the lognormal distribution. The portion of the distribution covered by this "geological success but economic failure" varies according to the geography of the well. In 3000 ft of water, that cut off could be 100 MMBO. In Kansas, it could be as little as 10 MBO, but in the Beaufort Sea, it could be 1 BBO or more.

Now let's find a way to combine two or more of these hybrid distributions each representing a prospective producing horizon in an exploration well. Let's do an example with two horizons, as in Figure 20. Let each of these horizons be like our previous example where we calculated a mean of 622 MBO and a variance of 1.4. One horizon has a chance of 20% and the other 10%. If we drill this well with only two prospective horizons (a big assumption), then we have four possible outcomes: both hit, A hits but B does not, B hits but A fails, or both

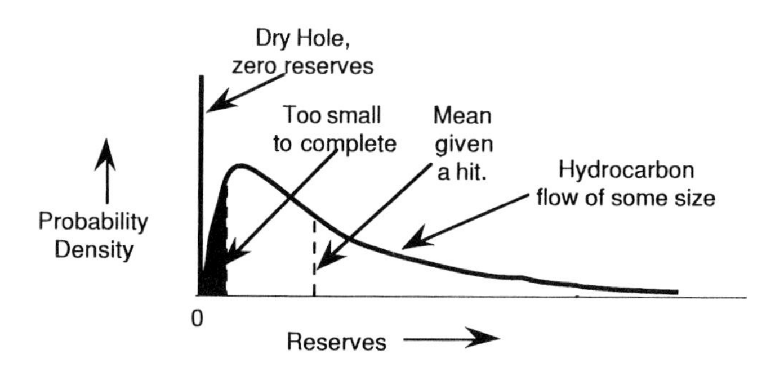

Figure 19. Hybrid binomial–lognormal distribution.

miss. Using the notation P(x) = probability of x, we have the following calculations:

P(A and B) = 0.2 x 0.1 = 0.02
P(A, not B) = 0.2 x 0.9 = 0.18
P(B, not A) = 0.1 x 0.8 = 0.08
P(not A, not B) = 0.8 x 0.9 = 0.72

What does this picture look like? Not much different from either horizon by itself except with different means, variances, and chances. The new "chance of nothing" is 0.72.

Strangely, the form of the "given a hydrocarbon flow" still approximates the lognormal closely enough to use that calculation model. Those readers familiar with the *central limit theorem* know that adding variables ultimately leads to normal or bell-shaped distributions. Multiplying variables leads to lognormal distributions.

Mean of mixed distribution
= 0.02[E(A + B)] + 0.18[E(A)] + 0.08[E(B)]

Since we add horizons, why not switch to the normal distribution? The central limit theorem says that as we add more terms, the distribution of the answer becomes *more nearly* like a normal distribution. For this example, in effect, we add one A to less than one-half of a B and one-ninth of an A + B. The sum still most closely resembles A. And even if A didn't dominate, we would find that highly skewed distributions do not yield quickly to the central limit theorem. You can demonstrate this to yourself by doing a Monte Carlo simulation of an A and B combination and plotting the results cumulatively on lognormal graph paper.

Now let's get a mean (M^*) and variance (V^*) for the sum of such hybrid distributions. That will enable us to have a mean and variance for any combination of horizons for either a single exploratory well or an entire drilling program. (See chapter appendix for the derivation.) Although a formidable hand calculation, it offers

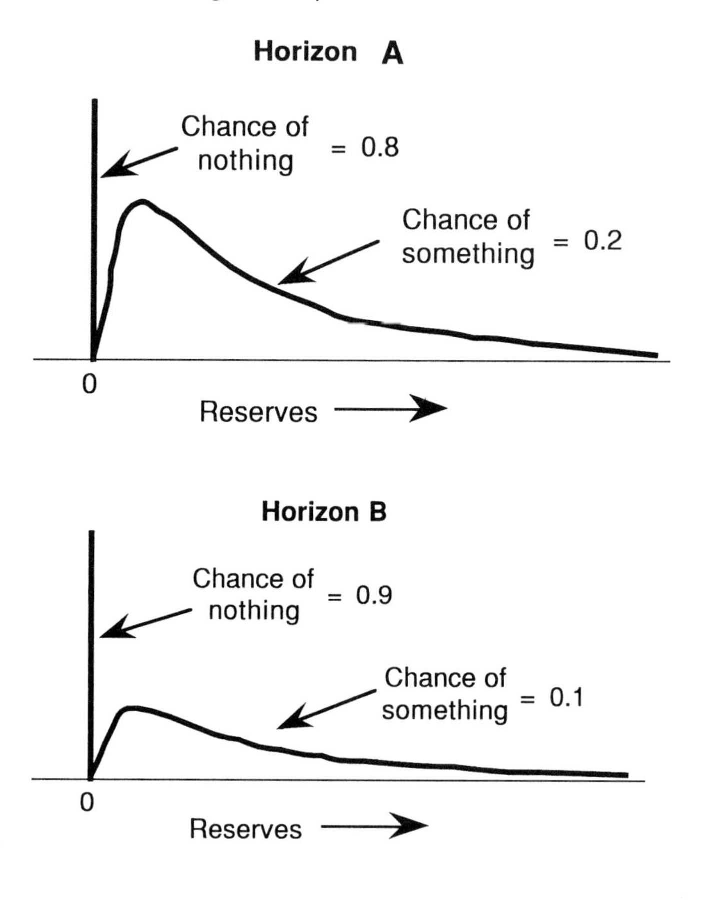

Figure 20. Two potential zones.

no challenge to a personal computer or even a programmable calculator. Also, it takes but a small fraction of the time a Monte Carlo simulation would take and gets a better answer if our assumptions of independence hold. Many Monte Carlo simulations, by the way, also make the independence assumption.

The two formulas are[1]

$$M^* = \left(\sum p_i M_i\right) / \left(1 - \prod q_i\right) \qquad (25)$$

$$V^* = \left[1/\left(1 - \prod q_i\right)\right]\left\{\sum p_i M_i^2 e^{\sigma_i^2} - \sum p_i^2 M_i^2 - \left[\prod q_i / \left(1 - \prod q_i\right)\right]\left(\sum p_i M_i\right)^2\right\} \qquad (28)$$

where
$\sum$ indicates the sum and $\prod$ indicates the product and
p_i = chance of success for each horizon or prospect
M_i = mean reserves for each horizon or prospect
q_i = chance of failure for each horizon or prospect

and the * on M and V signifies the condition of at least one discovery. For plotting on lognormal graph paper, you only need μ and σ^2 from Equations 10, 11, 21, and 22.

[1]Missing Equations 26 and 27 can be found in the Chapter Appendix.

WHERE'S THE MAP?

Strange things can happen when you composite reserves by any legal method. Let's look at an example to see the problem. Suppose that Horizon A has a mean of 15 MMBO and a chance of 0.10. Horizon B has a mean of 500 MBO and a chance of 0.4. Using Equation 25 for M^*, we have

$$M^* = \left(\sum p_i M_i\right) / \left(1 - \prod q_i\right)$$
$$= [(0.10 \times 15 \text{ MMBO}) + (0.4 \times 0.5 \text{ MMBO})]/[1 - (0.9 \times 0.6)]$$
$$= [1.5 + 0.2]/0.46 = 3.7 \text{ MMBO}$$

Note that the composite chance, 0.46, appears in the denominator.

How does the salesmanship work here? If we use only one horizon, we can excite the decision maker with talk of 15 MMBO, a good-sized discovery for the onshore lower 48. But when we put in the other horizon and composite, we find about 4 MMBO as a mean. No map can show z4 MMBO because it came from mathematical manipulation. When we composite from multiple horizons or even multiple wells, the units attached to our number may be reserves, but we are really calculating a betting odds number. It's a paradox—making decisions based on a number you cannot find on a map. That's why those making the decisions must understand the rules of their gambling game. A person would be a fool to engage in world champion poker without a solid working knowledge of betting odds and the probabilities of various poker hands. No less true in oil and gas exploration.

In the previous example, we can calculate risk-weighted reserves with and without Horizon B:

With B:	0.46 x 3.7 = 1.7 MMBO
Without B:	0.10 x 15 = 1.5 MMBO

The expected reserves, and consequently, any economics based on them, will be greater for the composite case. But the romance may not be the same.

TRUNCATIONS

Lognormal distributions start at zero and go to infinity on the high side. Neither boundary works satisfactorily for petroleum exploration. On the high side, we know of no explorationists who have yet found the infinite oil field. On the low side, it makes no sense to have 100 bbls of oil as part of the distribution since such small sizes do not warrant development. Figure 21 may help to clarify this. We have to find a way to prevent counting that part of the distribution

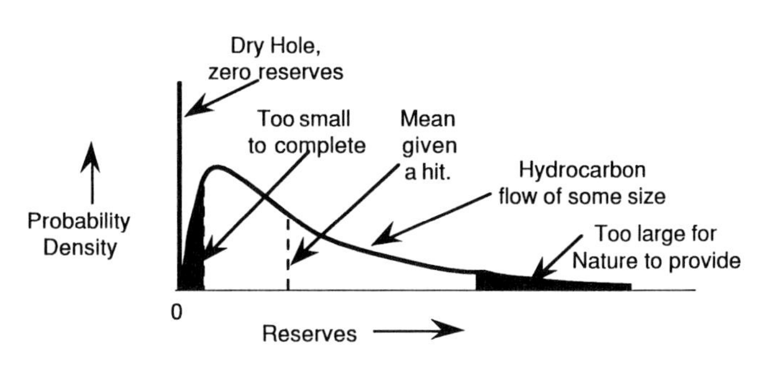

Figure 21. Hybrid binomial–lognormal distribution, double truncation.

that's too small to complete because those accumulations will turn out dry. Likewise, we may not wish to account for the possibility of 1 BBO for a shallow Kansas wildcat. We need the mean of that part of the lognormal distribution that lies between these two boundaries. For completeness, we include this correction. Most wells in the United States will not encounter a problem. However, in frontier areas, the minimum size to complete may become quite large and may greatly affect the answer.

To understand how the math works, look at Figure 22. In general, we would like a way to handle both the low-side and high-side truncations at the same time. You should always define the low side, L, and with a little knowledge of how much it takes to complete the well, this should not cause difficulty. The high side, H, deserves considerably more thought. Don't let your map fool you. As pointed out earlier, that map may not be anywhere close to what nature plans for you. Don't be too conservative. Leave yourself room for pleasant surprises on the high side.

We need to calculate $m_{L:H}$, the mean of that clear area between L and H. The math appears in the appendix.

$$m_{L:H} = M\{[F(z\#_{\ln H}) - F(z\#_{\ln L})]/[F(z_{\ln H}) - F(z_{\ln L})]\} \quad (29)$$
$$z_{\ln L} = (\ln L - \mu)/\sigma \quad (30)$$
$$z\#_{\ln L} = z_{\ln L} - \sigma \quad (31)$$
$$z_{\ln H} = (\ln H - \mu)/\sigma \quad (32)$$
$$z\#_{\ln H} = z_{\ln H} - \sigma \quad (33)$$

where the # symbol reflects the distribution with mean $= \mu + \sigma^2$.

$$F(z)_{\text{approx.}} = 1 - [1 + (0.644693 + 0.161984z)^{4.874}]^{-6.158} \quad (34)$$

Warning! Do not let $|z| > 3.9$. The approximation deteriorates rapidly for z outside that range.

Literally, the only numbers you need are the 10% and 90% points along with the two truncation points, L

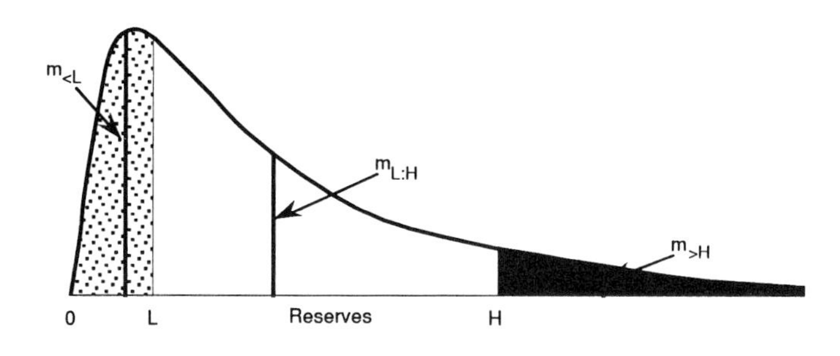

Figure 22. Truncated lognormal.

and H. To clarify this, let's do an example. Say we have a prospect in the Gulf (Figure 23) whose 10–90% reserve range is 5 MMBO to 50 MMBO. With less than 20 MMBO, we could not afford to set a platform from which to produce, so 20 MMBO has to be our minimum commercial size. On the high side, we might question how even God could stuff more than 75 MMBO into these fault blocks. Remembering the map's limitations and leaving room for high-side surprises, we'll cut off the high end at about 100 MMBO.

From our previous definitions, we know that

$$L = 20 \qquad H = 100$$
$$\ln L = 2.996 \qquad \ln H = 4.605$$
$$\ln 90\% \text{ point} = \ln 50 = 3.912$$
$$\ln 10\% \text{ point} = \ln 5 = 1.609$$
$$\mu = (3.912 + 1.609)/2 = 2.761$$
$$\sigma^2 = [(3.912 - 1.609)/(2 \bullet 1.28)]^2 = 0.809$$
$$M = e^{2.761+0.809/2} = 23.7$$
$$\text{Median} = e^{\mu} = e^{2.761} = 15.8$$
$$\text{Mode} = e^{\mu-\sigma^2} = 7.04 \qquad (35)$$

Now proceed to get the mean of the truncated form:

$$z_{\ln L} = (2.996 - 2.761)/0.899 = 0.261$$
$$z\#_{\ln L} = z_{\ln L} - \sigma = 0.2614 - 0.899 = -0.638$$
$$z_{\ln H} = (4.605 - 2.761)/0.899 = 2.051$$
$$z\#_{\ln H} = z_{\ln H} - \sigma = 2.051 - 0.899 = 1.152$$

From Equation 34, we have

$$F(z_{\ln L}) = 0.600 \text{ (calculate or read from graph)}$$
$$F(z\#_{\ln L}) = 0.261$$
$$F(z_{\ln H}) = 0.980 \text{ (calculate or read from graph)}$$
$$F(z\#_{\ln H}) = 0.878$$
$$m_{L:H} = 23.7[(0.878 - 0.261)/(0.980 - 0.600)]$$
$$= (23.7)(1.624) = 38.5 \qquad (29)$$

Without truncation, we see a mean of about 25 MMBO. With truncation, it increases to almost 40 MMBO. Did we just get a free lunch? Hardly. Reserves given a success go up dramatically, but chance goes down. Say we estimate a 15% chance of finding hydrocarbons that flow. We call that a geological chance of success, C_g. The chance that must ultimately go into the economics, however, is the chance that we complete the well. For this example, that means also setting a platform.

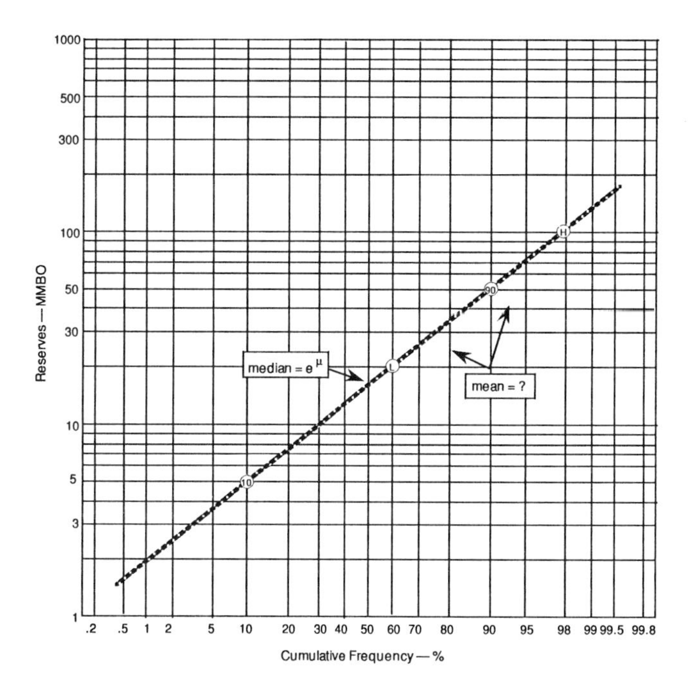

Figure 23. Truncations.

$$\text{Commercial chance} = C_c = C_g[1 - F(L)] = 0.15(0.4) = 0.06 \qquad (36)$$

C_g gives the probability that you're somewhere on the line that defines your particular lognormal distribution, and $[1 - F(L)]$ gives the chance that you'll have a commercial discovery (>20 MMBO) given hydrocarbons flow. Multiply to get the unconditional chance that you will have a commercial discovery.

Look at the difference in risk-weighted reserves.

Without truncation:	0.15(23.7) = 3.6 MMBO
With truncation:	0.06(38.5) = 2.3 MMBO

While chance and reserve changes tend to offset one another, you will still lose some value with truncation. More correctly, without truncation, you count value you shouldn't.

A few words about the term *commercial*. The meaning most consistent with the way the industry gathers and reports its data is this: a commercial oil or gas field is one that produces some oil or gas for sale.

Nothing in that definition says the field has to make money. The dictionary says that the word *commercial* does not necessarily imply profit, but just engaging in business (or "commercial") activities. So our wells are either dry or commercial. An expendable hole that discovers oil or gas should count as a commercial success even though that particular well may never produce.

OPTION PRICING

No, you haven't stumbled onto a typo—the heading *does* say option pricing. As lore goes, the "rocket scientists" took over financial analysis with their high powered and obscure mathematics. You may have heard of the Black–Scholes option pricing model that has found practical use on Wall Street. Some industry people are trying to apply these ideas more widely to project economics for lease purchases (Lehman, 1989). Bob Stibolt (of ARCO), who works proficiently in both financial and oil patch evaluation methodology, made the connection in 1989 that the Black–Scholes model was identical to what we've just discussed—getting the mean of a truncated lognormal distribution. So don't feel lonely if you decide to play "rocket scientist" with your prospect evaluations.

PROJECT PORTFOLIO

Notice that the multiprospect problem is identical to the multihorizon one. We have a number of prospects each with different reserves and chances. We combine them into a single distribution showing the chance of portfolio success and the distribution of those successes. The mathematics are almost exactly the same. For the prospect, we did not truncate the component horizons nor the area, pay thickness, or recovery. We truncated only at the answer or reserves stage. With portfolio analysis, we'll need to combine prospects that we may have already truncated, which means we'll need the variance of a truncated lognormal distribution. Since adding this variance to the analysis complicates things, let's do an example both with and without the correction so you can judge whether you need it.

Let's look at an example to see how to work a portfolio. First, we'll represent our three prospect portfolio on lognormal graph paper in the standard way using 10% and 90% points along with a truncation on one prospect because of a 5 MMBO lower limit for economic reasons (Figure 24). While this is a good prospect, even in our wildest dreams, we cannot see more than 1 BBO so we've also truncated the Cobalt prospect at that point. The other two have lower limits of 10 MBO, too small to worry about for affecting the mean or variance calculations. For these two, we've

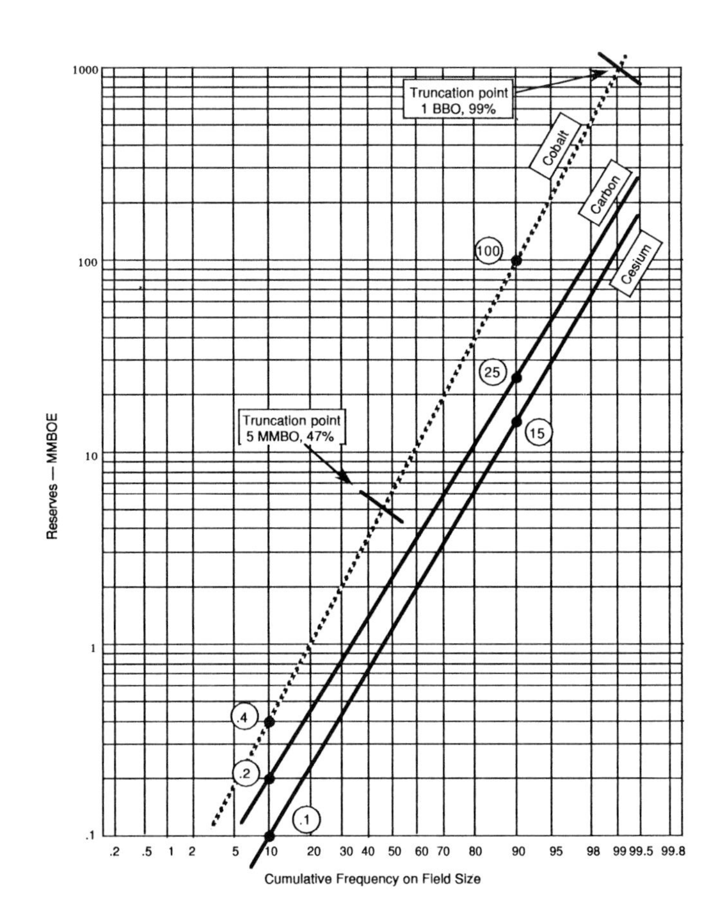

Figure 24. Exploration portfolio.

assumed no upper end truncation. We estimate the chances for the three prospects at 0.15, 0.25, and 0.30, respectively.

The only difference between multizone analysis and portfolio analysis is the way in which chance and truncation enter. For the prospect, we employ geological chance on the horizons and combine them, and after truncating, we end up with commercial chance and mean. In portfolio analysis, we start with the commercial (conditional) means and chances, and we need not truncate the answer.

First, we get the means and variances for all three prospects. Note that boldface type indicates input data.

Prospect Cobalt:

$$\sigma^2 = [(\ln 90\% \text{ point} - \ln 10\% \text{ point})/(2 \bullet 1.28)]^2$$
$$= [(\ln \mathbf{100} - \ln \mathbf{0.4})/2.56]^2 = (5.52/2.56)^2 = 2.157^2$$
$$= 4.65$$

$$\sigma = 2.16$$

$$\mu = (\ln 90\% \text{ point} + \ln 10\% \text{ point})/2 = (3.69)/2$$
$$= 1.84$$

$$M = e^{\mu + \sigma^2/2} = e^{1.84 + 4.65/2} = 64.4 \text{ MMBO}$$

C_g = **0.15** = geological chance of flowing hydrocarbons

Lower truncation = **5 MMBO**

Upper truncation = **1000 MMBO**

Prospect Carbon:

$\sigma^2 = [(\ln \mathbf{25} - \ln \mathbf{0.2})/(2.56)]^2 = (4.86/2.56)^2 = 1.89^2$
$= 3.56$

$\sigma = 1.89$

$\mu = (\ln 25 + \ln 0.2)/2 = (3.42)/2 = 1.71$

$M = e^{1.71+3.56/2} = 32.8$ MMBO

$C_g = \mathbf{0.25}$

Prospect Cesium:

$\sigma^2 = [(\ln \mathbf{15} - \ln \mathbf{0.1}]/(2.56)]^2 = (5.01/2.56)^2 = 1.95^2$
$= 3.80$

$\sigma = 1.95$

$\mu = (\ln 15 + \ln 0.1]/2 = (2.36)/2 = 1.18$

$M = e^{1.18+3.8/2} = 21.8$ MMBO

$C_g = \mathbf{0.30}$

The 11 boldfaced numbers show the only input data requirements!

Because of the truncation, we must rework prospect Cobalt. We must calculate some cumulative distribution functions as described earlier (Equations 30–34).

$z_{\ln 5} = (1.61 - 1.84)/2.16 = -0.1065$

$z\#_{\ln 5} = z_{\ln 5} - \sigma = -0.1065 - 2.16 = -2.27$

$z_{\ln 1000} = (6.91 - 1.84)/2.16 = 2.35$

$z\#_{\ln 1000} = z_{\ln 1000} - \sigma = 2.35 - 2.16 = 0.186$

$F(z_{\ln 5}) = 0.454$

$F(z_{\ln 1000}) = 0.990$

$F(z\#_{\ln 5}) = 0.012$

$F(z\#_{\ln 1000}) = 0.570$

The truncated mean for Cobalt is

$$m_{L:H} = M[F(z\#_{\ln 1000}) - F(z_{\ln 5})]/[F(z_{\ln 1000}) - F(z_{\ln 5})]$$
$$= 64.4(570 - 0.012)/(0.990 - 0.454)$$
$$= 64.4(0.558/0.536) = 67.0 \text{ MMBO}$$

So we have a new mean, M, which is slightly higher than before, and a new lower chance:

$$C_c = C_g[1 - F(5 \text{ MMBO})] = 0.15(1 - 0.454) = 0.0819$$

It may be helpful to put the key numbers into a table so we can easily refer to them during the compositing process (Table 5). Our plan is the same as when evaluating multiple horizons. We want to come up with a single distribution that describes the composite of the three prospects. What's the chance we would get nothing from our three wells? Given at least one of the wells hit, what are the probabilities of getting various amounts of oil or gas? We will not tie economics to this analysis, just reserves.

Table 5. Key Number for Portfolio Compositing

Prospect	M	σ^2	p	q	μ
Cobalt	67.0	4.65	0.082	0.918	1.84
Carbon	32.8	3.56	0.25	0.75	1.71
Cesium	21.8	3.80	0.30	0.70	1.18

$$M^* = \left(\sum p_i M_i\right) / \left(1 - \prod q_i\right)$$
$$= [(0.082)(67) + (0.25)(32.8) + (0.30)(21.8)]$$
$$/[1 - (0.918)(0.75)(0.7)] = 39.1 \text{ MMBOE}$$

$$V^* = \left[1/\left(1 - \prod q_i\right)\right]\left\{\sum p_i M_i^2 e^{\sigma_i^2} - \sum p_i^2 M_i^2 - \left[\prod q_i / \left(1 - \prod q_i\right)\right]\left(\sum p_i M_i\right)^2\right\}$$
$$= [1/0.518]\{(0.082)(4489)(104.6)$$
$$+ (0.25)(1076)(35.2) + (0.30)(475)(44.7)$$
$$- (0.00672)(4489) - (0.0625)(1076) - (0.09)(475)$$
$$- [0.482/0.518][(0.082)(67) + (0.25)(32.8)$$
$$+ (0.3)(21.8)]^2\}$$
$$= [1.93]\{38503 + 9469 + 6370 - 30.2 - 67.3 - 42.8$$
$$- [0.931](5.49 + 8.20 + 6.54)^2\}$$
$$= [1.93]\{54202 - [0.931](409)\} = 103875$$

$\sigma^{*2} = \ln(V^*/M^{*2} + 1) = \ln(103875/1529 + 1) = 4.23$

$\sigma^* = 2.06$

$\mu^* = \ln M - \sigma^2/2$
$= \ln 39.1 - 4.23/2 = 1.55$

$C_c^* = 1 - \prod q_i = 0.518$ (commercial chance)

Median $= e^{\mu} = 4.72$ MMBO

Mode $= e^{\mu - \sigma^2} = e^{1.55-4.23} = 0.0686$ MMBOE

10% point $= e^{\mu - 1.28\sigma} = e^{1.47-(1.28)(2.06)} = 0.311$

90% point $= e^{\mu + 1.28\sigma} = e^{1.47+(1.28)(2.06)} =$ 6[illegible]8

Now we have the composite portfolio distribution completely specified (and plotted) (see Figure 25). While we have carried three and sometimes more significant digits throughout the calculations, we know such accuracy far exceeds the precision of exploration's tools. With the kind of uncertainties prevalent in this business, even one-digit accuracy remains illusive. So while we may use the 39.1 MMBO mean in further calculations, we would simply report a program mean of "about 40 MMBO" and a program chance of "about 50%."

Does this mean that the program has a 50% chance of finding 40 MMBO? *No.* It has a *zero* chance of finding 40 MMBO! The language of probability is relatively precise; don't get caught being sloppy or you may mislead your investor. The program has a 50% chance of at least one discovery, and given some discovery, 40 MMBO represents the average of all that can happen. Sportscaster ignorance doesn't fit here; we don't

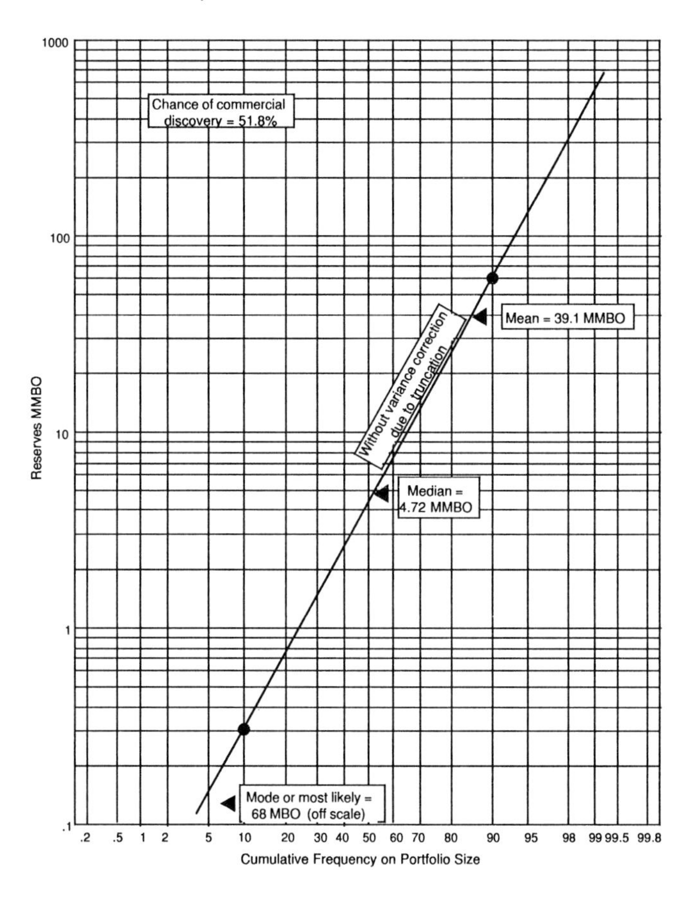

Figure 25. Three-well portfolio.

average 40 MMBO every time we drill a well, just as no running back "averages" 5 yards *every time* he carries the ball.

As we pointed out earlier, these calculations assume independence among the prospects, which even for prospects along the same trend probably will not cause a problem. Another assumption could cause trouble, however. Earlier in the discussion, we said that while we could find the mean of a truncated lognormal distribution, variance does not yield so quickly. Next let's present a way to deal with this difficulty that is easy to build into any spreadsheet calculation. We will also check to see how much difference this correction makes in the example just worked.

A truncated distribution must have a smaller variance than an untruncated one. Perhaps Figure 26 will show what is going on. The vertical or z axis shows the variance of the truncated distribution as a function of the beginning or untruncated variance (a northwest–southeast axis we'll call x) and the spot of the truncation (a northeast–southwest axis, our y). A few boundaries ought to be obvious. If the variance of the untruncated distribution equals zero (x axis), then it does not matter where you truncate. The truncated variance is zero also. So the entire surface of the

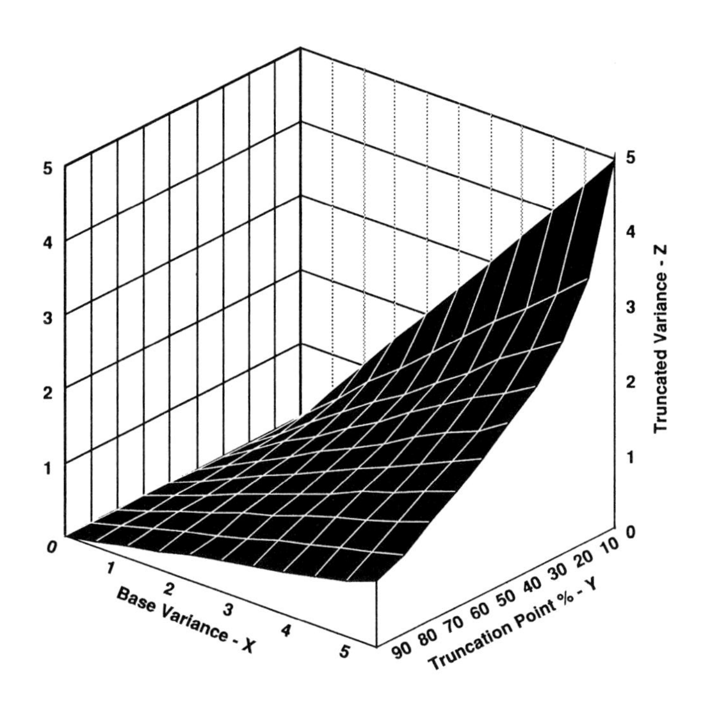

Figure 26. Variance of truncated lognormals.

variance function intersects the vertical left side plane at zero. Also, we know that if the truncation takes place at the 0% cumulative probability on the y axis, then the truncated variance will equal the untruncated variance since no effective truncation exists. Thus, you see on the right side vertical plane that the surface intersects on a 45° line.

The rest of the variance surface shows how variance decreases with the severity of the truncation. I calculated these variances using a Latin hypercube shortcut to sampling the truncated distribution using up to 1000 equal intervals. As you can see, the surface turns out to be fairly smooth. In fact, it becomes virtually flat if we transform the two axes showing change to the natural log of variance. Using that transformation and a multiple linear regression, it is possible to get a reasonable estimate of truncated variance using the x and y variables defined on the graph's axes.

$$\ln \text{truncated var.} = (-0.11219 - 0.018687)(\text{truncation point} + 1.001045)(\ln \text{untruncated var.})$$

We can rewrite this as

$$\text{Truncated var.} = (0.894)(\text{untruncated var.})e^{-0.0187(\text{truncation point})} \qquad (37)$$

(Remember that variance refers to the variance of the natural log of reserves, etc. The truncation point is in percent.)

The fit works well, giving a 0.997 correlation. Over the entire surface, 85% of the points on the regression surface come within 10% of the actual variance and no point errs by over 21%. Such accuracy far surpasses what our problem needs, so we can use Equation 37 with a feeling of security. Notice, however, that we have not included the upper end truncation because the lower one usually has the most effect. Severe upper end truncations may be caused by lack of imagination on the part of the explorationist!

Back to our example. Using the newly derived equation for variance of truncated distributions, let's recalculate variance for prospect Cobalt. The truncation point lies at the 45% point, and we previously calculated the untruncated variance as 4.65.

$$\text{Truncated var.} = (0.894)(4.65)e^{-0.842} = 1.79$$

Now, if we replace the 4.65 we used earlier with 1.79, we can get a new composite variance. (Look at the $e^{\sigma_i^2}$ term.) The boldfaced terms in the following are the ones affected in this new computation:

$$M^* = 39.1$$

$$\begin{aligned}V^* &= \left[1/\left(1-\prod q_i\right)\right]\left\{\sum p_i M_i^2 e^{\sigma_i^2} - \sum p_i^2 M_i^2 \right.\\ &\quad \left. - \left[\prod q_i/\left(1-\prod q_i\right)\right]\left(\sum p_i M_i\right)^2\right\}\\ &= (1/0.518)\{(0.082)(4489)(\mathbf{5.99})\\ &\quad + (0.25)(1076)(35.2) + (0.30)(475)(44.7)\\ &\quad - (0.00672)(4489) - (0.0625)(1076) - (0.09)(475)\\ &\quad - (0.482/0.518)[(0.082)(67) + (0.25)(32.8)\\ &\quad + (0.3)(21.8)]^2\}\\ &= (1.93)[\mathbf{2205} + 9469 + 6370 - 30.2 - 67.3 - 42.8\\ &\quad - (0.931)(5.49 + 8.20 + 6.54)^2]\\ &= (1.93)[\mathbf{18044} - (0.931)(409)] = \mathbf{34090}\end{aligned}$$

$$\sigma^{*2} = \ln(V^*/M^{*2} + 1) = \ln(\mathbf{34090}/1529 + 1) = \mathbf{3.15}$$

$$\sigma^* = \mathbf{1.77}$$

$$\mu^* = \ln(M) - \sigma^2/2 = \ln(39.1) - \mathbf{3.15}/2 = \mathbf{2.09}$$

$$C_c^* = 1 - \prod q_i = 0.518, \quad \text{commercial chance}$$

$$\text{Median} = e^{\mu} = \mathbf{8.08\ MMBO}$$

$$\text{Mode} = e^{\mu - \sigma_i^2} = e^{2.09-3.15} = \mathbf{0.346\ MMBO}$$

$$10\%\ \text{point} = e^{\mu - 1.28\sigma} = e^{2.09-(1.28)(1.77)} = \mathbf{0.839\ MMBO}$$

$$90\%\ \text{point} = e^{\mu + 1.28\sigma} = e^{2.09+(1.28)(1.77)} = \mathbf{77.9\ MMBO}$$

To see whether this correction has made a significant difference, we must construct our lognormal cumulative lines on the probability graph paper shown in Figure 27. This graph gives us much information. We've already talked about getting the median. We observe it by looking at the 50% point on the cumulative frequency or by calculating it. The median or 50% point often serves as a reality check for your intuition. Does it seem likely that there's an equal chance of getting less than or greater than 8 MMBO given at least one of the three wells is successful? However, do not use that number as a prediction for what you will get. Use the mean of 39.1 MMBO for that.

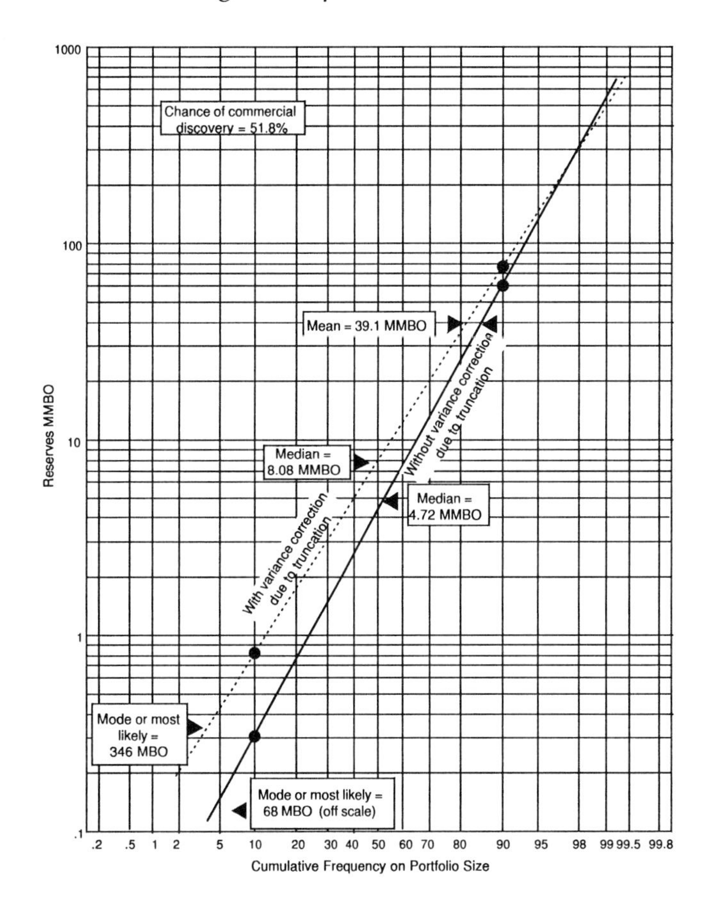

Figure 27. Three-well portfolio with truncation variance correction.

Although we cannot observe the mode (the most likely) from looking at a cumulative graph, we did calculate it as 0.346 MMBO. This number has little or no use in the decision process, but it does serve to remind us that in our business, we get lots of little ones and not many big ones. So while the mean of our portfolio promises about 40 MMBO, we will suffer through years of such three-well programs with only a few hundred thousand barrels to show for the effort. For those patient few who keep the faith (and perhaps get lucky), a 200 MMBO or greater bonanza will occur every so often (3% chance) to make people glad they became explorationists.

In Figure 27, we show both a dotted line and a solid line for the two ways of handling the variance for truncated lognormals. I recommend using the dotted line with the variance correction. For the example, you can see that, although the mean of each line is the same, the probability statements we make about various sizes show some differences and the calculation takes no longer.

So as an explorationist, what can we say about the portfolio? First, we know that we have about a 50% chance of hitting at least one of the three wells. Given a hit, we expect about 40 MMBO total, but because of the highly skewed nature of exploration results, we're much more likely to get a few hundred thousand barrels. With similar programs each year, only one in ten would give us less than 800 MBO, given we made any discoveries at all. On the positive side, we want to know if our program gives us some reasonable chance of hitting a big one. As we know, big is relative, but this analysis suggests an 8% chance of getting more than 100 MMBO. That means about 1 year in 24 when we factor in the 50% chance of getting anything at all.

If you are concerned about replacing production, just take the commercial chance C_c of 51.8% and multiply by the mean discovery size of 39.1 MMBO to get 20.2 MMBO. Thus, with this kind of portfolio, you can expect to find about 20 MMBO per year. If you produce 20 MMBO per year, then only about 1 year in 7 will you at least replace your reserves, and you may not survive long enough for the averages to bail you out. (Notice from the graph that the chance of exceeding 20 MMBO is 30%, given you make some commercial discovery, which only happens in half the years. That leaves you with a 15% chance of exceeding your "requirement.")

If you produce more than 20 MMBO per year and your production profits will fund no larger program, then you face a declining asset base. Best you know now while you can still try out some other strategies such as the use of higher technology for upgrading prospects prior to drilling or reducing expenses somewhere to find more drilling dollars.

The frequency distribution of large portfolios should begin to look more bell-shaped because of the addition process and the central limit theoren. You might ask how many wells a portfolio can have before we would want to switch away from the lognormal approximation and this relatively quick analytical approach. Surprisingly large. Say you have a 300-well portfolio with an average 30% success ratio and a 2 MMBO mean size. You really have 90 wells instead of 300 when you remove the dry holes. Now consider the dynamic range of sizes. With 90 successful wells, you might end up with 4 of them above 10 MMBO. Three-fourths would contain less than 1 MMBO. Out of the 180 MMBO expected reserves for the portfolio, over half would come from the largest 9 discoveries. A few large finds drive the shape of the distribution.

Using the 300-well example just described and an estimated σ^2 of 3, I ran a Monte Carlo simulation and found the portfolio reserves to be highly skewed—much closer to lognormal than normal. The worst difference I observed was at the 2% point. The Monte Carlo showed a 2% chance of the portfolio having less than 33 MMBO, while the lognormal approximation gave a 2% chance of less than 20 MMBO. Given the realm of exploration uncertainties, the difference between the two methods does not seem important. For any one 300-well program, the outcome could be as low as 25 MMBO or as high as 400 MMBO (using a 95% range).

Remember Malcom? Management wanted exactly the kind of information from him that was discussed in this chapter. Business needs the language of uncertainty and no explorationist should show up without the tools that provide such information. But neither should an explorationist promise to deliver the impossible. It's OK to be wrong. Since the competition performs no better in its ability to make accurate guesses, the economics that have grown up around our business allow for large errors—of the unbiased kind. If you have biased estimates, you need to find out quickly and act.

CONCLUSIONS

We set out to convince readers that explorationists have very large uncertainties to contend with and that those interested in the subject should keep estimate versus actual records on themselves to measure their AEF (average error factor), bias, and amount of correlation between actuals and estimates. We then provided the tools for plotting data on lognormal graph paper and explained why the lognormal distribution makes sense for those exploring for minerals. Next, we presented methodology for taking uncertainty estimates (10% and 90% points) on productive area, net pay thickness, and recovery in barrels per acre-foot (or MCF) and combining them to get a distribution of reserves—all without having to resort to Monte Carlo overkill. We also introduced how to truncate both ends of the distribution to conform with nature and the economics of the prospect and how to transform geological chance into economic chance with the proper low end truncation. From a single horizon prospect, we showed how to composite several horizons into one prospect reserves distribution. Finally, we took the horizon compositing lessons and applied them to a portfolio of prospects so a company could make probability judgments about its entire program at planning or budget time.

Earlier we noted that using a basin average for all reserves estimates would have been better for some companies. Then we laboriously laid out a method to make individual prospect estimates as if there would be some great value in that effort. Sounds almost paradoxical. But I think improving the estimates is possible, and these methods will help. However, explorationists must get more dynamic range into their estimates. If a prospect has some chance of being really big, then make sure it gets that credit by allowing for a more bullish 90% point. This allows for real selectivity and differentiation because not every prospect has significant upside potential.

Just as important is the realization that most discoveries are quite small and one's suite of estimates ought to reflect that fact. Not one of the Ten Commandments forbids the finding of small fields. If you know ahead of time that you'll be "blessed" with unexciting discoveries, then the least you can do is recognize that truth, start reducing your front end expenditures, and thus enhance your chance for profit.

References Cited

A short word about references and why there are so few. This material has been developed at ARCO starting in the mid-1960s without knowledge of similar activities going on elsewhere. For this reason, we've included more detail than we might have otherwise. If you find the material useful, we hope you will be able to duplicate the calculations for your own analyses.

Aitchison, J., and J. Brown, 1957, The lognormal distribution: Cambridge, Cambridge Univ. Press, p. 87.

Burr, I. W., 1967, A useful approximation to the normal distribution function, with application to simulation: Technometrics, v. 9, n. 4, p. 647.

Capen, E. C., and R. V. Clapp, 1974, The ratio: A possible clash between instinct and science: Journal of Petroleum Technology, v. 26 (May), p. 483.

Capen, E. C., R. V. Clapp, and W. M. Campbell, 1971, Competitive bidding in high risk situations: Journal of Petroleum Technology, v. 23 (June), p. 641.

Hoel, P. G., 1970, Introduction to mathematical statistics, 4th ed.: New York, John Wiley and Sons, p. 401.

Lehman, J., 1989, Valuing oil field investments using option pricing theory: Proceedings of the 1989 SPE Hydrocarbon Economics and Evaluation Symposium, Society of Petroleum Engineers Paper # SPE 18923.

Massey, F. J., Jr., 1951, The Kolmogorov–Smirnov test for goodness of fit: Journal of the American Statistical Association, v. 46, pp. 68-78.

Root, D. H., 1982, Historical growth of estimates of oil- and gas-field sizes, *in* Gaff, S. I., Oil and Gas supply modeling: National Bureau of Standards Special Publication 631, p. 350-368.

Chapter Appendices

A "ROAD MAP"

To aid readers in finding their way through perhaps unfamiliar territory, this appendix lists the important formulas numbered as they appear in the main text or in subsequent appendices. We've also provided a chart (Figure 28) that should help in connecting the various activities involved in uncertainty analysis with the correct formula.

Estimating performance measures:

$$\text{Bias} = (1/n)\sum \ln(E/A) \quad (1)$$

$$\text{Average error factor} = \text{AEF} = e^{(1/n)\sum|\ln(E/A)|} \quad (2)$$

Probability density functions for normal and lognormal:

$$f_N(x) = \left(1/\sigma\sqrt{2\pi}\right)e^{-(x-\mu)^2/2\sigma^2} \quad (3)$$

$$f_{\ln}(y) = \left(1/y\sigma\sqrt{2\pi}\right)e^{-(\ln y-\mu)^2/2\sigma^2} \quad (4)$$

$$\text{Mean of } x = \mu = \sum p_i x_i \quad (5)$$

$$\text{Sample average} = \bar{x} = \left(\sum x\right)/n$$

where n = number in sample (6)

$$\text{Sample variance} = s^2 = \left[\sum(\ln y)^2 - \left(\sum \ln y\right)^2/n\right]/(n-1) \quad (7)$$

Cumulative frequency, for goodness of fit test = (Point number) /(Total number of points) (8)

Cumulative frequency, for plotting = (Point number) /(1 + total number of points) (9)

$$10\% \text{ point} = e^{\mu-1.28\sigma} \quad (10)$$

$$50\% \text{ point} = e^{\mu} \quad (11)$$

$$90\% \text{ point} = e^{\mu+1.28\sigma} \quad (12)$$

$$\sigma = (\ln 90\% \text{ point} - \ln 50\% \text{ point})/1.28 \quad (13)$$

$$M = e^{\mu+\sigma^2/2} \quad (14)$$

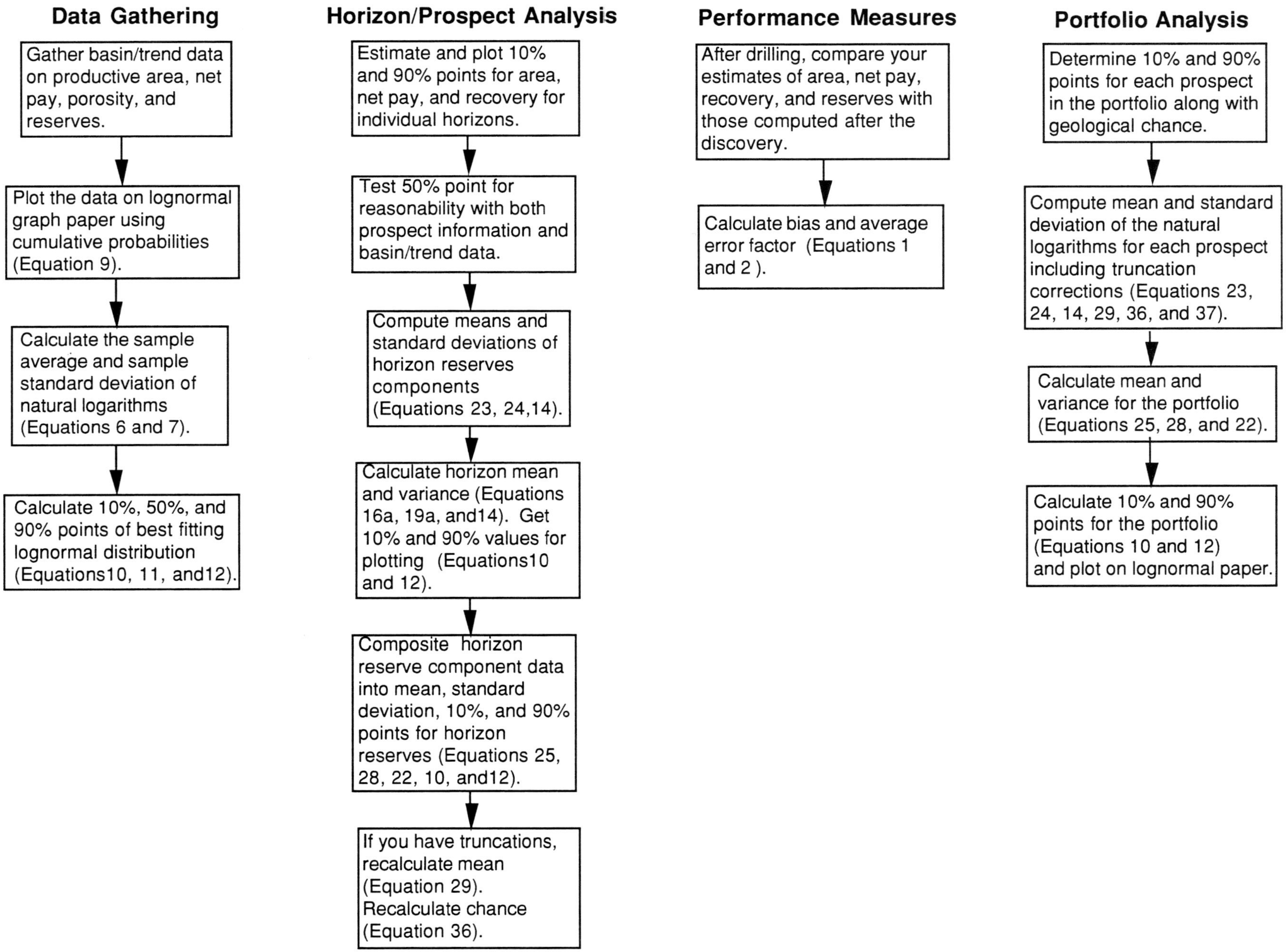

Data Gathering

Gather basin/trend data on productive area, net pay, porosity, and reserves.

Plot the data on lognormal graph paper using cumulative probabilities (Equation 9).

Calculate the sample average and sample standard deviation of natural logarithms (Equations 6 and 7).

Calculate 10%, 50%, and 90% points of best fitting lognormal distribution (Equations10, 11, and12).

Horizon/Prospect Analysis

Estimate and plot 10% and 90% points for area, net pay, and recovery for individual horizons.

Test 50% point for reasonability with both prospect information and basin/trend data.

Compute means and standard deviations of horizon reserves components (Equations 23, 24,14).

Calculate horizon mean and variance (Equations 16a, 19a, and14). Get 10% and 90% values for plotting (Equations10 and 12).

Composite horizon reserve component data into mean, standard deviation, 10%, and 90% points for horizon reserves (Equations 25, 28, 22, 10, and12).

If you have truncations, recalculate mean (Equation 29). Recalculate chance (Equation 36).

Performance Measures

After drilling, compare your estimates of area, net pay, recovery, and reserves with those computed after the discovery.

Calculate bias and average error factor (Equations 1 and 2).

Portfolio Analysis

Determine 10% and 90% points for each prospect in the portfolio along with geological chance.

Compute mean and standard deviation of the natural logarithms for each prospect including truncation corrections (Equations 23, 24, 14, 29, 36, and 37).

Calculate mean and variance for the portfolio (Equations 25, 28, and 22).

Calculate 10% and 90% points for the portfolio (Equations 10 and 12) and plot on lognormal paper.

Figure 28. A "roadmap" for uncertainty calculations.

$$\text{Swanson's mean} = 0.3(10\%\ \text{point}) + 0.4(50\%\ \text{point}) + 0.3(90\%\ \text{point}) \quad (15)$$

$$E(X + Y) = E(X) + E(Y) = \text{Sum of means always} \quad (16)$$

$$\mu_{\text{Res.}} = \mu_{\text{Pay}} + \mu_{\text{Area}} + \mu_{\text{Rec.}} \quad (16a)$$

$$E(XY) = E(X)E(Y) = \text{Product of means} \quad (17)$$

$$E(X/Y) = E(X)E(1/Y) \quad (18)$$

$$\neq E(X)/E(Y)$$

$$\text{Var}(X + Y) = \text{Var}(X) + \text{Var}(Y) = \text{Sum of variances} \quad (19)$$

$$\sigma^2_{\text{Res.}} = \sigma^2_{\text{Pay}} + \sigma^2_{\text{Area}} + \sigma^2_{\text{Rec.}} \quad (19a)$$

$$V = M^2 e^{\sigma^2 - 1} \quad (20)$$

$$\mu = \ln(M) - \sigma^2/2 \quad (21)$$

$$\sigma^2 = \ln(V/M^2 + 1) \quad (22)$$

$$\mu = \ln 10\%\ \text{point} + 0.5(\ln 90\%\ \text{point} - \ln 10\%\ \text{point}) = (\ln 90\%\ \text{point} + \ln 10\%\ \text{point})/2 \quad (23)$$

$$\sigma^2 = [(\ln 90\%\ \text{point} - \ln 10\%\ \text{point})/(2 \cdot 1.28)]^2 \quad (24)$$

$$M^* = \left(\sum p_i M_i\right) / \left(1 - \prod q_i\right) \quad (25)$$

$$\text{Var}(X) = E[(X - \mu)^2] = E(X^2) - [E(X)]^2 \quad (26)$$

$$E(Y^2) = e^{2(\mu + \sigma^2)} \quad (27)$$

$$V^* = \left[1/\left(1 - \prod q_i\right)\right]\left\{\sum p_i M_i^2 e^{\sigma_i^2} - \sum p_i^2 M_i^2 - \left[\prod q_i / \left(1 - \prod q_i\right)\right]\left(\sum p_i M_i\right)^2\right\} \quad (28)$$

Mean of doubly truncated lognormal:

$$m_{L:H} = M\{[F(z\#_{\ln H}) - F(z\#_{\ln L})]/[F(z_{\ln H}) - F(z_{\ln L})]\} \quad (29)$$

$$z_{\ln L} = (\ln L - \mu)/\sigma \quad (30)$$

$$z\#_{\ln L} = z_{\ln L} - \sigma \quad (31)$$

$$z_{\ln H} = (\ln H - \mu)/\sigma \quad (32)$$

$$z\#_{\ln H} = z_{\ln H} - \sigma \quad (33)$$

$$F(z)_{\text{approx.}} = 1 - [1 + (0.644693 + 0.161984z)^{4.874}]^{-6.158} \quad (34)$$

$$\text{Mode} = e^{\mu - \sigma^2} \quad (35)$$

$$\text{Commercial chance} = C_c = C_g[1 - F(L)] \quad (36)$$

$$\text{Truncated var.} = (0.894)(\text{untruncated var.})e^{(-0.0187)(\text{truncation pt.})} \quad (37)$$

THE SCIENTIFIC METHOD: TESTING HYPOTHESES ON RESERVES ESTIMATES

Well over a half century ago, statisticians were inventing basic tools to help scientists set up experiments and make probabilistic sense out of the results. The idea could hardly have been simpler. First set up a hypothesis you wish to accept or reject. Use your data to calculate a "statistic." Compare the statistic with a value in a table. If your calculated statistic is on one side of the value in the table, reject your hypothesis. If it's on the other side, accept the hypothesis.

One of these tools, the "*t*" statistic, allows management to weigh evidence for or against good estimating. Do the data support the contention by most explorationists that estimates and actuals are related? First set up a hypothesis:

*Predrill reserves estimates are **not** related to actual reserves discovered.*

It is hoped that the data will allow you to reject this uncomfortable statement! Calculate

$$t = (b - \beta_0)/s_b = b/s_b$$

where b = slope of the best fitting line through the data (we would like $b = 1$, meaning estimates = actuals); β_0 = slope of the line under the hypothesis = 0, that is, no relationship; and s_b = standard error of the slope b.

A large b compared to the uncertainty of its estimate s_b is evidence that the hypothesis does not hold true. Equivalently, a large t says to reject the hypothesis. How large? That is a function of how many data points you have and the chance of error you are willing to accept. You look up the values in tables found in statistics books. For this chapter, I have chosen to live with a 5% chance of being wrong when rejecting the hypothesis. Claiming that reserves estimates are correct when in fact they are not is a very costly error, so I would not want to make it very often. (The test must be "one-tailed" since presumably no one considers negative slopes for the best fit line. You have to know whether the test is one-tailed or two-tailed to look up the right number in the table.)

Now for the bias test. Agility is required! The bias test requires a "flip" in reasoning.

$$\text{Bias} = (1/n)\,\Sigma \ln(E/A) = 0 \text{ if there is no bias}$$

This time the hypothesis will be

*Reserves estimates show **no** tendency toward optimism or pessimism.*

Now you hope the data will *not* cause you to reject the hypothesis. Having a bias and not recognizing it becomes the most costly error, so we will set a much higher chance of being wrong when rejecting, say 40%. Because biases can go both ways, the test should be two-tailed. However, since most biases lean toward the optimistic side, we can do a 20% one-tailed test.

$$t = (\text{Bias} - 0)s_{\text{bias}}$$

where bias is defined as above and s_{bias} = standard error of bias. A large t is evidence of a significant bias and leads to rejection of the hypothesis.

DERIVATION OF MEAN AND VARIANCE OF BINOMIAL–LOGNORMAL DISTRIBUTIONS

Recall from the chapter that

$$\begin{aligned} E(XY) &= E(X)E(Y) \\ &= \text{Product of means if } X \text{ and } Y \text{ are independent} \end{aligned} \quad (17)$$

Now we'll use that result where X will become b, a binomial variable that takes the value of 1 with probability p and the value of 0 with probability q, where $q = 1 - p$. The mean of a binomial is p. Y will be a lognormal variable with mean M, variance V, mean of logarithms μ, and variance of logarithms σ^2.

$$E(bY) = E(b)E(Y) = pM$$

If we want $E(bY)$ given the zone flows, we simply divide by the success probability:

$$E(bY) \text{ (given } b = 1) = pM/p = M$$

which is an answer that should satisfy intuition. If we have n different zones, then

$$E(nbY) = npM$$

For example, say we have two zones (n) each with a 10% chance of success (p) and an exact size of 5 MMBO (M). Thus, $npM = (2)(0.1)(5) = 1$ MMBO. But we want the mean under the condition that at least one horizon hits:

$$\begin{aligned} \text{Chance of at least one hit} &= 1 - \text{Chance of no hits} \\ &= 1 - q^n \end{aligned}$$

For the example, $q = 0.9$, so

$$1 - q^n = 1 - 0.9^2 = 1 - 0.81 = 0.19$$

$$p^2 = 0.01 = \text{Chance of hitting both horizons}$$

Now we have all possibilities counted:

0.81 = Chance of nothing
0.01 = Chance of two hits
0.18 = Chance of one hit
1.00

Therefore, the mean given at least one hit (M^*) for n zones each with the same mean and chance is

$$M^* = npM/(1 - q^n)$$

Continuing with the example, $M^* = 1/0.19 = 5.3$ MMBO. Can we get some intuitive comfort here? Sure. The result says that, on average, given at least one successful horizon, we expect 5.3 MMBO. We know that most of the time (18/19) when we get at least one hit, we get *only* one hit and one hit has to be 5 MMBO. The only other possiblity is two hits for 10 MMBO, which occurs 1/19 of the time. Weighing these two sizes by their respective probabilities yields 5.3 MMBO.

If the zones do not have equal chances and reserves, we change the formula a little to accommodate this:

$$M^* = \left(\sum p_i M_i\right) / \left(1 - \prod q_i\right) \quad (25)$$

where the $\sum$ symbol means to sum all the products from $i = 1$ to $i = n$ and the symbol $\prod$ means to multiply all terms from $i = 1$ to $i = n$. You can put a red mark on that formula; you'll be using it often.

With independent horizons, the previous formulas give you the mean "exactly" and without regard to the form of the distribution. I've put the word *exactly* in quotes to remind you that in this business, this word has little value. Our estimating (guessing?) abilities about area, pay thickness, and recovery are not that good, so approximate formulas can also work.

If you wish to test the assumption of independence, you can run a sensitivity test and decide for yourself. Take those zones that you believe are not independent and just treat them as one. Add their thicknesses together for the net pay variable, if the productive areas are about the same. Average the recovery for the pair. Then reduce n by 1 in the formulas just given. Solve it again and see what you get. If you don't like either the assumption of complete independence or complete dependence, work it both ways and pick a reserves value somewhere in between.

Variances are tougher. Here we must *assume the form of the distribution,* and not surprisingly, we choose lognormal. From mathematical statistics, we know that

$$\text{Var}(X) = E[(X - \mu)^2]$$

which is the expected value of the squared differences between the values X and the mean of X. With a little algebra we have

$$\text{Var}(X) = E(X^2) - [E(X)]^2 \quad (26)$$

Remember that our variable is a hybrid made up of two other variables, b and Y.

$$\text{Var}(bY) = \text{E}(b^2Y^2) - [\text{E}(bY)]^2$$
$$= \text{E}(b^2Y^2) - (pM)^2$$

Recall that we used Equation 17 to evaluate the second term. The first term will take a bit more time. We again assume independence between b and Y—the chance that the zone will flow hydrocarbons is independent of the ultimate amount of hydrocarbons that flow:

$$\text{E}(b^2Y^2) = \text{E}(b^2)\text{E}(Y^2)$$

More jargon to come. Sorry. But for those who need to understand why, we want to provide that help. The rest of you can skip down to the answer and simply use it.

You may remember *moments* from a physics or engineering course. $\text{E}(Y^2)$ is the second moment about the origin, zero. The variance turns out to be the second moment about the mean. For a lognormal distribution,

$$\text{E}(Y^2) = e^{2(\mu+\sigma^2)} \quad (27)$$

where μ and σ^2 are the mean and variance, respectively, of the natural logarithms (Aitchison and Brown, 1957). Now we only need $\text{E}(b^2)$, or the mean of the square of a binomial. Another way to get the mean of any variable is to add up all the possible values of that variable weighted by their respective probabilities, $\text{P}(b)$:

$$\text{E}(b^2) = \sum\left[b^2\,\text{P}(b)\right] = (1)p + (0)q = p$$

So now we have

$$\text{Var}(bY) = \text{E}(b^2)\text{E}(Y^2) - (pM)^2$$
$$= p\,e^{2(\mu+\sigma^2)} - (pM)^2$$
$$= pM^2(e^{\sigma^2} - p),$$

which is the variance of a single horizon. Given independence among horizons, we can add variances to get the variance of the sum of these hybrid variables:

$$\text{Var}\left(\sum b_iY_i\right) = \sum p_iM_i^2(e^{\sigma_i^2} - p_i)$$

But if we want the variance given some sort of success, that is, a zone that flows, then we have to convert this unconditional variance to one conditional on at least one success. As before, we know to divide by

$$\left(1 - \prod q_i\right)$$

which is the chance that at least one horizon flows. But with the variance formula, we have a complication—we have the square of the mean as a term in the variance. We have to make that mean conditional also. As a reminder,

$$\text{Var}(bY) = \text{E}(b^2Y^2) - [\text{E}(bY)]^2$$

where the mean = $\text{E}(bY)$ has to be conditional on at least one successful horizon. Rather than bore the reader with more algebra, let's simply state the answer, where the * again denotes conditional on at least one success:

$$V^* = \left[1/\left(1 - \prod q_i\right)\right]\left\{\sum p_iM_i^2 e^{\sigma_i^2} - \sum p_i^2M_i^2 - \left[\prod q_i/\left(1 - \prod q_i\right)\right]\left(\sum p_iM_i\right)^2\right\} \quad (28)$$

To gain a measure of comfort, let's see if this monster collapses to something familiar. Say we have one horizon with a 100% chance of hitting. Removing the Σ's, Π's, i's, etc. gives you

$$V^* = M^2(e^{\sigma^2} - 1)$$

which is exactly what we started with earlier in Equation 20.

DERIVATION OF TRUNCATION EQUATIONS

We need to calculate $m_{L:H}$, the mean of that clear area between L and H (look back at Figure 22). Statistical theory gives us $m_{>H}$, or the mean of that part of the distribution above a cutoff such as H (or L). Recall that the mean, M, of the entire distribution is equal to the chance weighted means of its segments:

$$M = m_{<L}\text{F}(L) + m_{L:H}[\text{F}(H) - \text{F}(L)] + m_{>H}[1 - \text{F}(H)]$$

where the notation $\text{F}(L)$ is the standard cumulative distribution function, which means the probability under the distribution curve from the far left side up to the point L. The problem with this equation is that we don't know enough to solve it. We know M, and we can get $\text{F}(L)$, $\text{F}(H)$, and $m_{>H}$. We ultimately want $m_{L:H}$. That leaves us with another unknown, $m_{<L}$. So let's write it in terms of quantities that we do know.

$$M = m_{<L}\text{F}(L) + m_{>L}[1 - \text{F}(L)]$$

so

$$m_{<L} = \{M - m_{>L}[1 - \text{F}(L)]\}/\text{F}(L)$$

and we can find $m_{>L}$. Aitchison and Brown (1957) tell us how to get the mean of that part of the distribution above a cutoff point, such as $m_{>L}$ and $m_{>H}$.

$$m_{>L} = M\{[1 - \text{F}(L \mid \mu + \sigma^2, \sigma^2)]/[1 - \text{F}(L \mid \mu, \sigma^2)]\}$$
$$m_{>H} = M\{[1 - \text{F}(H \mid \mu + \sigma^2, \sigma^2)]/[1 - \text{F}(H \mid \mu, \sigma^2)]\}$$

where $\text{F}(L \mid \mu + \sigma^2, \sigma_2)$ is the probability of L or less

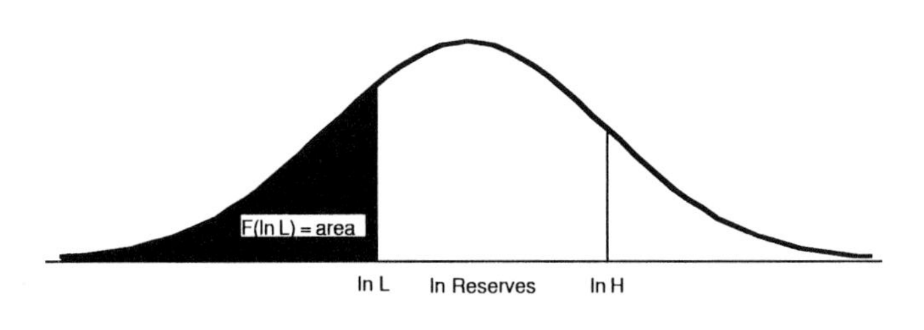

Figure 29. The cumulative distribution function.

given that the mean of the logarithms is $\mu + \sigma^2$ and the variance of the logarithms is σ^2. The meaning is identical to our use of F(*L*) earlier except we explicitly state the mean and variance of the distribution for clarity because the expression deals with two separate distributions differing in their means.

Now, if we go back to the formula that includes $m_{L:H}$ and rewrite it solving for the mean of this middle portion (and leaving out a few steps of algebra), we have

$$m_{L:H} = M\{[F(H \mid \mu + \sigma^2, \sigma^2) - F(L \mid \mu + \sigma^2, \sigma^2)] / [F(H \mid \mu, \sigma^2) - F(L \mid \mu, \sigma^2)]\} \quad (29)$$

How do we solve for all these F's? A look at Figure 29 may help us see what we need. We've slipped back to the normal form of the distribution for ease of calculation.

$$F(L)_{\text{lognormal}} = F(\ln L)_{\text{normal}}$$

Because they are areas under curves, the F's are integrals:

$$F(L) = \int_0^L f(x)dx$$

which is the area under the curve defined by $f(x)$ from 0 to *L*. We have a very good numerical solution to this integral to almost four standard deviation units or 99.9% of the distribution. That'll do for any exploration problem you'll likely run into. The solution, however, works for the standard normal distribution, with mean = 0 and variance = 1. That means we have to convert ours to this one to be able to use the formula. We usually denote the variable z as a standard normal and x as any other normal with mean μ and variance σ^2.

Table 6. Errors in F(*z*) Approximation

z	F(*z*)	$F(z)_{\text{approx.}}$	Difference
−3.0	0.001	0.001	0.0005
−2.5	0.006	0.006	0.0004
−2.0	0.023	0.024	−0.0008
−1.5	0.067	0.069	−0.0023
−1.0	0.159	0.160	−0.0010
−0.5	0.308	0.306	0.0017
0.0	0.500	0.496	0.0040
0.5	0.692	0.690	0.0019
1.0	0.841	0.843	−0.0021
1.5	0.933	0.935	−0.0022
2.0	0.977	0.978	−0.0008
2.5	0.994	0.994	0.0003
3.0	0.999	0.998	0.0004

Then,

$$z = (x - \mu)/\sigma$$

We have four different z values to calculate, two associated with ln *L* and two with ln *H*:

$$z_{\ln L} = (\ln L - \mu)/\sigma \quad (30)$$

$$z\#_{\ln L} = (\ln L - \mu - \sigma^2)/\sigma = z_{\ln L} - \sigma \quad (31)$$

$$z_{\ln H} = (\ln H - \mu)/\sigma \quad (32)$$

$$z\#_{\ln H} = (\ln H - \mu - \sigma^2)/\sigma = z_{\ln H} - \sigma \quad (33)$$

where the # symbol reflects the distribution with the mean = $\mu + \sigma^2$.

$$F(z)_{\text{approx.}} = 1 - [1 + (0.644693 + 0.161984z)^{4.874}]^{-6.158} \quad (34)$$

If you worry about approximations, Table 6 should put you at ease. It seems unlikely that anyone would do truncations beyond the 0.1% tails of the distribution, but even there the approximation works. It does break down for absolute values of $z > 3.9$, however (Burr, 1967).

Now we have all the pieces of the puzzle for calculating the mean of a double truncated lognormal distribution. For clarity, let's rewrite Equation 29 in a form that will be more useful:

$$m_{L:H} = M\{[F(z\#_{\ln H}) - F(z\#_{\ln L})]/[F(z_{\ln H}) - F(z_{\ln L})]\} \quad (29)$$

ED CAPEN retired as Distinguished Management Advisor for ARCO in Dallas in 1992. He began his career with the company as a student doodle bugger working out of Laredo, Texas, in the summer of 1956. Since joining the company's research lab in 1957 as a physicist, he has worked in seismic research, well logging research, as Director of Research Planning and Evaluation, as Director of Operations Analysis for ARCO's Production Division, and as Manager of Capital Administration and Expense Control for ARCO's corporate headquarters in Los Angeles. He has authored numerous articles on competitive bidding, economic analysis, and handling uncertainty and communication and has spoken at national meetings of AAPG, SPE, SEG, and API. He is a member of AAPG and SPE. In 1974 and 1987, Mr. Capen was a Distinguished Lecturer for SPE, and in 1989 he received the J. J. Arps Award from SPE for distinguished contributions to the field of petroleum economics and evaluation. He now has a teaching and consulting practice in Dallas.

Chapter 6

Estimating Prospect Sizes

Robert E. Megill
Consultant
Kingwood, Texas, U.S.A.

Oil at $100 per barrel will not close a contour that is not there, and gas at $10 per MCF will not take enough clay out of a sandstone's matrix to make it a reservoir.
—J. F. Bookout

HOW LARGE IS THE PROSPECT?

This chapter deals with estimating the size of potential prospect reserves. But before we can consider this topic, a brief review of lognormality is necessary. It is well established in the literature that a multiplicative process involving random data produces a lognormal distribution. It is therefore easy to understand why reserve sizes of oil and gas fields are approximately lognormal. We calculate the sizes (the estimated ultimate recovery) of fields discovered by multiplying several factors together to get a product. In the most simple case, we multiply net pay x recovery per acre-foot x area to get a prospect size in barrels or million cubic feet. In a sealed bid sale, individual bids by a number of bidders on the same tract result from multiplying several factors. Thus, it should not be surprising that they are approximately lognormal.

Plotting a Field Size Distribution

The following steps are involved in plotting a field size distribution.

1. From a list of fields (such as Table 1, column 1), produce a ranking of those fields with the largest field first, the second largest second, and so on (see Table 1, column 2). The ranking of fields is one of the variables used for plotting purposes to establish a cumulative frequency distribution of field sizes. A straight line on logprobability paper (Figure 1) means that the data plotted are lognormally distributed. This ranking produces a "greater-than" cumulative frequency distribution. An "equal-to-or-less-than" distribution is produced by ranking with the smallest field first. On logprobability paper, the less-than curve has a reverse slope from the greater than ranking.
2. Now calculate a fraction called a *fractile*. The denominator is the number of fields (n) plus 1. The numerator is the rank number of the field. Thus, the fractile plotting point for the first field is 1 divided by (n + 1). For the fields in Table 1, the plot point for the first field is 8.33%, derived by dividing 1 by (11 + 1), which equals 0.0833 or 8.33%. For the second-ranking field, the fractile is 2 divided by (11 + 1) or 0.1667 equals 16.67% (see Table 1, column 3).

A plot of the points from Table 1 is shown in Figure 1 with a straight line drawn through the points. In many trends, the points are slightly more scattered, but the general trend of lognormality is almost always apparent.

A field size distribution as shown in Figure 1 illustrates the uncertainty of *size* in a given trend. Preferably, a field size distribution should deal with fields in a specific formation or play. One would not plot salt dome fields along with low relief fields in the Outer Continental Shelf as part of the same distribution.

Information Gained from a Field Size Distribution

Because we have a cumulative plot on logprobability paper, we obtain probabilities. For example, the

Table 1. A Cumulative Distribution of Field Sizes

	Field Sizes (in BCF)	Field Sizes in Rank Order	Plot Points for Logprobability Paper (in %)[a]
1	150	320	8.33
2	20	150	16.67
3	40	90	25.00
4	5	70	33.33
5	15	55	41.67
6	320	40	50.00
7	10	30	58.33
8	30	20	66.67
9	90	15	75.00
10	55	10	83.33
11	70	5	91.67
Average		73	

[a]Percentage equal to or greater than.

median field size (in Figure 1) is 40 BCF. Thus, only one-half of the fields found in this trend are equal to or greater than 40 BCF. If the prospects to be drilled in this trend are similar to those already found, then the chance of getting a field of 40 BCF or larger is only 50%—if there is no risk. In this manner, the uncertainty associated with size is amply demonstrated. Size uncertainty can be combined with chance of success to ascertain the expected risk of a given size or larger. For example, if the trend in which these fields exist has had an average success rate of 20%, then the chance of getting a field of greater than 40 BCF is, on the average, only 50% x 20% = 10%. That is, only a one-in-ten chance exists of getting a field equal to or greater than 40 BCF. Obviously, a cumulative frequency distribution of fields found provides valuable information to the explorer.

In Figure 1, the arithmetic average of all fields is 73 BCF. Note how much larger the average is than the median. The difference indicates a lognormal distribution. The greater the difference between the average (mean) and the median, the greater the range between the largest and smallest field. A lognormal distribution always has a larger mean than median.

The average value takes into account the larger fields in the upper half of the lognormal distribution. One would not expect to find the average field or larger as often as the median size or larger. In fact, one would expect to find 73 BCF or more only one-third of the time. Yet the average is still the most representative reserve size for the entire suite of fields. To represent the distribution of these fields, the average should always be accompanied by a measure of dispersion, such as the standard deviation.

Applying the Concept of Field Size Distributions to Prospects

For reasons to be given later, prospect size ranges (i.e., prospect size distributions) are useful and necessary. The same concepts used to obtain field size distributions can be used to determine the range in prospect sizes. One important difference must be kept in mind, however. A field size distribution has a larger range of values than a prospect size distribution. A prospect size distribution is site specific. One may have a large or small prospect, but in any case, the range of possible sizes will not be as large as the size range for all fields found to date.

Consider the field size distribution shown in Table 1. The range is from 320 to 5 BCF. A prospect, in contrast, might range from 60 BCF at the 10th percentile to 40 BCF at the 90th percentile, with a median of about 50 BCF. These points are plotted along a dashed line in Figure 1. In the greater-than configuration, the prospect distribution has a much flatter slope than the field size distribution. Thus, its mean and median are much closer in value than those for the distribution of field sizes.

ESTIMATING THE RANGE IN RESERVE SIZE FOR PROSPECTS

Most people outside of the petroleum industry acknowledge that exploring for new oil and gas fields involves considerable risk. However, that thought is most often associated with the chance of finding hydrocarbons. Explorers also know that considerable differences of opinion can exist as to the reserve size of a prospect or prospects in a play. This chapter discusses the uncertainty in estimating potential prospect sizes.

First, why does one say *sizes*, not *size*? The answer is that the exact size is unknown and that it can be significantly different from a single estimate of size. In fact, the uncertainty is so great that a range of estimates is preferable to a single estimate. The decision maker must be able to see the entire spectrum of possible prospect sizes.

An Illustration

The need to consider more than one prospect size can be illustrated by the following example. Suppose we have a prospect mapped. Every map has some uncertainty; after all, we are dealing with unseen surfaces and thicknesses thousands of feet below the ground. Using net pay, recovery per acre-foot, and areal extent as the three critical reserve parameters allows an example of range for prospect sizes. Now

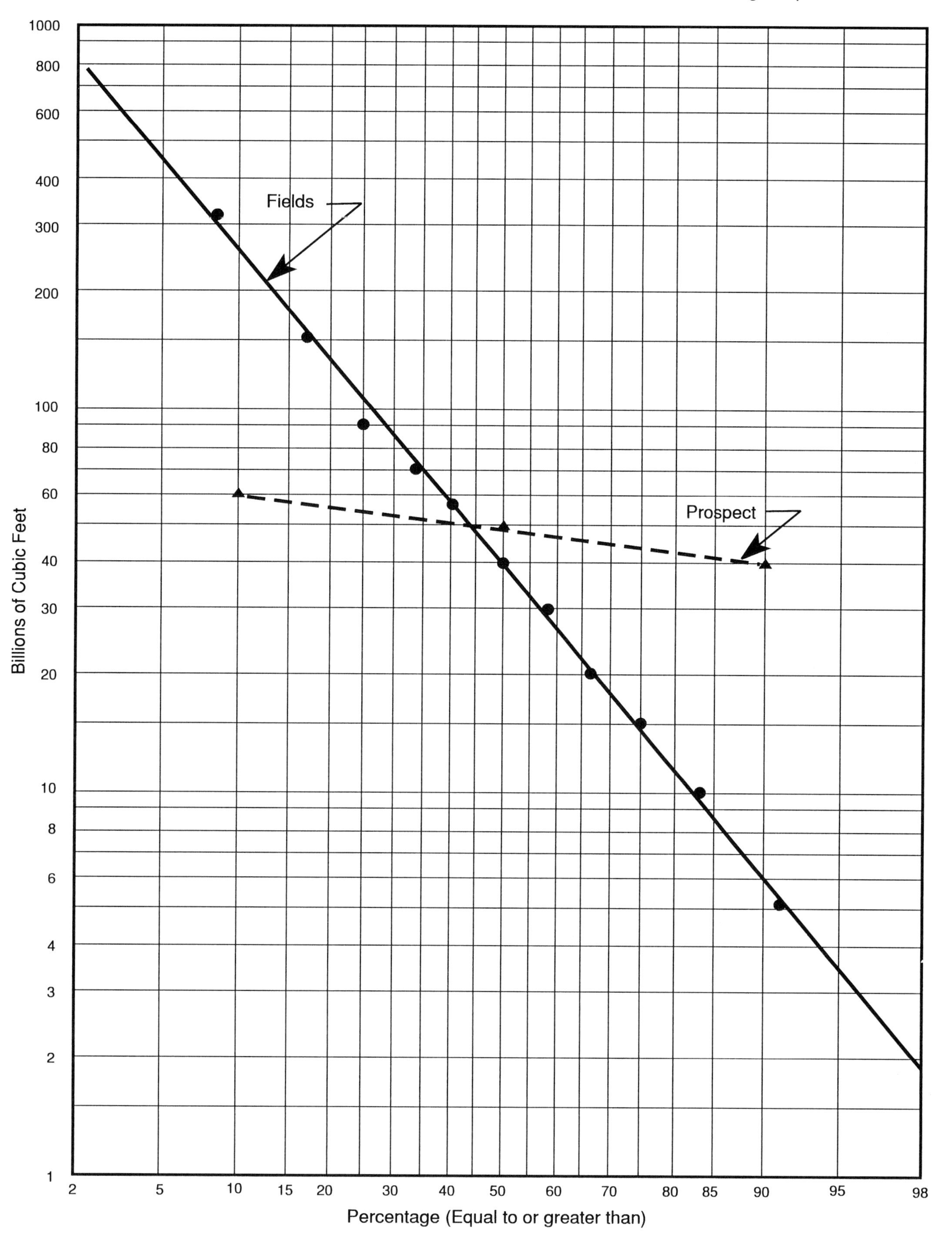

Figure 1. Cumulative distributions of field and prospect sizes.

suppose that in our prospect the net pay thickness varies from 12 to 16 ft over an areal extent ranging from 400 to 500 acres. The recoveries per acre-foot in the area range from 300 to 400 bbls. To get the full range of possible sizes, we first multiply 12 ft x 300 bbl/acre-ft x 400 acres to get the low-side possibility of 1.44 million bbls. The high-side possibility is obtained by multiplying 16 ft x 400 acre-ft x 500 acres to get 3.2 million bbls. The large value is 122% greater than the small value. The size of the prospect is thus estimated to range between 1.44 and 3.2 million bbls.

From this example, one can see the considerable uncertainty in the exact size of the prospect. Because of this uncertainty, several sizes should be examined. How many sizes is a function of the data, the computer programs available, and the management's familiarity and desire for simple or complex (Monte Carlo) solutions. Those familiar with Monte Carlo simulations (or other types of simulation such as the Latin hypercube) know that simulations provide a suite of answers *plus* cumulative probabilities dealing with probability of occurrence.

The Procedure

Some managers, however, still prefer to deal with specific examples rather than a suite of all possible answers. For such managers, alternate possibilities exist. One possibility, as outlined by Megill (1988a,b,c), involves the use of logprobability paper and a hand calculator. The steps involved are as follows:

1. High- and low-side values for pay thickness (limit of expected ranges) are plotted at the 90th and 10th percentiles. Note that Figure 2 uses the less-than configuration.
2. The median value is then checked for reasonableness; if deemed relevant, the projection of ranges is accepted.
3. The same procedure is applied to a plot of the expected limits for recovery per acre and areal extent of the prospect. In each case, after the 10th and 90th percentile values are plotted, the median value is checked for reasonableness. Three straight lines result from this effort (on logprobability paper).
4. To obtain the range of reserve sizes requires specific calculations. The reason is that the 10th percentile values of pay, recovery and area *cannot* be multiplied to get the 10th percentile value on the reserve plot; they, in fact, produce a plot point at about 1.3%. Likewise, the 90th percentile value on the reserve plot cannot be obtained by multiplying the 90th percentile values for pay, recovery, and area. Multiplying these values produces the 98.7% value. Because some types of logprobability paper do not contain values beyond 2% or 98%, another method must be used. (The only value on the three parameter plots that can be multiplied directly is the median value [50th percentile]; thus, one point on the reserve curve is always available directly.)
5. Megill (1988a,b,c) recommends Capen's "35/25" method to get the two other points on the reserve curve (E. C. Capen, pers. comm., 1988, 1989). It involves multiplying the 35th percentile values for pay, recovery and area, but plotting the result at the 25th percentile. The 25th percentile point plus the median point represent two points on a straight line. If a third plot point is desired, the 65th percentile values can be multiplied and plotted at the 75th percentile. The straight line can be extended to show what the 10th and 90th percentile values would be for the reserve range. (A second method for the two additional reserve points is to multiply the 23rd percentile points to be plotted at the 10th percentile for the reserve plot and the 77th percentile points to be plotted at the 90th percentile.)

   For each straight line, the median value is examined for reasonableness. The two-value approach (10th and 90th percentiles) has two advantages. First, an 80% range—a useful working range—is established. Second, because the limits were examined first, the estimator is not "locked" into a preconceived value for either the individual parameters or the final reserve estimate. The median values should be near those that would be mapped for the prospect. Figure 2 illustrates the 35/25 concept for the single parameters of pay, recovery, and area.
6. Figure 3 and Table 2 contain the plot points and calculations to determine the range of prospect values. The 35/25 method was used to obtain three plot points for the reserve curve. After the line is drawn through the three points at the median and the 25th and 75th percentile points using the 35/25 method, the 10th and 90th percentile points are spotted on the line.

If you are still involved in a process in which single estimates of prospect area, thickness, and recovery per acre are used to establish a single reserve size, you are not considering all possibilities. The easiest way to demonstrate this inexorable fact is to make a record of the predrill reserve sizes and compare them to the postdrill results. This exercise in humility will demonstrate the difficulty the explorer faces if he or she considers only a single number as *the* reserve size to be expected.

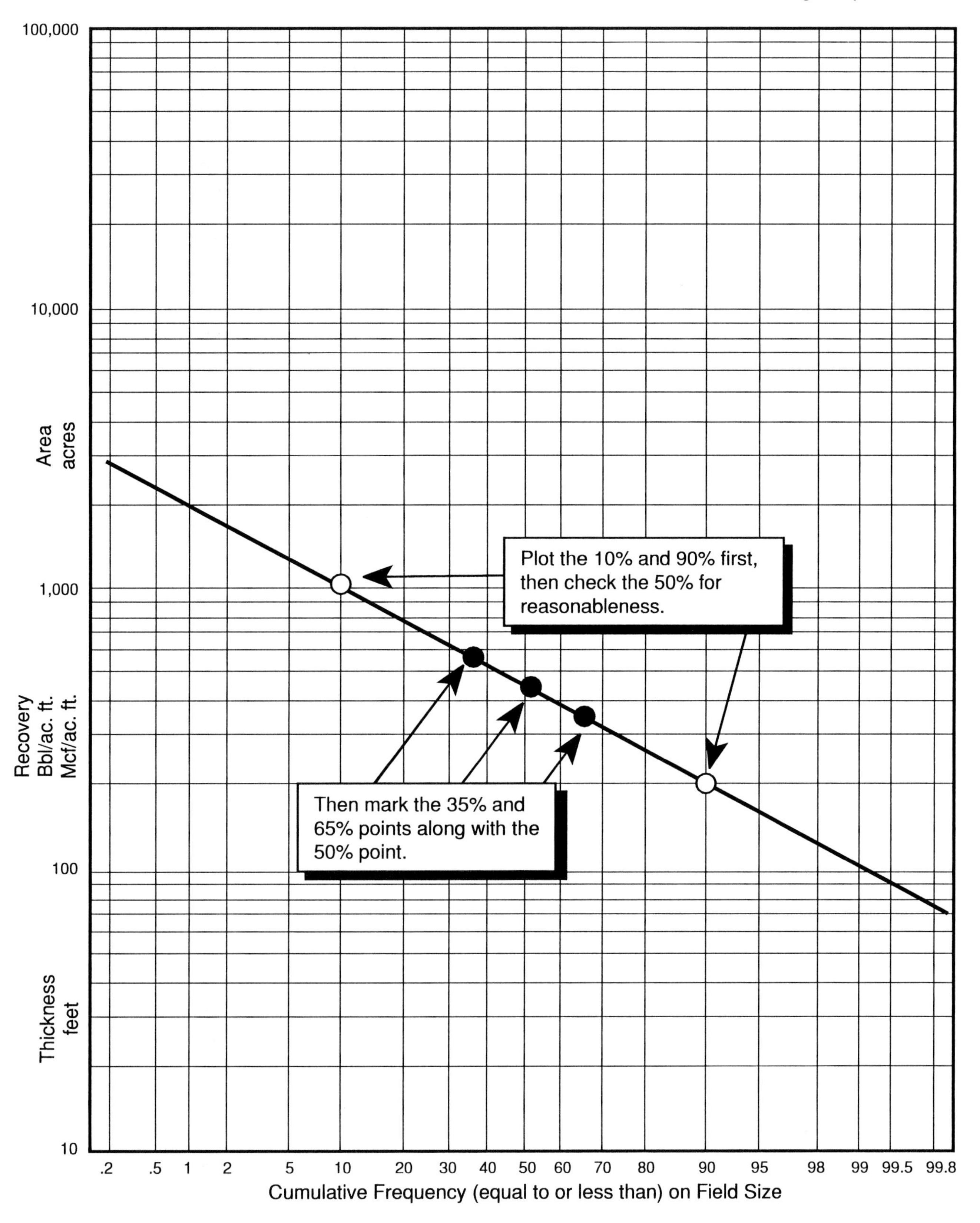

Figure 2. Prospect area, pay thickness, or recovery. (Courtesy of E. C. Capen.)

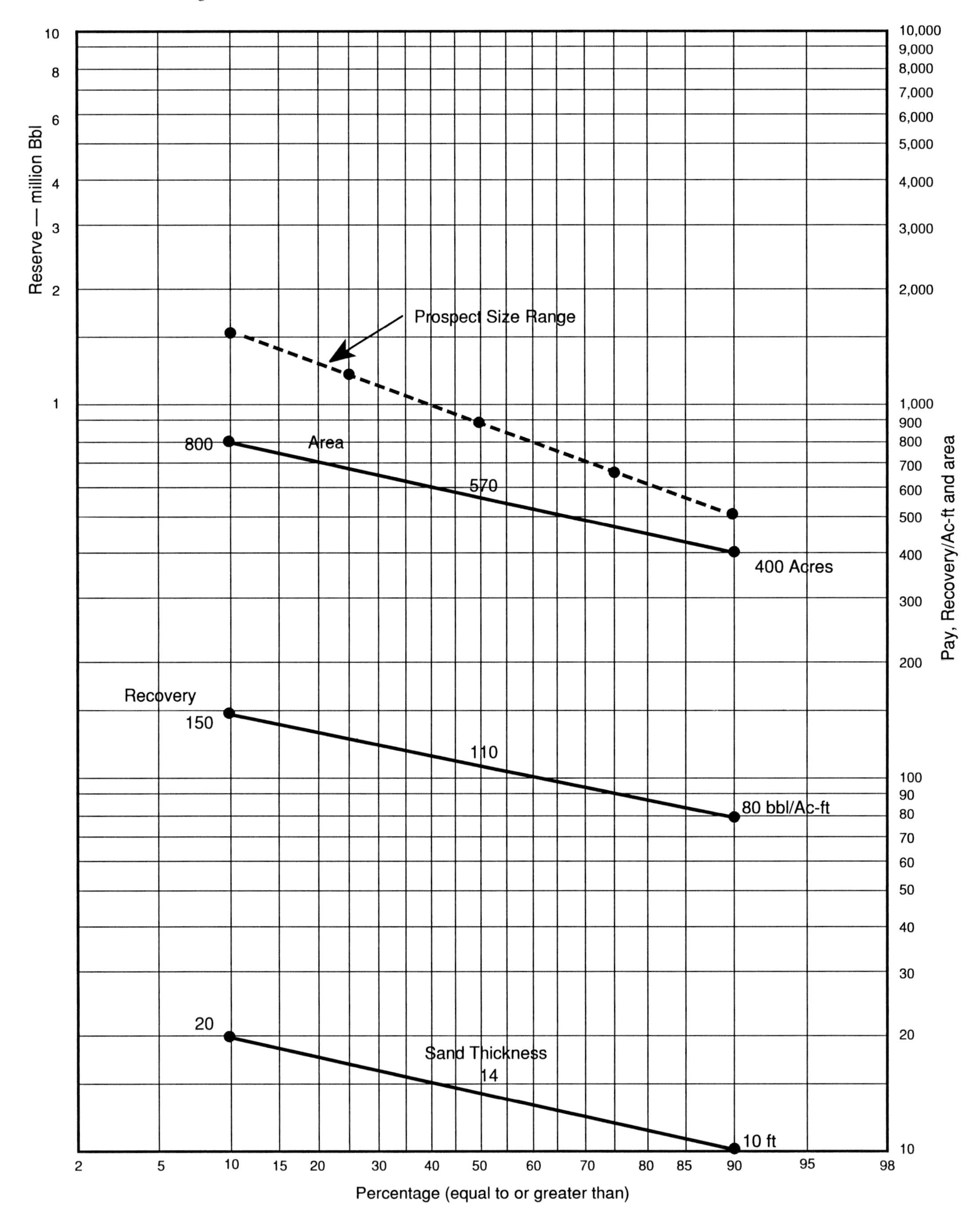

Figure 3. Estimating the range in prospect size.

Table 2. Estimating the Range of Prospect Sizes

	Percentiles			Plot Point Percentiles	
Parameter	10th	50th[a]	90th	25th[b]	75th[c]
Net pay: feet	20	14	10	17	12
Recovery: bbl/ac-ft	150	110	80	130	92
Area: acres	800	570	400	700	460
Reserves: million bbl	508[a]	878	1547[a]	1210	663

[a]Value determined from plot.

[b]Calculated using 35th percentile values.

[c]Calculated using 65th percentile values.

CONCLUSIONS

The process described in this chapter does not rule out the need for good maps. Structural maps, isopach maps, and cross sections will always be among the invaluable tools of the explorer. What the process does is to recognize that in an area of much uncertainty, a single answer does not supply the information needed by the decision maker. Where uncertainty exists, the full range of possibilities is the only valid information to give the people providing the funds. Anything else deprives the decision maker of the degree of uncertainty inherent in all input factors and the range of possibilities in any exploratory investment.

References Cited

Megill, R. E., 1988a, Charting the range of uncertainty: AAPG Explorer, v. 9, n. 10, p. 34-35.

Megill, R. E., 1988b, Estimating prospect reserves vital: AAPG Explorer, v. 9, n. 11, p. 18-19.

Megill, R. E., 1988c, Capen's compensation yields 35/25: AAPG Explorer, v. 9, n. 12, p. 21.

ROBERT E. MEGILL, a consulting geologist in Kingwood, Texas, received a B.S. in Geological Engineering from the University of Tulsa. He worked for Exxon, including predecessor companies Carter and Humble, from 1941 to 1984. He served as staff geologist, petroleum economist, reserve geologist, Head of Reserve Economics, Head of Exploration Economics, Manager of Planning, and Coordinator of Planning and Coordinator of Economic Evaluation. Mr. Megill has published numerous technical articles and is widely known for his books and courses on Introduction to Risk Analysis and Introduction to Exploration Economics. He is a member of AAPG and for the last 7 years has written a monthly column in The Explorer on "The Business of Petroleum Exploration."

Chapter 7

Chance of Success and Its Use in Petroleum Exploration

Peter R. Rose

Telegraph Exploration, Inc.
Austin, Texas, U.S.A.

The more uncertainty attends a given prospect, the more a systematic expression of subjective probability is needed.
—Peter R. Rose

INTRODUCTION

Originally, this chapter was to be titled "The Chance of Hydrocarbon Presence." In oil and gas ventures, however, what the prospective investor really wants to know is not whether hydrocarbons are present but rather the chance that the drill bit will encounter a reservoir containing sufficient petroleum to warrant completion of the well. Another important question relating to both reserve estimates and project profitability is this: if the well is completed, how much profit will it generate? These topics are addressed in other chapters of this book.

For a given prospect, geologists can assess the likelihood of key geological factors (reservoir rock, hydrocarbon charge, etc.) recognized as essential for an accumulation of petroleum to exist in the subsurface. This leads to estimating the chance of hydrocarbon presence—or *geological success*—as described here.

A second issue concerns the *quality* of the accumulation (reserve volumes and production rates) in relation to costs and wellhead revenues. For example, an onshore wildcat that discovers a million barrel oil field is completed in confident anticipation of a profitable new field, whereas a counterpart well offshore may be plugged and abandoned with hardly a second thought. Both wells encountered petroleum accumulations—geological successes—but only the onshore well was *completed*—the reporting measure of success. So the location or setting affects success rates.

However, many exploratory wells that are completed do not repay the costs of completing them, or even finding them. So a third consideration is the *commerciality* of the completed well and the accumulation it discovered. What is the chance that the well will repay the costs of completion and subsequent operation? The chance that it will also repay the cost of exploratory drilling? The further chance of returning a reasonable profit beyond that? Thus, *success* has many definitions. Techniques are available, however, to make useful predictions of success, but only if we use consistent and strict definitions.

Definitions

Geological success A well that encounters mobile and measurable reservoired hydrocarbons.

Completion success A well that is completed for production.

Incremental success A completed well that will return a profit only upon the investment of completing and operating it, ignoring (as sunk costs) the exploratory drilling costs required to find it.

Commercial success A discovery well that finds a field capable of paying for all subsequent costs of development drilling, completion, surface equipment, operation costs, and wellhead taxes, returning a reasonable profit margin plus the cost of an equivalent number of dry holes. Note that this definition allows the inclusion of a small proportion of commercial successes that do not also repay other exploration costs such as geological and geophysical (G & G), lease bonuses, and analogous exploratory dry holes.

The Historical Record

Drilling statistics gathered by AAPG's Committee on Statistics of Drilling (CSD) provide a general perspective on the probabilities of success associated with the several classes of wells drilled in the United States (Table 1). Again, it must be emphasized that *success* as used by the CSD means only that the well was completed—not that it was a profitable venture!

Although 13 to 18% of U.S. new field wildcat wells were "successful" during the 1980s, it appears that only about 1 to 2% discovered fields of a million oil-equivalent barrels or larger, the figure chosen by the CSD as significant with regard to reserve additions. Nehring (1990) disputes such low percentages, pointing out that, because of gradual development of discovered fields beyond the 6-year time frame used by the CSD, the actual success percentage for significant field discoveries should be higher—perhaps 4 to 6%. Indeed, the CSD has now suspended their annual reporting of this measure pending a reevaluation of their methodology and results. Even so, the chance of a domestic new field wildcat venture finding a significant accumulation is small.

In any case, such statistical reports blend drilling results throughout the United States, onshore and offshore, in different geological provinces and trends, to depths from a few hundred feet to more than 20,000 ft, by operators having highly diverse geotechnical capabilities, and in exploration theaters ranging from undrilled basins to maturely developed producing areas. Accordingly, such figures should not be used to project the chance of success for any specific drilling venture. Likewise, drilling success rates can be compiled for specific provinces, basins, or mature trends, but these data are most useful as broad reality checks. They also do not provide information specific to any particular prospect being considered for drilling.

Table 1. Generalized Success Rates of Various Well Classes Drilled in the U.S. Onshore and Offshore during the 1980s as Reported by the CSD

Well Class	Percentage Successful
Development wells	75–80%
All exploratory wells	20–30
Extensions (outposts)	40–45
In-field wildcats	25–35
New field wildcats	13–18

CHANCE OF SUCCESS AND EXPECTED VALUE

Expected Value Concept

Economic analysis of a contemplated oil and gas venture must be carried out on the assumption that the project is successful, with success often being expressed as one of several levels of profitability based on various ranges of the geotechnical and economic parameters that influence project commerciality. For example, a firm may generate economic runs on projected minimum, modal, median, mean, and maximum reserve sizes. (But note that the best single representation of the prospect reserve distribution is the mean [Megill, 1990].) However, many exploration ventures fail, and the consequences of such failure must also be considered in appraising the economic merit of a drilling prospect. *Expected value* (EV) includes probabilities as well as economic consequences of both success and failure:

$$\text{EV} = \text{Probability}_{\text{success}}(\text{Project present value}) + \text{Probability}_{\text{failure}}(\text{Costs of failure}) \quad (1)$$

Now imagine that you had the opportunity to participate in a simple game in which you try to call correctly the toss of a fair coin. If your call is correct, you win $20,000; if it is incorrect, you win nothing. If you were able to play this game free of charge, the expected value (EV) of each trial would be +$10,000. If you paid $10,000 each time you played, the EV would be zero, so that, statistically, you would just be trading dollars. If you were willing to invest $8,000 in one trial of this game, the EV would be +$2,000 (Table 2).

Faced with choosing among several options, all having positive EVs, the decision rule is to select the option having the highest positive EV (Newendorp, 1975). Obviously, when operators choose to participate in ventures having negative expected values, they are betting against the House, that is, they are gambling.

In exploratory ventures, prospect net present value (NPV) ordinarily includes all front-end exploratory costs. Cost of failure usually includes dry hole cost, cost of lease bonuses of the condemned leases, and some G&G costs. For especially large prospects, the cost of one or more confirmation–delineation wells may also be included. Newendorp (1975) thoroughly covers the subject of expected value for the reader who wishes additional background.

Calculating and Using Prospect Expected Value

The EV concept is extremely useful in analyzing oil and gas ventures because it enables us to *risk weight* the perceived value of prospects. It allows us to compare a prospect having large reserve potential but low

Table 2. Expected Value Examples (Coin Toss)

Trial	Outcome	Consequence – Cost	Profit/Loss x Probability =	Risked Result
Free trial	Correct call	+$20,000 – 0 =	+$20,000 x 0.5 =	+$10,000
	Incorrect call	0 – 0 =	0 x 0.5 =	0
			EV =	+$10,000
$10,000 trial	Correct call	+$20,000 – $10,000 =	+$10,000 x 0.5 =	+$5,000
	Incorrect call	0 – $10,000 =	–$10,000 x 0.5 =	–$5,000
			EV =	0
$8,000 trial	Correct call	+$20,000 – $8,000 =	+$12,000 x 0.5 =	+$6,000
	Incorrect call	0 – $8,000 =	–$8,000 x 0.5 =	–$4,000
			EV =	+$2,000

Table 3. Calculations and Comparisons of Expected Values of Two Prospects

Prospect	Calculations
Prospect A	Estimated AFIT[a] present value of 10,000 bbl mean reserves[b] = $30,000,000
	Estimated chance of success = 10%; chance of failure = 90%
	Estimated AFIT dry hole cost = –$2,500,000
	EV = 0.10($30,000,000) + 0.90(–$2,500,000) = **+$750,000**
Prospect B	Estimated AFIT present value of 10,000 bbl mean reserves[b] = $3,000,000
	Estimated chance of success = 40%; chance of failure = 60%
	Estimated AFIT dry hole cost = –$1,000,000
	EV = 0.40($3,000,000) + 0.60(–$1,000,000) = **+$600,000**

[a]AFIT = after federal income tax.
[b]Contains all project costs, including costs of exploratory well.

perceived chance of success with another prospect having modest reserve potential but higher probability of success. For example, guided only by EV, Prospect A in Table 3 would be preferable to Prospect B because it has a higher EV. However, the decision maker may in fact choose Prospect B, accepting a smaller reward in exchange for the substantially higher chance of success and lower front end costs. This would be an expression of risk aversion, as covered in Chapter 9. Even so, EV is valuable in detecting and coping with risk-averse behavior in this example.

Prospect inventories—aggregated into exploration programs of operating districts, divisions, regions, and entire corporations—depend upon EVs of prospects and are of substantial value in forecasting general results of such programs. Decision tree analysis also depends upon reliable EV figures. But you must have responsible estimates of chance of success (in addition to sound estimates of prospect reserves and present value, as well as dry hole costs) to calculate EV reliably.

GEOLOGICAL CHANCE FACTORS AND PROBABILITY OF SUCCESS

Requirements for a Petroleum Accumulation

Petroleum geologists generally agree that, for a subsurface accumulation of hydrocarbons to exist, there must be porous and permeable reservoir rock, hydrocarbons that have moved from a petroleum source rock to the reservoir rock, and a sealed closure or trap capable of containing the hydrocarbons (Landes, 1951; Dott and Reynolds, 1969). All three of these general requirements must be met for a hydrocarbon accumulation to form; if any one of them fails, no accumulation will be present. This paradigm becomes the fundamental basis for the utility of geological chance factors in estimating probability of geological success (Table 4). In its simplest form, each of the three geological chance factors is treated as an independent variable having a probability ranging from 0 to 1.0.

Serial multiplication of all three factors produces a decimal fraction equivalent to the probability that a hydrocarbon accumulation is present, which is the probability of geological success, or P_{sg} (and by subtracting P_{sg} from 1.0, we derive its counterpart, the probability of geological failure, or P_{fg}). These expressions, along with estimates of prospect value and dry hole cost, are needed to calculate the expected value of the venture.

Explorationists commonly ask whether one geological chance factor is more important than another, thereby deserving more weight. The answer is definitely *no*. They should be thought of as links in a chain: if any link breaks, the chain fails. By analogy, if any one of the geological factors is zero, the prospect will be dry.

However, in the real world, things are not so neatly binary (Figure 1, top). There are many reasons why an exploratory well might be dry. The objective reservoir may be absent, tight, saturated with water, or simply filled with inadequate concentrations of oil or natural gas. The latter are ordinarily recorded as dry holes with shows. Finally, if the exploratory well is completed, the new accumulation it has found may be either commercial or noncommercial.

Recorded Success Rates versus Geological Success Estimates

When compared with exploration drilling statistics, such as those reported annually by the CSD, P_{sg} (derived using geological chance factors) corresponds roughly, but not exactly, to the CSD's definition of success, that is, that the subject well was completed and did produce some hydrocarbons. This does not mean that the venture made a profit. In fact, the CSD's definition contains four possible successful outcomes (Figure 1):

1. The well was completed as the discovery well for a field in which average wells will generate sufficient production revenues to cover at least the cost to drill, complete, and operate them, plus a reasonable profit (and usually pay for the sunk costs to find the field, as well as its proportional share of equivalent exploratory dry holes). This is a *commercial success.*
2. The well was completed because anticipated future production will return a profit on the cost of completing and operating it, but not on the costs of exploratory drilling, which are thus viewed as sunk and not recoverable (Capen, 1991). Such a well is an *incremental success.* Ordinarily, no more wells would be drilled on such a prospect by the operator, assuming that subsequent events did not provide new encouragement to drill again.

Table 4. Example of Calculation of Simplistic Probability of Geological Success and Failure Using Three Geological Chance Factors

Geological Chance Factor	Probability
Reservoir rock	0.7
	x
Hydrocarbon charge	0.8
	x
Sealed closure	0.5
Product = probability of geological success (P_{sg})	0.28
Probability of geological failure (P_{fg}) = (1 – 0.28)	0.72

3. The well was completed as either a commercial success or an incremental success, but subsequent performance was inadequate to sustain operating costs, resulting in abandonment a short time later.
4. The well was completed only for business reasons, that is, to hold a lease position or to satisfy a contractual or regulatory obligation.

As defined, however, there is yet a fifth class that falls within the geological success category, but outside the CSD's definition of success. This is an exploratory well that discovers a petroleum accumulation too small to warrant the expense of completing and operating it (Figure 1). Such accumulations are commonly recognized as *good shows*. Practically speaking, however, we can eliminate this class by introducing concepts of minimum dimensionality or volumetrics into the definitions of the geological chance factors, using the U.S. onshore as an effective minimum standard. This accomplishes three important purposes:

1. It allows the geological chance factors to yield a product that corresponds effectively with the world's most liberal definition of chance of completion success, thus permitting comparison of conventional reporting measures of success with independently derived geological estimates of probability of success.
2. It provides a general and basic standard against which subsequent adjustments can be made for exploration projects having more demanding economic requirements, such as deep overpressured tests, offshore prospects, remote frontier ventures, or international contract areas with severe financial obligations.
3. It ensures essential compatibility between completion success and minimum reserve size, assuming that minimum but finite dimensions are required for area of accumulation, average net pay, and hydrocarbon recovery factor. The concept here is that some small but finite volume and minimum reservoir quality must exist for an accumulation

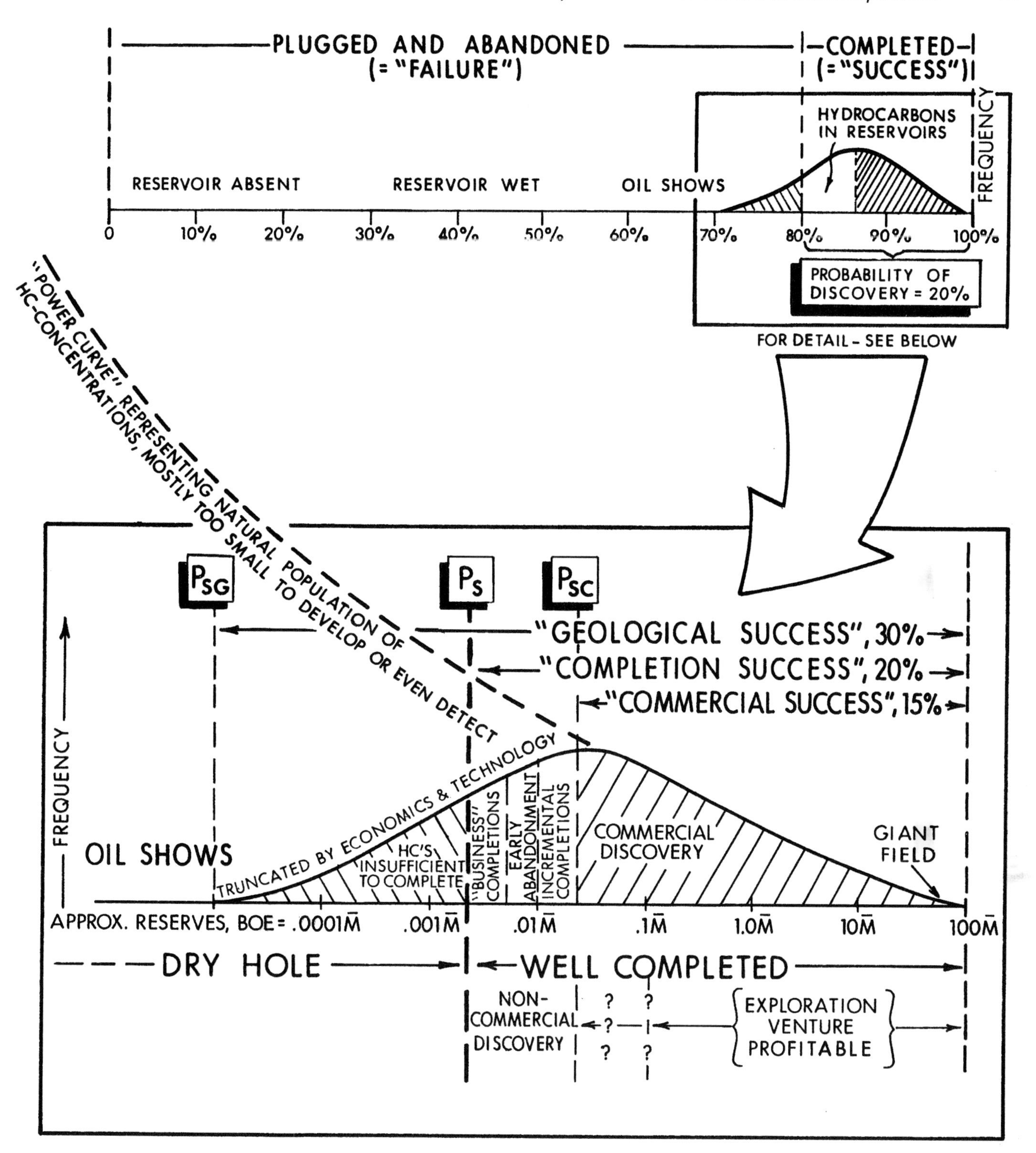

Figure 1. Different definitions of *success* in relation to probabilities of occurrence and general reserve levels using the U.S. onshore as a minimum standard. Note the log scale on the reserves distribution in the lower graph.

even to be *detected* by an operator. In other words, the lower limit of an accumulation, thus defined, is substantially larger than one barrel of oil!

Recommended System

For all exploratory prospects (including shallower pool, deeper pool, and extension wildcats), the geotechnical professional should independently express his or her confidence in four critical geological aspects of any prospect:

1. *What is the probability (= confidence) that **reservoir rock** is present of sufficient porosity and permeability to be productive and in some minimal thickness sufficient to contain detectable (i.e., measurable) quantities of mobile hydrocarbons, or to tempt a prudent onshore domestic operator to attempt a completion?*

One company uses the requirement of *flowability* to satisfy the need for some minimum standard here. In any case, what we seek is the geologist's confidence in the existence of at least a minimal reservoir—thickness, porosity, and effective permeability. Under this approach, encountering a wet, commercial quality sandstone would not be a failure in the reservoir category, but rather a failure in one of the other categories, such as an unexpected structural low, an absence of hydrocarbon charge, or a leaky trap. However, the presence of a 1-ft tight siltstone where a 10-ft porous sandstone objective had been predicted would be a reservoir failure.

2. *What is the probability (= confidence) that the **geological structure** of the reservoir objective is, in reality, essentially as represented on maps and cross sections?*

It is important to note here that we do not require a "structure," such as a domal anticline or a fault closure, but only that prospect maps and sections accurately depict the true structural configuration. For example, if only regular monoclinal south dip is required in the case of a stratigraphic trap prospect, then the geologist should express his or her confidence—as a probabilistic estimate—that the structure in the vicinity of the prospect actually is regular monoclinal south dip. If the map shows an antithetic fault closure, then what is the probability that such a structural configuration will actually turn out to be present?

This geological chance factor is formulated to apply to stratigraphic as well as structural traps and in tacit acknowledgment that the structural map is usually the single most important map involved in most prospects and many development projects. Also, errors in structural interpretation are a common reason for dry holes (Rose, 1987).

The geological structure chance factor, in combination with the reservoir requirement, focuses on the geometry of the envisioned oil or gas accumulation and on the volumes of fluids necessary to sustain a production test or prudent drill stem test.

3. *What is the probability (= confidence) that hydrocarbons are present in the subsurface geological environment such that the prospect has had access to them in some quantity to provide at least some modicum of **hydrocarbon charge**?*

This geological chance factor deals with such questions as the volumetric adequacy of petroleum source rocks, the generation of oil and/or gas, the migrational pathways to the site of the prospect and the concentration of hydrocarbons in the reservoir fluid (hydrocarbon saturation of at least 50% is required). The question of timing is *not* addressed here. In most frontier basins, confidence in hydrocarbon charge may be relatively low; in established basins and producing trends, however, this chance factor tends to be somewhat higher. For development projects, the hydrocarbon charge requirement has ordinarily been satisfied.

4. *What is the probability (= confidence) that a **sealed trap** exists, based on the lithological combinations and structural configurations depicted, and that the trapping configuration was already formed when hydrocarbons were migrating into the area of the prospect?*

Here we address three questions. First is the idea of *sealing capability* between reservoir and top seals, seat seals, and lateral seals (whether formed by stratigraphic contrasts or faults). Fluid viscosity, bed thickness, differential permeability, and fault history all influence the seal question. Second is the question of *timing* as noted previously in item (3): if the trapping configuration came into being after migration occurred, then "the gate has been shut only after the horse got out." The third question has to do with *preservation* from subsequent freshwater flushing or degradation of reservoired hydrocarbons. As used here, the term *trap* has no implications of geometry or configuration—only of containment and sealing. The troublesome issue of *fill-up*—best represented as a percentage—falls into this category. For most development wells, the sealed trap requirement has been satisfied.

In the U.S. onshore arena, serial multiplication of the four decimal fractions representing these four geological chance factors, *as defined*, yields a product that corresponds closely to the chance of completion, designated here as the *probability of success*, or P_s. From this we can derive the *probability of failure*, or P_f, where $1 - P_s = P_f$. Thus defined, P_s will correspond closely to completion success (= probability of discovery), as illustrated in Figure 1.

Multiple Objective Prospects

Prospects having multiple objectives tend to be more attractive than one-objective prospects because the chance for at least one of the two objectives to be productive is higher and thus the EV of the venture is higher (Table 5). However, a prospect having two objectives is not *twice* as likely to be successful as a one-objective prospect—nor does its probability of discovery equal the sum of the probabilities of discovery of the two multiple objective targets. If all the geological chance factors ($P_{reservoir}$, $P_{structure}$, P_{HCs}, $P_{seal/trap}$) for the two objectives are independent of one another, then $P_{discovery}$ and $P_{failure}$ can be derived by binomial expansion (see Table 6, case 1).

In most cases, however, some geological chance factors are independent and some are dependent (i.e., they are common to the prospect). In such cases, P_s and P_f are derived via a two-step process, as shown in Table 6, case 2. Discovery probability (and thus EV) for a multiple objective prospect having independent geological chance factors will usually be higher than for a multiple objective prospect having one or more dependent geological chance factors. Thus, the effect of dependent geological chance factors is to *reduce prospect discovery probability*. Dependent geological chance factors (i.e., common to the multiple objective prospect) most often include the structural aspect and the hydrocarbon charge aspect. The reservoir and trap/seal aspects tend to be independent.

An additional complication sometimes occurs when a given geological chance factor contains several subfactors, some independent and others dependent. The task here is to assign relative fractional weights to the subfactors so that their product equals the probability of the parent chance factor. This is done in the following example:

Probability of hydrocarbon charge		**0.7**
Subfactors:	Adequate organic richness	0.9
		x
	Thermal maturity	0.9
		x
	Migration/emplacement	0.9
	Product	= 0.73

On a multiple objective prospect, if the hydrocarbon charge chance factor is perceived to be higher for one exploration objective than for the other (even though they were charged from the same hydrocarbon source rock), the dependent subfactors are pulled out and combined with the other dependent primary chance factors in the two-step calculation (see Table 6, case 3).

Although having legitimate multiple objectives will indeed make a prospect more attractive economically, explorationists are often tempted to include secondary objectives that are only of marginal value. The improvement is illusory, however. The real effect is generally to provide only a bail-out so that a dry hole in the primary objective is compensated by a marginal or incremental completion in a secondary objective that may pay for pipe, maintenance, and some part of the drilling cost. For that reason, a multiple objective prospect having a high P_s should be viewed with caution.

Alternative Procedures

Many versions of this recommended procedure are used by the exploration industry today, ranging from a simple three-component system to one that employs 14 separate components. From a pragmatic viewpoint, a three-factor system seems to produce unrealistically high prospect probabilities. Also, each factor is so broad as to hinder subsequent dry hole analysis (Rose, 1987): too many subfactors exist in each factor. In contrast, systems having more than six geological chance factors tend to produce probabilities of success that are too low because of serial multiplication of decimal fractions. Moreover, a system having so many factors may imply unwarranted geological discrimination.

Since 1977, the writer has continuously used and tested the four-factor system previously outlined and has taught it successfully to more than 1500 professional explorationists. Nevertheless, it is not the only valid system, and procedures utilizing five or even six geological chance factors are certainly feasible. For example, the hydrocarbon charge factor could be divided into source rock and generation/migration components. Timing and/or preservation could be separated out of the seal/trap category.

All explorationists in a corporate unit should utilize the same system; otherwise, comparing their prospects would be like comparing apples and oranges. Also, to achieve confidence, the system should be consistent over time; once a procedure is adopted, stick with it. Unless an exploratory venture is truly a random drilling exercise, P_s should be site specific: it should be determined on the geological attributes and data density of the prospect, not on the basis of trend or region drilling success ratios.

Table 5. Geological and Economic Assumptions for Multiple Objectives Prospect

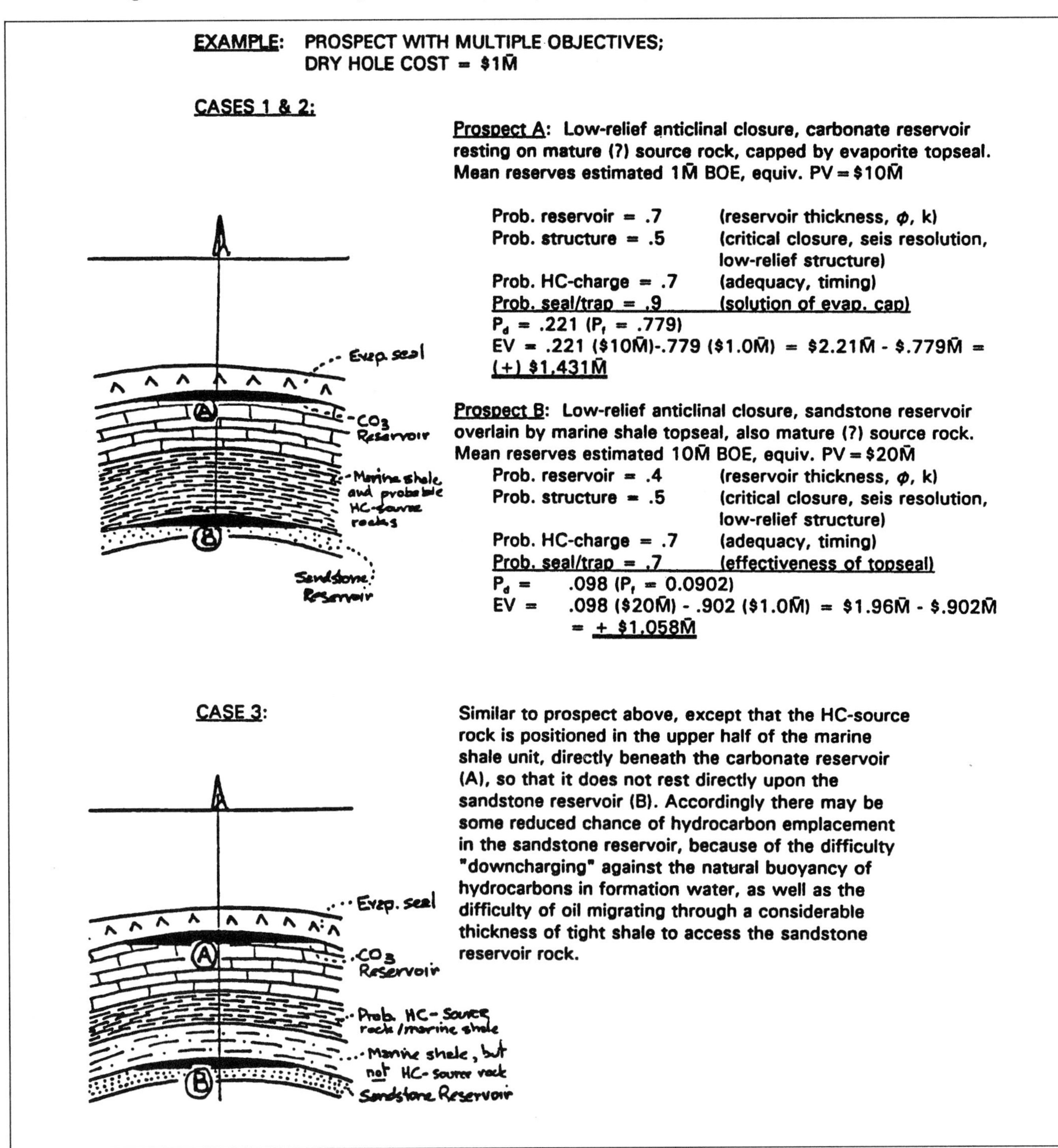

<u>EXAMPLE</u>: PROSPECT WITH MULTIPLE OBJECTIVES; DRY HOLE COST = $1M̄

<u>CASES 1 & 2:</u>

<u>Prospect A</u>: Low-relief anticlinal closure, carbonate reservoir resting on mature (?) source rock, capped by evaporite topseal. Mean reserves estimated 1M̄ BOE, equiv. PV = $10M̄

Prob. reservoir = .7		(reservoir thickness, φ, k)
Prob. structure = .5		(critical closure, seis resolution, low-relief structure)
Prob. HC-charge = .7		(adequacy, timing)
Prob. seal/trap = .9		(solution of evap. cap)

P_d = .221 (P_f = .779)

EV = .221 ($10M̄)-.779 ($1.0M̄) = $2.21M̄ - $.779M̄ = (+) $1.431M̄

<u>Prospect B</u>: Low-relief anticlinal closure, sandstone reservoir overlain by marine shale topseal, also mature (?) source rock. Mean reserves estimated 10M̄ BOE, equiv. PV = $20M̄

Prob. reservoir = .4		(reservoir thickness, φ, k)
Prob. structure = .5		(critical closure, seis resolution, low-relief structure)
Prob. HC-charge = .7		(adequacy, timing)
Prob. seal/trap = .7		(effectiveness of topseal)

P_d = .098 (P_f = 0.0902)

EV = .098 ($20M̄) - .902 ($1.0M̄) = $1.96M̄ - $.902M̄ = + $1.058M̄

<u>CASE 3</u>:

Similar to prospect above, except that the HC-source rock is positioned in the upper half of the marine shale unit, directly beneath the carbonate reservoir (A), so that it does not rest directly upon the sandstone reservoir (B). Accordingly there may be some reduced chance of hydrocarbon emplacement in the sandstone reservoir, because of the difficulty "downcharging" against the natural buoyancy of hydrocarbons in formation water, as well as the difficulty of oil migrating through a considerable thickness of tight shale to access the sandstone reservoir rock.

Table 6. Multiple Objectives

CASE 1: ALL GEOLOGIC CHANCE FACTORS ARE INDEPENDENT; DRY-HOLE COST = $1M

Productive Chance		A		B		Venture Prob.		Value		Risked Value
both = .022	a) A&B both productive	.221	x	.098	=	.022	x	(+)$30M	=	(+) .660M
only one = .275	b) A productive, B dry	.221	x	.902	=	.199	x	(+)$10M	=	(+) 1.990M
at least one = .297	c) A dry, B productive	.779	x	.098	=	.076	x	(+)$20M	=	(+) 1.520M
	d) A&B both dry	.779	x	.902	=	.703	x	(-) $1M	=	(-) .703M
						1.000		EV	=	(+)$3.467M

CASE 2: DEPENDENT (STRUCTURE, HC-CHARGE) AND INDEPENDENT (RESERVOIR, SEAL/TRAP) CHANCE FACTORS; DRY-HOLE COST = $1M. In this case, structural and HC-charge chance factors are both common to both prospects (i.e., not independent): if Prospect A has no closure, neither will Prospect B; also if Prospect A has no HC-charge, neither will Prospect B.

Step 1:

Prospect A = $P_{res} \times P_{seal}$ = .7 x .9 = .63 = P_d (and .37 P_f) are Independent factors
Prospect A = $P_{str} \times P_{HC}$ = .5 x .7 = .35 = P_d (and .65 P_f) are Dependent factors
Prospect B = $P_{res} \times P_{seal}$ = .4 x .7 = .28 = P_d (and .72 P_f) are Independent factors
Prospect B = $P_{str} \times P_{HC}$ = .5 x .7 = .35 = P_d (and .65 P_f) are Dependent factors

Step 2:

Productive Chance		Independent Factors: A x B = Prob.	Dependent Factors: Both A & B	Venture Prob.		Value		Risked Value
both = .062	a) A&B both productive	.63 x .28 = .176	.35	.062	x	(+)$30M	=	(+) 1.860M
only one = .195	b) A productive, B dry	.63 x .72 = .454	.35	.159	x	(+)$10M	=	(+) 1.590M
at least one = .257	c) A dry, B productive	.37 x .28 = .104	.35	.036	x	(+)$20M	=	(+) .728M
	d) A&B both dry	[1 - (.062 + .159 + .036)] = 1 - .257 =		.743	x	(-)$ 1M	=	(-) .743M
								EV = (+)$3.435M

CASE 3: DEPENDENT AND INDEPENDENT CHANCE FACTORS, AS IN CASE 2, BUT NOW INCLUDING ONE GEOLOGICAL CHANCE FACTOR WITH SUBFACTORS THAT ARE BOTH DEPENDENT AND INDEPENDENT; DRY-HOLE COST = $1M. Because of the position of the HC-source rock as noted in Table 5, there might be some reduced chance of hydrocarbon emplacement in the sandstone reservoir.

For Prospect A, we break down the HC-charge chance factor into three subfactors:

Probability of HC-charge = 0.7
Subfactors: HC-1) adequate organic richness = 0.9
HC-2) thermal maturity = 0.9 Product = 0.73
HC-3) migration/emplacement efficiency = 0.9

Subfactors 1 and 2 are common to Prospect B as well as Prospect A, but subfactor 3 (migration/emplacement efficiency) is more problematical in Prospect B, and is therefore assigned a probability of 0.5. All other geological chance factors remain as shown in Case 2 above.

Step 1:

Prospect A (Independent) = $P_{res} \times P_{seal} \times P_{HC3}$ = .7 x .9 x .9 = .56 = P_d (and .44 P_f)
Prospect A (Dependent) = $P_{str} \times P_{HC1,2}$ = .5 x .8 = .40 = P_d (and .60 P_f)
Prospect B (Independent) = $P_{res} \times P_{seal} \times P_{HC3}$ = .4 x .7 x .5 = .14 = P_d (and .86 P_f)
Prospect B (Dependent) = $P_{str} \times P_{HC1,2}$ = .5 x .8 = .40 = P_d (and .60 P_f)

Step 2:

Productive Chance	Independent Factors: A x B = Prob.	Dependent Factors: Both A & B	Venture Prob.		Value		Risked Value
a) A&B both productive	.56 x .14 = .078	.40	.031	x	(+)$30M	=	(+) .936M
b) A productive, B dry	.56 x .86 = .482	.40	.193	x	(+)$10M	=	(+) 1.930M
c) A dry, B productive	.44 x .14 = .062	.40	.025	x	(+)$20M	=	(+) .500M
d) A&B both dry	[1 - (.031 + .193 + .025)] = 1 - .249 =		.751	x	(-)$ 1M	=	(-) .751M
							EV = (+)$2.615M

CONVERTING COMPLETION CHANCE TO CHANCE OF COMMERCIAL SUCCESS

Probability of Commercial Success

For exploration ventures, the recommended method to assess the chance of commercial success is to first identify the minimum field size associated with the firm's definition of the threshold of commerciality and then to determine what proportion of such fields occur in the natural population of counterpart accumulations in the subject trend, play, or basin. This requires the geologist or engineer to construct a field size distribution.

Let's look at an example. For a given extension prospect having a predicted mean reserve size of 1,500,000 BOE, a geologist has concluded that the confidence of reservoir rock is 0.9, the structural confidence is 0.8, and the confidence of hydrocarbon charge is 0.9. The chief geological risk concerns whether a key fault will or will not seal, and the geologist assesses this as a 50/50 chance. Thus, the perceived chance of *completion success* (P_s) is 32%. However, construction of a field size distribution for the 20 analogous fault-separated fields in the trend reveals that only three-fourths of them exceed 200,000 bbls, which is the minimum economic field size in this trend for this firm (see Figure 1). Therefore, the chance of *commercial success* (P_{sc}) is 0.75 x 0.32 = 0.24. Calculated in this way, 0.24 represents the chance of finding a field of 200,000 bbls *or larger*.

The chance of incremental success (P_{si}) is determined similarly by first establishing the minimum sustained well deliverability that will pay for completion and operation (but not drilling) of a well. Then that figure is converted to equivalent ultimate field reserves. The probability associated with that reserve size is found on the appropriate trend field size distribution.

Technical and Mechanical Effects

Many variables other than geological chance factors affect exploration success. For example, firms that use state-of-the-art technology seem to have rates of success approximately two to four times greater than firms that drill without benefit of geotechnical guidance. This should be taken into account.

Also, mechanical chance factors should be considered, such as the chance of not getting the well down to the objective, the chance of incorrectly locating the well, and the chance of geotechnical errors in mapping, logging, or testing. However, if you anticipate that the well will be redrilled should such difficulties occur, you should include such trouble costs in the cash flow schedule for the project rather than as a chance of success factor. Generally, such considerations do not make a substantial difference except in economically marginal prospects.

Some authorities (Baker, 1988) suggest including a chance factor that deals with the probability that the exploratory well has been located and evaluated properly. It is certainly true that significant fields are occasionally discovered (or recognized) only after several penetrations. However, the writer's experience is that, for most prospects, this aspect can be covered within the geological structure or reservoir rock categories, that is, if the well turns out to be improperly located, it is commonly perceived as a failure in adequately assessing structural or reservoir risk. However, when dealing with frontier basins and plays, it is advisable to include a separate chance factor assessing the confidence that the well will be located properly and evaluated adequately.

Expressing Economic and Political Uncertainty

In assessing major projects that require large front-end investments or long elapsed time between expenditure and payout, the firm may wish to appraise the likelihood of a severe and extended drop in wellhead prices or operating costs. Basically, the procedure here is to identify what sustained low price levels—or elevated costs—would cause abortion of the project, then try to obtain estimates from knowledgeable petroleum economists as to the probability of such occurrences. The chance of commercial success (P_{sc}) is then multiplied by (1 – the chance of economic failure). Less severe price and cost fluctuations should simply be considered within the project cash flow model.

Political uncertainty can be expressed similarly. Again, knowledgeable, objective political experts should express their opinions as to the likelihood of a change of regulation, law, or regime severe enough to cause project abortion. Commonly, this probability can also be linked to a time span of 2 to 10 years or more, which can then be related to the cash flow model. The chance of commercial success (P_{sc}) can then be multiplied by (1 – the chance of political failure).

Some managers and economists prefer to use the chance of commercial success rather than the chance of completion in calculating the EV of prospects. This is perfectly acceptable, provided they recognize that the mean prospect reserves of the commercial group of fields will be substantially larger than for the population of all fields, commercial and noncommercial. Often, the difference between the EVs of completion successful versus commercially successful prospects will not be substantial. The reduced chance of success will be offset by the increased mean reserve size and thus increased NPV.

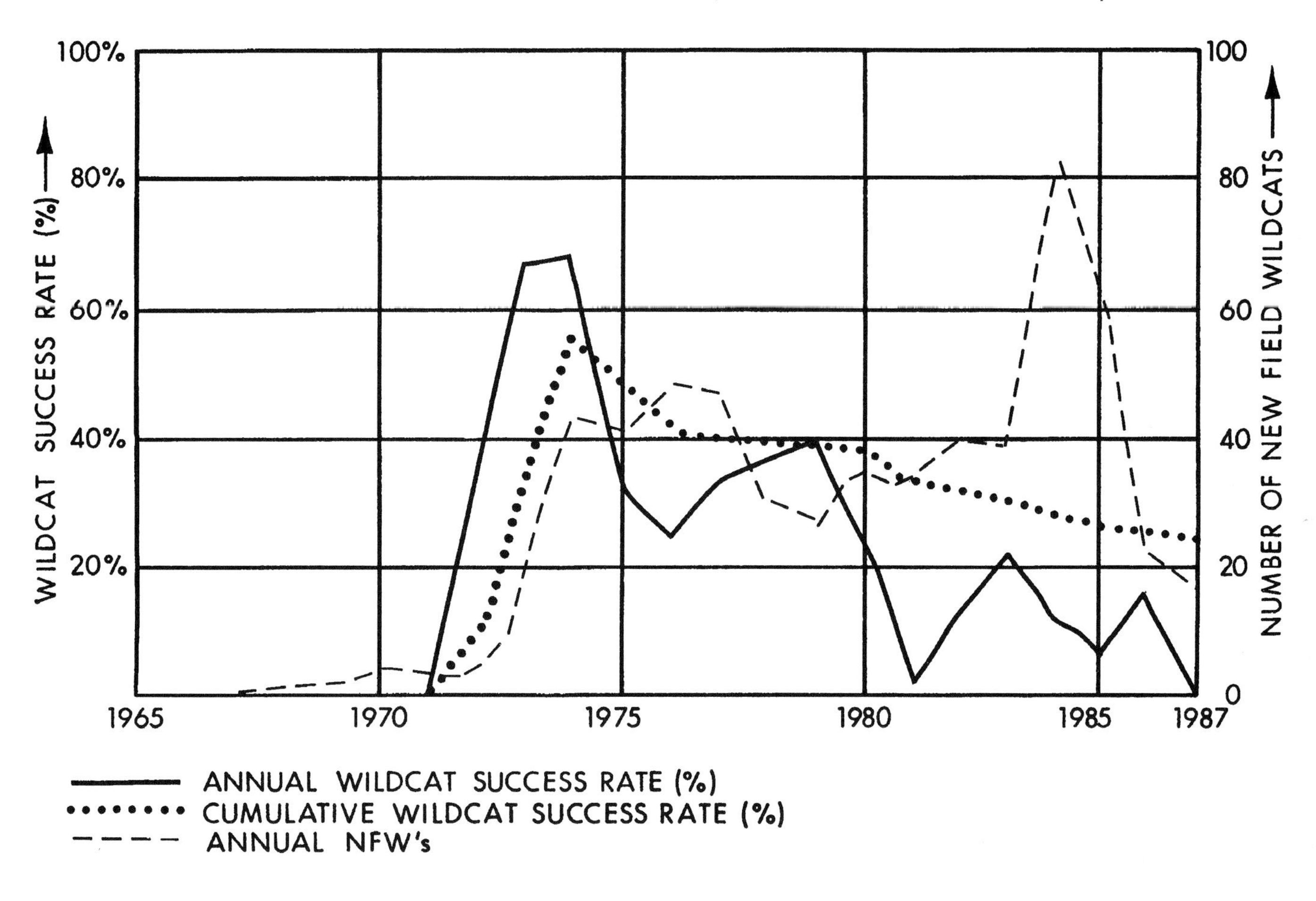

Figure 2. Annual and cumulative success rates compared with annual drilling, Niagaran Reef trend, Benzie and Manistee counties, northern Michigan.

Historical Trends in Discovery Probability

In most exploration theaters (basins, trends, plays, etc.) wildcat success rates change through time. Success rates are characteristically high during the early phases of exploration as larger and more evident fields are found, then decline as the industry searches for fields that are smaller and/or harder to find. Figure 2 shows actual data from a segment of the Niagaran (Silurian) Pinnacle Reef trend of northern Michigan. Although the discovery probability should be estimated based upon the geotechnical characteristics of the prospect, the prudent explorationist will also consider the trend's state of exploration maturity in arriving at a final estimate of P_S.

APPLICATIONS IN FRONTIER BASINS AND PLAYS

In assessing the probability of success for a well exploring a new play or lightly drilled basin, we must consider additional chance factors. Fundamentally, the procedure here is to estimate the probability (that is, the confidence) of the *play's existence*—in other words, the geologist's confidence that at least one such field of at least the minimum commercial size or larger is present. That probability is then combined with the chance of correct location or evaluation, as well as the chance of success for the specific prospect (as determined here). This procedure is well described by Baker et al. (1986) and tacitly recognizes that many unsuccessful new field wildcats are drilled in plays or trends that are clearly productive.

SUBJECTIVE PROBABILITY ESTIMATES IN EXPLORATION

Expert judgments of the probability of discovery of any drilling prospect are classic examples of subjective probability estimates, which some geoscientists resist or even reject. They claim it is just guessing. The record argues otherwise. Given a logical procedure, knowledgeable explorationists can generate such estimates with surprising consistency, agreeing not only on discovery probability but also on the relative certainty or uncertainty of the several geological risk factors in a given prospect.

Geoscientists have many reasons for their reluctance to estimate chance of success. One has to do with the traditions of geology as an observational and descriptive natural science. A second reason is that many professionals have never been trained in techniques of subjective probability estimating, nor have they been encouraged to examine the accuracy of their prior predictions. Third, a significant chance always exists that such predictions may turn out to be wrong, and our cultural and corporate values often associate scientific error with mediocrity or even moral turpitude. Yet a fourth reason is that geoscientists do not recognize the high degree of uncertainty involved in petroleum exploration, still believing that the secret to exploration success lies almost entirely in geotechnical skill and effort. Finally, we hear the very common excuse, "We don't have enough data to make an estimate." Unfortunately, explorationists *never* have "enough" information—this is inherent in the business! Moreover, the more uncertainty that attends a given prospect, the more a systematic expression of subjective probability is needed—not less. All that can be reasonably requested is the best objective estimates possible, given the time, skill, and budget available. This is indeed the professional explorationist's responsibility.

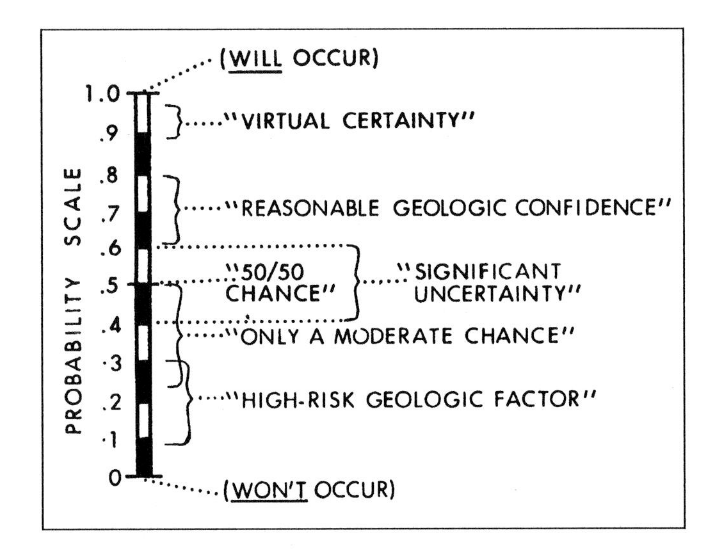

Figure 3. Geological chance factors: subjective expressions of probabilistic confidence.

Practical Aspects of Implementation

No substitute exists for actual experience in assessing and estimating confidence (= probability) in the various geological chance factors. Figure 3 relates various subjective phrases used by the writer in relation to a complete probability scale, and it may help the novice get started. One point should be emphasized here: do not use probabilities of 1.0 for any geological chance factor involved in a new field wildcat prospect. You simply cannot be that sure; supreme confidence is 0.95!

For development wells, the hydrocarbon charge and trap/seal factors have ordinarily been met. Only reservoir adequacy and the structural aspect remain as serious unknowns.

Analyzing and discussing real prospects with a peer group is an effective method for acquiring confidence in estimating geological chance factors and probability of discovery. This also helps explorationists standardize their definitions and procedures.

Naturally, accuracy of predictions on probability of discovery cannot be judged on a single prospect, or even two or three. Only the outcome of a *program* of exploratory ventures can provide a fair indication as to whether the assignment of discovery probability has been optimistic, objective, or pessimistic. Of course, if the program has involved many high-risk wells, a larger sample may be required.

However, results of even a relatively small program can be instructive with regard to correct identification of several geological chance factors. You should review predrilling projections of dry holes to see whether those geological chance factors identified as the high-risk factors did indeed correspond to the geological reasons why the hole was dry. Although dry holes are an inevitable aspect of petroleum exploration, the capable professional should usually find that he or she has correctly anticipated the real geological risks. If not, then the geological risk is not adequately understood or identified.

Explorationists can also gain experience by making estimates on wells drilled by competitors operating in the same trend or basin. Commonly, the geoscientist will have considerable knowledge about the subsurface conditions attending such wells, which thus provides expanded opportunities for developing a useful experience base for predictive confidence.

Finally, we cannot overemphasize the power of independent multiple judgments in assessing the geological chance factors and the probability of discovery. Firms are well advised to obtain several different opinions and to combine them into a final estimate of the probability of project success.

Virtues of Using Geological Chance Factors

By separating the various components of geological chance, they can be analyzed more thoroughly and objectively, leading to better geological understanding of the prospect. Also, the identification of a high-risk geological chance factor helps exploration management and staff focus on those items of greatest uncertainty. For example, if a prospect has a low probability of the

Table 7. A Model Prospect Inventory

PROSPECT	DRY HOLE COSTS ($1M)	DISCOVERY PROBABILITY	MEAN RESERVES (MBOE)	MEAN NPV ($M)	RISKED MEAN RESERVES (MBOE)	RISKED NPV ($M)	RANKING MEASURES: EXPECTED NET PV ($M)	RANKING MEASURES: PROFIT/' RISK	RANKING MEASURES: INVESTMENT ($M)
A	3.2	0.20	8.0	18.0	1.6	3.60	1.04	18.0 / 3.2 = 5.6	3.2
B	5.4	0.10	24.0	56.0	2.4	5.60	0.74	56.0 / 5.4 = 10.4	5.4
C	1.0	0.25	2.0	5.0	0.5	1.25	0.50	5.0 / 1.0 = 5.0	1.0
D	2.0	0.20	4.0	10.0	0.8	2.00	0.40	10.0 / 2.0 = 5.0	2.0
E	1.8	0.15	6.0	11.8	0.9	1.77	0.24	11.8 / 1.8 = 6.6	1.8
F	6.0	0.05	50.0	118.0	2.5	5.90	0.20	118.0 / 6.0 = 19.7	6.0
G	4.0	0.15	10.0	24.0	1.5	3.60	0.20	24.0 / 4.0 = 6.0	4.0
H	0.6	0.30	1.0	1.8	0.3	0.54	0.12	1.8 / 0.6 = 3.0	0.6
I	1.5	0.20	3.0	6.4	0.6	1.28	0.08	6.4 / 1.5 = 4.3	1.5
J	0.4	0.40	0.5	0.8	0.2	0.32	0.08	0.8 / 0.4 = 2.0	0.4
N = 10	25.9	2.00 (AVG = 0.2)	108.5	251.8	11.3	25.86	3.60		

NOTE: Annual overhead of this exploration organization is $8.0M̃.

ANTICIPATED RESULTS: THIS INVENTORY OF 10 EXPLORATORY WELLS IS A BALANCED PROGRAM INCLUDING THREE LOWER-RISK EXTENSION WELLS, FIVE MEDIUM-RISK TREND WILDCATS, AND TWO HIGH-RISK NEW FIELD WILDCATS. MOST PROBABLE OUTCOME OF THIS PROGRAM IS: TWO DISCOVERIES, TOTALLING 11.3 M̃BOE RESERVES, HAVING A NPV OF $25.86M̃. PROGRAM COST WILL BE $25.9M̃DHC + $8.0M̃ OVERHEAD = $33.9M̃ TOTAL COST; COST OF FINDING SHOULD BE APPROXIMATELY $33.9/11.3 = $3.00/BOE.

structural factor, an additional seismic line may be warranted. Conversely, when geoscientists see that additional data acquisition is not likely to increase materially the chance associated with a given geological factor, exploration becomes more cost effective, and usually more timely. But most important, explorationists cannot really analyze and thus improve their performance in predicting probability of discovery unless they systematically identify, forecast, and inspect the predictive results attending the several geological chance factors (Rose, 1987).

PROBABILITY OF DISCOVERY AND PROSPECT INVENTORIES

Program Forecasting

Prospect inventories (Table 7) are tabulations of prospects showing their respective costs, discovery probabilities, estimated mean reserves and NPVs, and risked reserves and NPVs (the latter are derived by multiplying estimated mean reserves and NPVs each by P_s). Prospect inventories are one of the most effective tools to improve exploration performance. Responsible professional estimates of individual prospect discovery probabilities, in combination with reliable forecasts of mean reserves (and thus prospect mean expected present value), provide the basis for predicting the following:

1. approximate number of discoveries from a given multiwell program;
2. approximate total new reserves added and their present value; and
3. approximate program cost-of-finding, using project cost forecasts.

Naturally, the accuracy of such program forecasts is keenly sensitive to the average discovery probability, the variance in reserve sizes, and the predictive skill. For a group of modest-sized prospects having an average P_s of 0.4, an inventory of 20 might provide an adequate sample for forecasting. However, for large-reserve, high-variance, high-risk (0.10) prospects, an inventory of 250 wells or more might be required (Schuyler, 1989).

Because of their inherent differences in scope and time frame, plays and prospects should not be included and compared in the same inventory. It is far better to maintain two separate inventories, one for plays and another for prospects.

Capital Allocation

Prospect inventories are also useful in optimizing the allocation of capital among competing exploration groups within the firm. To compare prospects, several different economic parameters can be employed (some form of profit/investment ratio or investment efficiency, expected net present value [ENPV], growth rate of return, or risk-adjusted value), resulting in multiple rankings. The 10 prospects in Table 7 are ranked on ENPV.

This is a form of economic high-grading, in which the procedure is to invest in the best projects and farm-out, sell, or drop the lower ranked ones, consistent with risk-spreading principles. Naturally, no prospect should enter the inventory unless it achieves established hurdle rates, usually a discounted cash flow rate of return (DCFROR) requirement, which should be set as low as possible so as to build the largest possible inventory, thus increasing the selective power of the inventory.

Again, for a prospect inventory to function properly requires responsible professional estimates of discovery probability, as well as mean prospect reserves. Biased or inept estimates will generate useless inventories.

Exploration Management

Probably the most cost effective use of the prospect inventory is at the operating level, where exploration funds are allocated to projects and plays as they are conceived and developed into drillable prospects (Figure 4). This is where the key exploration decisions are being made. Under this philosophy, provisional estimates of geological chance and reserves are made even as a concept emerges as a "lead," long before it becomes a documented prospect. Such leads and plays should be force-ranked and funded for exploratory study commensurate with their apparent ranking.

Professional Development

Exploration inventories encourage systematic improvement of individuals, teams, and operating units by providing an organized basis for comparison of predictions and results, that is, feedback. Naturally, this requires periodic postdrilling reviews of prospects or follow-ups by the professionals involved. To be effective, such exercises should be conducted in a constructive, rather than punitive, atmosphere by the professional geoscientists themselves, rather than by management (Rose, 1987). However, a brief report summarizing the main conclusions should be provided to all participants and to supervising management.

Requirements for Success of Prospect Inventories

Prospect inventories represent a powerful tool for any company that participates in multiple exploratory projects. Even smaller independents, who hold only fractional shares in drilling ventures, can benefit from the use of the prospect inventory. However, for a prospect inventory system to be successful, the following requirements must be met. First, a valid operating procedure must exist for evaluating prospect chance of success and mean reserves. Second, the system must be applied consistently and objectively in all exploration theaters. Third, economic and geotechnical prospect evaluation must be simple, quick, and reliable. Finally, the economic yardsticks employed must be meaningful and appropriate to the firm.

Properly organized and employed, the prospect inventory will inevitably set up internal competition for company funds. Although this leads directly to optimized capital allocation, some professionals express skepticism that competing groups will remain objective in generating geotechnical estimates. In fact, several effective ways exist to achieve consistency and equity among competing groups:

1. annual open comparison of predictions versus results;
2. a senior review team to review all prospects annually;
3. an annual exploration meeting in which each operating unit presents several recent prospects to peer groups from competing units; and
4. a management that does not reward over-estimating.

Acknowledgments *The writer thanks Elizabeth Huebner and Barbara Wiley for their conscientious work in helping to prepare the manuscript and illustrations.*

References Cited

Baker, R. A., 1988, When is a prospect or play played out?: Oil & Gas Journal, Jan. 11, p. 77-80.

Baker, R. A., H. M. Gehman, W. R. James, and D. A. White, 1986, Geologic field number and size assessments of oil and gas plays, *in* Rice, D. D., ed., Oil and gas assessment, methods and applications: AAPG Studies in Geology, n. 21, p. 25-31.

Capen, E. C., 1991, Rethinking sunk costs, *in* Hydrocarbon economics and evaluation symposium: Dallas, TX, Society of Petroleum Engineers, SPE 22017.

Dott, R. H., Sr., and M. J. Reynolds, 1969, Sourcebook for petroleum geology: AAPG Memoir 5, 471 p.

Landes, K. K., 1951, Petroleum Geology, 1st ed.: New York, John Wiley and Sons, 660 p.

Megill, R. E., 1990, Mean is best value to represent whole, *in* Business side of geology: AAPG Explorer, v.11, n 7.

Nehring, R., 1990, Let's get rid of dumb exploration: Houston Geological Society Bulletin, v. 32, n. 6, p. 26-30.

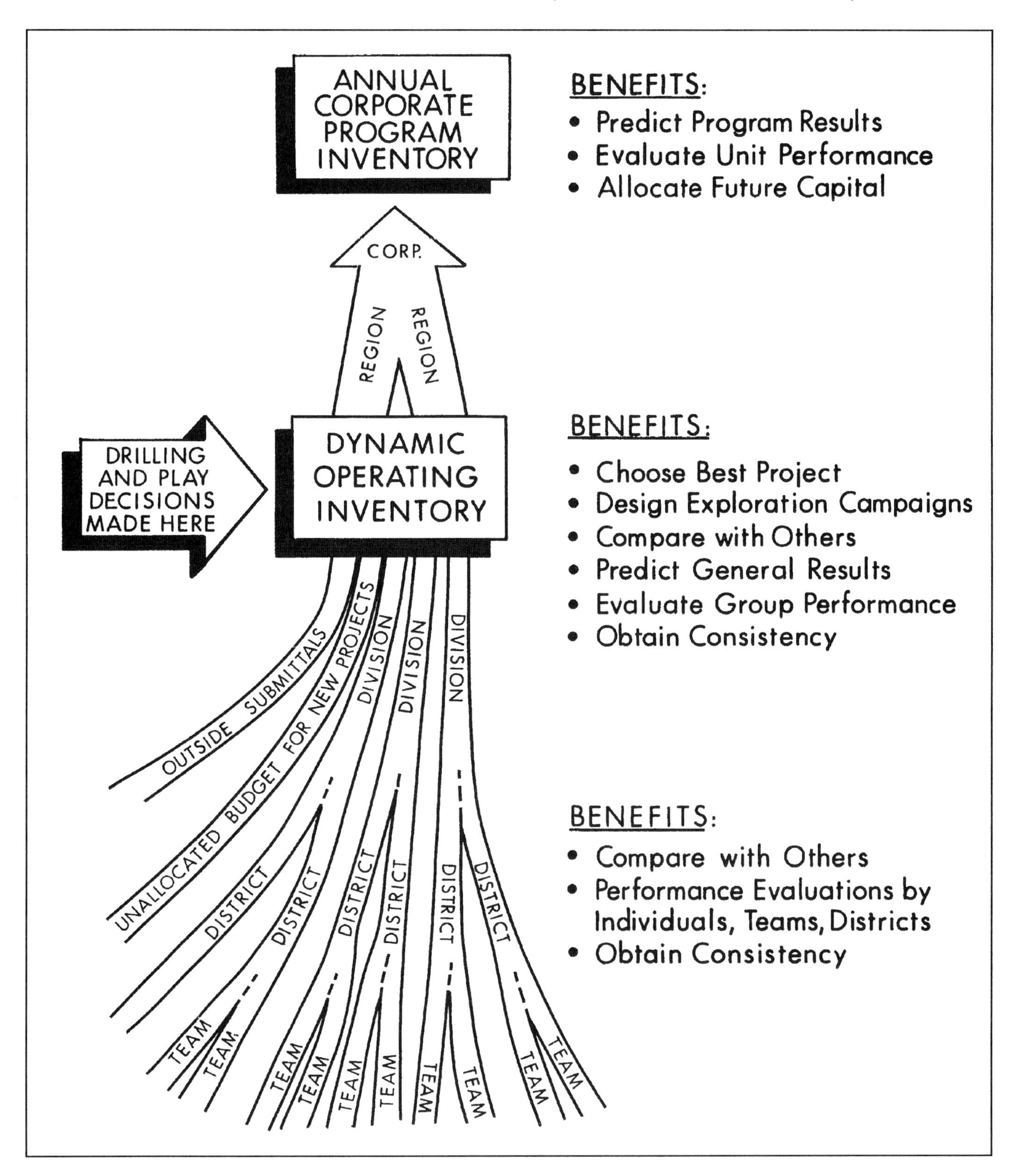

Figure 4. Generic representation of a functioning prospect inventory system in a corporate hierarchy and the benefits attending different organizational levels.

Newendorp, P. D., 1975, Decision analysis for petroleum exploration: Tulsa, OK, PennWell Books, 668 p.

Rose, P. R., 1987, Dealing with risk and uncertainty in exploration: How can we improve?: AAPG Bulletin, v. 71, n. 1, p. 1-16.

Schuyler, J. R., 1989, Applying expected monetary value concept: how many prospects is enough?: Oil & Gas Journal, Dec. 11, p. 87-90.

PETER R. ROSE, an AAPG member and Certified Petroleum Geologist, established his own independent oil and gas consulting firm, Telegraph Exploration, Inc., in Austin, Texas, in 1980. His clients include international exploration firms, most major U.S. companies, and many prominent independents. Prior to forming his own company, Dr. Rose was Staff Geologist with Shell Oil Company, Chief of the Oil and Gas Branch of the U.S. Geological Survey, and Chief Geologist and Director of Frontier Exploration for Energy Reserves Group, Inc. (now BHP Petroleum, Americas). He has explored for oil and gas in most North American geological provinces and has published and lectured widely on U.S. resource assessment, basin analysis, play development, prospect evaluation, and risk and uncertainty in exploration. He was an AAPG Distinguished Lecturer in 1985–1986 and continues to teach extensively at the corporate and professional level. Since 1990 he has been deeply involved in design and implementation of comprehensive exploration risk analysis systems for several major oil companies.

Chapter 8

Selecting and Assessing Plays

David A. White
Consultant
Austin, Texas, U.S.A.

The strengths of the assessment process lie in systematically documenting assumptions and in having multiple opportunities.
—David A. White

INTRODUCTION

Play analysis and mapping help in estimating risks and oil/gas volumes of prospects, as well as in assessing hydrocarbon potentials of plays, basins, and regions. A *play* is a group of prospects and any related fields having common oil or gas sources, migration relationships, reservoir formations, seals, and trap types. The prospects thus share any common elements of geological risk. This coherent relation to risk in regional perspective makes the play the key unit for geological analysis and mapping, regardless of whether the direct object of exploration is a prospect, a play, or a basin. Basins and larger areas are best assessed by aggregating their play assessments.

The chief exploration challenges in play analysis are (1) to develop and map correctly the interrelations of the critical geological controls of oil and gas occurrence at the play level; (2) to use effective methods for prospect and play assessment; and (3) to strive for new play concepts, looking for places where source, reservoir, and trap can be put together in unusual ways as well as in the usual ways. This chapter supplements work by Baker et al. (1984) and White (1987, 1988).

Of all the factors that go into an economic analysis of future exploration profits, the most critical are the geological estimates of the risks and potential oil and gas volumes of the prospects or plays. Unfortunately, such estimates must necessarily be made from limited information prior to drilling and are therefore subject to error. The main cause of error in prospect and play assessment is the complexity of the geological controls of oil and gas occurrence. Two dozen essential conditions (listed here in eight groups) must be met to create a commercial field:

- Adequate organic content, organic quality, thickness, and paleodrainage area of the source rock
- Enough time and temperature for maturation
- Adequate expulsion from the source and secondary migration to the trap
- Sufficient net thickness, porosity, permeability, and continuity of the reservoir rock
- Adequate closure area and height of the trap, with effective thickness and lithology of top and lateral seals
- Proper trap timing relative to migration
- Freedom from serious flushing, biodegradation, diffusion, or overcooking
- Adequate drive, concentration (not too dispersed or diluted), and oil viscosity for effective production

If any single one of these factors is absent or inadequate, the chance of commerciality is wiped out. Even if all these factors are adequate and a field exists, there are additional finding, environmental, political, producing, and economic risks.

Geological and economic assumptions should be kept as separate as possible. One way to do this is first to assess geologically the undiscovered resource base—the potentially recoverable petroleum volumes postulated to exist, regardless of present accessibility or economics. Then the assessor can sort out the undiscovered attainable potential—that part of the resource base deemed accessible and economically and technologically findable, producible, and marketable.

SELECTING AND MAPPING PLAYS

Established productive plays typically have the least risk because the effectiveness of the regional geological controls is known. However, success rates and field sizes are apt to decline with continued exploration (Meisner and Demirmen, 1981). Yet the more complex or more subtle the geological controls, the more likely it is that some relatively large field may lurk undetected in a new reservoir facies or an obscure trap. Witness the Mondak Mission Canyon field in the Williston basin (Parker and Hess, 1980), or the Beaverhill Lake pool at Caroline in the Alberta basin, whose discoveries lagged 20 to 30 years behind the first flush days of their plays. Mondak is a good example of a new discovery concept applied to an old area, where the collapse of the reservoir with other roof rocks over a deeper dissolved-out salt unit, areally juxtaposed with a mature source rock, created a fractured, productive fairway.

Frontier areas with the possibility of good source, reservoir, and trap, but with few data, typically have higher perceived geological risks but may also have higher potential rewards. Risks become even greater, of course, if the geological controls appear unfavorable.

Effective definition and delineation of a play sets the stage for assessment. The play as defined should consist of prospects having fundamentally the same elements of source, reservoir, and trap. Play boundaries coincide with any geological limits of adequacy or with any changes that affect group risks or potential field sizes. Boundaries can also be arbitrary, for example, international, concession, water depth, or other economic boundaries chosen for convenience. Assessors can handle any dependencies between such arbitrary subplays when aggregating assessment results (Gehman et al., 1981). Assessors may also separate different interspersed trap types or reservoir zones into different plays.

A play summary map superimposes the critical boundaries of adequacy of source, migration, reservoir, trap, timing, seal, preservation, and recovery conditions. It is the key to grading relative prospect chances of success and to locating the most favorable area within the play, thereby reducing exploration risks (White, 1988). Such a map integrates the adequacy boundaries from the conventional suite of exploration structure, isopach, and other maps. It also provides the complete inventory of known fields, tested dry prospects, and untested prospects that is essential for quantitative volume and risk assessment of the play. Explorers can start making rudimentary summary maps with very few data. Even if the early interpretations are not accurate, they at least document the reasons for the exploration decisions that have to be made at the time.

METHODS OF ASSESSING PLAYS

The chief methods of play assessment are (1) summing the assessments of the component prospects, (2) geologically estimating future field numbers and sizes with related risks, (3) discovery process modeling, and (4) using analogies such as other productive plays or basin and play classifications. The first method, summation of prospect assessments, provides the most information about the locations of potential fields but requires large doses of data and time.

The second and commonly most practical method is the geological analysis of future field potentials (Baker et al., 1984). The estimate of future field numbers normally comes from applying a success ratio to a count of the untested prospects, both known and postulated. The application of prospect grading to the field number estimate is discussed in the next section. The estimate of future field sizes comes from scaling prospect sizes or from look-alikes. Despite the uncertainties inherent in any forecasting, this geological method has profound advantages over more statistical discovery process approaches. First and foremost, it preserves the spatial relationships of the potentials and allows one to locate the most favorably assessed areas on the ground. It can be applied by explorationists themselves—by those who best know the geology and the future, commonly subtle, opportunities. The method can be used at any knowledge level and is fully compatible with prospect summation. All assumptions are documented and readily updatable, and the results are suitable bases for economic analyses of all hydrocarbon liquids and gases.

The third play assessment method, discovery process modeling (Adelman et al., 1983) as originally pioneered by Arps and Roberts (1958), statistically projects past discovery history by field number and size into the future. The process assumes that there are more little fields than big ones, that the bigger ones generally are found earlier, and that later discoveries fill in existing holes in some regular distribution (e.g., lognormal) of a finite, ultimate population sampled without replacement. In mature areas, this approach provides an interesting check on geological assessments. However, it does not work in sparsely drilled areas; it cannot anticipate subtle deposits or new play concepts. It will not tie the results to the ground, nor does it ensure that nature will fill in any holes in distributions.

Basin and play classifications, which compose the fourth method, have key uses in developing exploration concepts, but are not adapted to quantitative assessment. No classification, including the author's (White, 1980), can properly take into account the infinite variations in all of the two dozen critical controls of oil and gas occurrence, as noted by Bally and Snelson (1980). A basin is very complex, and each of its component plays is unique. Similarities to productive areas are often emphasized, but differences are ignored. The use of scaled analogies is necessary where data are limited, but as much as possible, assessments should be based on studies of all aspects of the target areas themselves.

GEOLOGICAL PLAY ASSESSMENT PROCEDURES

General Concepts

The three basic requirements for geologically assessing the undiscovered petroleum potential of a play are (1) a distribution of the number of future fields, usually derived from a prospect count and success ratio; (2) a distribution of future field sizes—prospect sizes—in recoverable barrels or cubic feet or cubic meters; and (3) an estimate of the play's chance of existence—that is, the chance of at least one significant field being present. The hand-calculated mean or average unrisked assessment equals the average number of fields multiplied by the average field size. The risked mean assessment equals the unrisked mean times the play chance. The distributions can be combined in a Monte Carlo computer simulation to provide the full range of possible answers in a probability curve (Baker et al., 1984). The mean values calculated by hand are just as accurate, however, as those derived from a computer simulation.

The first step in the assessment is to select the practical minimum field size to be included and counted. This minimum size can be the minimum economic size for the area or it can be of any size whose prospects are countable and significant to the assessor's company or nation. The selection of the minimum-size field to be counted affects every major factor in the assessment. It affects the prospect count because only prospects big enough to hold the minimum size should be included. It obviously affects the field size distribution and its mean, which would be drastically reduced by the inclusion of countless insignificant fields. It also affects the play chance, which is the chance of having at least one field of at least minimum size and which thus varies with the minimum size selected. (The total play chance reflects risks related both to regional geology and to limited numbers of prospects, as noted by Baker et al. [1984].)

This practical truncation at the minimum is absolutely essential to effective play assessment. Smaller and smaller fields come in progressively larger numbers that make up successively smaller effective proportions of the total resources. Their inclusion in the count destroys the focus of the assessment without practical gain, especially since a lump sum estimate of small field potentials can be added without difficulty. Even in very mature areas where all future fields are relatively small, the largest of these remaining fields will carry the bulk of the resource. Every known field size distribution is truncated on the small end by economics, and data for assessment below this truncation level are unreliable. Any play assessment method that does not truncate bogs down in these hazy numbers and loses touch with reality. The late Harry Gehman's sagacious dictum for deciding what field sizes to count was "to start at the largest and work down the list until you're tired and then quit."

In play assessment, the risk analysis is divided into two aspects represented by the success ratio and play chance. The success ratio is the conditional probability that the average prospect will be a field of at least minimum size, given that the play is successful—that the geological controls of oil and gas are all regionally adequate. The average future success ratio equals the average estimated number of future fields divided by the total number of untested prospects (known plus postulated) that are big enough to hold a field of at least minimum size. Or looking at it the other way, the average number of future fields equals the success ratio times the total untested prospects. The *success ratio* is a measure of the prospect-specific, individual, or independent risks within the play; the failure of one prospect from these causes would not condemn the others. The *play chance of adequacy*, in contrast, is the marginal probability reflecting the group or dependent risks affecting all prospects in the play. If one prospect fails, all will fail from the same cause if the risk analysis is correct.

The *average prospect chance of adequacy* in a play equals the success ratio times the play chance. The play chance of adequacy, just like the overall prospect chance, is the probability of exceeding the *minimum* assessed oil or gas volume. The probability of adequacy is 1.0 minus the risk of failure—the risk of not having even one significant (minimum) field. These are the great principles of risking. Play analysis helps sort out dependent versus independent risks and puts the chances of adequacy of different prospects in perspective. Particularly important is the tie between adequacy chance and the exceedence of the minimum value. Any assessor who assigns exceedence chances relative to the mean or most likely volume instead of to the minimum is apt to be guilty of seriously overrisking and therefore underestimating ultimate potentials.

Field Number and Success Ratio from Prospect Grading

Prospect grading by relative chance of adequacy is a quick way to tailor a success ratio to a play's local geology, if data permit. Where local data are very limited, success ratios must come from experience or from look-alike plays. Success ratios range from 0.0 to 1.0, and a typical value is 0.25. The problem with using look-alikes is that success ratios vary with the population chosen. The success ratio for the group of prospects in the heart of the oil patch of a play may approach 1.0. The success ratio may be much lower, however, if the surrounding poorer prospects are included in the count. The boundaries of the target and look-alike plays should therefore be defined similarly, which is not easy. Prospect grading gets around these difficulties.

The ultimate way to grade the qualities of individual prospects within a group is to assign chances of adequacy to each, as one would do if assessing all prospects individually and aggregating the results for the play assessment. A short-cut commensurate with our field number and size approach is simply to divide the prospects into three groups—the "A" group with the greatest chances of becoming fields, the "C" group with intermediate chances, and the "F" group with the least chances. In grading, the assessor considers all geological controls but focuses on the most critical one(s) for the play.

A major advantage of prospect grading is locating and documenting the most favorable area perceived within the play. Consider the simple play map of Figure 1. There are four untested prospects in each of three rows progressively farther from the mature source rock of the basin deep. Migration capability, reservoir quality, trap configuration, and seal effectiveness appear to be uniform over the whole area and adequate for all prospects. Source, interpreted to be of marginal quality, is the critical factor for the play. Prospects in the front row appear best and get A ratings. If the front row fills and spills, the second-row prospects, although having lower chances of adequacy for source, still qualify for C grades. Prospects in the back row have still lower source chances and receive F ratings.

From this grading, we can set up a triangular distribution of assessed future field numbers where the minimum as usual is 1, the most likely is 4 (summation of A's), and the maximum is 8 (summation of A's and C's). The average of this distribution is the sum of these three numbers divided by 3, or 4.33. The future average success ratio is 4.33 divided by the 12 prospects, or 0.36. Had there been 8 prospects in the front row, 4 in the second row, and 0 in the back row, we would estimate 1-8-12 fields, averaging 7.0, with a resulting success ratio of 0.58. If the traps looked adequate for only 2

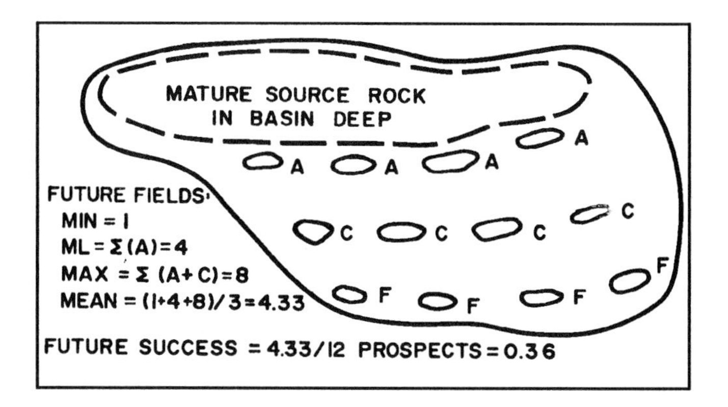

Figure 1. Map of prospects in a play, showing prospect grades (A, C, and F) and method of calculating possible minimum, most likely, maximum, and mean numbers of future fields and future success ratio.

prospects in the front row and 1 in the second row, and all others appeared likely to have fault leaks, we would estimate 1-2-3 fields, averaging 2.0, with a resulting success ratio of only 0.17. The numbers thus follow and document our current interpretation of the geology, and they provide an early rationale for leasing acreage or taking other required action. Although the interpretations can be wrong, the rewards are apt to go to those who can come up with the most realistic picture from the least data.

Field densities can help estimate numbers of future fields where prospects cannot be counted, either in frontier areas or in very mature areas having prospects too small to identify regionally. Densities, expressed as the ultimate number of fields (greater than the minimum size specified for the play) per thousand square miles or kilometers, come either from distant look-alikes or from more thoroughly explored parts of the same play. Often the assessor must postulate some additional fields in the control area to ensure a full measure in the assessment.

The next step is to estimate the fraction of the total assessed number of fields deemed to be oil fields. Gas fields make up the remainder. This tough estimate reflects interpretations of known fields and shows, source rock properties, maturation levels, differential migration and entrapment of oil versus gas, and possible preferential leakage, dissolution, or diffusion of gas. If we assume an oil fraction of 0.75 for our example, we would have 9 oil and 3 gas prospects, and 3.25 oil and 1.08 gas fields on average, if the play were productive.

Prospect Size Distributions

The estimate of future field sizes comes from representative prospect assessments, from processing ranges of the play's prospect volume factors in a Monte

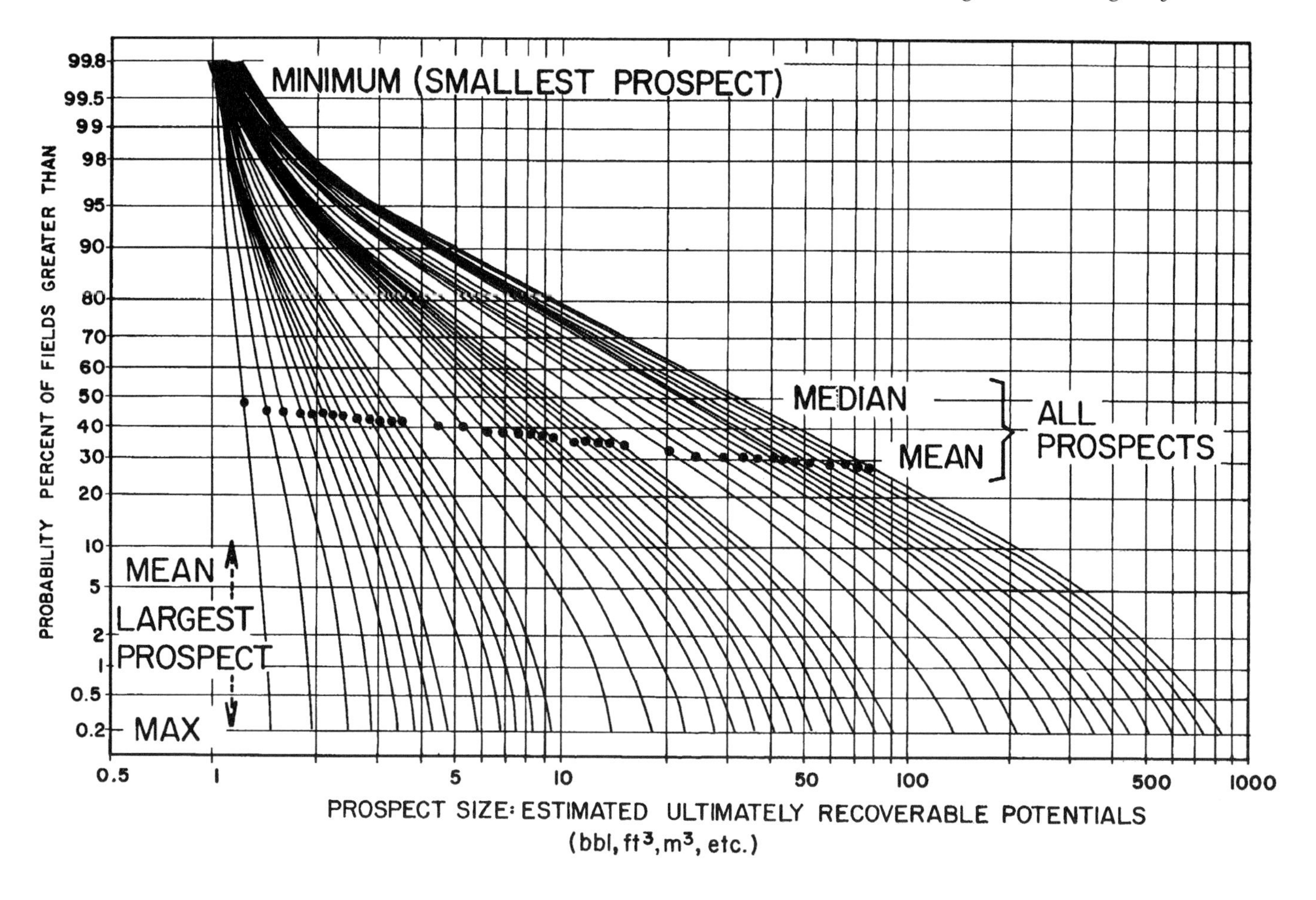

Figure 2. Prospect size distributions with a minimum of 1.0, showing the median and mean for all prospects, and the general location of the mean and maximum size for just the largest prospect within each distribution.

Carlo simulation, or from the ultimate reserves of known fields in look-alike plays (Baker et al., 1984). A fourth related approach is using the prospect size distributions described and tabulated as follows.

The ranges of potentially recoverable volumes of oil and gas in groups of prospects or future fields are conveniently represented as truncated lognormal distributions. As previously discussed, the truncation at the low end is essential. There must also be a truncation at the high end if the assessment result is to be realistic. This is an agonizing but unavoidable judgment. The inclusion of impossibly large fields in a distribution may cause a severe overestimate, the resultant loss of credibility for the assessor, and a wasted investment. However, too conservative a truncation may cause a severe underestimate and a missed exploration opportunity. There is no statistical way to resolve this dilemma, and discovery process modeling is no comfort. The answer rests entirely on the size of the largest remaining prospect and on our geological abilities to divine that size correctly. The most important consideration in selecting appropriate future field size distributions is the estimated largest size expected.

A family of prospect size distributions that can represent practically any distribution known in the world is shown in Figure 2. Each curve represents a log-triangular distribution truncated at a minimum of 1.0, as listed in Table 1. The most likely equals the median in each of these distributions. Successive curves are truncated at progressively larger maximums. Because all distributions are tied to the same minimum, the largest anticipated size essentially determines the distribution. The units can be barrels, cubic feet, cubic meters, or anything else. The curves and table can accommodate any minimum used. For a minimum of 5, for example, simply multiply every number on Table 1 or on the horizontal axis of Figure 1 by 5; for a minimum of 0.1, multiply every number by 0.1. To accommodate the largest field size distribution in the world, that of the Arab play in Saudi Arabia, multiply every number by 100 million bbl; the maximum of the largest distribution would then be 100 billion bbl, or about the size of the world's largest oil field, Ghawar. For a very mature area, one might choose the minimum-most likely-maximum of 1-10-100 multiplied by 0.05 million bbl, or 0.05-0.5-5.0.

There are two main ways to select a distribution for a

Table 1. Prospect Size Distributions for Minimum = 1.0

Minimum	ML	Maximum	Mean
1	1.22	1.50	1.23
1	1.41	2.00	1.43
1	1.58	2.50	1.61
1	1.73	3.00	1.78
1	1.87	3.50	1.94
1	2.00	4.00	2.08
1	2.12	4.50	2.22
1	2.24	5.00	2.36
1	2.45	6.00	2.63
1	2.65	7.00	2.85
1	2.83	8.00	3.08
1	3.00	9.00	3.31
1	3.16	10.00	3.51
1	3.87	15.00	4.50
1	4.47	20.00	5.39
1	5.00	25.00	6.13
1	5.48	30.00	6.91
1	5.92	35.00	7.60
1	6.32	40.00	8.13
1	6.71	45.00	8.89
1	7.07	50.00	9.58
1	7.75	60.00	11.00
1	8.37	70.00	11.80
1	8.94	80.00	12.70
1	9.49	90.00	13.70
1	10.00	100.00	15.10
1	12.20	150.00	20.50
1	14.10	200.00	24.30
1	15.80	250.00	29.20
1	17.30	300.00	32.80
1	18.70	350.00	35.80
1	20.00	400.00	41.40
1	21.20	450.00	43.40
1	22.40	500.00	46.90
1	24.50	600.00	52.40
1	26.50	700.00	60.60
1	28.30	800.00	65.40
1	30.00	900.00	72.00
1	31.60	1000.00	77.00

play assessment. The best way, if data permit, is to use the mean of the whole distribution. The estimate of the future mean can come from averaging the assessed unrisked means of representative prospects in the play, from the average of the ultimate reserves of look-alike fields larger than the specified minimum, or from the average size of recent discoveries (larger than the specified minimum) deemed to represent the future of a productive play. If such an overall mean were 5.4 million bbl and our assessment minimum were 1.0, we could run down the "Mean" column of Table 1 until we hit the closest number to 5.4, which is 5.39. The selected distribution would be 1-4.47-20. Before finalizing the selection, we should feel as comfortable as possible about the assumption that no future field could be larger than 20 million bbl. No fields smaller than the minimum or larger than the maximum can enter the assessment. If uncomfortable, we might try the next larger distribution with a maximum of 25, noting that its selection would drive the mean up to 6.13, an increase of 37%, which would have to be justified. Ultimately, we pick the mean and maximum that we and our management can live with. Distributions can be interpolated between those listed in Table 1, but this is usually not necessary.

The second more precarious but less data-dependent way to select a distribution is to use the curve on which the assessed size of the largest prospect plots on Figure 2. After all, the difference between distributions is fundamentally a question of how big the biggest prospect included is. But this approach is precarious because of the uncertainties in both of the plotting coordinates—assessed prospect size and its exceedence probability percentage. Volumetric prospect assessment is uncertain even when done by competent people using good methods with good data (Uman et al., 1979; Rose, 1987). The plotting percentage for an oil distribution depends on the number of oil prospects, which is generally uncertain. The procedure is to plot on Figure 2 the largest prospect's unrisked mean (from its prospect assessment curve) at the percent calculated by dividing 100% by the number of oil prospects plus 1. This is where the largest prospect would fall had we plotted the unrisked means of all the oil prospects. Thus, for our nine oil prospects of Figure 1, we would plot on Figure 2 the largest one's unrisked mean (say, 10.0 million bbl) at 100/9 + 1, or 10%. This point falls on the curve whose minimum-most likely-maximum is 1-4.47-20, and the mean for this whole distribution of all prospects (as distinct from the 10.0 assessed mean of the largest one only) is again 5.39 million bbl. The maximum of 20 million bbl represents the maximum possible allowed size of the largest prospect. The largest prospect should not be one-of-a-kind, but should tie to a roughly continuous distribution of smaller prospects down to the minimum. In spite of its uncertainties, this use of the largest prospect has the key advantage of tying the assessment specifically to the opportunities perceived in the target area itself.

Oil fields and gas fields can have different distributions with different units, different minimums, and different numbers of prospects. If desired, associated gas can be included in an oil field size distribution, and natural gas liquids can be included in a gas field size distribution, as energy equivalents.

Assessment using this family of distributions (Figure

2 and Table 1) reasonably represents nature, requires little or no plotting, is completely versatile as to units and minimum field sizes, handles oil versus gas effectively, makes optimum use of local data, and allows the assessor to see the sensitivities and impacts of different assumptions and choices easily.

Play Chance and Assessment Results

Play chance is the probability of having at least one field of at least minimum size. Thus, it is the marginal or dependent probability of the play's existence or adequacy as a productive entity, or the regional chance that the previous field number and size assumptions are correct. Because it involves only the group risks affecting all of the play's prospects and because the individual risks are already accounted for in the success ratio, the assessor must take care not to overrisk. For this reason, it is prudent to multiply only three chance-of-adequacy judgments, one each for source, reservoir, and trap, instead of the seven or more judgments commonly used for prospects—closure, reservoir facies, porosity, source and migration, seal and timing, preservation, and recovery. Still, the assessor should identify and evaluate within these three numbers any critical regional factor(s) among the two dozen essential geological controls. In our example of Figure 1, the only critical regional factor is source. If we assign a 0.8 chance that source will be adequate somewhere in the play (presumably in the front row of prospects) for at least one field of at least 1 million bbl, and if we see no problems within the reservoir and trap categories, the overall play chance would be 0.8 x 1.0 x 1.0 = 0.8.

Average play chances range from 0.15 to 0.60 and average 0.35 overall for the occurrence of at least one major field in several play groupings for a worldwide sampling of 1150 plays in 80 productive basins (White, 1980). Many plays are dry, their chances being 0.0. The chance for continued discovery in a known productive play is commonly 1.0 if the minimum field size is not too high.

The unrisked mean oil assessment is simply the mean number of oil fields times the mean oil field size; for our example play, it equals 3.25 fields times 5.39, which is 17.5 million bbl. The risked mean oil assessment equals the play chance of 0.8 times the unrisked mean of 17.5, or 14.0 million bbl. A similar calculation could be made for gas after we picked a gas prospect size distribution.

Note that a mean or *expected value* is only one possible answer located near the middle of a cumulative probability curve representing the range of many other possible answers. Also, the use of decimals in these calculations gives a false sense of precision to what are very tenuous numbers, but rounding is best left until after the assumptions and calculations are carefully documented.

The economics of the play's potential reward as a success is run on the unrisked mean, or on representative cases selected from the unrisked assessment curve, weighted by the play chance. The anticipated costs for a failure, such as geological, geophysical, land and dry hole expenses, carry the weight of 1.0 minus the play chance.

The risked mean alone is a useful but dangerous number when it comes to describing a play or a prospect. Risked means are useful in aggregating different assessments because they are the only numbers that can be directly added legitimately. But they give poor ideas of the sizes of the exploration targets, especially when chances of adequacy are low, because they are abstract combinations of two totally different populations—the successes and the failures.

PERSPECTIVES

Despite the impressiveness of fancy economic and engineering analyses, the biggest factor in the possible payoff of a prospect or play is the geological, judgmental estimation of hydrocarbon volumes and risks. It is sad but true that volume estimates may err either too high or too low by an order of magnitude (Uman et al., 1979; Rose, 1987) and that any assigned adequacy chance that is not 0.0 or 1.0 is bound to be wrong. After drilling, a field of at least minimum size either actually exists (probability 1.0), or it does not (probability 0.0). These are the only two possible outcomes of exploration. Yet not knowing the answer prior to drilling, the assessor must play the averages and guess in between, because he or she can neither be 100% sure of success nor afford to multiply respectable opportunities by 0.0. The trick is to put sufficiently high chances on the really good prospects and plays and to put low chances on the poor ones.

The cause is not lost just because each individual prospect or play assessment is doomed to be wrong to one degree or another, as the first and only law of resource appraisal states. The strengths of the assessment process lie in systematically documenting assumptions and in having multiple opportunities. If the assessor does a good long-term job of playing the averages for hydrocarbon volumes and risk levels, overestimates will balance underestimates in the long run, and the sum of the risked means over many trials should about equal the sum of discovered ultimate reserves.

References Cited

Adelman, M. A., J. C. Houghton, G. Kaufman, and M. B. Zimmerman, 1983, Energy resources in an uncertain future: Cambridge, MA, Ballinger Publishing Company, 434 p.

Arps, J. J., and T. G. Roberts, 1958, Economics of drilling for Cretaceous oil on east flank of Denver–Julesburg basin: AAPG Bulletin, v. 42, n. 11, p. 2549-2566.

Baker, R. A., H. M. Gehman, W. R. James, and D. A. White, 1984, Geologic field number and size assessment of oil and gas plays: AAPG Bulletin, v. 68, n. 4, p. 426-432.

Bally, A. W., and S. Snelson, 1980, Realms of subsidence, *in* Miall, A. D., ed., Facts and principles of world petroleum occurrences: Canadian Society of Petroleum Geologists Memoir 6, p. 9-75.

Gehman, H. M., R. A. Baker, and D. A. White, 1981, Assessment methodology: An industry viewpoint, *in* Assessment of undiscovered oil and gas: Bangkok, United Nations ESCAP, CCOP Technical Publication 10, p. 113-121.

Meisner, J., and F. Demirmen, 1981, The creaming method, a Bayesian procedure to forecast future oil and gas discoveries in mature exploration provinces: Journal of Royal Statistical Society, v. 144, n. 1, p. 1-31.

Parker, J. M., and P. D. Hess, 1980, The Mondak Mississippian oil field, Williston basin, U.S.A.: Oil & Gas Journal, Oct. 13, p. 210-216.

Rose, P. R., 1987, Dealing with risk and uncertainty in exploration: How can we improve?: AAPG Bulletin, v. 71, n. 1, p. 1-16.

Uman, M. F., W. R. James, and H. R. Tomlinson, 1979, Oil and gas in offshore tracts: Estimates before and after drilling: Science, v. 205, p. 489-491.

White, D. A., 1980, Assessing oil and gas plays in facies-cycle wedges: AAPG Bulletin, v. 64, n. 8, p. 1158-1178.

White, D. A., 1987, Conventional oil and gas resources, *in* McLaren, D. J., and B. J. Skinner, eds., Resources and world development: Energy and minerals, water and land: Chichester, U.K., John Wiley and Sons, p. 113-128.

White, D. A., 1988, Oil and gas play maps in exploration and assessment: AAPG Bulletin, v. 72, n. 8, p. 944-949.

DAVID A. WHITE, a consulting geologist in Austin, Texas, received a B.A. degree in Geology from Dartmouth College and M.S. and Ph.D. degrees in Geology from the University of Minnesota. He retired in 1986 from Exxon Production Research as a Senior Research Advisor after 32 years of work on source rocks, reservoir facies, global tectonics, and the development and application of assessment techniques. Dr. White served on assessment committees for the National Research Council and the National Petroleum Council, and teaches a course on Prospect and Play Assessment.

Chapter 9

Risk Behavior in Petroleum Exploration

Peter R. Rose

Telegraph Exploration, Inc.
Austin, Texas, U.S.A.

Most highly risk-averse people (and firms) never realize the high price they pay for their conservatism.
—Peter R. Rose

INTRODUCTION

Exploration and Risk Aversion

Like many other common four-letter words, *risk* is widely and variously used in petroleum exploration. Megill (1971, 1979) defined it as "an opportunity for loss." In this chapter, I use the term in the sense of *risk aversion*, which weighs the magnitude of investment against four factors: (1) size of available budget, (2) potential gain, (3) potential loss, and (4) probabilities of each outcome. It is not synonymous with uncertainty, which is defined as the range of probabilities that some condition may exist or occur—although uncertainty clearly affects risk behavior.

Modern petroleum exploration consists of a series of investment decisions on whether to acquire (1) additional technical data (geological, geophysical, engineering, drilling, or economic) and/or (2) additional mineral interests (Rose, 1987). Each decision should allow a progressively clearer perception of project risk versus reward and should support timely management action concerning the inferred accumulation. An idealistic definition of *exploration* could be "a series of investment decisions made under decreasing uncertainty." Every exploration decision involves considerations of both risk and uncertainty. Risk comes into play in deciding how much to pay for additional data or mineral interests and in considering the high impact of front-end costs on project profitability, as well as the substantial likelihood of ultimate project failure. Uncertainty is intrinsically involved in all geotechnical predictions with respect to the reserves contained in the prospective accumulation, the chance of discovery, and the costs to both find and develop it. Great uncertainty also attends forecasts of future oil and gas prices.

Companies searching for oil and natural gas make hundreds of such exploration decisions each year. So the problem in serial exploration decision making is twofold: (1) to be consistent in the way we deal with risk and uncertainty and (2) to perceive risk and uncertainty accurately and reduce them where possible. Risk aversion is not just a hypothetical nuisance. It causes explorationists to make inconsistent investment decisions, and it costs exploration companies millions of dollars annually in lost opportunities, bad choices, and wasted investment dollars.

Expected Value and Risk Aversion

The concept of *expected value* (EV) can be explained using the example of a simple game in which a coin is flipped one time (see Chapter 7). If you call the outcome (heads or tails) correctly, you win $20,000; if you call it incorrectly, you win nothing. If you were able to play this game free of charge, the EV of the game would be +$10,000. If you paid $10,000 each time you played, the EV would be zero. This can be shown as follows:

Free Trial: 0.5($20,000) + 0.5(0) = $10,000 EV
$10,000 Trial: 0.5($20,000 – 10,000) + 0.5(–$10,000) = $0 EV

The question is, how much would you, with your own unique financial status, obligations, and values, be willing to pay for one chance to play this game? Think about it and try to formulate an answer to this proposition. Most players would not pay $10,000 to play the game one time, but many would pay $1000 to $5000. A few risk-prone or affluent players might chance as much as $9500, whereas some highly conservative or fiscally burdened players might not be willing to hazard even a small loss on the venture.

This example illustrates the subtle, variable, but powerful human attribute called *risk aversion*—the proportion by which an investor discounts the EV of a venture. Another way of expressing this concept is to say that most prudent investors will not knowingly participate in a venture that has a negative or even neutral EV—they properly insist on a premium. The magnitude of that premium is a direct expression of their risk aversion.

Bernoulli was apparently the first to write about risk aversion, in about 1738 (Newendorp, 1975), although many thoughtful investors—but not many gamblers—had doubtless identified its effects long before!

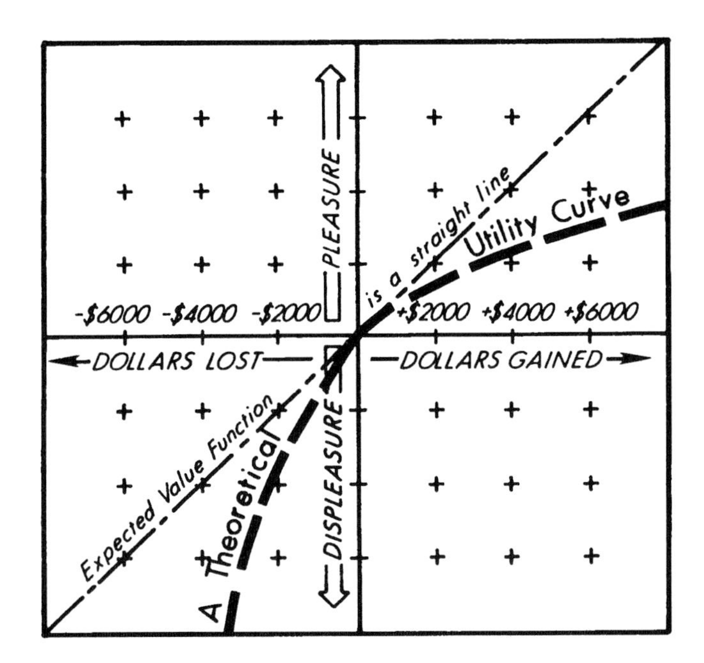

Figure 1. A theoretical utility curve. (From Rose, 1987.)

Utility or Preference Theory

The concept of utility (or preference) theory has been a controversial subject in petroleum exploration, primarily because it has been so difficult to apply operationally. However, an awareness of the psychological biases affecting our reaction to risk—and, we hope, a consistent (even quantified) approach to risk—clearly will improve investment strategies aimed at repeated fractional participation in a series of technically guided exploration ventures. A simplified summary of utility theory is as follows (after Tversky and Kahneman, 1981), illustrated here by the idealized utility curve in Figure 1.

1. The pleasure (utility) associated with winning $4000 is generally less than the *displeasure* of losing the same amount, that is, it hurts more to lose than it feels good to win. People will take a greater chance to avoid a loss than to make a gain of the same amount.
2. People feel more pleasure about gaining $10 going from, say, $10 to $20, than they do about gaining $10 going from $1500 to $1510.

Theoretically, by analyzing past decisions, we can construct a utility curve (and therefore, derive the corresponding utility function r) for an individual, a department, or an entire company, which might thereafter be used to adjust future investments so that they are consistent with previous behavior or current attitudes and expectations (Cozzolino, 1977). No single "correct" utility function can be determined for a company, however—only one that reflects risk attitudes of previous decisions. In addition, a firm's risk response will vary greatly depending on such factors as the overall economic climate, changing budgets, management philosophy, and company fortunes. A utility curve can also be constructed by proposing a series of hypothetical risk ventures to a decision maker, as shown by Grayson (1960) and Walls (1989).

Ubiquity of Risk Aversion

Risk aversion is a natural human attribute that attends almost all the significant decisions people make. Except for pathological gamblers, most of us are risk-averse, many strongly so. Although risk aversion may safeguard us against unwanted losses, it also imposes a subtle yet powerful penalty by significantly limiting our ability to realize optimal returns from our investments. Most highly risk-averse people (and firms) never realize the high price they pay for their conservatism. In addition, risk aversion makes us behave inconsistently with respect to serial decision making: bias inclines us to favor one project over another, not upon objective assessment of EV, but upon our risk-averse psychology.

In general, risk aversion affects all decisions involving elements that are precious to us—money, life and death, personal and professional status, personal relationships, and basic spiritual and/or philosophical standards. In corporate decision making, the concern for individual career status and the fear of criticism often outweigh the traditional concerns about corporation financial loss or gain.

Table 1. Biases Affecting Risk Decisions

Name of Bias	Common Example
Framing effects	Decision makers will take a greater gamble to avoid a loss than to make an equal gain.
Existence of a prior account	Decision makers are more inclined to take a risk at the beginning of a project than later in the project's life.
Maintaining a consistent frame of reference	Decision makers are more likely to invest during a run of good fortune and less likely to invest during a run of bad fortune.
Probability of success	A venture having a perceived high chance of success is preferred over a second venture having a low chance of success, even though the EV of the second venture is clearly superior.
Wrong action versus inaction	Because of fear of criticism or of losing career status, managers prefer to take a risk by not making a decision rather than taking action that could result in the same loss.
Number of people making decision	Groups are more prone to take risks than are individuals.
Workload and venture size	Large-volume ventures are preferred over smaller ones, especially when decision makers are busy.
Personal familiarity	The comfort bias: decision makers are more risk-prone in deals or environments with which they have had good experience.

BIASES AFFECTING RISK DECISIONS AND THEIR CONSEQUENCES

A considerable body of recent work in psychology, led by Daniel Kahneman and Amos Tversky, reveals many inconsistencies in the ways most people make risk decisions. Most of these inconsistencies fall into patterns of *heuristic bias* (systematic errors in pragmatic reasoning processes) and clearly affect exploration decision making (Table 1).

Framing Effects

Tversky and Kahneman (1981, p. 453) used the term *decision frame* to refer to "the decision-maker's conception of the acts, outcomes, and contingencies associated with a particular choice." The frame is controlled by both the formulation and presentation of the problem, as well as the personal attributes of the decision maker. In general, decisions framed in terms of avoiding a loss tend to be much less risk averse than those presented in terms of making a gain, although the EV of both ventures may be the same. To avoid a sure loss, we take a gamble on an even greater loss, but we seem to prefer a small "sure thing" to a gamble on a larger gain. These patterns seem to operate in all aspects of decision making: the actions involved, the outcomes that result, and the various contingencies that arise. Several variations on this theme apply directly to mineral exploration and production.

For example, suppose that the installment of an enhanced recovery project is being considered for a company producing property. Management will be more likely to approve the proposed venture if it is presented (that is, "framed") as an opportunity to avoid "losing" future production, rather than as an opportunity to "gain" additional reserves beyond projected conventional production limits.

Existence of a Prior Account

When we perceive that a given decision is related to a previous investment, the resulting choice tends to be more risk averse than if it had been made on its independent merits. For example, most people would pay another $50 for a theater ticket after having lost a $50 bill, but would not be willing to buy a second $50 ticket if they had acquired and then lost the first ticket (Tversky and Kahneman, 1981). By analogy, it would seem prudent to sell a five-well, stratigraphic pinch-out play as a "package deal", including all five tests, because investors are likely to become significantly more risk averse after several dry holes than when initially considering the venture. This could be described as the "good money after bad" syndrome.

Maintaining a Consistent Reference Frame

Apparently, our perception of risk is highly sensitive to ongoing events, which may incline us to take risks ordinarily shunned, or vice versa. For example, in pari-mutuel racetrack betting, bets on long shots are most popular on the last race of the day (McGlothlin, 1956). Most exploration geologists have vivid, even pungent, firsthand experience with the impact of current program success or failure upon management receptiveness to proposed new prospects.

Probability of Success

Kahneman and Tversky (1982) maintain that low-probability ventures are commonly overweighted (underrisked) with respect to certainty, whereas high-probability ventures are usually underweighted (overrisked). They point to the immense popularity of lottery tickets and flight insurance as practical examples of these trends with regard to choices by individuals. Whether these observations extend to organizational exploration decision making may be questioned. The experiences of many prospectors suggest that low-risk exploration ventures are more saleable than equally priced high-risk prospects, even when the EV of the low-risk deal is inferior. Research by Swalm (1966) tends to support this viewpoint, but more industry input on the question is needed.

Wrong Action Versus Inaction

One of the most interesting psychological patterns discussed by Kahneman and Tversky (1982) is the comparative regret associated with losses stemming from wrong action versus those that arise from inaction. Apparently, most people would prefer to take a loss by "standing pat", than to take action that turns out to be equally unprofitable. The source of this attitude lies in fear of criticism and possible loss of position or status. This bias exerts a powerful and debilitating influence on decision making, especially in larger firms, leading sometimes to virtual organizational paralysis. The most common manifestation of this bias is the frequently observed pattern, particularly in larger corporations, in which many decisions are in fact made by not making *any* decision.

Other Biases

Personal experience, as well as industry folklore, suggests that the number of people involved in a risk decision may influence the outcome. In general, groups may be more risk-prone than individuals (Rose, 1987). Although not verified by research, this observation offers support for firms that make exploration decisions by committee. Such risk-prone behavior may be caused by what has been delicately referred to as "the dilution of corporate accountability."

Another bias has to do with the interplay between our workload and the overall magnitude of the prospect. Commonly, a highly profitable but small-volume venture may be rejected in favor of a larger scope project that has a lower profit-to-investment ratio, especially when the prospect screener is pressed for time. In other words, "being busy" and "small scope" may produce a risk-averse tendency.

Undeniably, personal familiarity with the basic elements of a risk venture may prejudice decision makers, either because of past success with similar ventures or because of anticipated receptive attitudes of managers or potential partners. We can term this the *comfort bias*.

CHARACTERISTIC INDUSTRY PRACTICES TO DEAL WITH RISK AVERSION

Financial Risk-Averse Practices

When the risk versus reward balance of an otherwise attractive drilling prospect seems unfavorable with regard to the bankroll (or annual budget), the common approach is to diminish the amount of front-end cost at risk, usually by reducing working interest (that is, fractional participation). We thus reduce the potential reward, but we can still participate partially and thus "spread the risk." Alternatively, we may try to improve the probability of success by acquiring new data. We also might try to obtain a disproportionately larger share of the venture in relation to our share of the cost. Five common industry practices accomplish such aims:

1. Farming-out leased acreage in exchange for a drilling commitment.
2. Making bottom hole or dry hole contributions to encourage drilling on nearby competitor acreage.
3. Paying substantially less than the going price for a deal.
4. Acquiring a legal option allowing subsequent enlargement of mineral interest in the venture in the event of early success.
5. "Promoting" our partners (gaining a disproportionately large share of a joint venture) as the result of our early favorable lease position or unique technical capabilities, or their legal or financial disadvantage.

For ongoing ventures involving several partners, the occasional practice of *going nonconsent* (declining to participate in a specific subproject, while remaining in the venture as a whole) represents a tactic—or remedy—by which a partner can avoid an unacceptable front-end expenditure by accepting a greatly reduced share of subsequent revenues.

Organizational Risk-Averse Patterns

Fear of criticism or loss of personal status—and only secondary concern about the firm's profit or loss—is the force that drives the second category of risk aversion common to oil and gas companies, here termed *organizational risk aversion*. Exploration companies plagued by excessive organizational risk aversion are characterized by at least 11 common patterns of behavior or corporate culture:

1. Greater preoccupation with dry hole ratios than with reserve sizes of prospects.
2. Domination of project inventory by orthodox prospects and plays.
3. Not acting swiftly on decisive choices.
4. Frequent temporizing by shooting unneeded additional seismic in lieu of drilling.
5. A long, convoluted decision chain requiring excessive documentation.
6. Very conservative economic yardsticks.
7. A highly authoritarian and rigid management style.
8. A short-term reference frame for success criteria.
9. An orientation toward activity and data rather than goals and results.
10. Operating managers with little financial authority.
11. Corporate maintenance of geotechnical and economic apartheid.

Hidden Hurdles

Overly risk-averse exploration organizations commonly burden their internal procedure of prospect evaluation and ranking by *hidden hurdles*—undeclared screening parameters that lurk within the economic evaluation process and disqualify many exploration prospects, often for the wrong reasons. Hidden hurdles are typically imposed by well-meaning individuals trying to protect the firm from perceived irresponsible ventures and are not generally apparent to the explorationists responsible for generating the prospects. Hidden hurdles are classic examples of unaccountable risk-aversion and include the following:

1. *Unrealistically high corporate discount rates or minimum DCFRORs.* These are commonly imposed in the belief that those projects that do qualify will be the "best," when in fact those excluded will tend to be large-profit "cash cows"—just the kinds of ventures that build companies!
2. *"Gold-plated" field operations.* This is particularly common in firms in which newly discovered properties are transferred to the production department without any reverse transfer of funds or credits, so the property is seen to be arriving "free." As a result, there is less pressure to conduct truly cost-effective field development because there is no front-end cost to contend with. The result is excessive development cost estimates that may reduce the apparent value of an exploratory prospect.
3. *Overcautious drilling cost estimates on authority for expenditures (AFEs).* When firms make a practice of overestimating exploration drilling costs (to avoid cost overruns), the unseen effect is to burden exploration projects with excessive front-end costs, thereby reducing the estimated economic value of the project. The goal should be an equal number of drilling cost overruns and underruns each year.
4. *Overcautious wellhead price forecasts.* Exploration is inherently a long-term industrial activity plagued by short-term, overreactive economics. Rather than proceeding by inefficient fits and starts as oil prices fall and rise, a better course is to identify crude price levels and inflation rates that provide median historical and economic comfort and then proceed with systematic exploration.
5. *Excessive seismic programs relative to drilling expenditures.* Some companies find it easier to decide to shoot another seismic line than to make the hard decision to drill an exploratory well, even though managers may realize that the new data will not significantly improve the perceived risk to reward ratio.

Exploration Risk Versus Development Risk

In many firms, the prevailing folklore is that exploration ventures are generally riskier than development ventures. Exploration ventures are perceived to have relatively low chances of success, low capital investment requirements, and high ratios of project profit to exploration investment. Conversely, development ventures have high perceived chances of success, high capital investment requirements, and low ratios of project profit to development investment.

Common problems encountered in development ventures include underestimated chance of failure related to delays, higher costs, lower prices, process inefficiencies, and political or legal difficulties. When such problems are combined with the common large front-end capital requirements, plus relatively low earning rates, it is clear that development projects may be just as risky as exploration projects. They simply represent different combinations of risk.

Cost of Unwarranted Risk Aversion

Risk aversion is not just a theoretical phenomenon or a minor organizational nuisance. Large, financially powerful companies who manifest overly risk-averse behavior are paying a high price for their excessive conservatism, through inconsistent investment decisions as well as lost opportunities, especially those that involve unorthodox, large-reserve prospects having a characteristically low perceived chance of success. Such firms often report relatively high wildcat

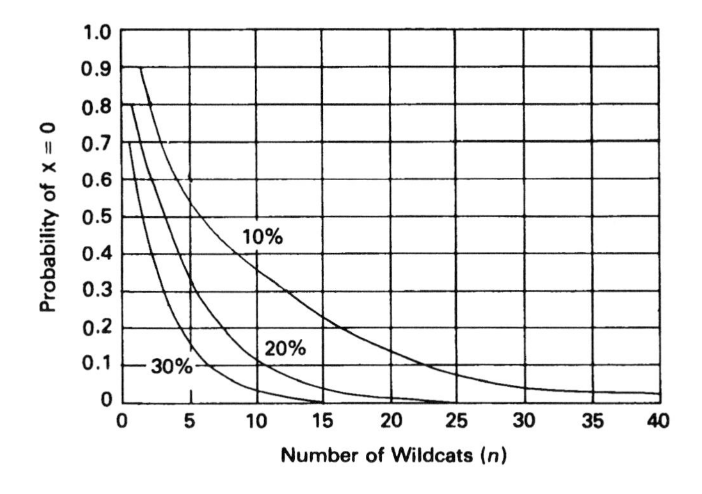

Figure 2. The probability of all dry holes using binomial probability. (From Megill, 1984.)

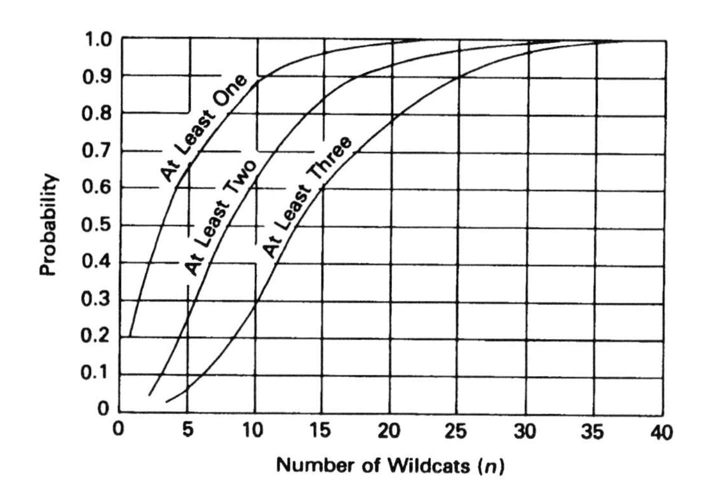

Figure 3. The chance of at least one, two, or three discoveries at a 20% success rate (*p*). (From Megill, 1984.)

success ratios, but the average reserves addition represented by each discovery is often quite small, less than 1 MMBOE.

Schuyler (1990) shows how to estimate the actual cost of excessive *financial* risk aversion to a company. One simulation involved a $2 million exploration program containing 20 ventures having an average success chance of 20%. The actual cost of risk aversion was about 15% of the $10 million program value as determined using the positive expected value decision rule. A second case simulated a $20 million development program containing 20 ventures having an average success chance of 80%. In this one, the actual cost of risk aversion was about 30% of the $70 million program value, as determined using the positive expected value decision rule. Schuyler (1990, p. 57) concludes, "When applied across many decisions, the impact of risk adjustments can be great, often higher than one expects or desires."

However, *organizational* risk aversion is probably the more debilitating influence affecting the efficient operation of exploration firms and—because it is hidden—may be much more difficult to measure. In particular, the loss of opportunity and profit incurred simply by unnecessary time delays is possibly the most insidious expression of organizational risk aversion.

WAYS TO IMPROVE RISK DECISIONS

Gambler's Ruin and Binomial Expansion

Clearly, no single risk aversion factor is "right"—only one that offers comfort and consistency to the user. The need for consistency is particularly apparent among companies that have fractional interests in many exploration ventures. The numerous operative biases provide additional reasons to strive for consistency in risk decisions.

Arps and Arps (1974, p. 77) defined *gambler's ruin* as "a situation in which a risk-taker with limited capital goes broke through a continuous string of failures that exhausts his/her funds." As Megill (1984, p. 87) points out, "the higher the risk (i.e., where chances of success are small), the greater the chance of going broke through a normal run of bad luck."

The problem, then, is to ensure that the firm has enough capital to ride out the eventual, inevitable runs of simple bad luck. Basically, this involves balance and consistency among (1) capital, (2) the number of wildcats planned, and (3) the chance of success or failure. The firm must choose an acceptable chance level for a normal run of bad luck, that is, 10%, 5%, 1%, or whatever the firm deems "normal."

Many managers would like to know the probable outcome of drilling programs—the chance of having all dry holes or one, two, or more discoveries from a multiwell campaign. Binomial probability tables (which can be found in any modern textbook on probability and statistics) provide the basis for assessing the general likelihood of success or failure, as well as predicting the probability of various numbers of successes from an exploration program of *n* independent prospects (Figures 2 and 3). This technique has certain limitations that suggest caution in its application to evaluating risk of exploration programs (Megill, 1984):

1. It assumes that the events (prospects) are independent. However, the outcome of one prospect is in fact often related to the outcome of another. We learn as we drill each prospect, so the odds change.

2. Binomial expansion does not fit the inventory principle (drilling the best prospects first) since no prospect is identified as being better than another.
3. It assumes "sampling with replacement," but exploration is clearly sampling *without* replacement. This is probably not a serious deficiency if the trends or plays have many prospects.

Graphical Risk Limits

Arps and Arps (1974) have developed a graphical method to judge the risk acceptability of any particular venture. Data required for analysis and solution include

1. available risk budget,
2. failure cost (dry hole cost plus land and seismic),
3. present value of successful outcome,
4. probability of success of the venture, and
5. minimum acceptable program success rate.

This method combines two aspects of risk aversion—gambler's ruin, and "breaking even in the long run," a form of EV. The procedure involves cross-plotting a point on a nomogram-like chart relative to the location of certain "isorisk" contours. If the plotted point lies above the index contour, the project represents unacceptable risk to the firm.

This method was far advanced for its time, being both statistically sound and practical. However, it provides only a binary answer to the acceptable risk question. It does not yield an answer that can be compared usefully with another project. Nor does it indicate the investment level that is optimal and consistent with respect to budget, chance, reward, and risk.

Risk-Adjusted Value

Cozzolino (1977) has advocated a formula to determine *risk-adjusted value* (RAV) using the risk aversion function r, as follows:

$$\text{RAV} = \frac{-1}{r}\ln\left[p \times e^{-r(R-C)} + (1-p) \times e^{rC}\right] \qquad (1)$$

where
R = gross reward (in millions of dollars),
C = cost (in millions of dollars),
p = probability of success,
r = risk aversion function (in millionths).

Note that this equation combines both expected value (EV) and risk aversion.

The risk aversion function r, however, is not easy to determine and, it will change according to company philosophy, economic climate, and other factors. Instead, as a "quick and dirty" method, Cozzolino (1977) has suggested using the reciprocal of the annual exploration budget, in millionths, as a first approximation of r. By applying Cozzolino's method to past ventures, we can assess the consistency of risk decisions. Of course, each venture will have a different RAV, but the proportion of the RAV to the EV of each venture—the percentage discount—will lie at or above a certain level for all ventures if risk decisions have been made consistently. Naturally, ventures in which the participant made a particularly favorable deal will show percentage discount values well above the desired discount level.

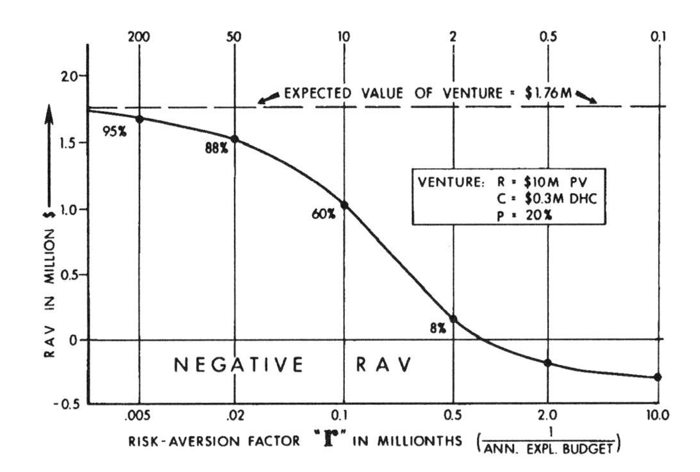

Figure 4. Risk-adjusted value (RAV) diminishes geometrically with diminishing budget size. R = gross reward, C = cost, p = probability of success, M = million, PV = present value, DHC = dry hole costs. (From Rose, 1987.)

Figure 4 illustrates the impact of diminishing budget level on risk-adjusted value. For a company having a $200 million annual exploration budget, the RAV is 95% of the $1.76 million EV of the venture. As budget decreases to $50 million, $10 million, and $2 million, the RAV of the venture declines to 88%, 60%, and 8% of the EV, respectively, and thereafter to negative values.

Figure 5 illustrates how reducing interest in a venture enables a small-budget company to participate. The hypothetical deal is the same as in Figure 4. For a company with a $500,000 annual budget, this venture has a negative RAV at share levels greater than about 26%. However, at the 10% share level, the RAV is 37% of EV, and at the 1% share level, the RAV is 88% of EV, which is relatively as attractive to the small company as 100% of the venture was to the company having an annual budget of $50 million.

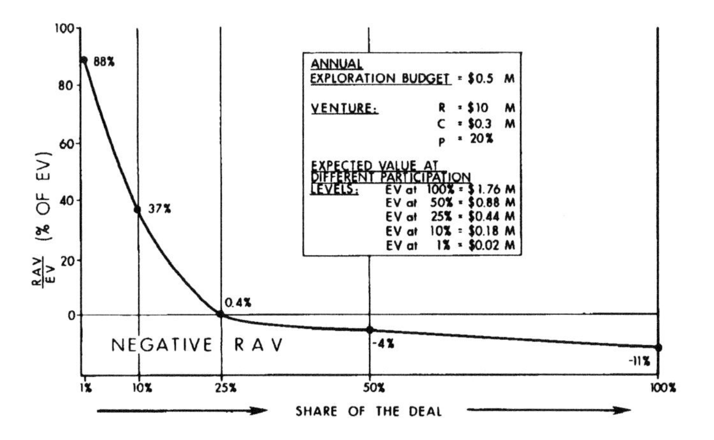

Figure 5. Reducing interest in a venture allows smaller firms to participate because RAV/EV (= percentage of EV) increases geometrically as the share is reduced. EV = expected value. Other abbreviations are as in Figure 4. (From Rose, 1987.)

Establishing Consistent Participation Levels

One of the more difficult tasks facing the modern exploration decision maker is deciding how large a share to take in each of a number of qualifying and competing ventures. In other words, how should the firm invest appropriately and consistently in serial exploration ventures, given its strategies and budgets? Walls (1989) offers a practical method by which the *r* factor—the firm's utility function—can be determined and then applied so as to indicate a consistent and justifiable participation level.

Walls (1989) provides a series of model drilling ventures, each involving different investment levels, returns, and chances of success. The decision maker evaluates and selects the share that represents the maximum level of comfortable participation. This approach was also used by Grayson (1960).

Discussion

In trying to improve the ways in which we make exploration risk decisions, we should strive to be consistent. In other words, our choices should reflect similar risk-aversion attitudes. Two approaches seem to be promising (Rose, 1987).

1. Remember that many natural and subtle psychological influences tend to bias decision making in risk ventures. Use of the expected value (EV) concept may serve as a useful yardstick against which to view propositions. Pay attention to the terms in which a deal is presented—that is, its decision frame. Also, periodically reviewing decisions made by management inaction may prove illuminating (but not necessarily reassuring).
2. Use some consistent form of risk adjustment when considering exploration deals. The approach of Arps and Arps (1974) will indicate levels beyond which additional participation is hazardous. Use of Cozzolino's (1977) risk-adjusted value (RAV) may be more attractive because it can indicate optimal participation level. Until a company's *r* function can be determined, the reciprocal of its annual exploration budget can serve as a rough check on consistency. The goal should be to show that, for all ventures, proportional discounts from the EV of the company's share lie consistently at or above a certain percentage.

Although all decisions on exploration ventures may be made by "headquarters," prospect generators and deal screeners employed by the company should be thoroughly familiar and in tune with the organization's risk philosophy. One of the most damaging aspects of the risk aversion phenomenon occurs when a prospect screener's risk aversion is much higher than the firm's. The result is that the screener tends not to show managers high-risk/high-potential ventures they may well be seeking. Conversely, if the management's risk-aversion level is much higher than that of the deal screener or prospector, they may not have the opportunity to review enough low-risk, close-in ventures that they prefer.

Formalized Risk Analysis and Organizational Risk Aversion

Corporations that have adopted formalized, integrated procedures of exploration risk analysis—that is, firms that have "institutionalized" risk analysis—have done so with good reason. They have recognized that the inherent concern of professional employees for career security sometimes works against the true interests of the firm, by intimidating them so that they do not recommend projects or prospects in which the firm should indeed properly participate—ventures that are unorthodox or otherwise daunting, but which promise rewards that justify the risks. Formal risk analysis programs allow professional employees the comfort of going by the book, while at the same time making risk decisions that are proper for the strategies, values, and budgets of the firm. The use of RAV as one of several economic yardsticks employed in prospect inventories is recommended as a simple and effective measure to bring consistency to serial exploration investing.

RISK-ADJUSTED VALUE, FAIR MARKET VALUE, AND DEAL MAKING

In connection with negotiated property settlements, condemnation actions, and lawsuits, it is often important to establish the fair market value of oil and gas properties. For producing properties, the procedures are well established. You determine the *net present value* (NPV) of the property by performing a cash flow analysis at an appropriate discount rate and compare the NPVs of the subject property and counterpart properties that have been sold recently. If a comparable sale cannot be documented, recalculate a second NPV for the subject property using a discount rate roughly double that used in the first NPV. The second NPV should provide a general guide for an appropriate purchase price.

The logic here is that if the property is worth, say, $2 million discounted at 10%, and $0.8 million at 20%, then purchase at about the lower figure should provide an adequate cushion to protect the buyer from unanticipated negative surprises in the future, while still returning at least 10% on the purchase price. Such methodology is commonly used, but it should be recognized as pragmatic in the extreme, inasmuch as the discount rate itself has nothing to do with chance of commercial success.

But what about exploration properties—tracts under which oil and gas reserves are unproven and therefore speculative? How can the value of such tracts be estimated objectively? Here the RAV concept can be useful. Required data include

1. the mean net present value based on estimated mean reserves, production rates, and costs;
2. the chance of commercial success;
3. the estimated dry hole costs; and
4. the seller's risk aversion function r, based on his or her recent decisions, or a calculated r obtained by dividing 1 by the annual exploration budget, as a last resort.

The calculated RAV represents an approximation of the value of the property to the owner and therefore provides a general upper indication of fair market value. To obtain an alternative value for purposes of comparison, the property's RAV can also be calculated using the buyer's r function.

RAV calculations are also useful in deal making by indicating what a certain venture might be worth to both prospective parties, purchaser and seller. Keep in mind, however, that many firms seem to use r factors substantially higher than indicated when using the generic value calculated as 1/annual exploration budget. That is, firms often tend to behave in a more risk-averse manner than we might expect (M. R. Walls, pers. comm., 1990).

Finally, the prospective deal seeker should always bear in mind the simple remedy for otherwise attractive prospects that are too risky. Simply recalculate the RAV of the venture at a smaller share. Commonly, a project that is too risky at a working interest of 50% will fit your firm's risk limits at 25% or 12.5%.

Acknowledgments *The writer thanks Elizabeth Huebner and Barbara Wiley for their conscientious work in helping to prepare the manuscript and illustrations.*

References Cited

Arps, J. J., and J. L. Arps, 1974, Prudent risk-taking: Journal of Petroleum Technology, v. 26, p. 711-715.

Cozzolino, J. M., 1977, The difficulty of assessing uncertainty: Journal of Petroleum Technology, v. 28, p. 843-850.

Grayson, C. J., 1960, Decisions under uncertainty: Harvard Univ., Div. of Research, Graduate School of Business Administration, 402 p.

Kahneman, D., and A. Tversky, 1982, The psychology of preferences: Scientific American, v. 246, n. 1, p. 160-173.

McGlothlin, W. H., 1956, Stability of choices among uncertain alternatives: American Journal of Psychology, v. 69, n. 6, p. 604-615.

Megill, R. E., 1971, An introduction to exploration economics, 1st ed.: Tulsa, OK, Petroleum Publishing Co., 159 p.

Megill, R. E., 1979, An introduction to exploration economics, 2nd ed.: Tulsa, OK, PennWell Books, 180 p.

Megill, R. E., 1984, An introduction to risk analysis, 2nd ed.: Tulsa, OK, PennWell Books, 274 p.

Newendorp, P. D., 1975, Decision analysis for petroleum exploration: Tulsa, OK, PennWell Books, 668 p.

Rose, P. R., 1987, Dealing with risk and uncertainty in exploration: How can we improve?: AAPG Bulletin, v. 71, n. 1, p. 1-16.

Schuyler, J. R., 1990, Decision rules using the EMV concept and attitude toward risk: An improved approach: Oil & Gas Journal, Jan. 15, p. 55-57.

Swalm, R. O., 1966, Utility theory: Insights into risk taking: Harvard Business Review, v. 44, n. 6, p. 123-136.

Tversky, A., and D. Kahneman, 1981, The framing of decisions and the psychology of choice: Science, v. 211, p. 453-458.

Walls, M. R., 1989, Assessing the corporate utility function: A model for the oil and gas exploration firm: South Texas Geological Society Bulletin, v. 4, p. 13-27.

PETER R. ROSE, an AAPG member and Certified Petroleum Geologist, established his own independent oil and gas consulting firm, Telegraph Exploration, Inc., in Austin, Texas, in 1980. His clients include international exploration firms, most major U.S. companies, and many prominent independents. Prior to forming his own company, Dr. Rose was Staff Geologist with Shell Oil Company, Chief of the Oil and Gas Branch of the U.S. Geological Survey, and Chief Geologist and Director of Frontier Exploration for Energy Reserves Group, Inc. (now BHP Petroleum, Americas). He has explored for oil and gas in most North American geological provinces and has published and lectured widely on U.S. resource assessment, basin analysis, play development, prospect evaluation, and risk and uncertainty in exploration. He was an AAPG Distinguished Lecturer in 1985–1986 and continues to teach extensively at the corporate and professional level. Since 1990 he has been deeply involved in design and implementation of comprehensive exploration risk analysis systems for several major oil companies.

PART III

ECONOMIC ASPECTS OF THE BUSINESS

The five chapters in this part concentrate on money-making aspects of the business. They cover exploration economics, development economics, oil and gas property evaluation, computer software for economic evaluations, and databases. Chapters 10, 11, and 12, on exploration and development economics and property evaluation, form a logical progression. Before a play or prospect is drilled, its economic outcomes and thresholds must be estimated. If the exploratory well is successful, then development economics are needed to determine whether or not to proceed with appraisal or development drilling. Next, as exploration opportunities diminish in mature plays and production of oil and gas fields goes into natural decline, it becomes increasingly important to evaluate producing properties as a reserve replacement opportunity for all companies and independents. Finally, Chapter 13 on computer software reviews a variety of programs available to perform economic evaluation, while Chapter 14 on databases offers some guidance to anyone thinking of building a well history data file. A synopsis of each chapter follows.

Chapter 10, "Exploration Economics," by Robert E. Megill, deals entirely with determining the economic viability of a play or prospect. It includes (1) cash flow analysis, (2) present value concepts, (3) various economic yardsticks, (4) predrilling reality checks, and (5) postdrilling, postplay analysis. The author prefers play over prospect analysis because, being the sum of several prospects, play economics give a broader view of the investment opportunity. He states that play and prospect economics are always slightly optimistic because they seldom include all charges that ultimately will be borne by successful prospects.

Chapter 11, "Development Economics," by Field Roebuck, addresses the economic evaluation of investment opportunities that occur after a discovery well is successfully completed. It considers (1) the development phase of an exploration project, (2) mutually exclusive or alternative investment opportunities, and (3) improved and enhanced recovery projects. The author admonishes investors to remember that two separate prospects, with equal economic values based solely on their exploration phase, can have quite different values when their individual development potentials are considered. Also, explorationists might be able to enhance the economic attractiveness of their prospects when development potential is included in the evaluation.

Chapter 12, "Oil and Gas Property Evaluation," by Gene B. Wiggins III, presents the basic methods most commonly used today by the industry in determining the value of oil and gas properties for acquisition or divestment. This chapter goes into extensive detail using worksheets keyed to the text. The first half reviews techniques for estimating and forecasting oil and gas reserves, including discussions of the volumetric and performance methods. The second half presents economic methods used for analyzing and valuing future production streams once reserves have been determined and forecasted.

Chapter 13, "PC Software Considerations for Economic Evaluation of Investments in the Petroleum Industry," by John M. Stermole, has three objectives: (1) to identify what users can expect to find in the subject PC software, (2) to illustrate many salient considerations in evaluating such PC software, and (3) to create a level of awareness of specific considerations that may require a vendor to explain how a particular program handles various calculations. The author also alerts us that the software requirements of each user can be very different and need to be thought out completely. Included is a list of software vendors and their available software packages based on the results of a questionnaire. The intent of this list is to provide a general understanding of what is available and where to look for more information.

Chapter 14, "Databases: Basic Considerations Using Well History Data as an Example," by Robert W. Meader, is not simply a list of items to include in a well history database. Rather, it raises the more critical issues of geoscience database development and those philosophical questions users and managers should ask before building a database. Examples of some issues addressed include (1) *ownership*—"Many databases are developed with one mind set, but used by a second."; (2) *geological tops*—"An implied permanence is placed on a geological top once it is entered into machine storage, a finality that was not so with traditional practices."; (3) *core descriptions*—"What are the anticipated geoscience questions the database may be reasonably expected to answer?"; and (4) *longevity*—"Archival storage is neither trivial in pursuit nor inconsequential in storage." The message is, do not rush into building a well database without a detailed understanding of its purpose and future use.

Chapter 10

Exploration Economics

Robert E. Megill
Consultant
Kingwood, Texas, U.S.A.

. . . first look at the size of the prize and the risk involved. Geology first, economics second.
—Robert E. Megill

INTRODUCTION

Chapter 6 discusses the estimation of prospect sizes, while Chapter 7 discusses the estimation of the chance of hydrocarbon presence. These two aspects of exploration economics are of primary concern to the geologist. It is here that his or her technical talent is brought to the fore. Furthermore, they are the vital ingredients to the third aspect: will it make money?

This chapter deals entirely with the third aspect—determining the economic viability of the play or prospect. At the outset, one point is important. Preexploration economics are important because they enable geologists to see if their assumptions will prove profitable. Their assumptions must consider the full range of possible outcomes, even if only some portion of that range may contain prospects or plays that are estimated to be profitable. Play economics are preferable to prospect economics because, being the sum of several prospects, they give a broader view of the investment opportunity. Finally, remember that play and prospect economics are always slightly optimistic. They seldom include all of the exploration and overhead charges that must ultimately be borne by the successful prospects.

IT ALL BEGINS WITH CASH FLOW

In Chapter 12 on oil and gas property evaluation, the building of a cash flow is demonstrated. Any investment evaluation begins with a cash flow analysis. A cash flow analysis measures the money coming into and going out of the corporate treasury (or a personal bank account). It can be negative or positive in a given year, but it must be positive in total for an investment to be economically viable.

Any investment evaluation takes into account all of the expenditures and revenues related to the opportunity. Exploration evaluations include development economic calculations. The development portion of a prospect evaluation is identical to what would be made for drilling wells to develop a new discovery. What is different for the prospect or play evaluation is that two additional items are added to the development well cash flow stream—exploration expenditures and exploration risk. Because these two items are "add ons," the returns for prospect are lower than those for the development wells. Thus, any economic measure for a development well (or wells) will be greater (larger) than the same measure for the entire prospect (Figure 1).

Example Cash Flow for an Exploratory Opportunity

Following the steps outlined in Chapter 6 and using the mean prospect size, the cash flow stream shown Table 1 was calculated for an exploratory play (or prospect). Note that, as is true for most exploratory investments, the first few years are negative. This example cash flow has three negative cash flows and eight positive ones. The first negative one (cash outlay) is in the year zero—defined as the first significant investment in the play or prospect. By the end of year one, a total of $1.5 million has been spent. By the end of year two, the total cash outlay has totaled $2.0 million. Thereafter, the negative cumulative cash flow

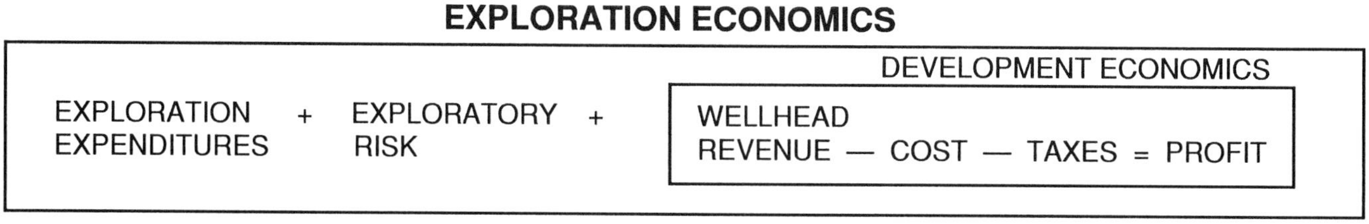

EXPLORATION ECONOMICS

EXPLORATION EXPENDITURES	+	EXPLORATORY RISK	+	DEVELOPMENT ECONOMICS: WELLHEAD REVENUE — COST — TAXES = PROFIT

Figure 1. Exploration economics is development economics with exploratory risks and expenditures added.

diminishes and becomes positive during the fourth year. The actual payout is 3.67 years. The total cumulative net cash flow is $4.7 million and occurs at the end of the tenth year.

Three Yardsticks from Cash Flow Analysis

All investment analysis begins with a cash flow stream. The net cash flow can be calculated before Federal income tax (BFIT) or after it (AFIT). Deal makers presenting a prospect often deal with BFIT cash flows because they do not know the tax situation for a prospective buyer of their deal. Major companies almost always deal in AFIT cash flows.

Three useful investment "yardsticks" emerge from cash flow analysis. They are noted in Figure 2, which is a plot of the cumulative net cash flow for the West Wind basin play. The first yardstick is the *maximum negative cash flow*. It occurs at the end of year two. The second is the *payout period*, and the third is the *total net cash flow*, which is the total amount of profit to be made from the projections of production, revenue, cost, taxes, etc. The maximum negative cash flow (seldom equivalent to the dry hole cost) is an estimate of the total drain on the corporate treasury to make the play. As such, it is a key investment yardstick. If all other yardsticks are identical, take the project with the smallest maximum negative cash flow. The maximum negative cash flow is, in essence, an estimate of the cost to make the play. It is sometimes used as the investment (I) in profit to investment ratios (P/I).

The payout period is calculated from the cumulative cash flow for the play. It is an estimate of the number of years for the play to pay back the initial investment. If all other yardsticks are identical, take the investment with the quickest payout.

The net cash flow, $4.7 million for this play, is the total amount of money to be made from the play. It should be a risk-adjusted value for evaluation purposes. (Risk-adjusted values and their calculations are discussed later in this chapter.) One should always be interested in the total amount of money to be made from an investment. If all other yardsticks were identical, you would take the investment with the largest net cash flow.

Review of Cash Flow

The mean (or average) reserve of a prospect distribution is first converted to a production stream based on the producing characteristics of the formations found in the prospects of the play. This production stream is then turned into a revenue stream by applying the price forecast to annual production rates. Next, exploration and development costs, operating expenses, and taxes are deducted to arrive at the net cash flow from the play. It was the final calculation, the net cash flow, that was just illustrated. From the cumulative net cash flow three yardsticks were illustrated: the maximum negative cash flow, the payout period, and the total net cash flow. The first segment of this chapter briefly demonstrated the conversion of a prospect or play reserve estimate into a cash flow stream. Such a conversion must be made to calculate not only the three yardsticks just discussed but also numerous other yardsticks used to compare one investment to another.

PRESENT VALUE CONCEPTS

A significant addition to investment analysis is made from calculations of present value for a specific cash flow stream. From present value concepts come additional and valuable yardsticks to compare one investment to another. Several yardsticks contain a combination of both compounding and discounting and both will be explained.

Most people understand the concept of compound interest. An interest rate, an investment amount, and the frequency of compounding are the three ingredients involved. Table 2 shows an example of compounding using a 10% interest rate. From this table, you can see that the value of $1.00 invested at 10% grows to a value of $2.59 in 10 years. The key ingredients are annual compounding, a 10% interest rate, and a $1.00 investment. Examining the table illustrates the growth of any investment. The future value (FV) of any investment is equal to $1.00 plus the interest rate (expressed as a decimal fraction) taken to the power of the number of years invested at that interest rate. It can be expressed

Table 1. Annual and Cumulative Net Cash Flow, West Wind Basin Play

	Cash Flow ($1000)	
Year	Actual	Cumulative
0	–500	–500
1	–1000	–1500
2	–500	–2000
3	1000	–1000
4	1500	500
5	1200	1700
6	1000	2700
7	800	3500
8	600	4100
9	400	4500
10	200	4700

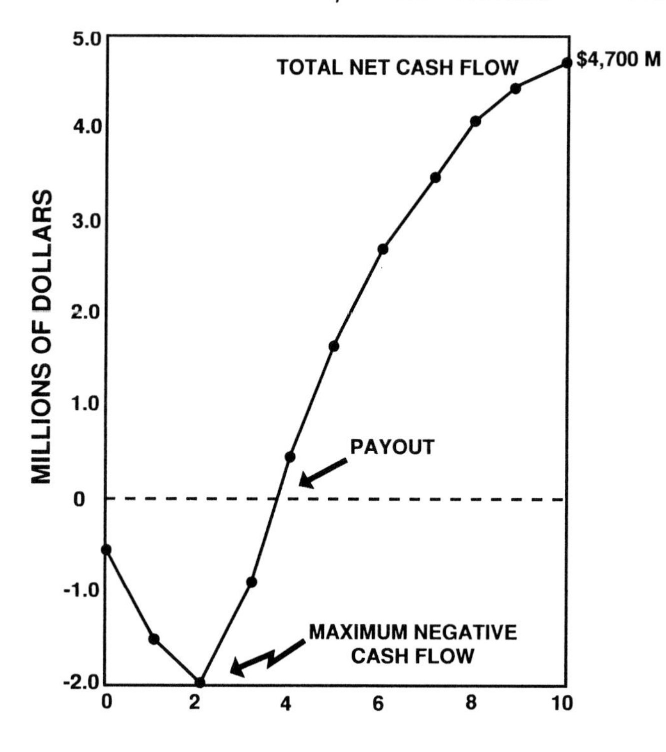

Figure 2. Three valuable investment yardsticks are shown by a plot of cumulative net cash flow for the West Wind basin play.

in the following equation:

$$\text{FV of I} = (1 + i)^n$$

where
I = investment,
i = interest rate, expressed as a decimal fraction,
n = number of years I is invested at rate i.

An interesting rule of thumb comes from this relationship called the "rule of 72." Divide the number 72 by the percentage interest rate and you will get a good approximation of the number of years it takes to double the initial investment. For example, 72 divided by 10 is 7.2. If you invest money at 10%, it will double in value in about 7.2 years.

One way to compare investments is to pick a year in the future and compound the positive revenues to that year, subtract the negative cash flows, and then see which investment produces the largest amount of money.

What do you do if given two or more cash flow streams to compare? What might you measure if the cash flow streams are of different lengths (life spans) and the annual flows are not similar? What is the value, today, of money to be received 5 or 10 years from today? Discounting of cash flow streams was introduced to answer these questions. Discounting is always related to an interest rate. The rate used should be related to the average reinvestment rate possible for the money flowing in (as represented by the annual cash flows projected). If you expect to be able to reinvest future cash flows at 10%, then 10% is the interest rate you should use for discounting.

The equation to calculate the discount factor for each year is the reciprocal of the equation for future value (FV):

$$\text{Discount factor} = 1/(1 + i)^n$$

Each individual annual cash flow has a different discount factor.

An example calculation, shown in Table 3, is based on the previous net cash flow for the West Wind basin play. The discount factors used in this table are end-of-the-year factors. Many computer models use mid-year discount factors. To calculate mid-year discount factors, 0.5 is subtracted from the value n for each year. For example, for the sixth year, the factor would be $1/(1 + i)^{5.5}$. The factor would thus equal 0.592, not 0.564 as shown in Table 3. Mid-year factors are larger and thus produce larger discounted cash flows (and larger yardstick values) for the same cash flow stream than do end-of-the-year factors.

The first column in Table 3 shows the year, including year zero, while the second column shows the net cash flow, as given in Table 1. The third column contains the individual discount factors for 10%, and the final column gives the *present value* (also called *present worth*) discounted at 10%. The last column is obtained by multiplying the individual discount factors times the annual net cash flow for the particular year of the factor. Note that in the tenth year, the discount factor is only 0.386. One dollar received 10 years from now has a present value of only 38.6 cents. This occurs because of the loss of interest at 10% for a period of 10 years. The factor for each individual year reflects the loss of interest for $1.00 for the number of years indicated.

Table 2. Future Value of One Dollar

Year	Value of $1 Invested at 10%
0	1.00
1	1.1 x 1.00 = 1.10
2	1.1 x 1.1 = 1.21
3	1.1 x 1.21 = 1.33
4	1.1 x 1.33 = 1.46
5	1.1 x 1.46 = 1.61
6	1.1 x 1.46 = 1.77
.	.
.	.
.	.
10	1.1 x 2.36 = 2.59

Table 3. Discounting a Cash Flow Stream, West Wind Basin Play

Year	Net Cash Flow ($1000)	Discount Factor (10%)	Present Value Discounted at 10% ($1000)
0	–500	1.000	–500
1	–1000	0.909	–909
2	–500	0.826	–413
3	1000	0.751	751
4	1500	0.683	1025
5	1200	0.621	745
6	1000	0.564	564
7	800	0.513	410
8	600	0.467	280
9	400	0.424	170
10	200	0.386	77
Totals	4700		2200

Another way of looking at the present value of tomorrows' dollars is as follows: the discount factor at some time in the future is the amount that you would have to invest *today,* at the interest rate related to the factor, to yield $1.00 at that time. For example, at 10% interest you would have to invest $0.386 today to have $1.00 in 10 years.

The final column in Table 3 totals $2.2 million. It is the present value (present worth) discounted at 10%. In other words, the West Wind play is estimated to yield 10% *plus* $2.2 million. One could spend another $2.2 million in year zero to make the West Wind play and still make 10%.

Constructing a Present Value Profile

In Table 4, the net cash flow of the West Wind play is given additional discounting. The discounted cash flow for 20%, 30%, and 40% is shown. Remember that for each discount rate, the factors (end-of-the-year) are calculated using the equation $1/(1 + i)^n$. For example, the discount factor for 30% in the sixth year is calculated as $1/(1.3)^6$. The result is 0.207, which when multiplied by the net cash flow for year six ($1 million) yields $207,000 dollars.

The term *present value* must always be accompanied by a specific discount rate. One can calculate as many present values as there are discount rates. So this term is not complete unless you know the discount rate applied to the annual net cash flows.

However, a graph can be constructed, on which many discount rates are shown, to determine a specific discount rate. In Figure 3, a present value *profile* is shown for the West Wind basin play. The x axis has discount rates from 0% to 50%. The scale for the y axis has both plus and minus amounts for present value.

Figure 3 was constructed by plotting the *total* discounted net cash flow for the following discount rates: 0%, 10%, 20%, 30%, and 40%. The result is the curved line shown in Figure 3. Where this curve crosses the zero present value line, the discount rate is given a special name. It is called the *discounted cash flow rate* (DCFR). It is also known as the *discounted cash flow rate of return* (DCFROR), the *internal rate of return* (IRR), or the *investor's interest rate* (IIR). To calculate the DCFR, one discounts the net cash flow at increasingly larger and larger rates until the sum of the discounted cash flow becomes negative. In our example for the West Wind basin play, a rate of 40% discounts the net cash flow to a negative $216,000.

In Figure 3, the discount rate that discounts the net cash flow to zero is 34.1%. Thus, for this play, the DCFR is 34.1%. The DCFR has been used extensively as a key yardstick during the past 15 years. However, it has fallen out of favor in some companies because of its numerous problems and limitations. For example, the DCFR will not always give the proper ranking of investments. Also, it often assumes unreasonable reinvestment rates. In the last section of this chapter, the best yardsticks are illustrated and recommendations are made about those best suited to rank prospects.

Review of Present Value

From present value concepts come several useful yardsticks by which to evaluate and compare investment opportunities. In all instances, present value concepts for prospects or plays begin with a risk-weighted net cash flow stream. It is the stream of annual cash flows that is discounted. From present value concepts, we get the net present value at various discount rates (including the company's minimum guideline return), the present value profile and the discounted cash flow rate (DCFR). One key use of present value is the present value at the company's minimum guideline return divided by the *discounted* investment.

Table 4. Discounting a Cash Flow, West Wind Basin Play

Year	Net Cash Flow ($1000)	Discounted at 20%	30%	40%
0	–500	–500	–500	–500
1	–1000	–833	–769	–714
2	–500	–347	–296	–255
3	1000	579	455	364
4	1500	723	525	390
5	1200	482	323	223
6	1000	335	207	133
7	800	223	127	76
8	600	140	74	41
9	400	78	38	19
10	200	32	15	7
Totals	4700	912	199	–216

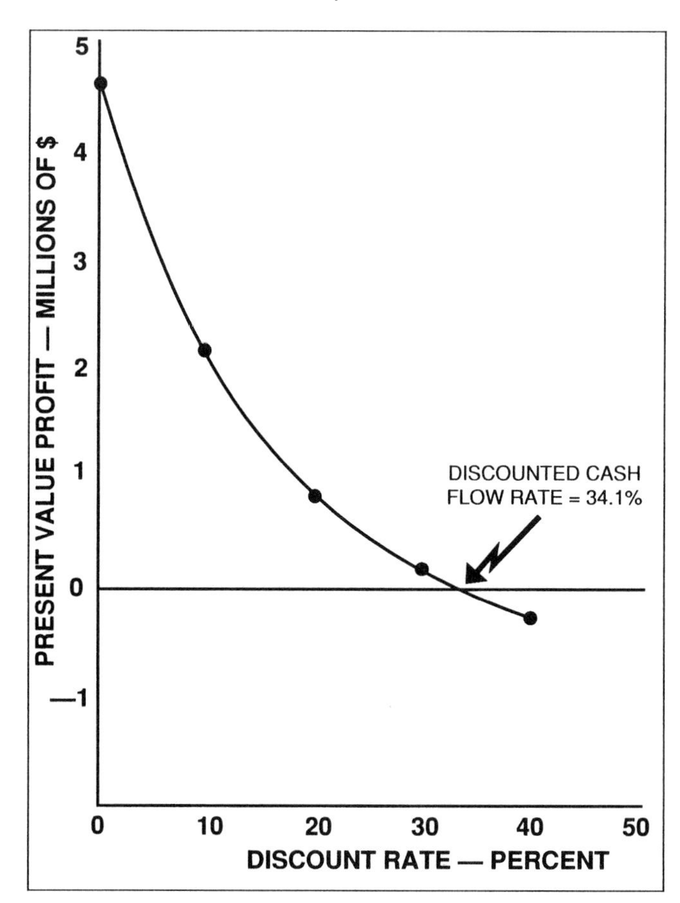

Figure 3. Present value profile, West Wind basin play.

YARDSTICKS OBTAINED FROM CASH FLOW AND PRESENT VALUE CONCEPTS

It is from cash flow analysis and present value concepts that most yardsticks to compare investments are derived. This section covers the yardsticks one can obtain for our example of the West Wind basin play. First, let's review what we mean by a risk-weighted cash flow stream. In the example shown in Table 5, a short cash flow stream for a productive prospect is combined with a dry case (which can be a dry hole or holes, more seismic, etc.). The *dry case* (often called the *failure case*) is defined as the amount of money spent before the property is abandoned. The prospect is given a 30% chance of success, which means that the dry case (failure case) has a 70% probability of occurrence.

The cash flows shown in Table 5 are assumed to be discounted at an interest rate of X%, as indicated in the table's title. The productive prospect would have a discounted net cash flow of +$3.4 million. The dry case totals –$1.1 million. The last column represents the combination of these cash flows with their probability of occurrence. There is only a 30% chance of getting the $3.4 million, but a 70% chance of getting the –$1.1 million. The combined result—the risk-weighted net cash flow—is $250,000. The last column, totaling $250,000, is where all yardsticks begin, and all yardsticks are calculated from it.

You do not have to use the annual cash flows to show the risk-weighted net cash flow. You can, in fact, use the totals as shown in Table 6. The total of the risk-adjusted value is $250,000 and is called the *expected monetary value* (EMV). The EMV is not a "real-world" number. Table 6 clearly shows that the real-world prediction is either a gain of $3.4 million or a loss of $1.1 million. Then what does the $250,000 mean? It is a theoretical value that takes into account the *probability* of each event occurring. The discovery value is expected to occur only 30% of the time; the dry case is expected to occur 70% of the time. Thus, the EMV is the combination, with probabilities, of both outcomes. It is a *repeated trial* number. If you had numerous investment opportunities similar to the $3.4 million discovery, you would expect to gain $250,000 on average for each investment, despite the fact that the most probable outcome for each single investment is the dry case.

In the calculation of the EMV, the economic result of each possible outcome is always shown as a present value (present worth) at the minimum acceptable guideline rate of return. The minimum return is the interest rate used to discount future cash flows. A few writers have stated that the discount rate should be the same as the average return expected on future investments. Because this value varies from investor to investor, an X is shown in Table 6 to indicate the discount rate.

Table 5. A Risk-Weighted Net Cash Flow Discounted at *X*%

Year	Productive Prospect Cash Flow ($1000)	Probability	Dry Case Cash Flow ($1000)	Probability	Risk-Adjusted Net Cash Flow ($1000)
0	−1000	0.3	−800	0.7	−860
1	1800	0.3	−300	0.7	330
2	1200	0.3	—	—	360
3	700	0.3	—	—	210
4	400	0.3	—	—	120
5	200	0.3	—	—	60
6	100	0.3	—	—	30
Totals	3400		−1100		250

Yardsticks of the West Wind Basin Play

Having examined cash flow analysis and present value concepts, we can now look at the economic yardsticks that describe the West Wind basin play. From cash flow analysis, we obtained three yardsticks: total net cash flow, payout period, and maximum negative cash flow. For the West Wind basin play, these values are

Total net cash flow:	$3.4 million
Payout period:	3.67 years
Maximum negative cash flow:	$2.0 million

As shown in Table 1, the maximum negative cash flow occurs at the end of year two. The total net cash flow for an exploratory investment should be a risk-weighted number. A risk-weighted number shows the average net profit per project if a number of similar investments could be made (repeated trials).

Yardsticks that relate to time include the project life, the discounted cash flow rate (DCFR, IIR, or IRR), the present value at various discount rates, and the present value profile. For the West Wind basin play, these values are

Project life:		10 years
Discounted cash flow rate:		34.1%
Present value at	10%:	$2.2 million
	20%:	$912,000
	30%:	$199,000
	40%:	−$216,000

One measure of an investment's value is the profit to investment ratio (P/I). The name sounds simple, but in actual practice, many combinations of things pass for the P/I. The ratio should be the net cash flow divided by the investment. However, several definitions of *investment* exist. As a consequence, there are several different ways of calculating the P/I.

In the 1920s and 1930s, the term *investment* usually meant the investment placed on the corporate books. With the emergence of cash flow analysis, the investment after Federal income tax (AFIT) was used by corporations. The newest yardsticks use variations of the early negative cash flows in the cash flow stream as the investment. The maximum negative cash flow is one such definition of investment. One yardstick, the *growth rate of return*, or GROR, uses the discounted negative cash flow as the investment. All profit to investment ratios attempt to provide a measure of investment efficiency, that is, how much profit you are getting per dollar invested.

The best way to rank investment opportunities is to use some measure of investment efficiency. The best measure of investment efficiency is a ratio in which the discounted net cash flow is in the numerator and the discounted investment is in the denominator. The discount rate should be the expected average investment rate for future investments. (Remember that for exploratory investments, the discount rate is applied to a risk-weighted cash flow stream.) The GROR is a measure of investment efficiency and will rank projects in their proper order. The DCFR will not always give the proper ranking. The net cash flow divided by the maximum negative cash flow is also a measure of investment efficiency.

A full description of the GROR is given by Capen et al. (1976). The GROR requires the choice of a target year to which some compounding is done. The idea behind the GROR (and an earlier yardstick called percentage gain on investment) is to pick a future year (target year) and compound all income to that year and see which investment yields the largest amount of money. If there are cash flows beyond the target year, you "sell" them at the discount rate related to the return expected on future investments. The discounted portion is added to the compounded portion to obtain a measure of "profit." Varying the target year will vary the precise value of the GROR, but will not affect its proper ranking of investments. For the West Wind

Table 6. An Expected Monetary Value

Event	Probability	Present Value @ X% ($1000)	Risk-Adjusted Value ($1000)
Discovery	0.3	3400	1020
Dry Case	0.7	−1100	−770
Totals	1.0		250

basin play, a target year of four yields a GROR of 34.1%, the same as the DCFR. As the target year increases, the GROR decreases; for example, a target year of seven yields a GROR of 23.2%. Obviously, using GROR as a yardstick requires a common target year for all investments compared.

What Economic Yardsticks To Use

Any investment opportunity should be measured by how much present value it gives per dollar invested. For the West Wind basin play, the discounted negatives (discounted at 10%) for the years zero, one, and two total $1.822 million. With this as the denominator (the investment) and the present value at 10% as the numerator ($2.2 million), one investment efficiency yardstick is 1.21. It is often called the *investment efficiency index*. When comparing one investment with another, the one with the highest investment efficiency index is the best opportunity. Always use discounted values for both profit (net cash flow) and investment to get the best ranking of prospects.

The DCFR is still a widely used investment yardstick. It has severe limitations in ranking investments and will not always give the same ranking of projects as the investment efficiency index. It is best used as a screen to eliminate those investments below a certain discount rate.

No accounting yardstick is adequate to make individual investment decisions. This broad statement applies to booked rate of return, return on assets, return on capital employed, and return on equity. These accounting standards do not provide the proper ranking of investment opportunities. They do not reflect what is actually happening in the corporate treasury. You have to begin your judgments about investment opportunitites with cash flow analyses that measure cash inflows and cash outflows.

Returns on assets, equity, and capital employed are published in all corporate annual reports, but they should not be used to choose the best investment opportunities.

Concluding Remarks on Investment Yardsticks

Most of us cannot make a go or no-go decision on the basis of a single number. This fact is particularly true of an exploratory venture because of the significance of other subjective factors such as risk and prospect potential. We have already established that there is no single answer to the question of prospect size and chance of hydrocarbon presence. Therefore, exploratory investment decisions should deal with a range of possibilities. Single values can be shown, but the range of possibilities is equally relevant.

Many existing computer programs allow the use of Monte Carlo simulation by which a full suite of values can be shown, from the most optimistic to the most pessimistic. Various yardsticks can be shown over the range of possible outcomes, with a most probable value given. An 80% range can be used—the values within which 80% of expected values should fall—to provide a quick bracket on the investment opportunity. Monte Carlo is only one type of simulation; there are others, such as the Latin hypercube. The value of simulation lies in its ability to show the range of all variables.

For single values, the best yardsticks to use are those that measure investment efficiency. The DCFR can be used as a screen. Do not use any yardstick that does not begin with a properly risked cash flow stream (which eliminates all accounting yardsticks). *Do* consider the total quantity of money to be gained by the investment. Explorers are the providers of future income to the corporation. They provide the goods on the shelf for tomorrow's production.

The dry case, as used in this chapter, is not just a dry hole. It may include several dry holes, additional seismic, and even more acreage purchases. It is the amount of money that would be spent before a property is abandoned.

If your computer program uses mid-year discount factors, you will generate larger present value profits and also higher DCFR values than with end-of-the-year discount factors—for the *same* annual net cash flows.

Exploratory ventures should begin with a look at the potential, the risk, and then the economic yardsticks. It is important to know how much money you could make and what the yardsticks infer, but you must first look at the size of the prize and the risk involved. Geology first, economics second.

PREDRILLING REALITY CHECKS

Reality checks should be used in any predrilling analysis. Such a practice is instinctive for the seasoned explorer. What is a *reality check?* It is the review of all that is stored in your mental reservoir of facts and relationships to apply to what you are analyzing. For example, suppose you have just finished risking all prospects in a play and the estimated average chance of success is 35%. An immediate reality check is the average success rate of all wells drilled in the area. If past drilling has produced a success rate of only 20%, then you would want to reexamine your risk assumptions (the geological controls governing the presence of hydrocarbons) to see why you are so much more optimistic than past history. History is one readily available reality check.

Reality checks are involved in both geological and economic phases of the analysis. In the geological phase, reality checks should be used to examine geological controls and the parameters used to estimate the range in prospect sizes. In the economic phase, the reality checks involve price forecasts, estimates of drilling and development costs, estimates of inflation, production rates, and so on.

A reality check tells the explorer if his or her estimates are in the right "ball park." Prospects and their parameters are site specific, but prospects normally fit within the general framework of the fields discovered in the area in which they are drilled. That makes history a valuable reality check.

Reality checks are a predrilling exercise. They encompass all available related facts to measure the approximate validity of assumptions made. Reality checks enable you to check experience against assumptions for logical fits. They are made as data are compiled and after compilation produces a result or an answer. They are a necessary and vital ingredient of any analysis in which much uncertainty exists.

POSTDRILLING, POSTPLAY ANALYSIS

After drilling has been completed, an after-the-fact reality check can be made—the postplay analysis. One could use the term *postprospect* analysis, but a *postplay* analysis is more meaningful than the review of a single prospect. Here all the new data from the play results are examined; they are now compared to the assumptions made before the play was initiated. The validity of predrilling assumptions can now be tested. Valuable information comes from the postplay analysis. It is information that enables the explorer to check on his or her technical capability. For the explorer, the postplay analysis should center around the estimates of the chance of success and the estimates of prospect sizes.

Estimates of the Range of Prospect Reserve Sizes

In the two-step process of estimating the chance of hydrocarbon presence and the range in prospect sizes, the sizing should be done first. Why? The basic reason to size first is that one of the definitions of the chance of hydrocarbon presence is the chance of finding the minimum size *or more*. If you have not sized first, you do not have in mind the minimum size as your reference point.

Chapter 6 had a discussion on estimating the range in prospect sizes using the 35/25 method. Three elements in reserve estimation were used: net pay, recovery per acre-foot, and areal extent. Thus, the first examination in the postplay analysis is to check the data developed from the play with the predrill estimates. One would hope that the actual value was within the range estimated. If not, why? This comparison relates to the three parameters (pay, recovery, and area) for all prospects. Here are some important questions to ask:

1. If actual data were not within the estimated range, is there some discernible reason why they fell outside the range?
2. On average, were actual estimates near the mean of the range of prospect values?
3. Is there some particular parameter (e.g., net pay) that was consistently in error (had a sharp deviation from the mean)?
4. What values were consistently well estimated?
5. Given the full range of comparisons, what must be done in future plays to improve your predrilling estimates?

The main purpose of the postdrilling review of size estimation is to enable you to do a better job *next time*. To neglect to examine what you have estimated, when the facts are in, is to ignore an important data set that can improve your technical skills. The postdrilling analysis points out the need to keep good records for predrilling estimates.

Estimates of the Chance of Hydrocarbon Presence

After sizing, the chance of success can be estimated. *Chance of success* (chance of hydrocarbon presence) in this instance means the chance of finding a field within a given range of prospect sizes—or the chance of finding the minimum size or more. The minimum size is determined from your estimate of the range in prospect size. For practical purposes, the minimum size can be taken as the 90th percentile on the range of sizes, with the 50th percentile as the median and the 10th percentile as the high-side value.

To estimate the chance of hydrocarbon presence, the geological controls governing the presence or absence of hydrocarbons must be isolated. Estimators use from four to seven controls. Four seems to be the ideal number to reflect the sensitivity needed to describe a control's uncertainty. For our example, assume that the following four controls can adequately describe the presence or absence of hydrocarbons: structure, reservoir, hydrocarbon charge, and trap. Each control must be carefully defined to ensure that all explorers will be estimating the same entity. Estimates of probability of adequacy are assigned to each geological control. A 1.0 indicates absolute certainty (do we ever have this possibility?), and 0.0 indicates no chance of occurrence.

Suppose our estimates of chance of adequacy for the four parameters are as follows: 0.5, 0.9, 0.8, and 0.7. Then our chance of hydrocarbon presence is 25.2%, or about 25%. Each control is initially assumed to be independent, and all four are multiplied to derive the 25%. Dependent controls can be handled, and the methodology for common (dependent) controls is well described in the literature (Gehman et al., 1975) (see also Chapter 7).

In the postplay analysis, the geological controls as initially assigned are tested against the new data from drilling the play. Wells by other operators also add data for the analysis. From the postplay analysis, the explorer is trying to determine if he or she was consistently wrong in estimating the adequacy of one particular parameter (such as the trap). Was an adequate reservoir assumed too readily? Did the structure form in time to trap the migrating hydrocarbons? Again, the postplay analysis enables the explorer to gain experience to improve technical skills.

Most prospects are dry. Therefore, on most prospects, there is at least one parameter whose adequacy was overestimated. Because of the inherently low chance of success, it will take more than one play to test a geologist's skills in estimating the chance of success. One prospect is akin to one roll of the dice. It doesn't tell you much about the game.

In review, estimate your sizes first, then estimate your chance of adequacy second, and finally, for every play, do a postplay, postdrilling analysis. The explorer is paid to find prospects. Prospects involve size estimates and estimates of the chance of hydrocarbon presence. From the postplay analysis, the explorer has a chance to review his or her own track record and to adjust accordingly.

A postplay analysis involves a little recordkeeping. But it is the kind of recordkeeping that the explorer should be doing anyway. We are all interested in enhancing our technical skills, and the postplay analysis is a vital ingredient in this process for the explorer.

Economic Phase

In setting up a cash flow analysis, many estimates are made. Each of these should be checked in a postplay analysis. If there is no success, then analysis is easy. If discoveries are made, a postplay analysis can be made to check estimates. The economic analysis should follow some time after discovery and the establishing of production, especially if you want to check estimates of development and operating costs. The review process is the same as for the geological phase. Check what was estimated before drilling with what actually occurred. From that comparison, learn what you did well and what you did poorly and adjust accordingly. The postdrill analysis is just as important for the economic phase as it is for the geological phase.

References Cited

Capen, E. C., R. V. Clapp, and W. W. Phelps, 1976, Growth rate: A rate of return measure of investment efficiency: Journal of Petroleum Technology, May, p. 521.

Gehman, H. M., J. R. Kyle, and D. A. White, 1975, Prospect risk analysis, *in* Probability methods in Exploration: AAPG Research Symposium, Stanford Univ., Preliminary Report, p. 16-20.

Additional Reading

Capen, E. C., 1976, The difficulty of assessing uncertainty: Journal of Petroleum Techonology, v. 28, p. 843-850.

Gehman, H. W., R. A. Baker, and D. A. White, 1981, Assessment methodology: An industry view, in Assessment of Undiscovered Oil and Gas: Proceedings of the Seminar at Kuala Lumpur, Malaysia, March 3–8, 1980, p. 113

Megill, R. E., 1984, An introduction to risk analysis, 2nd ed., Tulsa, OK, PennWell Books, 274p.

Megill, R. E., 1988, An introduction to exploration economics, 3rd ed.: Tulsa, OK, PennWell Books, 238p.

Megill, R. E., 1988, The business side of geology: Explorer, Oct., Nov., and Dec. issues.

Rose, P. R., 1987, Dealing with risk and uncertainty in Exploration: How can we improve?: AAPG Bulletin, v. 71, n. 1, p. 1-16.

ROBERT E. MEGILL, a consulting geologist in Kingwood, Texas, received a B.S. in Geological Engineering from the University of Tulsa. He worked for Exxon, including predecessor companies Carter and Humble, from 1941 to 1984. He served as staff geologist, petroleum economist, reserve geologist, Head of Reserve Economics, Head of Exploration Economics, Manager of Planning, and Coordinator of Planning and Coordinator of Economic Evaluation. Mr. Megill has published numerous technical articles and is widely known for his books and courses on Introduction to Risk Analysis and Introduction to Exploration Economics. He is a member of AAPG and for the last 7 years has written a monthly column in The Explorer on "The Business of Petroleum Exploration."

Chapter 11

Development Economics

Field Roebuck

Roebuck Associates, Inc.
Dallas, Texas, U.S.A.

The game is to make money on your investments, rather than just finding oil and gas at any cost.
—Vincent Matthews III

INTRODUCTION

Here the term *development economics* refers to the economic evaluation of investment opportunities that occur after the discovery well is drilled and completed, with specific regard to the techniques used and the economic yardsticks available for investment decisions. Three potential situations are considered in this chapter: (1) the incorporation of development wells into the outcomes of the original exploration project, (2) mutually exclusive or alternative investment opportunities, and (3) the installation of improved or enhanced recovery projects during or at the end of the primary producing life of a property.

For purposes of this discussion, it is assumed that the ultimate goal of the investor is to maximize wealth. Hence, the standard yardsticks of economic value can be applied. However, it should be recognized that any investor can have another, more immediate goal in mind, such as maximizing reserves or diversifying an inventory of properties for control of risk. In such cases, the same methods and the same logic of evaluation would apply, but different yardsticks might be used.

DEVELOPMENT PHASE OF AN EXPLORATION PROJECT

The techniques used for the evaluation of an exploration play or prospect were discussed in Chapter 10 on exploration economics, but primarily in terms of the investments and cash returns from the exploration phase of the project. Too often, an investor fails to consider that two separate prospects, which yield very nearly equal economic values when evaluated solely on the basis of the exploration phase, can have quite different values when their individual development potentials are considered. Likewise, entirely different conclusions about the relative merits of several exploration prospects will often result when the respective development potentials are included as part of the overall evaluation process. Finally, explorationists often can enhance the economic attractiveness of their prospects for potential investors when development potential is included in the evaluation.

The incorporation of subsequent development wells into an exploration prospect requires the definition of that potential. Additional investment requirements and net cash flow streams must be generated, and cash amounts must be discounted at an appropriate discount rate and according to some realistic time schedule. Also, the final outcomes must be expressed on a risk-weighted basis. Thereafter, the process of evaluation is fundamentally the same as for any other investment opportunity.

To illustrate the overall process, consider the example exploration/development project summarized in Table 1. An exploration well is to be drilled with a 30% chance (probability) of success. If the well is a dry hole, the project will be abandoned, and the overall present value cost is estimated to be $1.1 million, discounted at 10%.

However, if the well is successful, it will have a present value investment cost of $1.3 million and will yield a future discounted net cash flow of $3.4 million. Also, a successful well introduces the potential for

Table 1. Example Exploration/Development Project (in Present Value Dollars)

	Investment ($1000)	Net Cash Flow ($1000)
Exploration phase:		
Dry hole	1100	−1100
Producer	1300	+3400
Development phase:		
1. Single dry hole	455	−455
2. Single producer	650	+3635
3. Two producers	1240	+6940

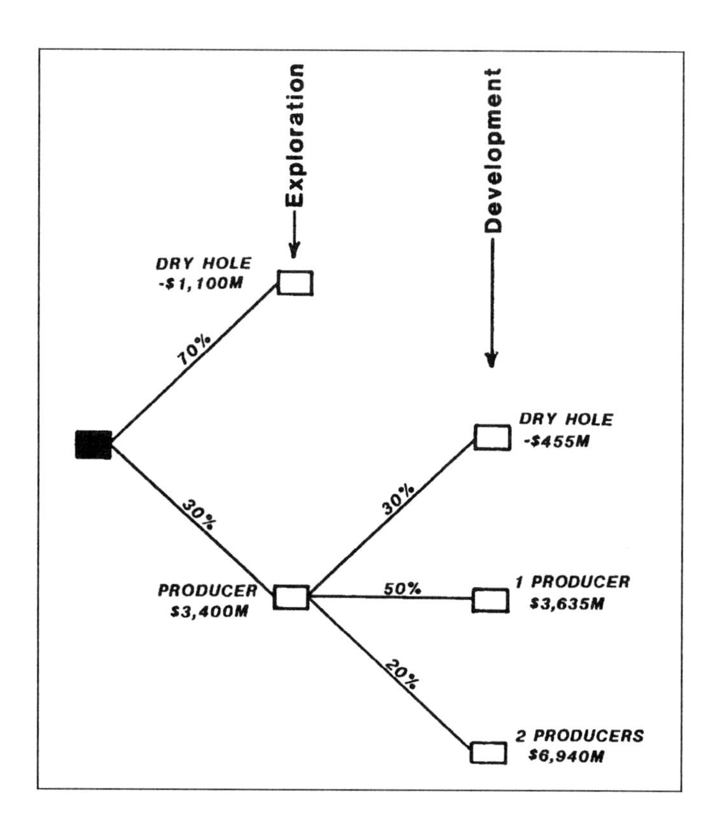

Figure 1. Exploration well with development phase.

additional development on the property, which can be assigned three possible outcomes, as shown in Figure 1.

First, it is estimated that there is a 30% chance that a single stepout well will be drilled, but it will be a dry hole, and with a 10% discount rate, it will cost $455,000 in present value terms. Second, there is a 50% chance that one successful development well will be drilled. It will have a discounted present value investment cost of $650,000 and will generate a present value net cash flow of $3,635,000. Third, there is a 20% chance that two successful development wells will be drilled for a present value investment cost of $1,240,000 and with a total present value net cash flow of $6,940,000.

Note that the net present values of the two-well development case are less than twice the present values of the single-well case because the second development well will be drilled a year after the first, and hence, both the investment costs and the net cash flows are discounted for an additional year.

As demonstrated in Chapter 10, the risk-weighted net present value (the expected value in discounted cash flow terms) for the exploration phase of the project can be calculated in thousands of dollars, as follows:

Success:	0.3 x $3400 =	$1020
Failure:	0.7 x −$1,100 =	−770
	Expected value =	$ 250

Using the same method, the risk-weighted investment costs for the exploration well would be $1,160,000, and the discounted profit to investment ratio (PIR) would be 0.216, which is obtained in the same manner as described in Chapter 10.

In Figure 1, all the defined outcomes for the overall exploration and development project have been diagrammed, along with the present value and the probability associated with each outcome. Such a diagram can assist in the calculation of the overall expected value of the redefined project, as illustrated in Table 2, where all monetary amounts are again in thousands of dollars.

Using this same method and the same probabilities, the total risk-weighted investment cost for the project can be calculated (Table 3) as $1,373,000. Therefore, the 10% discounted PIR for the total project, on a risk-weighted basis, is now 0.853, a ratio almost four times that obtained by evaluating only the exploration phase of the project. Also, this procedure yields a more realistic picture of the overall investment opportunity.

MUTUALLY EXCLUSIVE AND OTHER ALTERNATIVE INVESTMENTS

An alternative to any investment opportunity almost always exists, assuming that investment monies are available in the first place. For example, the investment funds could be placed in a (theoretically) safe and secure savings account paying some known rate of interest. That or some other type of interest-bearing account can always be considered as a viable alternative to any oil and gas investment. This is one line of logic that has long been used to establish a minimum hurdle rate for the investment yardstick commonly known as the internal rate of return (IRR). This means that, if some minimum rate of interest can be earned in a relatively safe investment, then any alternative oil and

Table 2. Calculation of Overall Expected Value

Outcome	Present Value ($1000)	Probability	Extension ($1000)
First well dry hole	−1100	0.7	−770.00
First well success:			
Development dry hole	3400 − 455	0.3 x 0.3	265.05
One development well	3400 + 3635	0.3 x 0.5	1055.25
Two development wells	3400 + 6940	0.3 x 0.2	620.40
		Expected Value	1170.70

Table 3. Calculation of Expected Value Investment

Outcome	Present Value ($1000)	Probability	Extension ($1000)
First well dry hole	1100	0.7	770.00
First well success:			
Development dry hole	1300 + 455	0.3 x 0.3	157.95
One development well	1300 + 650	0.3 x 0.5	292.50
Two development wells	1300 + 1240	0.3 x 0.2	152.40
		Expected Value	1372.85

gas project must return at least that rate of interest. And the oil and gas investment must yield a positive net present value, on a risk-weighted basis, when discounted at that minimum interest rate. This type of evaluation is straightforward.

However, having decided on the oil and gas investment, additional alternatives often present themselves. And some of these alternatives can be mutually exclusive, requiring a different approach—or at least a different line of logic—for proper evaluation. We confront mutually exclusive alternatives on a daily basis: to drive the car or ride the bus, to watch television or read a book, to use a credit card or pay cash, and on and on. It's an either/or situation; choose one or the other, not both.

The potential oil and gas investor is confronted with mutually exclusive alternatives when deciding between drilling or farming out a prospect, infill drilling or additional stimulation for rate acceleration, purchasing or leasing equipment, working interest participation or an overriding royalty interest, and so on. Such mutually exclusive alternatives should be analyzed with the concept known as incremental economics. Whether or not the arithmetic is actually used, the logic of incremental decision making must be there if a correct decision is to be made.

This logic can be stated simply: "If I can get this return with little or no investment, then I cannot credit the alternative with the total return. I can only credit the additional investment with the additional return it generates."

For example, consider the economics calculated for a drilling prospect, as shown in Table 4. This well can be drilled with 100% of the working interest for an investment of $600,000 and will return a present value net cash flow of $1,326,000.

As an alternative, consider that a farm-out arrangement (with no investment) is available. The farm-out terms provide for a one-eighth overriding royalty interest until payout (with a 20% provision for overhead) reverting to a 50% working interest thereafter. As shown in the second column of Table 4, this returns a present value profit of $761,000.

Since neither a rate of return nor any PIR can be calculated for the farm-out case (with no investment, all such yardsticks are infinite), the decision must be made solely on the basis of present values. And the drill alternative has the greater net present value. But it should not be considered that the drill case investment of $600,000 actually generates a net present value of $1,326,000. Of that amount, $761,000 can be obtained without making any investment at all. Therefore, the drilling investment only generates the incremental net present value difference of $565,000 (that is, $1,326,000 minus $761,000). Still, if an appropriate discount rate has been used, this is a positive (additional) value, and drilling the well would be the proper economic decision.

At times, still other alternatives should be considered. For example, if farming out this prospect and making the $600,000 investment elsewhere results in a total net present value greater than $1,326,000, the farm-out would be the proper decision.

This same logic can be extended to any type of project that includes a no-investment alternative, a circumstance that always precludes the use of the PIRs

Table 4. Drill versus Farm-out Economics

	Drill Case	Farm-out Case
Gross production:		
Oil (bbls)	196,788	196,788
Gas (MCF)	127,912	127,912
Net production:		
Oil (bbls)	141,687	58,965
Gas (MCF)	92,097	38,327
Net revenue:	$3,361,187	$1,437,756
Investment	600,000	0
Operating costs	382,790	179,388
Net cash flow	$2,378,397	$1,258,368
PV_{10} net cash flow	$1,326,482	$761,453
Profit/investment ratio	3.964	—
Discounted PIR @ 10%	2.211	—
Internal rate of return	53%	—

Table 5. Rate Acceleration and Incremental Economics

Year	Net Cash Flow (in $1000) Existing Well	Two Wells	Incremental
0	0	−150	−150
1	237	473	236
2	211	353	142
3	166	186	20
4	130	92	−38
5	101	39	−62
6	78	9	−69
7	59	0	−59
8	45	0	−45
9	33	0	−33
10	23	0	−23
11	16	0	−16
12	10	0	−10
13	5	0	−5
14	1	0	−1
Totals	1115	1002	−113
PV_{10} NCF	833.5	850.2	16.7

and rate of return yardsticks to assist in the decision. In some cases, it is actually more appropriate to evaluate the incremental cash flow stream rather than the individual alternatives.

For example, consider the case of an infill well that could be drilled solely for rate acceleration purposes. Suppose that an existing well is producing at its allowable producing rate and has been doing so for 2 years. It is expected that this well will continue to produce at its allowable for another 12 months and that its rate will then decline at 20% per year. This will give it a remaining life of 14 years, during which time it will generate a net cash flow of $1,115,000, as shown in Table 5. The present value of this future net cash flow, with mid-year discounting at 10% per year, amounts to $834,000.

Now consider that an offset well can be drilled at a cost of $150,000. The two wells together will produce twice the single-well allowable rate for a year before declining, recovering only a little more total oil but shortening the producing life to 6 years. As shown in Table 5, the total net cash flow from the property will be reduced to $1,002,000, but the net present value of the combined cash flow streams will be increased to $850,000.

A proper decision could be made solely on the basis of the respective present values, but an investor usually wants to determine the payout, the internal rate of return, and the PIRs. However, none of these yardsticks can be calculated for the case of the existing well because there is no investment on a time-forward basis. Note also that the additional investment for the second well actually loses money, but it generates an increase in the total present value of the property. Table 5 also includes the incremental cash flow stream, which is calculated by subtracting the existing single-well cash flow from the accelerated two-well cash flow, that is, by subtracting the case with the minimum investment from the alternative case.

The investment yardsticks can be calculated for the two-well case and for the incremental cash flow stream. The payouts can be calculated, as follows:

$$\text{Two-well payout} = 150/473 = 0.317 \text{ years}$$
$$\text{Incremental payout} = 150/236 = 0.636 \text{ years}$$

Likewise, the undiscounted and discounted PIRs can be calculated (with discounting at a minimum required rate of return, such as 10%) as follows:

$$\text{Two-well undiscounted PIR} = 1002/150 = 6.680$$
$$\text{Two-well discounted PIR} = 850.2/150 = 5.668$$

$$\text{Incremental undiscounted PIR} = -113/150 = -0.753$$
$$\text{Incremental discounted PIR} = 16.7/150 = 0.111$$

Although this is not the strongest project ever to come down the pike, the investment in the offset well generates an incrementally positive present value, with an internal rate of return greater than 10%. It is also possible that more porosity will open up and that it will turn out to be better than a pure rate acceleration project.

But what is the internal rate of return for the incremental cash flow case? Something interesting happens when the present value profile (Table 6) is calculated and the results are plotted, as shown in Figure 2. Since the incremental investment actually loses money, the net present value at a zero discount rate (undiscounted) is a negative amount. With

Table 6. Present Value Profile of the Incremental Net Cash Flow

Discount Rate (%)	Net Present Value ($1000)
0	−113.0
5	−33.5
10	16.7
15	48.5
20	68.7
30	88.5
40	93.8
50	92.4
60	87.7
70	81.7
80	75.1
90	68.5
100	62.0

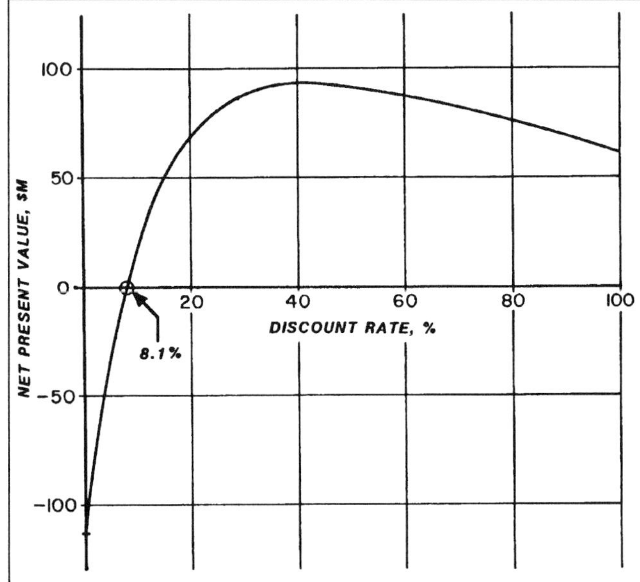

Figure 2. Present value profile for a rate acceleration case.

increasing discount rates, the net present value increases. It has a value of zero at a discount rate of about 8.1%, continues to increase to a value of approximately $93,900 at a discount rate of 42% and then begins to decrease. At a discount rate of 100%, the net present value is down to $62,000, and it is safe to assume that it will equal zero again at some greater discount rate.

So, what is the internal rate of return of the incremental cash flow stream? Some will argue that it is 8.1%, and some will choose the greater unknown (from this plot) rate. Others will argue that there are at least two rates of return for this single investment—an interesting if curious argument since it is almost inconceivable that a single investment can be earning two different rates of interest simultaneously.

Informed logic requires that the internal rate of return is indeterminate for situations such as this. However, it can be assumed that the rate of return is some rate between the two cross-over rates, provided the average interest rate at which the monies are reinvested is within that range. This is only one of several problems encountered when the internal rate of return is used as the chief yardstick for investment decisions. However, as before, the proper investment decision can be made solely on the basis of the discounted PIR and the respective present values of the alternative cash flow streams.

The second well increases the present value of the net cash flow from the lease, when discounted at the minimum required internal rate of return, which was assumed to be 10% for this example. Hence, the incremental cash flow stream yields a positive discounted PIR. So the proper decision is to make the investment and drill the second well. From this analysis, it can also be seen that any investor with a reinvestment rate capability or requirement of less than 8% could not properly choose to drill the second well.

IMPROVED AND ENHANCED RECOVERY PROJECTS

Any type of injection project designed to increase the ultimate recovery from a property can be evaluated in the same manner as an exploration prospect or a simple development project. The same logic and requirements apply: an investment is being made to generate a positive future net cash flow and a positive present value. The income tax treatments of the investments may differ slightly from those of exploration expenditures, but the same evaluation methods can be used.

For example, consider that a producing property is nearing the end of its primary life and that a water injection project is being considered. Two options exist: (1) the injection project can be started immediately, or (2) the project can be installed at the end of the primary life. Table 7 includes the estimated, future, after federal income tax (AFIT) net cash flows that would result from the remaining primary life of the property and from each of the two options, along with the net present value and internal rate of return for each case, using continuous mid-year discounting.

It is of interest to note that an immediate start for the injection project yields the greater net present value at a 10% discount rate, yet a greater internal rate of return would be achieved by the delayed project. This is another example of the problems that can result when the internal rate of return is used to rank projects in the order of their economic attractiveness.

It should be noted that the investment of $750,000 for the early injection case results in an incremental net

Table 7. Estimated Future Net Cash Flows

Year	Remaining Primary	Waterflood Now	Waterflood Later
0	—	–$750,000	—
1	$87,600	89,400	$87,600
2	40,900	119,500	40,900
3	19,700	617,200	19,700
4	7,300	520,700	7,300
5	—	393,900	–808,000
6	—	246,500	77,600
7	—	181,800	586,400
8	—	97,800	503,200
9	—	14,900	387,600
10	—	—	245,900
11	—	—	181,600
12	—	—	97,500
13	—	—	12,700
Totals	$155,500	$1,531,700	$1,440,000
PV_{10} NCF	$139,017	$ 826,355	$ 571,232
IRR (%)	—	32.7	>100

Table 8. Injection Project Timing Analysis

	Waterflood Now	Waterflood Later
Net present value @ 10%	$826,355	$571,232
Internal rate of return (%)	32.7	>100
Payout (from year 0) (years)	2.9	7.0
PIR	2.042	1.641
Discounted PIR @ 10%	1.102	1.021

present value of only $687,338 because the remaining primary net present value of $139,017 can be obtained with no investment. Likewise, the delayed case investment, which is $877,400 (with a present value of $559,455) yields an incremental net present value of only $432,215. (The delayed injection case has a greater investment requirement because of cost inflation.)

With a required minimum internal rate of return of 10%, and with the same methods as used before, the investment yardsticks can be calculated and summarized as shown in Table 8. In this example, the early water injection project is the more attractive case. It yields a greater total net present value and results in a greater increase in net present value wealth per present value dollar invested. Choosing between the two alternatives on the basis of the internal rate of return would result in an improper economic decision.

All of these measures can also be calculated from the incremental net cash flow streams if so desired. However, since the remaining primary is included in both of the injection cases, the conclusions would not change. But if the only consideration was the mutually exclusive choice between a continuation of primary production and an early start to the water injection project, the decision would be made on the basis of the incremental economics.

It is also important to note that, for either straightforward or incremental economic evaluations, the investment decision is often dependent on the discount rate selected for the net present value calculations. A plot of the net present value profiles can be used as an illustration. A present value profile can be calculated for each of the two alternative waterflood projects, as was done for the incremental net cash flow of the rate acceleration example (Table 6). If this is done and the results are plotted, the two curves in Figure 3 will result.

From this plot, it can be seen that selecting a discount rate greater than 32.7% will rule out the early startup alternative because it will no longer have a positive net present value. But more importantly, for any discount rate greater than 27.5%, the delayed project will have the greater net present value and would be the proper choice.

The delayed project alternative of this example indicates that an injection project to be installed at the end of primary life, in lieu of plugging and abandonment, can be evaluated in the same manner as any other investment project. However, one additional aspect must be considered. When evaluations are made on an AFIT basis, there can be a positive value to plugging and abandonment since any remaining tax deductions can be taken immediately when a property is abandoned. This AFIT value, if any, should be deducted from the value generated by the proposed investment—another case of incremental economics.

Also, the salvage value of a property (either AFIT or BFIT), along with any abandonment costs and cost liabilities, should be included in a cash flow stream when they occur near enough in the future to have a significant present value. This situation is most commonly encountered with improved or enhanced recovery projects, which can entail large investments for equipment. If the project has a relatively short life, the used equipment can still have appreciable market value at the end of the project, and the discounted (present value) market value can be significant.

RISK ANALYSIS AND MONTE CARLO SIMULATION

Any economic evaluation should ideally be made on a risk-weighted basis and with expected value outcomes, and this procedure was included in the first example of an exploration prospect followed by a development phase. The only significant differences in risk-weighting cash flow streams for exploration and development projects lie in the derivation and the values of the probabilities assigned to the various outcomes.

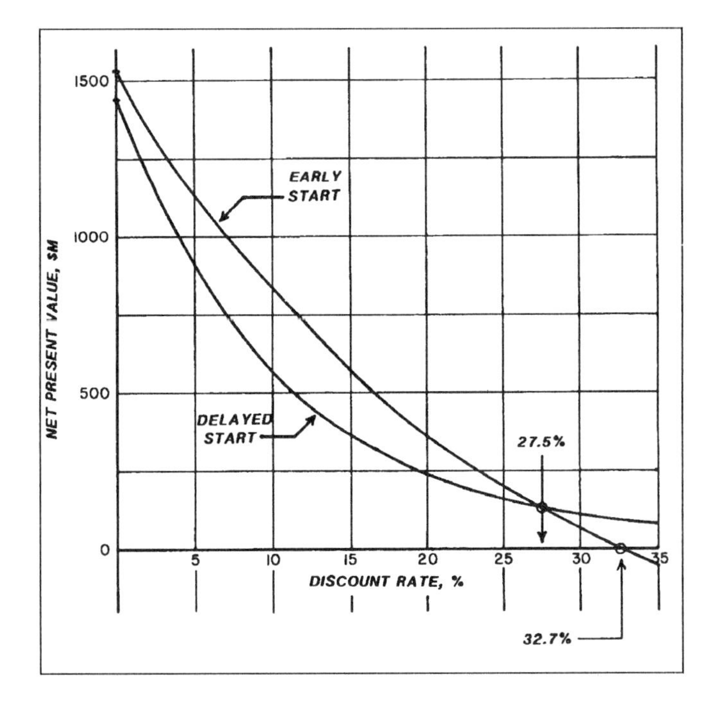

Figure 3. Present value profiles for a water injection project.

Probabilities assigned to development projects tend to be greater than those applicable to exploration prospects because of the more extensive data available and because of the known presence of commercial quantities of hydrocarbons. Also, formal techniques can be used more often for the assessment of probabilities for development projects because of the availability of factual data. Likewise, more precise methods can be used for predicting future recoveries and production performance under various operating programs.

Monte Carlo simulation is sometimes used for development projects, just as it is for exploration prospects. It can be a valuable tool when properly applied and when representative probability distributions can be defined for the individual variables. However, when the model is not properly assembled, when less than representative distributions are used, or when independent and dependent relationships are not properly defined, a Monte Carlo model is a dangerous tool.

A Monte Carlo model will calculate a probability distribution for the case defined. After enough computer processing, the result will almost always be a very believable distribution. But this is one situation where the old computer cliché holds true: garbage in, garbage out! Therefore, this type of risk analysis should be limited to those cases or situations in which it can be used properly and with a reasonable degree of confidence.

INVESTMENT YARDSTICKS

The yardsticks used for investment decisions for development projects are fundamentally the same as those used for exploration plays and prospects. Therefore, it follows that an investor should develop a consistent strategy that enables decisions to be made on the same relative basis for all investment opportunities that may arise. Opportunities can be screened with either the internal rate of return or the growth rate of return (GRR) to ensure that each project meets or exceeds some minimum standard, and this is commonly done. However, the same end can be accomplished simply by calculating the net present value at a discount rate equal to the required minimum rate of return. This can be done for any cash flow stream, even one with only positive values, a situation that precludes the calculation of a rate of return.

Still, most investors want to see the rate of return, as well as the payout and the undiscounted PIR. However, like Linus' security blanket, these yardsticks should be comfort factors only. The shorter the payout or payback period and the greater the undiscounted PIR, the more comfort there is with a project. Also, for any single class of projects, previous experience can often dictate minimum comfort standards for these yardsticks.

By the same token, an investor requiring a minimum rate of return of 12%, for example, is more comfortable with an investment showing a 48% rate of return than with one that shows a rate of return of only 14%. Such an opinion is generally valid because the first project has more margin for the inevitable error in projections of future rates, recoveries, and economics. Also, the first project will usually, but not always, yield the greater net present value per present value dollar invested.

This is the fundamental reason for ranking projects in the order of their relative attractiveness and for making ultimate investment decisions on the basis of the ratio of net present value to the present value of the investment, however the investment is defined. When there is no investment for a mutually exclusive alternative, such as a farm-out option, this yardstick can be applied to the incremental cash flow stream. This discounted PIR will always rank projects in the proper order of economic attractiveness—when the cash flow stream is constructed correctly and when the goal of the investor is to maximize wealth.

Annotated Bibliography

Through the years, many papers have been published in oil and gas periodicals on the subject of economic evaluation, with and without specific regard to development and production projects. Several books presenting the fundamentals of risk analysis and oil and gas evaluation are also available. Selected reading from these many publications will provide the investor with a wealth of information to develop a better understanding or to supplement the fundamentals of economic evaluation presented in this chapter.

However, for the beginner or the uninitiated, where to begin can be a real quandary. Too, all authors do not agree about all aspects of the subject, and ferreting out those whose opinions have withstood the test of time can be a difficult task. Fortunately, some published articles have become classics, and reading and understanding these would be worthwhile for every investor and student of economic evaluation. Following is a brief list of papers, which can be recommended for further reading.

Dean, J., 1954, Measuring the productivity of capital: Harvard Business Review, v. 32, n. 1, p. 120.
This paper includes a discussion of yardsticks of investment worth and the fallacies encountered in analyzing capital proposals, with emphasis on rates of return. It was the vehicle for the acceptance of the internal rate of return by the oil and gas industry.

Glanville, J. W., 1957, Rate of return calculations as a measure of investment opportunities: Journal of Petroleum Technology, v. 9, n. 6, p. 12.
This is an excellent discussion of the internal rate of return and its positive and negative characteristics. This paper was the immediate impetus for the widespread use of the internal rate of return by the oil and gas industry.

Herts, D. B., 1964, Risk analysis in capital investment: Harvard Business Review, v. 42, n. 1, p. 95.
This is a classic paper that describes Monte Carlo simulation and the methods for displaying and interpreting the results.

Northern, I. G., 1967, Risk, probability, and decision-making in oil and gas development operations: Journal of Canadian Petroleum Technology, v. 6, n. 4, p. 150.
This paper discusses the need for risk-weighted evaluation in production operations with example results from 76 wells. It includes estimates of reserves, rates, investments, and profits.

Seba, R. D., 1987, Determining project profitability: Journal of Petroleum Technology, v. 39, n. 3, p. 263.
This is an excellent discussion of the concepts and aspects of investment evaluation, including incremental economics, rate acceleration projects, ranking measures, the effects of taxes and inflation, sunk costs, opportunity costs, and investment selection criteria, with calculation methods and examples.

Silbergh, M., and F. Brons, 1972, Profitability analysis: Where are we now?: Journal of Petroleum Technology, v. 24, n. 1, p. 90.
This paper presents a concise discussion of the history and application of profitability yardsticks and criteria, including probabilistic analysis.

Smith, M. B., 1974, Probability estimates for petroleum drilling decisions: Journal of Petroleum Technology, v. 26, n. 6, p. 687
This presents a discussion, with examples, of probability estimates for various types of exploration and development projects.

FIELD ROEBUCK is President of Roebuck Associates, Inc., an independent consulting firm in Dallas. He received B.S. and M.S. degrees in Petroleum Engineering from the University of Texas at Austin. Mr. Roebuck has more than 35 years of internationxal experience as an oil and gas consultant and petroleum educator, including 5 years on the Petroleum Engineering Faculty of the University of Texas and 20 years in supervisory and management positions with the Engineering & Consulting Department of Core Laboratories, Inc. He also spent 2 years as Vice-President and General Manager of Engineering Numerics Corp. He is author of numerous publications on reservoir engineering, improved oil recovery, mathematical reservoir simulation, petroleum economics, risk analysis, and statistical estimation of reserves. He is a member of SPE, AAPG, SIPES, and the Dallas Geological Society. He is a Registered Professional Engineer, a graduate of the Southern Methodist University Institute of Management, and an Adjunct Professor of Geological Sciences at SMU.

Chapter 12

Oil and Gas Property Evaluation[1]

Gene B. Wiggins III
Atwater Consultants, Ltd.
New Orleans, Louisiana, U.S.A.

Supposing is good, but finding out is better.
—Mark Twain

INTRODUCTION

Exploration opportunities are declining domestically due to maturing producing basins and limited potential for major oil and gas discoveries coupled with increasing environmental constraints. As this occurs, acquisition and divestment of producing properties are going to continue to play important roles in the reserve replacement strategies of integrated producers and independents. Opportunities that exist in acquiring producing properties include (1) acquisition of immediate cash flow, (2) enhancement of property economics by means of production acceleration and cost reductions, and (3) new exploration and development opportunities in existing fields as a result of new technology and "taking another look." Factors contributing to increased activity in domestic property acquisitions and divestments include unstable world oil supplies and prices due to continuing hostilities in the Middle East coupled with a growing inventory of marginally economic properties operated by the large integrated producers.

This chapter presents methods in oil and gas property evaluation as they are generally applied in the evaluation, acquisition, and divestment of oil and gas properties. The first half of the chapter presents techniques for estimating and forecasting oil and gas reserves, including discussions of the volumetric method and performance methods. Definitions for oil and gas reserves as approved by the Board of Directors of the Society of Petroleum Engineers on February 27, 1987, are presented in the Chapter Appendix. These reserve definitions should be strictly adhered to for reserve reporting purposes. A further guide for reserve definitions and reserve reporting is found in a publication prepared by the Society of Petroleum Evaluation Engineers entitled *Guidelines for Application of the Definitions for Oil and Gas Reserves: Monograph 1*.

The second half of the chapter presents the economic methods most commonly used for analyzing and valuing future production streams once reserves have been determined and forecasted. The basic elements of the net cash flow are presented, which include future production rates, income from oil and gas production (including oil and gas prices), expenditures (including costs and severance and ad valorem taxes), and income taxes. Topics covered under federal income taxation include depreciation, depletion, and intangible drilling costs (IDCs). After the net cash flow has been determined on both a "before tax" and an "after tax" basis, the next step is application of profitability measures to evaluate the net cash flow. Measures of profitability presented include net present value, rate of return, payout period, profit to investment ratio, and discounted profit to investment ratio. Methods of fair market value determination are presented along with a discussion of how to incorporate risk and uncertainty into the evaluation process. A section on oil and gas deals is also presented along with a discussion on how deals should be considered in the evaluation process.

Tables 1 and 2 show the basic before tax and after tax cash flow models as arranged in typical computer output formats. The before tax model in Table 1 reflects that net production is multiplied by prices to give total revenue. Costs are then deducted from total revenues to give cash flow. The after tax model in Table 2 shows the income tax treatment for the various capital costs shown in Table 1. These costs are "recovered" in the form of deductions from operations cash flow to arrive at taxable income. Taxes payable are calculated from

[1]This chapter is dedicated to the memory of Gene B. Wiggins, Jr., AAPG.

taxable income by application of the appropriate tax rate. Cash flow after tax is determined by deducting taxes payable from cash flow before tax. To aid in understanding the cash flow model both before tax and after tax, the circled numbers above the columns in Tables 1 and 2 correspond to the boldfaced numbers in the text below each section heading where these components of the cash flow model are discussed.

TYPES AND USES OF EVALUATIONS AND APPRAISALS

An oil and gas property evaluation is generally limited to a determination of the cumulative undiscounted future net cash flow for a future production stream as well as the present worth of the future net cash flow discounted at a given discount rate. A fair market value appraisal, however, generally assigns a lesser value than the present worth to the future production stream using a combination of the methods discussed in the section entitled Fair Market Value. *Fair market value* is generally defined as the price at which a property would change hands between a willing buyer and a willing seller, neither being under any compulsion to buy or sell and both having reasonable knowledge of the relevant facts.

Property evaluations and appraisals have numerous applications. Requests for evaluations and/or appraisals are generally made by a lending institution for loan purposes, by the Internal Revenue Service for tax matters such as successions, by a bankruptcy court, by a seller and/or prospective purchaser of a property, or in court cases. Additional uses for evaluations and appraisals include investor reporting, choosing among investment alternatives, and annual company reporting of reserves and economics. The methodology used in evaluations and/or appraisals is often adjusted depending on the purpose of the report. For example, prices and costs may be held constant in an evaluation prepared for the Securities and Exchange Commission (SEC), whereas prices and costs are generally escalated in most other cases. This chapter will attempt to present the basic components and mechanics of a property evaluation in a manner applicable to all of these uses.

DATA REQUIRED FOR EVALUATION OF OIL AND GAS PROPERTIES

The property evaluator must assemble, organize, process, and assimilate a tremendous volume of material and data in the preparation of an oil and gas property evaluation. General categories of data collection include geological, geophysical, engineering, and economic data (Wiggins and Tearpock, 1985). Geological and geophysical data include all subsurface well logs (including correlation and detail sections), directional surveys, core analyses and descriptions, paleontological reports, base maps, seismic velocity surveys and profiles, and any previously generated maps. Required engineering data include production data of all produced fluids by well and by perforations, fuel and flare volumes, well tests, laboratory analyses of all produced fluids, pressure-volume-temperature (PVT) analyses of all recovered fluid samples, formation pressure (static bottomhole, buildup, and drawdown surveys), formation temperature, and well histories. Required economic data include the working and net receiving interests being evaluated, oil and gas prices, operating costs, appropriate severance and ad valorem taxes, and any anticipated capital costs. If an after tax case is run, knowledge of the interest owners qualifying status (i.e., integrated company, royalty owner, or independent producer) and tax rate are required. In cases where critical data are missing or unavailable to the evaluator, the evaluation report should state what is missing and assess the overall impact of the missing data on the final results.

DETERMINATION OF GROSS PRODUCTION (RESERVES)

1*

General Discussion

Methods in oil and gas reserves estimation (Arps, 1956) generally include analogy, volumetric methodology, and performance techniques. In most applications, more than one method is used for reserves determination as a means of cross-checking one estimate against another. Analogy or the comparative approach generally applies during the predrill or exploration phase of an oil or gas property. During this phase, subsurface data are limited. Reserve estimates are generally based on experience from similar fields, reservoirs, or wells in the same area. As a consequence, the range of reserve estimates tends to be subject to significant error.

Once one or more wells have been drilled and a property has been proved productive, subsurface data such as well logs, core analyses, and bottom hole pressure and fluid data become available. Using these data, volumetric estimates can be made based on subsurface maps and assumptions can be made regarding recovery efficiencies of in-place hydrocarbons based on the type of reservoir drive mechanism. Reserve estimates based on volumetric methodology tend to be more accurate than using an analogy or the

*Boldfaced numbers centered under headings throughout chapter correspond to numbered columns in Tables 1 and 2.

AS OF JANUARY 1, 1991

-END- MO-YR	GROSS PRODUCTION (1) OIL, MBBL	GROSS PRODUCTION GAS, MMCF	NET PRODUCTION (2) OIL, MBBL	NET PRODUCTION GAS, MMCF	PRICES (3) OIL $/B	PRICES GAS $/M	NO. OF WELLS	OPERATIONS, M$ (4) OIL REVENUE	GAS REVENUE	OTHER REVENUE	TOTAL REVENUE
12-91	226.300	171.988	169.725	128.991	22.50	1.80	1.0	3818.813	232.184	0.000	4050.997
12-92	91.250	69.350	68.438	52.013	23.85	1.91	1.0	1632.246	99.241	0.000	1731.487
12-93	36.500	27.740	27.375	20.805	25.28	2.02	1.0	692.067	42.078	0.000	734.145
12-94	14.600	11.096	10.950	8.322	26.80	2.14	1.0	293.437	17.841	0.000	311.278
12-95	5.840	4.438	4.380	3.329	28.41	2.27	1.0	124.417	7.565	0.000	131.982
12-96											
12-97											
12-98											
12-99											
12- 0											
S TOT	374.490	284.612	280.868	213.460	23.36	1.87	0.0	6560.980	398.909	0.000	6959.889
REM.	0.000	0.000	0.000	0.000	0.00	0.00	0.0	0.000	0.000	0.000	0.000
TOTAL	374.490	284.612	280.868	213.460	23.36	1.87	0.0	6560.980	398.909	0.000	6959.889

-END- MO-YR	OPERATIONS, M$ (5) SEV + ADV TAXES	(6) WF PROFIT TAXES	(7) NET OPER EXPENSES	(8) OPERATIONS CASH FLOW	CAPITAL COSTS, M$ (9a) TANGIBLE COSTS	(10a) INTANG. COSTS	(11a) LSEHOLD COSTS	(12) SALVAGE VALUE	(13) CASH FLOW BTAX, M$	(19a) 10.00 PCT CUM. DISC BTAX, M$
12-91	490.251	0.000	38.160	3522.586	400.000	600.000	2500.000	0.000	22.586	-141.346
12-92	209.232	0.000	40.450	1481.805	0.000	0.000	0.000	0.000	1481.805	1143.059
12-93	88.589	0.000	42.877	602.679	0.000	0.000	0.000	0.000	602.679	1617.961
12-94	37.512	0.000	45.449	228.317	0.000	0.000	0.000	0.000	228.317	1781.516
12-95	15.885	0.000	48.176	67.921	0.000	0.000	0.000	0.000	67.921	1825.748
12-96										
12-97										
12-98										
12-99										
12- 0										
S TOT	841.469	0.000	215.112	5903.308	400.000	600.000	2500.000	0.000	2403.308	1825.748
REM.	0.000	0.000	0.000	0.000	0.000	0.000	0.000	0.000	0.000	1825.748
TOTAL	841.469	0.000	215.112	5903.308	400.000	600.000	2500.000	0.000	2403.308	1825.748

Table 1. Example of "Before Tax" Reserves and Economics Schedule

AS OF JANUARY 1, 1991

	(8)	(9b)	(11b)	(10b)	(14)	(15)	(16)	(17)	(18)	(19b)
-END- MO-YR	OPER CASH FLOW, M$	DEPR. EXP., M$	DEPL. EXP., M$	INTANG. EXP., M$	INTEREST EXP., M$	TAXABLE INCOME M$	INVEST. CREDIT M$	TAXES PAYABLE M$	CASH FLOW ATAX, M$	10.00 PCT CUM. DISC ATAX, M$
12-91	3522.586	57.144	1510.721	600.000	0.000	1354.721	0.000	460.605	-438.019	-580.516
12-92	1481.805	97.960	609.162	0.000	0.000	774.683	0.000	263.392	1218.413	475.585
12-93	602.679	69.968	243.665	0.000	0.000	289.046	0.000	98.276	504.403	873.047
12-94	228.317	49.980	97.466	0.000	0.000	80.871	0.000	27.496	200.821	1016.905
12-95	67.921	124.948	38.986	0.000	0.000	-96.013	0.000	-32.644	100.565	1082.396
12-96										
12-97										
12-98										
12-99										
12- 0										
S TOT	5903.308	400.000	2500.000	600.000	0.000	2403.308	0.000	817.125	1586.183	1082.396
REM.	0.000	0.000	0.000	0.000	0.000	0.000	0.000	0.000	0.000	1082.396
TOTAL	5903.308	400.000	2500.000	600.000	0.000	2403.308	0.000	817.125	1586.183	1082.396

(20a) BTAX RATE OF RETURN (PCT)	74.50
(21a) BTAX PAYOUT YEARS	0.99
BTAX PAYOUT YEARS (DISC)	1.11
(22a) BTAX NET INCOME/INVEST	1.69
(23a) BTAX NET INCOME/INVEST (DISC)	1.52
PRODUCTION START DATE	1/ 1/91
MONTHS IN FIRST LINE	12.00
GROSS WELLS	1.00
MAX. OIL PRICE ($/B)	28.41
GROSS OIL WELLS	1.00
CUMULATIVE OIL (MBBL)	0.000
REMAINING OIL (MBBL)	374.490
ULTIMATE OIL (MBBL)	374.490
INITIAL W.I. FRACTION	1.00000000
INITIAL NET OIL FRACTION	0.75000000
INITIAL NET GAS FRACTION	0.75000000

(20b) ATAX RATE OF RETURN (PCT)	45.66
(21b) ATAX PAYOUT YEARS	1.36
ATAX PAYOUT YEARS (DISC)	1.55
(22b) ATAX NET INCOME/INVEST	1.45
(23b) ATAX NET INCOME/INVEST (DISC)	1.31
PROJECT LIFE (YEARS)	5.00
DISCOUNT RATE (PCT)	10.00
PRIOR DEPL. BASIS (M$)	0.000
PRIOR DEPR. BASIS (M$)	0.000
MAX. GAS PRICE ($/M)	2.27
GROSS GAS WELLS	0.00
CUMULATIVE GAS (MMCF)	0.000
REMAINING GAS (MMCF)	284.612
ULTIMATE GAS (MMCF)	284.612
FINAL W.I. FRACTION	1.00000000
FINAL NET OIL FRACTION	0.75000000
FINAL NET GAS FRACTION	0.75000000

----PRESENT WORTH PROFILE-----		
DISC RATE	PW OF NET BTAX, M$	PW OF NET ATAX, M$
0.0	2403.308	1586.183
2.0	2275.044	1473.972
5.0	2095.405	1317.134
8.0	1929.495	1172.614
10.0	1825.748	1082.396
12.0	1727.016	996.650
15.0	1587.542	875.698
18.0	1457.461	763.074
20.0	1375.491	692.193
25.0	1185.424	528.089
30.0	1014.002	380.371
35.0	858.529	246.615
40.0	716.808	124.856
45.0	587.021	13.484
50.0	467.667	-88.839
60.0	255.395	-270.607
70.0	72.046	-427.428
80.0	-88.175	-564.362
90.0	-229.605	-685.174
100.0	-355.536	-792.717

Table 2. Example of an "After Tax" Reserves and Economic Schedule

comparative approach given sufficient data. However, volumetric estimates need to be refined continually as additional wells are drilled, brought on production, and begin to exhibit production decline trends.

As an oil or gas property begins to exhibit production and/or pressure decline trends, performance techniques such as decline curve analysis and material balance methods can be applied to compare with previous volumetric estimates and assumptions. Given sufficient performance data, reserve estimates can be refined and accuracy increased over volumetric or analogy estimates. A high level of confidence for reserve estimates can be reached where good agreement is attained between volumetric reserve estimates and performance estimates. The reserve estimator should always consider the degree to which the geological "picture" does or does not fit the performance "picture."

Volumetric Method

Determination of Reservoir Bulk Volume

The first step in a volumetric estimate is preparation of geological maps, including structure maps, fault contour maps, and isopach maps. Structure and fault contour maps are used to determine the areal extent or productive area of the reservoir. The bulk volume of an oil or gas reservoir is determined from isopach maps that are based on the net effective saturated sand thickness in each well (Wharton, 1948). The bulk volume of a reservoir in acre-feet is usually determined from an isopach map by means of planimetric survey. Methods of volume calculations from planimeter measurement include the vertical slice and horizontal slice methods. Both methods are equally accurate, and selection of a method is usually a matter of choice. In the case of directionally drilled wells and steeply dipping beds, it is important to remember in construction of isopach maps that the calculated true vertical thickness should be used and not the measured thickness of a well simply corrected to vertical. Significant error can result in isopach construction if the true vertical thickness of a well is not used.

Volumetric Calculations

Once the isopach map is constructed and planimetered, the next step is calculation of the amount of oil and/or gas originally in place. This is followed by the application of an appropriate recovery factor and the planimetered bulk volume to determine the amount of oil and/or gas that is estimated to be ultimately recovered. It is important to remember that all reservoir engineering calculations are performed using values of absolute pressure (where psia = psig + 15) and absolute temperature (where °R = °F + 460). Here is the equation for determination of recoverable oil from a reservoir.

Recoverable oil in stock tank barrels

$$= \frac{7758(\phi)(1 - S_w)}{B_o} \text{ x R.F. x acre-feet}$$

where

ϕ	=	porosity, decimal
S_w	=	connate water saturation, decimal
B_o	=	oil formation volume factor, reservoir barrel/stock tank barrel
7758	=	the volume of barrels per acre-foot (43,560 feet^2/acre ÷ 5.615 feet^3/bbl)
R.F.	=	recovery factor, decimal
acre-feet	=	reservoir bulk volume from planimetric survey

The oil formation volume factor, B_o, accounts for the fact that under most conditions, liquid volumes in the reservoir shrink to lesser volumes as reservoir pressure declines or as the oil is produced from the reservoir to the surface and into the stock tank. Therefore, this factor is the critical component in volumetric calculations for converting reservoir barrels of oil to stock tank barrels of oil. Sources for oil formation volume factor include laboratory analyses of oil samples and correlations such as Standing's correlation (Standing, 1952), which are found in standard reservoir engineering textbooks.

The equation for determination of recoverable gas from a reservoir is as follows:

Recoverable dry gas in thousands of cubic feet (MCF)

$$= 43.56(\phi)(1 - S_w)\left(\frac{PT_{sc}}{P_{sc}TZ}\right) \text{ x R.F. x S.F. x acre-feet}$$

where

ϕ	=	porosity, decimal
S_w	=	connate water saturation, decimal
P	=	reservoir pressure, psia
T	=	reservoir temperature, °R (= °F + 460)
P_{sc}	=	pressure, standard conditions (14.65, 14.73, or 15.025 psia depending on the required pressure base)
T_{sc}	=	temperature, standard conditions, 60°F or 520°R
Z	=	gas deviation factor (compressibility factor)
43.56	=	number of thousands of cubic feet per acre-foot
R.F.	=	recovery factor, decimal
S.F.	=	shrinkage factor, decimal
acre-feet	=	reservoir bulk volume from planimetric survey

The gas volumetric calculation can be simplified depending on the pressure base being used, as follows:

Recoverable dry gas in thousands of cubic feet (MCF)
= 1507.57(ϕ)(1 – S_w)(P/TZ) x R.F. x S.F. x acre-feet for a 15.025 psia pressure base

= 1546.16(ϕ)(1 – S_w)(P/TZ) x R.F. x S.F. x acre-feet for a 14.65 psia pressure base

= 1537.76(ϕ)(1 – S_w)(P/TZ) x R.F. x S.F. x acre-feet for a 14.73 psia pressure base

Note that for Texas, Oklahoma, Alabama, and Kansas, pressure base equals 14.65 psia. For Colorado, Louisiana, Nebraska, Mississippi, Montana, New Mexico, and Wyoming, pressure base equals 15.025 psia. For California and Federal Offshore, pressure base equals 14.73 psia.

The gas deviation factor (Z) or compressibility factor, as seen in the gas volumetric calculation, accounts for the fact that hydrocarbon gases are nonideal contrary to Boyle's and Charles' laws. Whereas fluids such as oil and water exhibit little or no volume change with pressure, gas exhibits significant volume change with pressure and temperature and must be corrected for nonideal behavior for use in engineering calculations. The first step in determining the gas deviation factor is determination of the full well stream gas gravity (i.e., gravity of the gas at reservoir temperature and pressure). The full well stream gas gravity can be determined from laboratory data or it can be determined from the following calculation if only the separator gas gravity is known:

Specific gravity of the full well stream (15.025 psia pressure base)

$$= \frac{G_s + \dfrac{\overset{(1)}{4493} \times G_c}{R_c}}{1 + (V_c / R_c)}$$

where G_s = specific gravity of the separator gas
G_c = specific gravity of the separator condensate
= 141.5 ÷ (°API + 131.5)
R_c = gas condensate ratio, feet^3/barrel
V_c = condensate vaporizing volume ratio
= $\overset{(2)}{(2938.81)}$ x (1.03 - G_c)

Note that the calculations can easily be adjusted with the following factors depending on the pressure base being used.

	(1)	(2)
15.025 psia pressure base	4493	2938.81
14.73 psia pressure base	4583	2997.65
14.65 psia pressure base	4608	3014.02

After determining the specific gravity of the full well stream, the pseudocritical pressure and temperature can be determined from a chart of pseudocritical properties of condensate well fluids and miscellaneous natural gases prepared by Brown et al. (1948), which can be found in standard reservoir engineering textbooks. The pseudocritical pressure and temperature values obtained from the Brown et al. chart are then divided into the actual reservoir pressure and temperature, respectively, to give dimensionless values of pseudoreduced pressure and temperature. These pseudoreduced values of pressure and temperature are then used to enter a chart of compressibility factors for natural gases prepared by Standing and Katz (1942) to determine the gas deviation factor (Z) or compressibility factor. The chart by Standing and Katz can also be found in standard reservoir engineering textbooks.

The shrinkage factor (S.F.) as used in the gas volumetric calculation is similar to the oil formation volume factor. When a full well stream gas is produced from the reservoir to the surface, it goes through a condensation process whereby the condensate liquids separate from the upper-end gas components (methane, ethane, propane, butane, etc.). For purposes of volumetric calculations, it is necessary to reduce the calculated recoverable gas by a factor that accounts for the shrinkage of separator gas from full well stream gas. This factor can be determined from published correlations or can be calculated using the following equation (see specific gravity of the full well stream calculation for description of variables):

$$\text{Shrinkage factor (S.F.)} = \frac{R_c}{R_c + V_c}$$

In almost all oil and gas property evaluation applications, it is necessary to estimate secondary phase production (solution gas and condensate) as well as primary phase production (oil and gas). This is usually the case because income is derived from secondary phase production. When only limited performance data are available, solution gas or dissolved gas can be determined by multiplying the solution gas to oil ratio by the calculated recoverable oil, as shown in the following equation:

$$\text{Recoverable dissolved gas (MCF)} = (RN)(R_{si})/1000$$

where RN = recoverable oil, STB
R_{si} = solution gas to oil ratio, feet^3/bbl

Solution gas is the amount of gas dissolved in the oil. As pressure falls below the saturation pressure, or "bubble point", of the oil, the amount of gas in solution in the oil decreases rapidly with pressure decline, and conversely, the amount of free gas associated with the oil increases. The solution gas to oil ratio can be determined from correlations (Standing, 1952) found in standard reservoir engineering textbooks if laboratory analyses of the reservoir fluids are not available. In many cases, the initial solution gas to oil ratio is sufficiently close enough to the producing gas to oil ratio of the initial production so that they can be used interchangeably. It is important to note that the previous equation for calculating recoverable dissolved gas applies primarily to oil reservoirs in which reservoir pressure is maintained at a relatively high level as a result of aquifer encroachment and/or gas cap expansion. In the case where the drive mechanism is primarily by dissolved gas, solution gas recovery may significantly exceed the product of the solution gas to oil ratio times the calculated oil recovery when reservoir pressure drops below the saturation pressure. In this case, it is necessary to use other methods for forecasting secondary phase recovery.

For purposes of estimating secondary phase recovery in gas reservoirs under conditions of limited performance, the recoverable condensate is generally determined by multiplying the average condensate yield in barrels per million cubic feet (MMCF) of gas by the recoverable dry gas, as shown:

Recoverable condensate (bbl)
= Recoverable dry gas (MMCF)
x Condensate yield (bbl/MMCF)

Drive Mechanisms and Recovery Factors

The final step in the volumetric method is application of recovery factors to the calculated volumes of oil or gas originally in place. Knowledge of the type of reservoir drive mechanism is essential in determining the appropriate recovery factor to be applied, as recovery efficiencies are dependent on the drive type. The following table presents a general guideline for estimated ultimate recoveries from oil and gas reservoirs for a given drive mechanism:

Reservoir Type	Drive Type	Ultimate Recovery of HCs Originally in Place (%)
Oil	Dissolved gas	5–20
Oil	Gas cap	20–40
Oil	Water	30–60
Oil	Gravity drainage	25–80
Gas	Gas expansion	50–90
Gas	Water	40–75

Note that it is common for more than one of these drive mechanisms to occur at one time in a producing reservoir. For example, a partial water drive mechanism can exist in a producing oil or gas reservoir where limited water influx from an associated aquifer and expansion of dissolved gas occur simultaneously.

As the ranges of recovery factors suggest, knowledge of the drive type is essential to accurate reserve estimation. An incorrect assumption of the type of drive in a volumetric estimate of reserves can result in significant error. In the early producing life of an oil or gas property, it is often difficult, if not impossible, to define the type of drive in a reservoir. In those cases, a reserve estimator is obligated to assume the least efficient drive mechanism when assigning proved reserves (that is, dissolved gas drive for oil reservoirs and water drive for gas reservoirs).

For an oil reservoir, a *water drive* occurs when water from an adjacent aquifer expands into a portion of the reservoir as the reservoir is produced. The rate at which water expands into the reservoir is a function of the size of the aquifer and the permeability of the formation. In most cases, pressure decline is arrested to some degree, and oil recoveries are increased as the water expands and "pushes" oil out of the reservoir rock. The recovery factor calculation for determination of recoverable oil from oil originally in place for a strong water drive is

$$R.F. = \frac{(1 - S_w - S_{or})}{1 - S_w} \times S.E$$

where S_w = connate water saturation, decimal
S_{or} = residual oil saturation, decimal
S.E. = sweep efficiency of water drive, decimal

Sweep efficiency (S.E.) is defined as the ratio of the volume of the reservoir swept by encroaching water at any time to the total volume of the reservoir subject to invasion. Residual oil saturation (S_{or}), expressed as a fraction of the pore volume, is the fraction of oil remaining after water displacement. Residual oil saturation can be measured from log or core data, or a correlation can be used between reservoir oil viscosity, average reservoir permeability, and residual oil saturation as determined by Craze and Buckley (1945) and Arps (1956).

A *gas cap drive* in an oil reservoir occurs when a primary gas cap expands into the underlying oil zone as a result of declining reservoir pressure due to production. Increased oil recoveries occur as the expanding gas cap pushes more oil out of the oil zone than could be recovered by expansion of dissolved gas in the oil alone (Morgan, 1967). The effectiveness of a gas cap drive is dependent upon the size of the gas cap and the absence of a water drive, which tends to stabilize pressure and halt expansion of the gas cap. A large gas cap relative to the size of the oil zone is more effective in maintaining pressure in the oil reservoir and increasing oil recoveries. Estimation of recovery efficiencies in gas cap drive oil reservoirs is not a trivial procedure. Often it is necessary to use reservoir simulation methods to estimate recovery efficiency with any accuracy.

Dissolved gas drives (also referred to as *solution gas drives* or *depletion drives*) in oil reservoirs are characterized by the release of dissolved gas from the oil as the reservoir pressure drops below the saturation pressure (Morgan, 1967). Oil is driven to the well bore by the movement and expansion of the released gas. As the reservoir pressure continues to drop, more and more gas is released from the oil in the reservoir. As the gas saturation increases in the reservoir, the flow of gas in the oil zone increases. The produced gas to oil ratio will increase steadily, reaching a peak after about one-half of the ultimate oil recovery has been produced. The peak gas to oil ratio is generally several times greater than the original solution gas to oil ratio. Once the peak ratio has been reached, the gas to oil ratio begins to decrease.

Generally, a dissolved gas drive is the least efficient drive possible in an oil reservoir. A dissolved gas drive should be assumed for proved reserve estimation purposes if performance data do not indicate a more efficient drive mechanism. The recovery factor calculation for determination of recoverable oil from oil originally in place for a dissolved gas drive is as follows:

$$R.F. = 1 - \frac{(1 - S_w - S_{ga})(B_{oi})}{(1 - S_w)(B_{oa})}$$

where
S_w = connate water saturation, decimal
S_{ga} = gas saturation at abandonment, decimal
B_{oi} = initial formation volume factor, reservoir barrel/stock tank barrel
B_{oa} = abandonment formation volume factor, reservoir barrel/stock tank barrel

The recovery factor for an oil reservoir above the saturation or bubble point pressure that produces by means of liquid expansion of the oil only is as follows:

$$R.F. = \frac{(B_{ob} - B_{oi})}{B_{ob}}$$

where
B_{oi} = initial formation volume factor, reservoir barrel/stock tank barrel
B_{ob} = formation volume factor at the bubble point pressure, reservoir barrel/stock tank barrel

A *gravity drainage drive* in an oil reservoir is the movement of oil downward due to the effects of gravity as oil is withdrawn from the lower portion of the reservoir (Morgan, 1967). Economic recovery of oil from a gravity drainage drive will occur only when the vertical force of gravity on the oil exceeds the effect of capillary retention and any other force and obstacles resisting the movement of oil in the formation. Dissolved gas in a gravity drainage drive reservoir moves upward as it is released from solution. The gas accumulates in the gas cap or forms a secondary gas cap, which results in a pressure maintenance effect in the reservoir similar to that produced by a gas cap. This explains the high ultimate recoveries sometimes associated with gravity drainage reservoirs. As with gas cap drive reservoirs, estimation of recovery efficiencies in gravity drainage reservoirs is not a trivial procedure. Often it is necessary to use reservoir simulation methods to estimate recovery efficiency with any accuracy.

Recovery efficiencies in combination drive oil reservoirs can be difficult to quantify with accuracy. One approach for a partial water drive oil reservoir would be to calculate separately the recovery factors for effective water drive and dissolved gas drive. The reserve estimator would then have to make a judgment, based on reservoir performance and/or analogy, of the degree to which each mechanism was influencing recovery from the reservoir. For example, if the calculated recovery factors for dissolved gas drive and effective water drive were 10% and 50%, respectively, and it was the reserve estimator's judgment that each mechanism was expected to contribute 50% each to the total recovery efficiency, then the adjusted recovery factor for combination drive would be (0.10 x 0.50) + (0.50 x 0.50) = 0.30, or 30%.

The two basic drive mechanisms by which gas reservoirs are produced are gas expansion drive and water drive. A *gas expansion drive* in a gas reservoir is characterized by the expansion of compressed gas remaining in the reservoir as the pressure declines when part of the gas is produced (Morgan, 1967). As the reservoir pressure decreases, the productivity of wells completed in a gas expansion drive reservoir also declines. Ultimate recovery from this type of reservoir is a function of the reservoir pressure at which wells cease to produce at economical rates. Original gas in place can be determined from an extrapolation of the pressure versus cumulative curve to zero pressure once the pressure data has been adjusted for compressibility by dividing by the gas deviation factor (Z). The recovery factor calculation for determination of recoverable gas from gas originally in place for a gas expansion drive is as follows:

$$R.F. = \frac{P_i / Z_i - P_a / Z_a}{P_i / Z_i} \times C.F.$$

where P_i/Z_i = initial P/Z, psia
P_a/Z_a = abandonment P/Z, psia
C.F. = conformance factor, decimal

A conformance factor is used in the formula to account for heterogeneous formation conditions that may hinder reservoir communication.

A *water drive in a gas reservoir* is analogous to a water drive in an oil reservoir. Reservoir pressure maintenance results as encroaching water fills the part of the reservoir space originally occupied by gas. Recovery efficiencies in water drive gas reservoirs tend to be less than in gas expansion drive reservoirs due to trapping of the residual gas at higher pressures. Also, gas under pressure has characteristics similar to those of light oil and tends to resist complete displacement by water from the pore space. Finally, the efficiency of water drives is hindered by the wider spacing of wells in gas reservoirs as opposed to oil reservoirs which allow for the encroaching water to bypass more gas. The recovery factor calculation for an effective water drive in a gas reservoir is as follows:

$$R.F. = \frac{(1 - S_w - S_{gr})(S.E.)}{1 - S_w}$$

where S_w = connate water saturation, decimal
S_{gr} = residual gas saturation, decimal
S.E. = sweep efficiency of water drive, decimal

Residual gas saturation, expressed as a fraction of the pore volume, is the fraction of the gas remaining after water displacement. Residual gas saturation can be determined from laboratory measurements or from relationships of residual gas saturation versus porosity, as presented by Katz et al. (1966) and others.

The recovery factor calculation for a partial water drive with bottom hole pressure decline in a gas reservoir is as follows:

$$R.F. = \frac{(1 - S_w)(P_i / Z_i) - S_{gr}(\overline{B}_g)}{(1 - S_w)(P_i / Z_i)}(S.E.) + \frac{(P_i / Z_i) - (P_a / Z_a)}{(P_i / Z_i)}(1 - S.E.)$$

where S_w = connate water saturation, decimal
S_{gr} = residual gas saturation, decimal
P_i/Z_i = initial P/Z, psia
P_a/Z_a = abandonment P/Z, psia
$\overline{B}_g = \frac{(P_i / Z_i) + (P_a / Z_a)}{2}$
S.E. = sweep efficiency of water drive, decimal

Performance Methods

It has long been recognized that a producing hydrocarbon reservoir is a depleting resource that will exhibit declining production and/or pressure at some point in its producing life. Estimates of remaining recoverable hydrocarbons can be made from analyses and extrapolation of these declining performance trends.

The performance evaluation phase follows after sufficient actual performance data on the property have been available to make a check of previous volumetric estimates against decline curve trend analyses. Also, the pressure behavior may now make material balance work possible and give insight into the type of production mechanism, thereby giving a more accurate determination of the ultimate recovery. The performance methods most often used in estimating reserves of oil and gas recoverable under primary methods are as follows:

- Production decline curve methods
- Cumulative production curve methods
- Material balance methods
- Simulation methods

Each of these methods depends on a projection of past production trends and operating conditions into the future. The choice of a performance evaluation method is generally a function of the reliability of the data. For example, material balance methods for oil are not commonly used due to the lack of reliable pressure and fluid data. Production decline curves, however, are commonly used because the only requirement is monthly production. Reservoir simulation is another method that generally requires reliable data.

Production Decline Curve Methods

Any production decline method must be used with a degree of caution and understanding when predicting reserves and ultimate recoveries. Accurate production records are essential for accurate reserve estimating. Factors that can affect changes in rates of production and must be considered when using decline curve methodology are listed here:

- Completion of offset wells
- Gas lift installation

- Workovers
- Pump installation
- Proration
- Change in choke size
- Effect of reservoir drive mechanism

The reserve estimator should attempt to account for the causes of these irregularities in the reserve estimating process. More accurate reserve estimates using decline curve methods will result from extrapolation of periods of regular decline in which few or no changes in operating conditions have occurred.

The four types of decline curves shown in Figure 1 include constant percentage decline (exponential), hyperbolic decline, linear decline, and harmonic decline. Most production rate time graphs or decline curves have been found to exhibit either exponential decline or hyperbolic decline. Some well performance over an extended production life, in fact, can exhibit a combination of decline type curves. That is, an exponential decline may give the best fit over the first half of the producing life of a well, whereas harmonic decline gives the best fit over the last half.

The *exponential decline curve* is plotted on semilogarithmic graph paper with production rates plotted on the logarithmic scale versus time (generally, in months) plotted on the Cartesian scale. With exponential decline, the drop in production per unit of time is a constant fraction of the production rate. The relevant equations for exponential decline analysis are as follows:

Reserves (MCF or bbl) on decline

$$= \frac{\dfrac{12\,(q_i - q_{el})}{\ln(q_i / q_{el})}}{L}$$

Life (years) on decline

$$= \ln(q_i / q_{el}) \times \frac{R}{12\,(q_i - q_{el})}$$

Annual decline rate (%)

$$= 1 - e^{\frac{-\ln(q_i/q_{el})}{L}} \times 100$$

where
- q_i = initial production rate, MCF or bbl per month
- q_{el} = final or economic limit production rate, MCF or bbl per month
- L = life (years) on decline
- R = reserves on decline, MCF or bbl
- ln = natural logarithm

An example of decline curve analysis of a well exhibiting exponential or constant percentage decline is

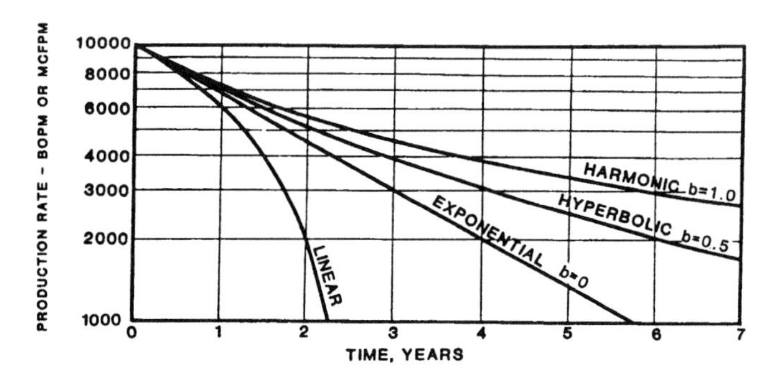

Figure 1. Semilogarithmic production rate versus time decline curve plots.

shown in Figure 2.

The exponential decline method is probably used more often than any other method due to its ease of use. Also, reservoirs generally exhibit exponential decline during the early stages of their declining productive lives. Therefore, forecasting by means of exponential extrapolation is straightforward and less subject to error.

Hyperbolic decline curves resemble hyperbolas when plotted on a semilogarithmic plot, as shown in Figure 1. The hyperbolic curve tends to flatten out with time. Recall that with exponential decline, the drop in production per unit of time is a constant fraction of the production rate. With hyperbolic decline, however, the drop in production per unit of time as a fraction of the production rate is proportional to a fractional power of the production rate. Areas of production that exhibit hyperbolic decline include the Medina Formation in New York, the Sprayberry and Austin Chalk trends in Texas, and the Denver–Julesburg basin in Colorado. The calculated reserve on decline for hyperbolic performance is as follows:

Reserves on decline (MCF or bbl)

$$= \left[\frac{q_i^b}{(a_i)(1-b)}\right]\left(q_i^{1-b} - q_{el}^{1-b}\right)$$

where
- q_i = initial production rate, MCF or bbl per month
- q_{el} = final or economic limit production rate, MCF or bbl per month
- a_i = the rate of decline (slope) at the beginning of decline
- b = hyperbolic exponent, the value necessary to fit the data during the hyperbolic portion of decline

The biggest problem with hyperbolic forecasting and analysis is the determination of the hyperbolic exponent

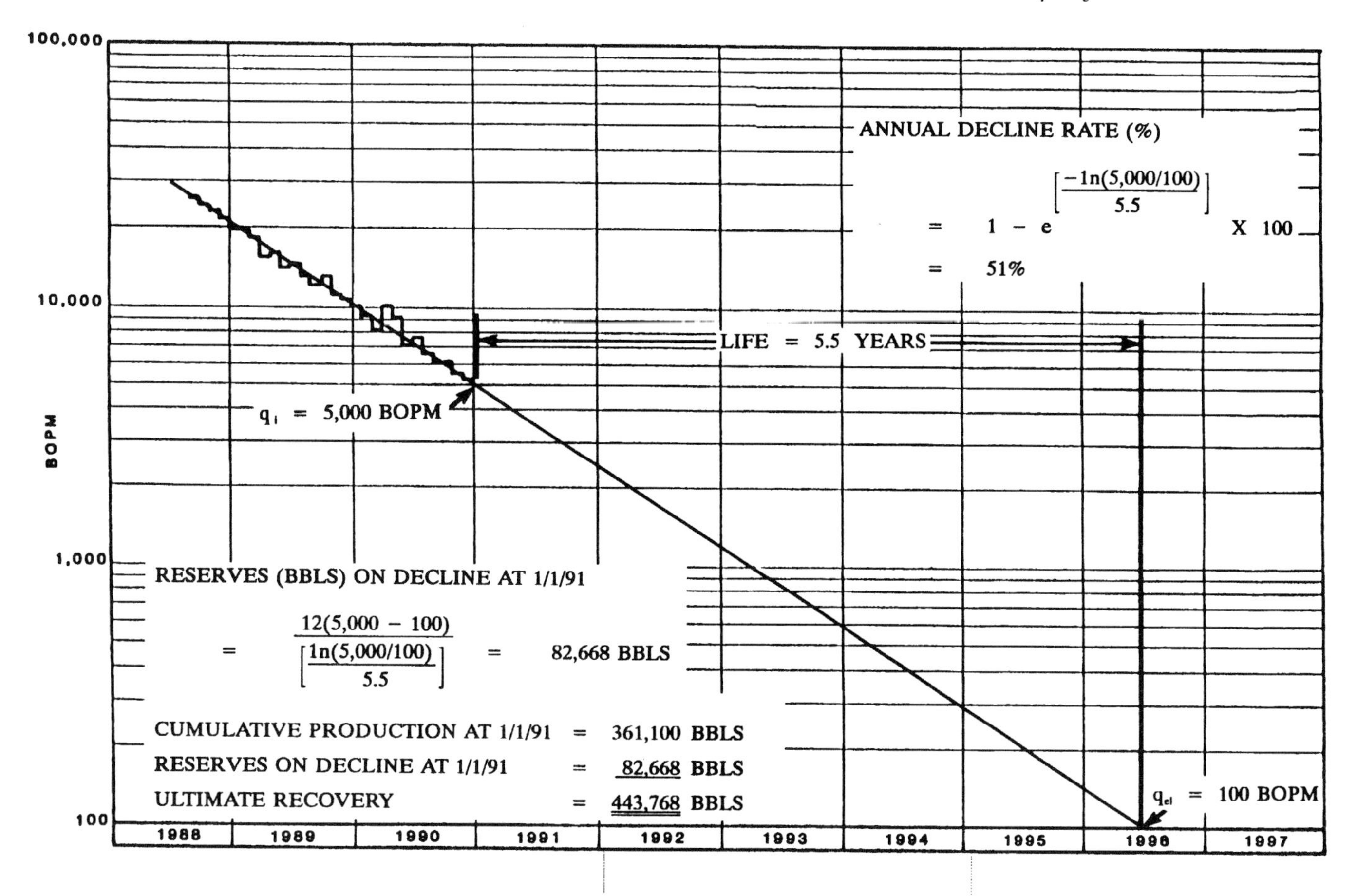

Figure 2. Example of exponential decline curve analysis.

b. The value of *b* can be different for different reservoirs and can vary widely over the life of a given reservoir as production changes occur. Estimates of *b* can be made by fitting existing data or using a value based on analogy. Although several techniques are available for evaluating *b*, the technique most commonly used is various iterations on *b* to evaluate the equation of the decline curve. Frequently, an evaluator will start with an arbitrary value of *b* that falls between 0 and 1, such as 0.5. Despite the difficulty of using hyperbolic decline analysis, it is true that most reservoirs follow some form of hyperbolic decline, particularly in the latter stages of depletion.

Another form of production behavior is *linear decline*, which exhibits a straight line when plotted on Cartesian coordinates as a function of time. With linear decline, the drop in production per unit of time is constant. Such behavior can sometimes be attributed to gravity drainage or an effective water drive. Water production upon completion is often associated with this kind of performance. The calculated reserve for linear decline performance is as follows:

$$\text{Reserves on decline (MCF or bbl)} = \frac{q_i + q_{el}}{2} \times L \times 12$$

where

q_i = initial production rate, MCF or bbl per month

q_{el} = final or economic limit production rate, MCF or bbl per month

L = life (years) on decline

For *harmonic decline*, the decline rate is proportional to the producing rate. The calculated reserve on decline for harmonic performance is as follows:

$$\text{Reserves on decline (MCF or bbl)} = \frac{q_i}{a_i} \ln \frac{q_i}{q_{el}}$$

where

q_i = initial production rate, MCF or bbl per month

q_{el} = final or economic limit production rate, MCF or bbl per month

a_i = the rate of decline (slope) at the beginning of decline

ln = natural logarithm

Table 3. Economic Limit Calculation for an Oil or Gas Well

Factor	Amount
Oil Economic Limit	
1. Oil Value	_____$/bbl
2. Product Value	_____$/bbl
3. Gas Value ($/MCF) x (MCF/Bbl)	_____$/bbl
4. Total Product Value (1) + (2) + (3)	_____$/bbl
5. Production Tax	_____$/bbl
6. Windfall Profit Tax	_____$/bbl
7. Value After Tax (4) – (5) – (6)	_____$/bbl
8. Net Value After Royalty (7) x (1 – RI)	_____$/bbl
9. Operating Expense	_____$/well/mo
10. Economic Limit (9) ÷ (8)	_____bbl/well/mo
Gas Economic Limit	
1. Gas Value	_____$/MCF
2. Product Value	_____$/MCF
3. Condensate Value ($/Bbl) x (Bbl/MCF)	_____$/MCF
4. Total Product Value (1) + (2) + (3)	_____$/MCF
5. Production Tax	_____$/MCF
6. Windfall Profit Tax	_____$/MCF
7. Value After Tax (4) – (5) – (6)	_____$/MCF
8. Net Value After Royalty (7) x (1 – RI)	_____$/MCF
9. Operating Expense	_____$/well/mo
10. Economic Limit (9) ÷ (8)	_____MCF/well/mo

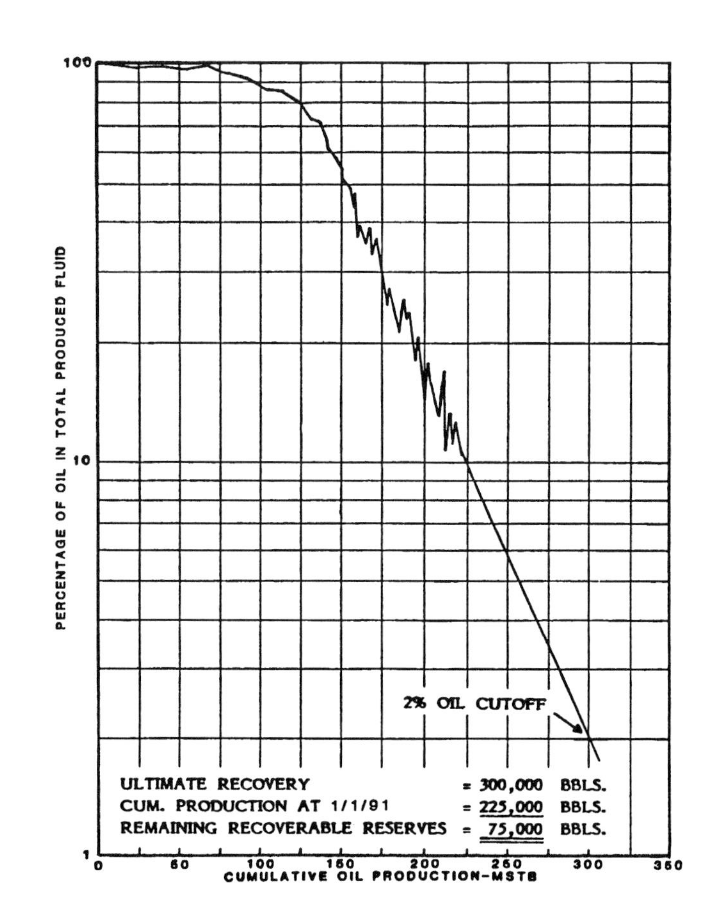

Figure 3. Example plot of percentage of oil in total produced fluid versus cumulative production.

Harmonic decline is exhibited by some reservoirs such as those involving gravity drainage. The Diatomite reservoir in the Belridge Field, Kern County, California, under primary depletion is an example of harmonic decline. Generally, however, harmonic decline is not recognized until late in the productive life of a reservoir. Therefore, harmonic decline analysis is not widely employed. Characteristic of harmonic decline is the straight line relationship between cumulative production and production rate on semilog paper.

As indicated in the decline curve equations previously presented, production decline curves are generally extrapolated to a cut-off point that represents an economic or physical limit. The economic limit is considered to be the point at which the operating costs are equal to the income from production. Oil wells are generally abandoned due to the costs of overall lifting, clean out, and/or treatment exceeding the value of produced oil. In addition, abandonment of a well may be desirable if an offset well can be more strategically located to produce the oil of a marginal producer. The calculations for estimating the economic limit of a producing oil or gas well are shown in Table 3.

Cumulative Production Curve Methods

Cumulative production curves from wells, leases, and fields versus oil percentage in total fluid or production rate are convenient methods for estimating ultimate recoveries graphically. Estimates of remaining recovery can also be made by subtracting current cumulative production from the estimate of ultimate recovery. As with production decline curve analysis, extreme care should be taken to understand fully the factors that are contributing to the declining production rate.

Plotting the percentage of oil in the total produced fluid versus cumulative production on semilog paper is a popular way for calculating ultimate recoveries in water drive reservoirs. The example plot in Figure 3 shows how a well producing from a water drive reservoir frequently exhibits a declining oil percentage as the water sweeps through the portion of the reservoir from which the well is producing. This declining oil percentage trend can be extrapolated to an "economic limit" percentage cutoff point for determination of ultimate recovery. This percentage cutoff point represents the lowest oil percentage, combined with the fluid producing capacity of the lease, at which income from production equals operating costs.

Figure 4 shows a plot of production rate versus cumulative production. The straight line behavior plotted on Cartesian coordinates is indicative of

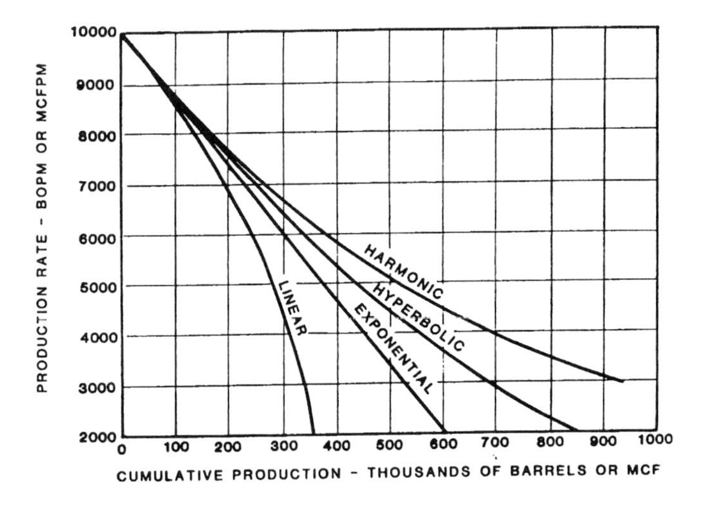

Figure 4. Coordinate production rate versus cumulative production decline curve plots.

constant percentage decline and can be extrapolated to an economic limit for determination of ultimate recovery. Straight line behavior exhibited on semilogarithmic coordinates with production rates on the logarithmic axis, as shown in Figure 5, is indicative of harmonic decline. Extrapolation of this straight line behavior gives estimates of remaining and ultimate recoveries for harmonic decline. Hyperbolic decline exhibited on production rate versus cumulative production curves can be adjusted to a straight line for extrapolation proposes using specially prepared graph paper. Also, preconstructed idealized hyperbolic decline curves plotted on transparent paper can be used for evaluating hyperbolic declines.

When using cumulative production curves, it is probably best to plot all decline curves on semilog graph paper to identify rate time changes easily. As additional production data becomes available, the curves should be updated and the reserve estimates adjusted to incorporate the current identifiable production trends.

Finally, cumulative production curves can be used to estimate ultimate recoveries of the secondary phases of production, including dissolved gas and condensate. The log of cumulative liquid production (oil and condensate) is plotted versus the log of its associated cumulative gas production (dissolved gas or nonassociated gas), and the established trend is extrapolated. With knowledge of ultimate recovery of the primary phase (oil or nonassociated gas) from volumetrics or performance evaluation, the extrapolated trend of the cumulative production curve can be used to forecast the ultimate secondary phase recovery (dissolved gas or condensate). This method is subject to significant error during the early productive period of a reservoir and should be used cautiously.

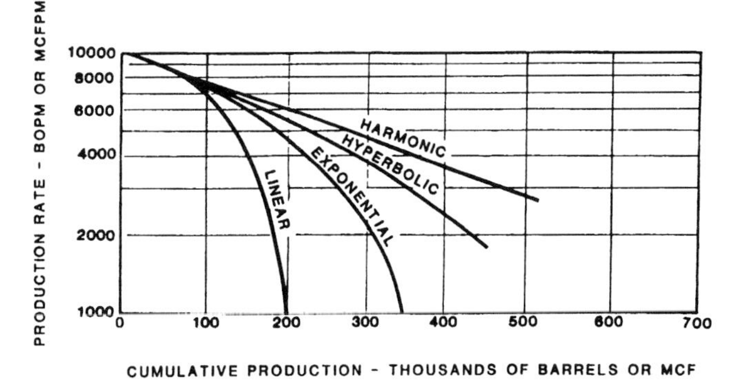

Figure 5. Semilogarithmic production rate versus cumulative production decline curve plots.

Material Balance Methods

The material balance method is another means of estimating original oil and gas in place in a reservoir to which a recovery factor must be applied to estimate ultimate recovery. The principle of material balance represents an application of the conservation of matter in a reservoir, which states that the volume of reservoir fluids produced plus the amount remaining in a reservoir must be equal to the amount present initially. The material balance equations are complicated, and discussion of the many applications of the methodology is beyond the scope of this chapter. Suffice it to say that the material balance method is highly dependent upon the amount and relative accuracy of the data required to perform the calculations. An essential data requirement for a reservoir study is knowledge of the amount of water encroachment at various times in the producing life of the reservoir being studied. Frequently, it is difficult to estimate water encroachment with any great accuracy. Therefore, this can be one area that affects the material balance and raises questions about the validity of the results.

One of the methods for analyzing data from a nonassociated gas reservoir is called the *pressure decline method*. It is based on a gas material balance and makes use of a P/Z versus cumulative gas production curve. The first step in this analysis is measurement of shut-in bottom hole pressure. The bottom hole pressures are then adjusted for nonideal gas behavior using the appropriate compressibility factor for each pressure. Again, Z factors are calculated using the specific gravity for raw or full well stream gas as opposed to dry or separator gas.

The pressures (adjusted for compressibility) are plotted versus cumulative production on Cartesian coordinate paper. For depletion drive reservoirs, the pressure decline trend can be extrapolated to zero pressure to give an approximation of initial gas in place, as shown in Figure 6. This initial gas in place value

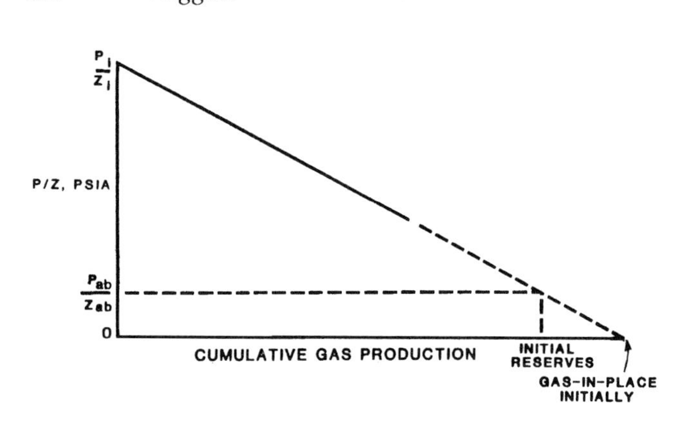

Figure 6. P/Z versus cumulative production plot.

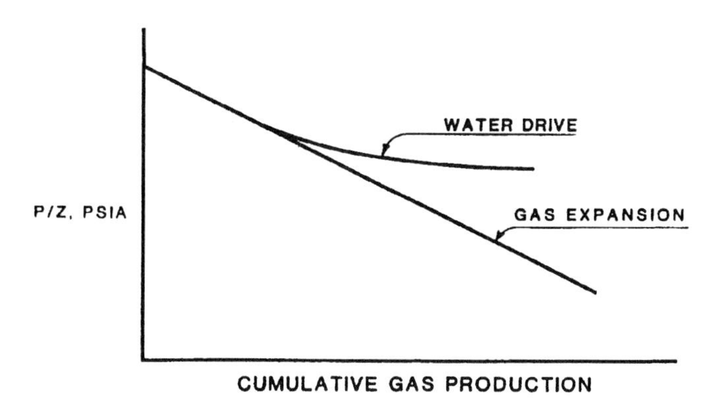

Figure 7. Comparison of gas expansion drive versus water drive on a P/Z versus cumulative production plot.

should compare closely with the volumetrically calculated initial gas in place, assuming that sufficient well control is available to define the reservoir limits. Ultimate recovery from a depletion gas reservoir is restricted by surface conditions (availability of compression) and generally ranges from 50 to 90% of gas initially in place. Abandonment P/Z should be based on actual field conditions (that is, surface line pressure and whether or not compression is available).

Figure 7 shows the effect of a water drive on a P/Z versus cumulative production curve. Water encroachment into the reservoir tends to maintain the reservoir pressure so that it does not drop as rapidly as it would in a gas expansion drive reservoir. Significant error can result from basing reserve estimates on P/Z extrapolation by assuming a depletion drive reservoir during the early productive life of a reservoir when, in fact, a part of the drive mechanism is water encroachment. The volumetric method should generally be relied upon for water drive reservoirs. Other material balance methods exist for calculating gas in place for a water drive; however, a discussion of their use is beyond the scope of this chapter.

For abnormally high pressured reservoirs, the extrapolation of a P/Z versus cumulative production curve can also lead to significant error. A P/Z curve for an abnormally high pressured reservoir can exhibit two slopes during the productive life of the reservoir. The first slope of the curve extends from the initial abnormal pressure to a pressure approximating a normal gradient. This is due to the compaction of the bulk volume and expansion of the sand grains during the production period above normal pressure, which influences reservoir behavior. Below normal pressure, the P/Z curve exhibits a much steeper slope. Significant error can result from extrapolation of the initial P/Z slope. Hammerlindl (1971) and others have presented methods for adjusting a P/Z versus cumulative production curve to determine reserve estimates for abnormally high pressured reservoirs that exhibit the two-slope phenomena previously discussed.

Scheduling of Reserves

The principle means for forecasting future production rates is decline curve analysis. However, decline curve analysis is not nearly as straightforward as it first appears, and care should be exercised to ensure proper use of decline curve methodology. Deliverability calculations and plots of the log of cumulative gas production versus the log of cumulative liquid production are other useful techniques for forecasting future production rates of gas production and secondary streams (dissolved gas or condensate).

Well head deliverability calculations are a method for estimating the capacity of a well to deliver gas into a pipeline under specified conditions. The usual well head deliverability testing method calls for shutting in the well for a sufficient period of time to allow for determination of the static well head pressure. After the static well head pressure is recorded, the well is opened to flow at the lowest test flow rate. When the well head flowing pressure has stabilized, the necessary data is recorded and the well is opened to the next higher flow rate. This procedure is repeated until data have been recorded for four flow rates. The well head deliverability curve is plotted from the measured data on logarithmic–logarithmic graph paper, as shown in Figure 8 using the following empirical equation:

$$q_g = C(P_{ts}^2 - P_{tf}^2)^n$$

where

q_g = flow rate at standard conditions, MMCF/day

P_{ts} = shut-in pressure at wellhead, psig

P_{tf} = flowing pressure at wellhead measured on a flowing column of gas, psig

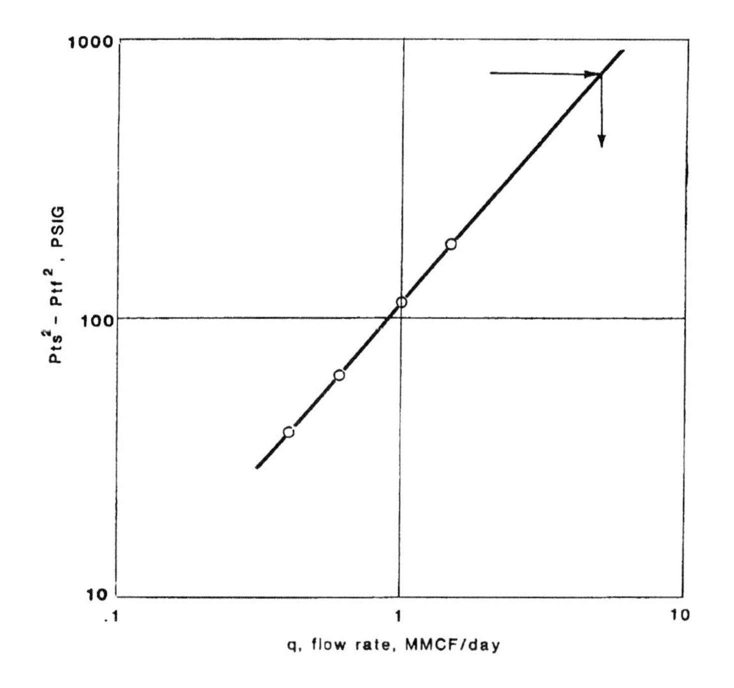

Figure 8. Multipoint test showing well head performance.

C = performance coefficient
n = performance exponent

The maximum well head flow rate against a given back pressure is read by reading the extrapolated curve through the measured data points at a point corresponding to the square of well head shut-in pressure minus the square of the back pressure. In most cases, this back pressure is the pipeline pressure that the flowing pressure of the producing well must overcome for production to enter the pipeline. It is important to note that the application of these well head deliverability calculations is limited to gas wells that are producing little or no liquids.

In the early phase of a producing property before declining production has commenced, it is necessary to make assumptions of when declining production will commence and at what rate of decline. Analogy to performance from adjacent and/or similar properties can often be used to assist in the forecast. A production schedule should reflect reasonable scheduling of production over the life of the field or property reserves. Properties with forecasted lives in excess of 20 years should be reviewed closely because they may represent candidates for acceleration projects at some future time. Again, the importance of accurate production forecasting cannot be emphasized enough. As shall be seen, time value of money concepts are extremely sensitive to cash flows in the first few years of production. A significant error in a production rate forecast during the early life of production can translate into significant error in the economic analysis.

DETERMINATION OF NET PRODUCTION (RESERVES)

2*

The production schedule consists of forecasted annual production rates for oil, gas, and/or natural gas liquids. These forecasted rates represent gross production rates and must be adjusted to account for the amount of production reserved for the royalty interest, an overriding royalty, or any other structured arrangement as set out in the lease agreement. For example, if the lease agreement sets out a 1/8 royalty to the landowner, then the total working interest production would be 7/8 of the gross production.

The net revenue interest is the particular economic interest being evaluated, which could include a working interest (adjusted for royalty), a royalty interest, an overriding royalty, or a carried interest. A *working interest* is an interest in minerals in place that bears the development and operating costs of the property. A *royalty interest* is a right to oil and gas or minerals in place that entitles its owner to a specific fraction, in kind or in value, of the total production from the property, free of development and operating expenses. The net revenue interest (NRI) for a working interest being evaluated is calculated by the following equation:

$$\text{NRI} = \text{Working interest} \times (1 - \text{Royalty interest})$$

Gross production as shown in Column 1 of Table 1 is multiplied by the net receiving interest to give net production.

The net production columns for oil and gas should also reflect adjustments for shrinkage and field use. Thermal shrinkage of oil and basic sediment and water (BS & W) can account for as much as 7% of the original fluid produced after primary separation. Additional shrinkage of gas after primary separation can occur as a result of further processing of gas to remove ethanes, propanes, and butanes in addition to contaminants such as hydrogen sulfide and carbon dioxide. Finally, volumes of oil and/or gas are often consumed on lease or in the field as fuel for production equipment and facilities. All shrinkage and field use must be accounted for in the net production columns in order to differentiate between production and sales. In summary, the net production column should represent

*Boldfaced numbers centered under headings throughout chapter correspond to numbered columns in Tables 1 and 2.

the projected sales volumes of oil and gas that are net to the economic interest being evaluated.

PRICES
3

Once the net revenue interest production has been forecasted, the next step is to determine the revenue to be derived from future production estimates. This is done by multiplying the net revenue interest production by the appropriate crude oil and/or natural gas value.

Crude oil prices are effected by the API gravity and sulfur content of the oil. The API gravity is usually determined by the use of a hydrometer, which can read either specific gravity or °API. The specific gravity of an oil is defined as the density of the oil divided of an equal volume of water at standard temperature. The American Petroleum Institute (API) has adopted a dimensionless specific gravity relationship for crude oil known as the API gravity. The gravity relationship is

$$^{\circ}\text{API} = \frac{141.5}{\text{Oil specific gravity}} - 131.5$$

For API gravity, the standard temperature is 60°F.

In general, all oil measurements and sales are required to be reported at standard conditions, which is 60°F. It is often the case, however, that the quantity of oil being measured is not at 60°F. The American Society for Testing Materials (ASTM) has published tables that give correction factors to adjust the volume of oil at a measured temperature to the standard temperature of 60°F. The most accurate method for determining the correction factor for a specific crude, however, is a laboratory test.

Crude oil prices are published by oil purchasers in price bulletins whenever posted prices change. Price bulletins reflect adjustments to oil prices based on gravity and sulfur content. Generally, price deductions are made for crude oils that have a gravity of less than 40°API and also for crude oils containing sulfur. In many cases, price deductions are made for crude oils and condensates that have gravities exceeding 45°API. In addition to adjustments for gravity and sulfur content, deductions in oil prices may be required to account for any transportation costs.

For purposes of forecasting future oil prices for a property evaluation or appraisal, the current posted price is generally escalated at an annual rate consistent with current and anticipated future market conditions. In general, the price is escalated to a ceiling such as $40 per bbl and held constant thereafter. If an evaluation is prepared for Securities and Exchange Commission (SEC) purposes, year-end prices (both oil and gas) are required to be kept constant with no escalation.

Natural gas prices are determined by federal regulations, the quality of the produced gas, and the terms of delivery as established by a gas contract or agreement. As a rule, gas contracts specify the minimum acceptable heating value for the gas, which is usually measured in British thermal units (BTUs) per cubic foot of gas. Gas prices are subject to adjustment upward or downward to reflect the actual BTU content by multiplying the gas price by the ratio of the actual BTU content to 1000. Thus, if the actual BTU content of a given gas is 1100, then the gas price would be adjusted upward by a factor of 1.1. In addition to BTU content, gas contracts specify the maximum water vapor content of the gas in pounds of water vapor per MMCF of gas. The purchaser's acceptance of the gas is also subject to restrictions of its content of oxygen, carbon dioxide, hydrogen sulfide, and total sulfur. Additional adjustments in gas prices may be needed to account for dehydration, compression, gathering, and/or transportation costs.

For purposes of forecasting future gas prices, the current price is generally escalated at an annual rate consistent with current and anticipated future market conditions. The exceptions to this would be those cases in which contracts or agreements specify the escalation factor to be used and those categories of gas that are still regulated by the Natural Gas Policy Act of 1978. Gas prices are generally escalated to a point in time that corresponds to the price ceiling for oil (say, $40 per bbl) and then held constant. Note that in the example cash flow problem in Table 1, oil and gas prices have been escalated at a rate of 6% per year.

Oil and gas pricing has become as important a variable in oil and gas property evaluation as the determination of reserves and the forecasting of production. The property evaluator must spend considerable time in solving the difficult tasks of price forecasting as well as devoting considerable time to understanding the myriad of government regulations as they relate to oil and gas prices. An oil and gas property evaluation with accurate estimates of reserves and forecasting of production is meaningless if the forecasts of future product prices are completely out of line with economic reality. It is interesting to note that economic experts in the early 1980s were projecting oil to escalate at 10% per year from $38 per bbl to over $100 per bbl by the early part of the 1990s. Many of the past and current problems faced by the oil and gas industry can be traced to these pricing projections.

REVENUE
4

The revenue to be derived from future production

Table 4. Severance Tax Calculation for Example Cash Flow Problem (see Table 1)

Year	Net Gas (MCF)	Gas Sev. Tax ($)	Net Oil (bbl)	Oil Price ($/bbl)	Oil Sev. Tax ($)	Total Sev. Tax ($)
1991	128,991	12,899	169,725	22.50	477,352	490,251
1992	52,013	5,201	68,438	23.85	204,031	209,232
1993	20,805	2,081	27,375	25.28	86,508	88,589
1994	8,322	832	10,950	26.80	36,680	37,512
1995	3,329	333	4,380	28.41	15,552	15,885

estimates is equal to the net revenue interest oil and gas production times the corresponding crude oil and natural gas prices.

SEVERANCE AND AD VALOREM TAXES 5

Production and conservation taxes based on production volume, production value, or both and levied by state governments are called *severance taxes.* Property taxes levied by state, county, and municipal governments are called *ad valorem taxes.* Ad valorem taxes may be based on reserves in the ground, actual average daily production rates, the assessed values of equipment and property, or other factors. A working interest owner pays severance and ad valorem taxes only on his share of the working interest in production or in the property. Severance and ad valorem taxes are subtracted from the gross revenue in calculating the net operating income before federal income tax. Production and state and local taxes are always paid whether or not a profit from production is realized. Severance and ad valorem tax rates vary widely from state to state and from one area to another. In addition, many states have complicated methods of calculation, exemptions, limits, or sliding scales. It is therefore recommended that the appropriate state and Internal Revenue office be contacted for specific details.

Severance tax calculations as shown in Column 5 of the example cash flow problem (Table 1) are based on $0.10 per MCF for gas and 12.5% of the gross value for oil. The calculation is shown in Table 4.

WINDFALL PROFIT TAX 6

The Windfall Profit Tax Act of 1980 was repealed on August 23, 1988, with signing of the Omnibus Trade and Competitive Act of 1988 by President Reagan. The tax raised about $77 billion during its 8-year life, but in the years prior to repeal, it had not been collecting any revenues for the federal government due to low oil prices. The original intent of the tax was to tax any gain realized from the decontrol of oil. The tax was applied to each barrel of domestic oil or condensate liquid removed and sold from a lease. Oil or condensate liquids used on a lease from which they were produced were not subject to the tax.

Windfall profits tax rates varied by tax tier and by the status of the producer as an integrated oil company (retailer or refiner) or an independent producer. An independent producer received reduced rates on the first 1000 BPD of qualifying production. The windfall profit tax computation is summarized in the following calculation:

Windfall profit
= Removal price – Adjusted base price
– Severance tax adjustment

Severance tax adjustment
= (Removal price – Adjusted base price)
x Severance tax rate

Windfall profit tax
= Windfall profit x Windfall profit tax rate

Removal price, in dollars per barrel, was the amount for which the oil was sold including any adjustments that were made after the initial sale, such as BS & W and/or gravity corrections. Adjusted base price, in dollars per barrel, was the base price adjusted by an inflation adjustment factor. Base prices for a barrel of oil for each tier were established by rules set by U.S. Treasury regulations and were published on a quarterly basis. Once a base price was established for a category of oil produced from a property, the base price did not change even though the gravity or the quality of oil might have changed.

The windfall profit per barrel of oil subject to tax was limited to 90% of the net income per barrel from each separate property for the taxable year. This was referred to as the net income limitation. Therefore, the windfall profit tax was actually equal to the lower of windfall profit or the net income limitation times the windfall profit tax rate.

NET OPERATING EXPENSE
7

Operating costs are those expenditures incurred by a producing oil or gas well after it has been drilled and completed. Operating costs differ from investments in that operating costs are those expenditures that will only benefit the period in which they are made. Total operating costs are the sum of direct operating costs and taxes (severance, ad valorem, and windfall profit taxes) excluding income taxes (Ikoku, 1985). Direct operating costs include all costs except taxes directly chargeable to a well or lease. Examples of direct operating costs include labor and material costs for operation, maintenance, and repair. Other examples include workovers and the costs associated with gas-processing plants, salt water disposal systems, and gas- or water-injection plants, along with an appropriate share of the overheads. Another example of operating costs would be any fixed charge per well per month as set out in the operating agreement.

Operating costs and their treatment can have a significant impact on future cash flow projections. It is appropriate to treat operating costs on a total cost basis (dollars per month) rather than on a per unit basis (bbl or MCF). The application of unit costs during a curtailment period or after a significant decline in production would result in a distorted picture of actual costs, as it generally costs about the same whether a well produces 100 bbl per day or 10 bbl per day. Escalation factors used for future operating costs in cash flow projections should also be carefully considered to ensure reasonableness. For the example cash flow problem shown in Table 1, operating costs were escalated at the rate of 6% per year. Sources for operating cost data for newly discovered production include expenditures from areas in which production processes or applications are similar.

The calculation of the economic limit or the producing rate at which the revenues exactly equal the costs has been presented in the reserves determination section of this chapter. It is important to always consider, however, the relevant costs and revenues in the analysis. For example, if a pumper will continue to be retained in an area of production whether or not a specific well under study is abandoned, then the portion of the pumper's salary allocated to the specific well would not be considered in the analysis because it is not a relevant cost. Only those revenues and costs that cease when a well is abandoned should be considered in the analysis.

OPERATIONS CASH FLOW
8

Operations cash flow is equal to total revenue minus severance and ad valorem taxes, windfall profit taxes, and net operating expenses.

TANGIBLE AND INTANGIBLE COSTS
9a and 10a

An *investment* is an expenditure that will generate benefits for more than one period (usually 1 year). An example of a capital expenditure or investment would be the cost of a machine that has a useful life of 5 years and would help generate revenue over the next 5 years. Capital expenditures or investments as they relate to oil and gas property evaluation would include all expenditures required to drill and complete the well and to provide the well with facilities to lift production from the reservoir and transport it from the well head to the field terminal. Also included are all future expenditures other than operating costs required to keep the well on production.

Investments are classified as either *tangible* or *intangible expenditures* to allow for proper handling in income tax calculations. Expenditures not represented by physical equipment are generally classified as intangible expenses and are not capitalized for income tax purposes. Tangible expenses are costs of items of physical equipment and are tax deductible only by depreciation. Labor costs for installing surface equipment are generally classified as intangible expense.

Investment costs used in oil and gas property evaluations should represent the most accurate information currently available. Escalation factors used for future investment costs in cash flow projections should also be carefully considered to ensure reasonableness. As will be seen, discounted cash flow projections are sensitive to estimates of investment costs particularly in the early stages of the cash flow projection.

LEASEHOLD COSTS
11a

Leasehold costs are the costs of exploring for and/or acquiring an oil and gas property. These items include geological and geophysical costs, legal costs, assessment costs, mineral rights, acquisition costs, lease bonuses, and capitalized intangible drilling and development costs (IDCs) not represented by physical property.

SALVAGE VALUE
12

The remaining value of an asset if it is not used to the exhaustion of its useful life is called the *salvage value*, which is the amount of money that will be realized upon the sale of the asset once it has reached the end of its useful life.

Table 5. MACRS Recovery Period, Depreciation Method, and Example Property Type

Recovery Period (years)	Depreciation Method	Example of Property Type
3.0	200% declining balance/straight line	Special tools, handling devices
5.0	200% declining balance/straight line	Cars, light vehicles
7.0	200% declining balance/straight line	Oil and gas equipment
10.0	200% declining balance/straight line	Petroleum refining equipment
15.0	150% declining balance/straight line	LNG plants, pipelines
20.0	150% declining balance/straight line	Utility plants
27.5	Straight line	Residential rental real property
31.5	Straight line	Nonresidential real property

CASH FLOW BEFORE TAX
13

Cash flow before tax is equal to operations cash flow less all capital costs, including tangible costs, intangible costs, leasehold costs, and salvage value.

FEDERAL AND STATE INCOME TAXATION

Net revenue as determined by subtracting total expenditures from total revenue is subject to both federal and state income taxation. Therefore, the impact of federal and state income taxation on the future income to be derived from an oil or gas property is an important part of an oil and gas property evaluation. State income taxes are determined by multiplying the net revenue before income tax by the appropriate state income tax rate. As a rule, state income taxes are lower than the federal income tax. Some state income taxes are based on percentages of the federal tax rate. Others are based on percentages of net income, which in this case is defined as revenue minus taxes and operating expenses but prior to capital investments. Still others are based on a combination of both. For state income taxes based on a percentage of the federal tax rate, the following formula can be used to calculate the effective tax rate:

$$\text{Effective tax rate} = [\text{Federal tax rate} \times (1 - \text{State tax rate})] + \text{State tax rate}$$

The assumption is that the state tax is deductible for the federal tax calculation.

Federal income taxes are obtained by multiplying the taxable income after all deductions by the corporate tax rate. Taxable income is determined by deducting depreciation expense, depletion expense, intangible expense, and interest expense from operations cash flow.

DEPRECIATION EXPENSE
9b

Depreciation is defined as a deduction in calculating taxable income that allows for the recovery of the cost of tangible items used in a trade or business (Thompson and Wright, 1985). Depreciable items generally include both well equipment (tangible drilling and development costs) as well as surface equipment (Ikoku, 1985). As a rule, items that retain a usable or reusable value over a period of time can be handled for tax purposes by means of depreciation. The purpose of depreciation is to allow a firm or business to recover the cost of a depreciable asset that is subject to wear and tear and obsolescence during its estimated useful life.

As a result of the Tax Reform Act of 1986 (TRA 86), the cost of most tangible depreciable property placed in service after December 31, 1986, is recovered using the Modified Accelerated Cost Recovery System (MACRS) methods (Stermole and Stermole, 1987). Cost recovery methods are the same for both new and used properties. Salvage value is not considered in calculating the depreciation deduction. Depreciation methods for MACRS for personal property include a 200% declining balance switching to straight line (at such time as to maximize the deduction) and a 150% declining balance switching to straight line. Straight line depreciation may be elected over the other "accelerated" depreciation methods previously discussed on an irrevocable basis. For residential rental real property or nonresidential real property purchased after 1986, straight line depreciation is required. Table 5 shows the recovery period, the depreciation method, and an example of property type for each class of recovery property both personal and real.

Another aspect of depreciation treatment under TRA 86 is timing conventions that affect the allowable depreciation in the first year (Stermole and Stermole, 1987). A half-year convention modified by a mid-quarter convention is specified for recovery property other than residential rental and nonresidential real property. Under the half-year convention, the assets are considered to have been placed in service at mid-year no matter when a particular asset was actually

Table 6. Depreciation Expense Calculation for Example Cash Flow Problem (see Table 2)

Year	Unrecovered Cost ($)		Depreciation Rate		Annual Depreciation Allowance ($)	Cumulative Depreciation Allowance ($)
1	400,000	x	(1/7 x 2)/2	=	57,144	57,144
2	342,856	x	1/7 x 2	=	97,960	155,104
3	244,896	x	1/7 x 2	=	69,968	225,072
4	174,928	x	1/7 x 2	=	49,980	275,052
5	124,948		124,948/3.5	=	35,699	310,751
6	89,249		124,948/3.5	=	35,699	346,450
7	53,550		124,948/3.5	=	35,699	382,149
8	17,851		124,948/3.5 x 2	=	17,851	400,000

Year 1: 200% declining balance with half-year timing convention
Years 2–4: 200% declining balance
Years 5–8: Straight line

placed in service. Therefore, only one-half of the year's normal depreciation is allowed in the year that the property is placed in service. If more than 40% of the assets are placed in service during the last quarter of the year, however, the mid-quarter convention applies. Under this convention, all assets placed in service in the last quarter would receive the fraction 1.5/12 times the calculated full first year depreciation as the allowed first year depreciation deduction. Assets placed in service during the first, second, and third quarters would receive first year deduction factors of 10.5/12, 7.5/12, and 4.5/12, respectively, times the calculated first year depreciation as the allowed first year depreciation deduction. For residential and nonresidential real property, a mid-month first year convention applies.

In the example cash flow problem in Table 2, depreciation expense is calculated as shown in Table 6. Note also in the example cash flow problem that, since the property ceases producing in year five as reflected in the reserves and economics schedule, the remaining unrecovered cost is deducted in the year in which the well goes off production.

DEPLETION EXPENSE
11b

Under current tax laws, the economic interest owner of oil and gas reserves (includes royalty owners but not shareholders) is allowed a deduction in calculating taxable income for the depletion of the natural resource. *Depletion* is defined as the exhaustion of a natural resource as a result of its extraction (Ikoku, 1985). In view of the fact that hydrocarbons are a nonrenewable resource, the economic interest owner is allowed to get back not only the cost of extracting the resource but also the original cost of attaining it. For oil and gas, the depletion deduction allowed is the greater of cost depletion or percentage depletion. The amount that may be deducted from taxable income is known as *allowable depletion.*

Cost Depletion

The costs of exploring for and/or acquiring an oil and gas property are included in the cost depletion basis for minerals. Items included in the depletable basis are geological and geophysical costs, legal costs, assessment costs, mineral rights acquisition costs, lease bonuses, and capitalized IDCs not represented by physical property.

Cost depletion is calculated by dividing the number of units of reserves at the beginning of the taxable year into the adjusted basis of the property for that year. This value is then multiplied by either the number of units for which payment is received during the tax year (if the cash receipts and disbursements method is used) or the number of units sold (if an accrual method of accounting is used). Reserves and production are treated on a barrels of oil equivalent (BOE) basis. The adjusted basis of the property is the original cost of the property less all of the depletion (cost and percentage) allowed on the property since its acquisition by the taxpayer. The formula for calculating cost depletion is

$$CD = B\left(\frac{P}{R+P}\right)$$

where CD = annual cost depletion allowance
B = the "adjusted basis" of the property
P = the number of units of production sold or for which payment was received during the tax year
R = the number of "recoverable units" of production remaining at the end of the tax year

Table 7. Cost Depletion Calculation for Example Cash Flow Problem (see Table 2)

Years	Oil Prod. (bbl)	Gas Prod. (MCF)	BOE	Total BOE	(Annual BOE ÷ Total BOE) x $2.5 million
1	226,300	171,988	28,665	254,965	$1,510,721
2	91,250	69,350	11,558	102,808	609,162
3	36,500	27,740	4,623	41,123	243,665
4	14,600	11,096	1,849	16,449	97,466
5	5,840	4,438	740	6,580	38,986
Totals				421,925	$2,500,000

In the example cash flow problem, cost depletion is calculated based on a gas–oil equivalency factor of 6 MCF per bbl of oil, as shown in Table 7. A brief comment should be made about gas–oil equivalency. In some cases, it is convenient to combine oil quantities and gas quantities. In general, equivalence is based on a BTU equivalence as opposed to a price equivalence. One barrel of 40°API oil is usually assumed to be equivalent to 5.7 million BTUs. Assuming 1 million BTU per thousand standard cubic feet of gas, 5.7 MCF is equivalent to 1 bbl of oil. Gas–oil equivalence is usually reported as barrels of oil equivalent (BOE). When equivalence is reported on the basis of prices, the price of a barrel of oil is divided by the price of 1 MCF of gas. If oil is sold for $20 per bbl and gas for $2 per MCF, it would take 10 MCF to equal 1 bbl.

Percentage Depletion

Percentage depletion allows for the deduction of a certain percentage of gross income from a mineral property. The original intent of percentage depletion was to provide operating owners with a relatively inexpensive source of financing for further investment in exploration and development. With the 1975 Tax Reduction Act and the 1986 Tax Reform Act, percentage depletion was essentially repealed with the exception of certain categories of natural gas production and certain independent producers and royalty owners.

Qualified independent producers and royalty owners are allowed percentage depletion on a limited amount of production. An independent is defined as a producer whose refinery runs do not exceed 50,000 bbl per day on any given day during the taxable year and whose retail sales do not exceed $5 million during the taxable year. Independent producers are allowed a depletion deduction equal to 15% of the gross income from qualifying production. The deduction for percentage depletion cannot exceed a specified percentage of taxable net income from the property after all deductions except depletion have been taken ("net income limitation"). With passage of the Omnibus Budget Reconciliation Act of 1990, the net income limitation on oil and gas percentage depletion was increased from 50% to 100% of the net income from the property for taxable years beginning after December 31, 1990 (KPMG Peat Marwick, 1990).

Further modifications of percentage depletion under the Omnibus Budget Reconciliation Act of 1990 relate to marginal production and repeal of the transferred proven property rule. For qualifying marginal production (stripper oil and heavy oil) for independent producer and royalty owners, the statutory percentage depletion rate of 15% was increased after December 31, 1990, by 1% for each whole dollar that the average domestic well head price of crude oil for the immediately preceding calendar year is less than $20 per bbl. This increase in percentage rate is limited to a maximum increase of 10%.

The transferred proven property rule did not generally allow for percentage depletion for properties that were transferred after December 31, 1974, by one taxpayer to another if, at the time of the transfer, the principal value of the property had been demonstrated by prospecting, exploration, or discovery work. With passage of the Omnibus Budget Reconciliation Act of 1990, taxpayers can claim percentage depletion on transferred oil and gas properties for transfers occurring after October 11, 1990.

Depletion Allowance

The depletion allowance is calculated by first determining cost depletion using the cost depletion equation. The next step is to determine percentage depletion, which is the smaller of gross depletable income times the statutory rate or gross taxable income (not gross depletable income) before depletion times the 100% net income limitation. Allowable depletion in a given tax year is the greater of cost depletion or percentage depletion.

INTANGIBLE DRILLING AND DEVELOPMENT COSTS

10b

Intangible drilling and development costs (IDCs) are defined as those costs incurred by a company in the drilling and development of new crude oil and natural gas reserves. IDCs include such items as the costs of labor, fuel, repairs, hauling and supplies used to prepare the drill site, drill and complete the well, and construct surface facilities as needed for production from the completed well. IDCs also include the cost of installation of tangible equipment into the well. However, the equipment itself is capitalized and depreciated.

IDCs can be deducted as an expense in the current tax year or can be capitalized. An expenditure that may be capitalized is an item such as land or equipment that can help generate revenue in future periods. Thus, when the cost is capitalized, the item becomes an asset to the firm. An expense is the cost of an asset that is assumed to expire during an accounting period in an attempt to generate revenues during the period. Generally, all taxpayer's elect to expense IDCs because of the immediate tax savings.

Prior to 1987, integrated producers were required to amortize 20% of the productive well IDC on a straight line basis over a 36-month period. Under the Tax Reform Act of 1986 (TRA 86), integrated producers were required to amortize 30% of the productive well IDC on a straight line basis over a 60-month period (Gallun and Stevenson, 1986). Productive well IDCs on wells outside the United States could no longer be deducted under TRA 86, but must either be amortized on a straight line basis over 10 years or be recovered as cost depletion. Small producers could continue to enjoy a full IDC deduction for costs of domestic wells.

INTEREST EXPENSE
14

Interest on a loan relating to property acquisitions or investments is treated as a tax deduction for federal income tax calculations.

TAXABLE INCOME
15

Taxable income is equal to operations cash flow minus deductions for depreciation, depletion, intangible drilling and development costs, and interest expense.

INVESTMENT TAX CREDIT
16

Prior to January 1, 1986, federal tax laws allowed for an investment tax credit (ITC) for tangible personal property or real property used as an integral part of production. The ITC calculation is as follows:

(Taxable income x Tax rate) – ITC = Taxes payable

For properties placed in service through 1982, the ITC was generally 10%. For properties placed in service after 1982 and subject to Accelerated Capital Recovery System (ACRS) depreciation, the ITC rates are shown in the following table:

ITC Rate (%)	Comments
10	ACRS depreciation equals cost minus 1/2 ITC taken.
8	ACRS depreciation is not adjusted.
6	Applies to 3-year ACRS property only. ACRS depreciation equals cost minus 1/2 ITC taken.
4	Applies to 3-year ACRS property only. ACRS depreciation is not adjusted.

Note that ACRS depreciation (predecessor to MACRS depreciation previously discussed) applies to most properties placed in service between January 1, 1981, and December 31, 1986, and is calculated by multiplying the unadjusted basis for a property (typically the cost of the property) times the appropriate ACRS depreciation rate as determined by the class of the property (3-, 5-, 10-, or 15-year). The class of properties and the ACRS depreciation rate to be used are found in tables in the tax code.

ITCs could be applied to both new and used properties (Thompson and Wright, 1985). For new properties, the ITC was limited to $25,000 plus 90% of any remaining tax liability for 1982 and 85% of any remaining tax liability after 1982. For used equipment, the ITC was limited to $125,000 in a given year. An unused ITC could be carried back 3 years or carried forward 15 years for a new property. However, unused credits for used property could not be carried forward or back. A portion of the ITC taken must have been recaptured if an ACRS property was sold before the end of its ACRS recovery period. The portion of the ITC recaptured was added to the tax liability in the taxable year the recapture occurred.

The ITC was repealed under the Tax Reform Act of 1986 effective January 1, 1986, except to certain binding contracts on that date. ITC carryovers to taxable years beginning after June 30, 1987, were reduced by 35% because of the reduced regular tax rates.

TAXES PAYABLE
17

Taxes payable are calculated using the following formula:

(Taxable income x Tax rate) – ITC = Taxes payable

Corporate income tax rates are summarized in the following table:

Taxable Income	Tax Rate (%) Effective July 1, 1987	Prior Law
Not over $25,000	15	15
Over $25,000, but not over $50,000	15	18
Over $50,000, but not over $75,000	25	30
Over $75,000, but not over $100,000	34	40
Over $100,000	34	46

CASH FLOW AFTER TAX

18

Cash flow after tax is determined by subtracting taxes payable from cash flow before tax.

MEASURES OF PROFITABILITY

General Discussion

The next step in the valuation of an oil and gas property is the application of profitability measures to evaluate the net cash flow stream both before and after tax. The principle "no-risk" measures of profitability include net present value, rate of return, payout, and profit to investment ratio.

No single measure of profitability considers all of the factors or dimensions of a net cash flow stream that are pertinent to the decision maker in the evaluation process. It is necessary to use a combination of profitability parameters that are consistent with a firm's philosophy and manner of doing business. A good measure of profitability (Newendorp, 1975) should provide a basis for comparing overall project profitability to minimum investment criteria as established by the firm, such as the firm's cost of capital or an average earnings rate. A profitability parameter should be suitable for the ranking and comparison of projects within an inventory of investment opportunities. A good profitability parameter should include quantitative assessments of risk and uncertainty. Finally, a good measure of profitability should consider the "time value of money" with respect to an investment opportunity.

It is important to note that the profitability measures discussed in this chapter use only direct investments, gross revenues, direct taxes and royalties, and operating costs in the computations. In most project analyses, it is the practice to keep the investment decision apart from the finance decision. This is sufficient for day-to-day routine cash flow analysis and is a generally accepted business practice. For investments that require very large sums of capital, however, the sources and costs of funds become important considerations. It becomes necessary to consider such financial parameters as sources and costs of capital, leveraged transactions, debt to equity ratios, and depreciation schedules and to factor financing charges into the analysis.

Measures of profitability discussed in this section include the following:

- Net present value profit
- Rate of return
- Payout
- Profit to investment ratio
- Discounted profit to investment ratio

Before discussing these profitability measures, it is necessary to talk in general about time value of money concepts.

Time Value of Money Concepts

The time rate patterns of cash flow resulting from an investment opportunity are generally related to some measure of profitability by means of *time value of money* concepts, which include both compounding and discounting. The compound interest equation is as follows: $P(1 + i)^n = F$. It relates the future value, F, to a principle amount of money today, P. The term $(1 + i)^n$ is called the *compound interest factor*. The present value equation (annual compounding and year-end cash flows) is a modified form of the compound interest equation:

$$P = \frac{F}{(1+i)^n} = F\left[\frac{1}{(1+i)^n}\right]$$

The interest rate *i* is often referred to as the *discount rate* in the previous equation, and *n* is the year in which the cash flow is received or disbursed. The term in the brackets is called the *discount factor*. The following calculation using the compound interest equation shows what $10 will appreciate to after 5 years at a rate of 10% compounded annually:

$$\begin{aligned} F &= P(1 + i)^n \\ &= \$10(1 + 0.10)^5 \\ &= \$10(1.611) \\ &= \$16.11 \end{aligned}$$

Thus, the compound interest equation states that $10 today will become $16.11 in 5 years. It can also be said that receiving $16.11 in 5 years has a *present value* or present worth of $10. As a rule, we are generally concerned with comparing cash flows at the present time or time zero. Therefore, the present value equation is more frequently used. Discount factors for annual compounding and year-end cash flows are calculated using the following equation:

$$\text{Year-end discount factor} = \frac{1}{(1+i)^n}$$

In cash flow analyses, various types of compounding (annual, monthly, daily, and continuous) and patterns of receipts (year-end, mid-year, monthly, and uniform) can be used. The important thing is that the type of compounding and the pattern of receipts used in the analysis should be representative of how cash flows in and out of a firm's treasury and consistent with the cash flow being analyzed. In general, oil properties are evaluated on a yearly basis using mid-year discounting. Discount factors for annual compounding and mid-year cash flows are calculated using the following equation:

$$\text{Mid-year discount factor} = \frac{1}{(1+i)^{n-0.5}}$$

Mid-year more closely approximates monthly cash flows better than year-end. In general, however, the change in present value between year-end and mid-year discounting is minor when compared to inaccuracies in production and price forecasts.

Net Present Value

19a and 19b

The net present value (NPV) calculation for a net cash flow before and after tax uses a single discount rate for all economic analyses. The single discount rate is often referred to as the *average opportunity rate* or the *opportunity cost of capital.* It is called the opportunity cost because it is the return foregone by investing in the project rather than investing in securities. An example of a NPV calculation is shown in Table 8, assuming annual compounding and mid-year cash flows.

In this case the discount rate is 10%, which is based on the firm's average opportunity rate. The positive NPV tells us that the investment of $1,500,000 to drill a well will earn a rate of return equal to 10% plus an amount of cash totaling $827,700 as of time zero. Another way of stating the results shown in Table 8 is that the investment of $1,500,000 and the associated positive net cash flow has a net present value or net present worth of $827,700. If the NPV equaled zero in this example, then the investment would yield a rate of return equal to the discount rate used. If the NPV was negative, the investment would yield a rate of return less than the discount rate. If the average opportunity rate of the firm's ability to invest capital is realistic, then it follows that an investment opportunity having a negative NPV should be rejected. As stated earlier, a positive NPV represents the present value cash worth

Table 8. Example of NPV Calculation Assuming Annual Compounding and Mid-Year Cash Flows

Year	Net Cash Flow	Discount Factor (10%)	10% Discounted Cash Flow
0	–$1,500,000	1.000	–$1,500,000
1	+1,000,000	0.953	+953,000
2	+800,000	0.867	+693,600
3	+600,000	0.788	+472,800
4	+200,000	0.716	+143,200
5	+100,000	0.651	+65,100
Total			$827,700 = NPV @ 10%

in excess of making a rate of return equal to the discount rate. Stated differently, a positive NPV is the amount of additional money that could be invested in the project and still realize a rate of return equal to the discount rate.

The biggest disadvantage of NPV is that it is independent of the absolute size of cash flows, as shown in the following example:

	Project A	Project B
NPV of revenues	$1,600,000	$250,000
Minus initial investment	–1,500,000	–150,000
NPV profit	$100,000	$100,000

In this case, it is apparent that the objective of maximizing NPV profit is not a completely adequate profitability criterion if there are limitations on the availability of capital. As stated earlier, the decision maker must use a combination of profitability measures such as NPV, payout, and profit to investment ratio to adequately evaluate a net cash flow stream.

An important consideration in the use of NPV is that cash flows received early in a project life are weighted more heavily than later cash flows. Cash flows received or disbursed late in the life of a project (after 15–20 years) have very little effect on the computed NPV. Also, the computed NPV is very sensitive to errors in estimating initial investment and the early cash revenues. A small variation in the initial investment can sometimes cause a much larger variation in the resultant NPV. All of these factors need to be considered when NPV is used to evaluate a net cash flow stream.

The process of determining the discount rate to be used in NPV computations is not an easy procedure, and several factors need to be taken into consideration. If a firm is operating on borrowed capital, then the discount rate should at least exceed the interest rate being paid on the loan. If the source of capital comes from several sources, such as internally generated funds, short- and long-term debt, and equity sources,

Table 9. NPV Calculation for Example Cash Flow Problem (see Tables 1 and 2) Assuming Annual Compounding and Mid-Year Cash Flows

Year	Net Cash Flow	Discount Factor (10%)	10% Discounted Cash Flow	Cum. Discounted Cash Flow
Before Tax				
0	–$3,500,000	1.000	–$3,500,000	–$3,500,000
1	+3,522,586	0.953	+3,358,654	-141,346
2	+1,481,805	0.867	+1,284,405	+1,143,059
3	+602,679	0.788	+474,902	+1,617,961
4	+228,317	0.716	+163,555	+1,781,516
5	+67,921	0.651	+44,232	+1,825,748
Total			+$1,825,748	
After Tax				
0	–$3,500,000	1.000	–$3,500,000	–$3,500,000
1	+3,061,981	0.953	+2,919,484	–580,516
2	+1,218,413	0.867	+1,056,101	+475,585
3	+504,403	0.788	+397,462	+873,047
4	+200,821	0.716	+143,858	+1,016,905
5	+100,565	0.651	+65,491	+1,082,396
Total			+$1,082,396	

then the discount rate should be based on a weighted average cost of capital (Brealey and Myers, 1984), which is calculated using the following equation:

$$r^* = r_D(1-T_C)\frac{D}{V} + r_E\left(\frac{E}{V}\right)$$

where
r^* = the adjusted cost of capital
r_D = the firm's current borrowing rate
T_C = the marginal corporate income tax rate
r_E = the expected rate of return on the firm's stock (a function of the firm's business risk and its debt ratio)
D,E = the market values of currently outstanding debt and equity
$V = D + E$ = the total market value of the firm

Finally, the discount rate used should reflect the corporate growth objectives that management has set for the growth of the firm's total assets.

At present, most firms are using "nonrisked" discount rates in the range of 9 to 11% for evaluating petroleum investments. It is often argued that the risk of oil and gas exploration and production is greater than less risky investments such as chemical processing, refining, and retail marketing and therefore a higher discount rate such as 25 to 30% should be used to account for risk and uncertainty. The recommended approach, however, is to incorporate an explicit analysis of risk and uncertainty into the net cash flow prior to discounting and then discount at the 9 to 11% nonrisked rate.

Calculation of NPV for the example cash flow problem in Tables 1 and 2 on both a before tax and after tax basis are shown in Table 9, assuming annual compounding and mid-year cash flows.

Rate of Return

20a and 20b

Rate of return is defined as the discount rate that makes the present value of net receipts equal to the present value of the investments. Stated differently, it is the discount rate that makes the NPV of a project or investment opportunity equal to zero. The calculation of rate of return is a trial and error process involving the use of various discount rates in the discounting process until NPV = 0 is determined. A rate of return calculation is shown in Table 10. Discount factors are based on annual compounding and mid-year cash flows.

Thus, the idea is to find an interest rate that makes the present value of net receipts equal to the present value of the investments. In Table 10, the expenditure of $1,500,000 to purchase the future series of cash revenues has a profitability equivalent to investing $1,500,000 in a savings account or any other type of investment that pays 52.8% interest compounded annually. It is important to note that rate of return is being defined as the earnings rate of the initial investment.

In selecting the interest rate for the trial and error computation of rate of return, a higher rate should be used if the sum of the discounted cash flow for the first

Table 10. Example of Rate of Return Calculation

Year	Net Cash Flow	Discount Factor (52.8%)	52.8% Discounted Cash Flow
0	–$1,500,000	1.000	–$1,500,000
1	+1,000,000	0.809	+809,000
2	+800,000	0.529	+423,200
3	+600,000	0.346	+207,600
4	+200,000	0.227	+45,400
5	+100,000	0.148	+14,800
Total			$0 = NPV @ 52.8%

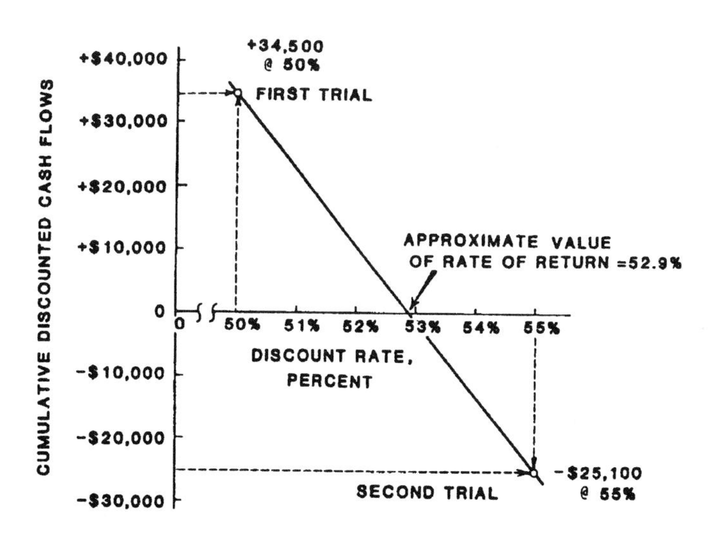

Figure 9. Rate of return approximation by means of graphical solution (see Table 11).

trial was positive. If it was negative, a lower discount rate should be used. A graphical approach to approximating the value for the rate of return of a series of cash flows is shown in Figure 9 and Table 11. This technique helps to minimize the number of trial and error iterations necessary to derive the rate of return for a given project.

In Figure 9, note the close agreement between the rate of return approximation of 52.9% shown in the graph and the actual computed rate of return of 52.8%. It is important to remember that this is only an approximation to aid in determination of the rate of return value. The actual correlation between the discount rate and the sum of the discounted cash flows is nonlinear. However, over relatively limited ranges of discount rate, the correlation will provide a good estimate for the third trial.

Rate of return has several advantages as a measure of profitability. Like NPV, rate of return considers the time value of money concept and is a useful measure of the relative profitability of investments having approximately the same total life and cash flow patterns. Rate of return is also a convenient measure of profitability to compare with a "minimum," such as a cost of capital or a corporate objective of a specified annual growth rate.

As with NPV, a disadvantage to rate of return is that it is independent of the magnitude of the cash flows. Stated another way, one cash flow could be in millions of dollars while another could be in thousands of dollars and yet the rate of return would be the same for both. The disadvantage here is that the magnitude of the investment is important to consider in the evaluation process. Therefore, other measures of profitability such as NPV, payout, and profit to investment ratio must be used in addition to rate of return to evaluate a net cash flow stream adequately.

Another consideration in the use of rate of return analysis is that cash flows received early in a project life are weighted more heavily than later cash flows, as is the case with NPV. Cash flows received or disbursed late in the life of a project, however, have very little effect on the computed rate of return. Also, the computed rate of return is very sensitive to errors in estimating initial investment and the early cash revenues. A small variation in the initial investment can result in a much larger variation in the resultant rate of return. All of these factors need to be considered when rate of return is used to evaluate a net cash flow stream.

Two last considerations need to be taken into account when using rate of return analysis. The first consideration concerns the case when a cash flow has more than one discount rate that satisfies the definition of rate of return. Such multiple rates of return can result from rate acceleration projects and projects requiring a large expenditure at a later point in the life of the project. It is important to note that a second sign reversal in the cumulative net cash flow represents a necessary condition to multiple rates of return. Multiple rates of return are also dependent on when the second sign reversal occurs and the magnitude of the negative cash flows causing the reversal. The important thing to remember is that there is no way to establish a true rate of return in cash flows having multiple rates of return. When multiple rates of return exist, other profitability criteria should be used such as net present value at a specified discount rate.

The second consideration in rate of return analysis concerns the reinvestment assumption and its effect on rate of return. It is argued by some that all cash flows from an investment opportunity must be reinvested at the computed rate of return when received so that the initial expenditure will have the earning power at the rate of return given. Thus, if a firm does not have unlimited reinvestment opportunities at a rate of, say, 40% for the next 10 years, rate of return will not be a realistic measure of true profitability for the firm. Others take the position that fulfillment of the reinvestment assumption is not critical for rate of return to be meaningful as a profitability measurement.

Table 11. Rate of Return Approximation by Means of Graphical Solution (see Figure 9)

		First Trial		Second Trial	
Year	Net Cash Flow	Discount Factor (50%)	50% Discounted Cash Flow	Discount Factor (55%)	55% Discounted Cash Flow
0	–$1,500,000	1.000	–$1,500,000	1.000	–$1,500,000
1	+1,000,000	0.817	+817,000	0.803	+803,000
2	+800,000	0.544	+435,200	0.518	+414,400
3	+600,000	0.363	+217,800	0.334	+200,400
4	+200,000	0.242	+48,400	0.216	+43,200
5	+100,000	0.161	+16,100	0.139	+13,900
Totals			+$34,500		–$25,100

Despite this potential weakness in rate of return, many decision makers still rely on rate of return analysis as a measure of profitability. Typically, firms will set a minimum rate of return expectation for various types of projects on the order of 15 to 35%. If the projected net cash flow fails to generate a rate of return in excess of the established minimum, then the project being analyzed is rejected. It is important to be aware of the strengths and weaknesses of rate of return as well as the other measures of profitability so that the profitability measure is applied in an appropriate manner that gives the best results.

Up to this point, the discussion of the rate of return concept has not formally considered risk in the analysis. There is, however, an approach for considering risk in a rate of return calculation (Newendorp, 1975). This can be done by increasing the initial investment to include monetary losses for the case in which no cash revenues are received from the project due to unsuccessful efforts. A "pseudoinvestment" as of time zero can be determined for use in a risk-weighted rate of return calculation using the following equation:

Pseudoinvestment at time zero for risk-weighted rate of return

$$= C_0 + \left[\frac{(1-p)}{p}\right] C_D$$

where

C_0 = investment at time zero that generates the future cash flow

C_D = investment loss if no future cash flow is generated

p = probability that the future cash flow will occur (decimal fraction)

$(1-p)$ = probability of no future cash flow (decimal fraction)

This approach has been widely used in the oil industry because it allows a decision maker to incorporate dry hole risks into a rate of return calculation. An important weakness of the pseudoinvestment in a risk-weighted rate of return calculation, however, is that it cannot consider more than two possibilities (that is, success or failure). In many oil field applications, it is often the case that more than two possibilities exist in any given consideration, such as the discovery of varying levels of oil and gas reserves. Risk-weighted rate of return calculations are not unreasonable approximations in considering the element of risks. However, more realistic ways exist to incorporate the element of risk in the property evaluation process, some of which are discussed in the section entitled Risk and Uncertainty.

Payout

21a and 21b

Payout is defined as the length of time required for the cumulative net revenues to equal the initial investment or, more simply put, the length of time required to get the invested capital back. Payout is widely used in the petroleum industry as a measure of profitability, but it has several disadvantages. The primary disadvantage is that it tells nothing about the timing of returns that occur before or after payout, and it does not consider the total profitability of the investment. Therefore, payout as a profitability measure is by itself generally insufficient to evaluate an investment opportunity and must be used with other profitability measures such as NPV, rate of return, and profit to investment ratio to evaluate a net cash flow stream properly. The obvious primary consideration in payout time analysis is how quickly the investment can be returned out of the generated net revenues. For most oil and gas investments, a generally acceptable payout period is in the range of 2 to 5 years. The payout calculation in the following example uses the net revenues shown, which are generated from an investment of $1,500,000:

Year	Net Cash Flow
1	$1,000,000
2	800,000
3	600,000
4	200,000
5	100,000

The net cash flow in the first year is estimated to be $1,000,000. Therefore, the unrecovered portion of the investment of $1,500,000 after the first year is $1,500,000 minus $1,000,000, or $500,000. The fraction of the second year required to recover this remaining balance is $500,000/ $800,000, or 0.625. Therefore, the payout time is 1.625 years.

As stated previously, payout is commonly used in the industry for evaluating oil field investments. Again, the inherent problem with payout (that is does not consider the timing of returns or the total profit from the investment project) should be kept in mind when using this measure of profitability.

Profit to Investment Ratio

22a and 22b

Profit to investment ratio is defined as the ratio of the total undiscounted net profit to investment. Unlike payout, the profit to investment ratio is a measure that reflects total profitability, as follows:

Undiscounted profit to investment ratio

$$= \frac{\text{Undiscounted net profit}}{\text{Investment}}$$

$$= \frac{\text{Total net cash revenues} - \text{Investment}}{\text{Investment}}$$

The use of the net income to investment ratio as opposed to the undiscounted profit to investment ratio is another widely used measure of profitability (see Table 2).

Net income to investment ratio

$$= \frac{\text{Net Income}}{\text{Investment}}$$

A profit to investment ratio calculation is given in the following example using the net revenues (shown in the following table) to be generated from an investment of $1,500,000.

Year	Net Cash Flow
1	$1,000,000
2	800,000
3	600,000
4	200,000
5	100,000
	$2,700,000

Undiscounted profit to investment ratio

$$= \frac{\$2,7000,000 - \$1,500,000}{\$1,500,000}$$

$$= 0.80$$

Net income to investment ratio

$$= \frac{\$2,7000,000}{\$1,500,000}$$

$$= 1.80$$

Profit to investment ratio is often used with payout to evaluate routine investment opportunities. A common industry rule of thumb for an acceptable investment opportunity is one in which the investment pays out in 2 or 3 years or less and yields a net income to investment ratio of greater than 2 to 1. As with payout, however, the major weakness with profit to investment ratio is that it does not reflect the rate of earnings of income from the project. For example, two investment opportunities could have the same payout time and profit to investment ratio and yet one investment opportunity returns total income in one-half the time compared to the other. Obviously, the investment opportunity that returns total income in half the time would be more desirable. Therefore, profit to investment ratio must be used with other profitability measures such as NPV, rate of return, and payout time to evaluate a net cash flow stream properly.

Discounted Profit to Investment Ratio

23a and 23b

It was previously stated that one problem with NPV is that it is independent of the absolute size of the cash flows. The *discounted profit to investment ratio* (DPR) is obtained by dividing NPV by the present value of the investment, and it represents the amount of discounted net profit generated in excess of the average opportunity rate per dollar invested. In terms of ranking investment opportunities, DPR is useful in maximizing profit per dollar invested under conditions of limited investment capital constraints. Calculations for DPR and discounted net income to investment ratio are shown in the following example:

Discounted net profit = $827,700 (at 10%)

Initial investment = $1,500,000

Discounted profit to investment ratio

$$= \frac{\$827,700}{\$1,500,000} = 0.55$$

Discounted net income to investment ratio

$$= \frac{\$827,700 + \$1,500,000}{\$1,500,000} = 1.55$$

As previously discussed, NPV is generally preferred to other profitability measures because it uses a single discount rate that gives a more realistic measure of true profitability. Since it is generally the case that investment capital is limited, it is clear that investment opportunities should be chosen on the basis of maximizing profitability per dollar invested. Sound investment strategy would dictate, therefore, that projects should be chosen that maximize the discounted profit to investment ratio. Many consider DPR to be the most representative measure of true earnings potential of an investment for this reason.

FAIR MARKET VALUE

As has been shown, an oil and gas property evaluator calculates reserves, forecasts production, and forecasts hydrocarbon prices and expenses to arrive at an annualized net cash flow stream from a producing property as of a prescribed date. The annualized net cash flow stream is then discounted to determine the present worth value of a property both before and after federal and state income taxes. Often the property evaluator is asked to state the fair market value of the property. Such requests are generally made by a lending institution for loan purposes, by the IRS for tax matters such as successions, by a bankruptcy court, by a seller and/or prospective purchaser who seek(s) a third party opinion, or in court cases. It is sometimes argued that the present worth value of a property represents the fair market value. However, experience shows that the present worth value often only represents the value to a specific individual or group of individuals who have an economic interest in the property. The value to a prospective buyer may be something entirely different, and the calculated present worth value becomes nothing more than a dollar figure that may be used as a basis for bargaining. It is important to remember that present worth values are based on certain assumed conditions and that buyers and sellers may evaluate the variables differently. Lending institutions and courts have viewpoints that are still different.

Fair market value (FMV) is generally defined as the price at which a property would change hands between a willing buyer and a willing seller, neither being under any compulsion to buy or sell and both having reasonable knowledge of the relevant facts. Many methods exist for determining fair market value, but most determinations are based in one form or another on one of the following methods (Campbell et al., 1978; Eggleston, 1964; Garb, 1985, 1990):

- *Rate-of-Return Method*: This method calculates FMV as that purchase price which provides for an acceptable rate of return on investment both before and after taxes.
- *Payout Time Method*: Using this method, the FMV would be equal to the cumulative undiscounted future net cash flow before tax for the first 2 to 5 years after the property is purchased. A rule of thumb for the maximum time length considered in this valuation method is no more than one-third of the remaining life.
- *Profit to Investment Ratio Method*: This method calculates FMV by dividing the expected profit by the purchase price of the property. Purchasers would typically seek a ratio of two or three to one or better.
- *Specified Fraction of the Present Worth of Future Net Income Method*: FMV is estimated by use of a specified percentage of present worth. A common rule of thumb in using this method is two-thirds of present worth.
- *Price per Barrel of Reserves in the Ground Method:* For this method, gas volumes are converted to equivalent barrels of oil on either a heating value or price ratio basis. A rule of thumb is that oil reserves in the ground are worth one-third the current market value.
- *Present Worth at a Specified Discount Rate Method:* FMV is considered to be the present worth of the investment and its future cash flow using a discount rate between 15 and 25%.
- *Going Unit Price Method:* This method is based on a percentage of the future undiscounted revenue. FMV for this method generally ranges from 40% of the future undiscounted revenue of newer properties to 25% for older properties.
- *Price per Barrel per Day of Producing Rate Method:* This method uses a specified price per barrel per day of the producing rate for the first year of the remaining life of the property.

As a rule, a combination of these methods is used to determine FMV. For example, an acceptable FMV may be one in which the rate of return is 15 to 20% after federal income taxes, the payout is in 2 or 3 years, and the profit to investment ratio is not less than two or three to one. Gustavson (1985) recommends a FMV that is an arithmetic average of at least three FMV methods having values within 10% of each other. His methods of FMV include value per barrel of oil equivalent, risked present worth, 3-year cumulative cash, and present worth at 25%. In addition to the FMV methods just discussed, an obvious source of FMV data is the sale price of similar properties similarly situated, when such information is available.

In the case of appraisals for bank loans, the amount that can be borrowed is generally slightly lower than the fair market value. An appraisal for loan purposes is used to satisfy two basic concerns of the lending institution—total reserves and annual cash flow capability. A property must have sufficient value to repay the loan if it were sold on the open market. At the same time, it is generally a requirement that the property have sufficient revenue-producing ability to pay operating costs and taxes as well as provide reduction of principal and interest. After establishing a fair market value, lending institutions will generally perform loan soundness tests, such as the following (Campbell et al., 1978):

1. 50% of the future undiscounted revenue is still remaining at the anticipated payout of the loan;
2. 50% of reserves are still in place at payout; or
3. a test to determine the effect on the loan if a 50% reduction in payments occurs due to a reduced price, curtailed production, or loss of production.

As is probably evident by now, the fair market value established by individuals or institutions for a property can vary quite significantly. Despite the fact that most everyone uses the same general rules and approaches, opinions of fair market value vary considerably due to the degree of uncertainty that always exists in studies of this nature. The best approach for the property valuation engineer under these conditions is to consider all of the data made available to him or her, apply the appropriate appraisal procedures in a correct manner, and fully document the appraisal procedures and assumptions made.

RISK AND UNCERTAINTY

The degree to which risk and uncertainty exist in reserve estimates, production forecasts, and forecasts of hydrocarbon prices makes it incumbent upon the property valuation engineer to formally consider and incorporate risk into the property appraisal process. There are several ways that risk can be considered in a property appraisal:

1. Adjustment of the required rate of return, present value, or other profit indicators in some manner to reflect the relative risk involved.
2. Use of present product prices without escalation and conservative production forecasts in order to minimize risk.
3. Use of formal probability methods that quantify risk and uncertainty. An example is Monte Carlo simulation techniques, which calculate the results to be expected, weighted by probability estimates.
4. Multiple executions of the appraisal process in which a combination of the reserve, property life, product prices, tax assumptions, capital investments, and operating methods are varied.
5. Application of a single risk adjustment factor to the evaluation.

In the case of the single risk adjustment factor (Garb, 1985), a best estimate case of reserves and cash flow is first prepared. If several reserve categories exist, such as proved developed, proved undeveloped, and probable reserves, then a separate weight factor is applied to the individual reserve projection and cash flow associated with each category. As a rule, there is generally little or no risk associated with proved developed reserves and a factor of 1.0 would be assigned. Where risks are involved relative to proved developed reserves such as one-well properties, it is recommended that risk be taken into account in the FMV determination. The risk factors corresponding with other reserve classifications would generally fall within the following ranges:

	Risk Factor
Proved undeveloped	0.60–0.90
Probable	0.30–0.70
Possible	0.00–0.15

The risk associated with each predicted cash flow is dictated by the confidence one has in the basic data and the reserve projections derived from these data. After the separate weight factors are applied to the individual cash flows, the cash flows can be accumulated to give a risk adjusted total. The risk adjusted total cash flow can then be discounted to determine its present worth, and the appropriate measures of profitability applied to determine fair market value.

In summary, the important thing is to incorporate risk formally into the property appraisal process regardless of which method is used. Appraisals that do not formally incorporate risk and uncertainty are not representative of the true value of a property.

OIL AND GAS DEALS

When evaluating oil and gas properties, it is important to understand how oil and gas deals are structured so that the correct net receiving interest is used in the cash flow analysis. Remember that the net receiving interest is equal to the working interest times the quantity (1 – royalty interest). In many oil and gas deals such as drilling funds, however, the relationship between the net receiving interest and the working interest is much more complex. On some occasions, the oil and gas property evaluator must spend more time trying to understand the structure of the deal than he or she spends on estimating reserves and preparing the cash flow analysis. Important questions to be answered include the following:

1. For the interest being evaluated, do any fees such as management fees need to be considered in the analysis in addition to operating costs?

2. Are there overriding royalty interests that need to be considered?
3. Does the interest being evaluated change after payout or some percentage of payout due to back-in or reversionary interests?
4. Is the deal burdened by a net profits interest?
5. Is there a penalty arrangement and/or a reduced back-in net revenue interest for nonparticipation in the drilling of a well?

Oil and gas wells can be drilled on either a promoted or nonpromoted ("heads up") basis when investors are involved. In the past, a common oil and gas industry promoted deal was often referred to as a "third for a quarter." In such a deal, three hypothetical investors would each pay one-third of the costs to drill a well to the casing point. Thus, they would each own a one-third working interest in the well to the casing point. If the well was completed, then the individual who promoted the deal would come back into the deal and the promoter and the three investors would share equally (25% each) in the costs to complete and operate the well and in the net revenue to be derived from production. Thus, the promoter in such a deal would be carried free of charge to the casing point. It is important to remember that the net revenue received by each party in such a deal would be the 25% working interest adjusted for royalty and any overriding royalty interests that might burden the deal.

In today's environment, the standard third for a quarter deal is generally no longer the basis for deals found in most domestic exploratory and development drilling ventures. In many cases, today's typical drilling deal is closer to a heads-up deal than in years past. This is mainly due to the less attractive economics and higher risk associated with domestic exploratory drilling projects versus other readily available investment alternatives such as oil and gas property acquisitions and international exploration and development opportunities, which are perceived by investors as having more favorable risk to reward relationships. An example of a common deal today that typifies the change in the basic structure of domestic drilling deals is one in which the promoter is paid a specified amount of cash plus an overriding royalty interest as compensation for putting the deal together. At payout, the overriding royalty interest for the promoter would convert to a small working interest plus possibly a reduced overriding royalty interest.

When reviewing oil and gas deals from a property evaluation standpoint, the important thing is to determine what percentage of the total revenue is to be received by the net revenue interest being evaluated. Drilling funds often operate on the basis of an arrangement in which the costs are divided among the limited and general partners who share all costs, for example, on a 90%–10% basis, while the net revenue is divided on a 75%–25% basis. Other examples of arrangements between limited and general partners include cost and revenue splits that change after payout. An important thing to be aware of in such deals is whether payout in a multiwell deal is defined on a well by well basis or on a project as a whole basis. The manner in which payout is defined in such deals can have a significant impact on cash flow for the net receiving interest being evaluated.

Oil and gas deals other than promoted deals that an oil and gas property evaluator needs to be aware of are listed here (Gossett, 1984):

- *Farm-out:* A farm-out is an agreement that assigns a certain interest from a party owning a lease(s) (farmor) to another party (farmee) for the drilling of a well at a specified location to a certain objective. For the farmor, the farm-out serves to keep the lease in force and have the area tested when the farmor either lacks funds to drill a well or believes the probability of a dry hole is high. For the farmee, the agreement offers an opportunity to establish production by the drilling of a well and earn a portion of the leasehold. The most common farm-out agreement is the undivided type in which the farmee will generally take an assignment of full interest in the drill site. The farmor generally retains an overriding royalty in the drill site with the option to convert to a predetermined working interest at payout. A less common farm-out agreement is the divided type that is commonly called a "checkerboard" farm-out. In this case, the lease(s) is divided into a checkerboard configuration that is worked out on the basis of a specified amount of acreage (for example, 640, 320, 160, or 40 acres). The farmor generally retains every other tract in the checkerboard while farming out the offset tracts to the farmee. The farmor will generally retain 100% working interest in his or her tracts and an overriding royalty interest in the farmed out tracts. Depending on the language of such agreements, the farmee may or may not be required to drill a well on each tract that is productive to earn the drill site tract. Farm-outs may be simple or complex. In either case, the property evaluator needs to understand how the interest being evaluated is affected by the farm-out agreement.
- *Overriding Royalty:* A royalty that is reserved by the farmor in addition to the royalty paid to the landowner is called an *override.* The overriding royalty can have the option to convert to a working interest after payout or can remain a straight override without conversion. The decision to convert an override is dependent on the results of the well. Good wells are generally

converted, while wells that are subject to long payout periods, reworking operations, or projected short lives may not be converted. When a property is burdened by an override, the oil and gas property evaluator must make a decision based on the available data as to whether he or she thinks the override will be converted at some point in the future (at payout). It is clear that a convertible override can have a significant effect on the future cash flow for a particular interest being evaluated.

- *Back-In:* A back-in deal allows for the farmor to come back into the deal for a specified working interest after payout. A common back-in deal is referred to as a 75%–25% deal. In this case, the farmee pays 100% of the initial well costs to earn 100% of the working interest until payout. At payout, the farmor comes into the deal for a 25% working interest and the farmee is reduced to a 75% working interest in the well. The costs of any subsequent wells are generally borne proportionately between the farmor and the farmee. The biggest problem for the property evaluator in this type of deal is estimating when payout will occur.
- *Net Profit Interest:* A net profit interest shares in a percentage of the profits of the working interest and not in the revenue as does a royalty or overriding royalty interest. *Profit* is defined as the gross revenue derived from production minus royalty; any overriding royalties; and all costs of production, equipment, operating expenses, taxes, and any other deductions authorized under the agreement from which the net profit interest was created. The holder of a net profit interest has no obligation for exploration or operating costs, nor does he or she participate in decision making or operations relative to the property burdened by the net profit interest. It is important to remember that a net profit interest shares only in the profits of the working interest. If a well has not paid out or is not profitable, then the net profit interest owner does not receive income. An oil and gas property evaluator must become familiar with the portion of the written agreement covering net profits because there is often confusion about what constitutes payout under a net profit agreement.
- *Production Payments:* A type of deal that is less common than those just discussed involves the substitution of a production payment for an override in a farm-out agreement. In this case, the farmor farms-out his or her acreage and reserves a specified sum of money to be paid out of a specified percentage of production as opposed to an override. An example of a hypothetical production payment would be $2,000,000 out of 1/8 of 7/8 of production.

The preceding discussion has been a brief, simplified summary of the subject of oil and gas deals. Many deals are often more complex and varied than those discussed (Thompson and Wright, 1985). It is important for oil and gas property evaluators to be fully aware of the deal and its impact on the property evaluation. A common problem with property evaluations is the realization after the fact that a property is burdened by a convertible overriding royalty interest or net profit interest that has inadvertently not been considered in the property evaluation. Oil and gas property evaluators need to have a thorough and up-to-date understanding of deals and an ability to interpret the written agreements that cover the various elements of the deal.

APPRAISAL OF UNDEVELOPED LEASEHOLD

Undeveloped leases associated with a producing oil and gas property should always be considered in the appraisal process. Undeveloped leasehold interests are often evaluated by one of two methods (Garb, 1985). The discounted cash flow method, heavily risk adjusted, is used if reserves can be estimated by analogy or by definition of the prospect area with geology or geophysics. If analogy is not applicable or if no actual drilling prospects are defined, then the undeveloped leasehold interests can be valued using a comparative approach based on the costs for obtaining similar leases in the same area.

APPRAISAL OF EQUIPMENT, FACILITIES, AND PLANTS

In many instances, appraisals of surface equipment, facilities, and plants are an important part of the oil and gas property evaluation process. This is particularly true in the case of buying or selling producing properties because equipment, facilities, and plants may add considerable value to a producing oil or gas property. In some cases, a review of surface equipment and production facilities may indicate that the equipment is in such deteriorating condition that substantial additional investment in equipment will be required in the future to support continuing production from the property or field over its remaining life. In those cases, it may be necessary to revise the reserves and economic cash flow projections for a property evaluation so as to include the additional investment requirements.

As a rule, experts who are knowledgeable about the appraisal of equipment, facilities, and plants are asked to review the properties in question and submit valuation reports. The results of these reports are then combined with the reserves and economics report prepared by the geologist and engineer in an effort to establish total fair market value for the oil and gas property. The principal methods for appraisal of plants, refineries, terminals, and other related facilities include (1) capitalization of future earnings, (2) depreciated replacement cost, and (3) comparable sales.

ENVIRONMENTAL COST CONSIDERATIONS

Other costs that should be considered in the evaluation of oil and gas properties are environmental and abandonment costs (Russell, 1989). Increasingly, federal and state environmental agencies are requiring operators to incur additional costs in connection with such things as closing of open pits associated with production and disposal of produced water and hazardous waste material. An environmental audit of a property may be required to ensure that the property is not negatively affected by the existence of hazardous substances or detrimental environmental conditions. With respect to abandonments costs, property evaluators in many cases in the past have assumed that the salvage value of equipment and materials would offset abandonment costs for onshore producing properties. It was also argued that in many cases, abandonment costs occurred so far in the future that the effect of discounting on the present worth or present value of the abandonment costs resulted in a cost so small that it could be ignored for all practical purposes. Consequently, little or no consideration was given to abandonment costs. Increasingly, however, abandonment costs are an important consideration in oil and gas property evaluation. For example, there is a significant cost difference between whether an operator is required by a lessor or state regulatory body to recover completely all buried flow lines from a lease versus simply cutting flow lines off at the surface. In the offshore area, abandonment costs can be very significant (an average of $2 million or more per platform depending on water depth) because operators are required to abandon production platforms and facilities completely and clean up any debris on the ocean floor that was associated with development of the property.

THE REPORT AND DUE DILIGENCE

In many cases, a written report is required as the final step of an oil and gas property evaluation. Although report formats vary, most report styles tend to contain the following standard sections: (1) introduction, (2) conclusions, (3) recommendations, (4) discussion, and (5) appendices. An evaluation report should be prepared with the understanding that it may be subject to due diligence review. Van Sandt (1989, pp. 1–2) defines the *due diligence process* as the

> . . . systematic review and verification of the information contained in an oil and gas evaluation report, prospectus, or other documents used by one or more parties involved in the sale of stock, bonds, or interests in a property, an acquisition, a merger, or bank loan and reporting those findings to the client as well as reporting any other factors which were found that may impact the proposed transaction.

Therefore, the report should be clear and forthright with respect to any factors or hidden liabilities that might impact the results of the report, either positively or negatively. In addition, the property evaluator should make full disclosure of any present or contemplated future interest that he or she might have in the subject property being evaluated that might tend to influence or bias the results of the evaluation.

ACQUISITIONS

Acquisition of oil and gas properties can provide opportunities to the integrated producer or independent who seeks a balanced approach to reserve replacement. As stated at the beginning of the chapter, opportunities that exist in acquiring producing properties include acquisition of immediate cash flow, enhancement of property economics by means of production acceleration and cost reductions, and new exploration and development opportunities in existing fields as a result of new technology and taking another look.

Oil and gas property evaluations (projections of reserves and cash flow) play a fundamental role in determining the value at which a property changes hands between a buyer and a seller. However, perceptions of the buyer and seller with respect to future price expectations also have a significant impact on establishing value and whether or not a property changes hands. A review of 96 transactions by Diggle and David (1987) between 1974 and 1987 indicated that reserves sell for about 50% of posted price during times of higher future price expectation (such as late 1976 to mid-1979) and at about 25% of posted price during times of lower future price expectation (such as late 1981 to early 1986). The average over the period reviewed by Diggle and David was 30%, which corresponds to the rule of thumb that oil reserves in the ground are worth approximately one-third of the current market value (discussed in the section entitled Fair Market Value).

During the 1980s, a decision criterion was emphasized in choosing among investment opportunities which was based on minimization of the replacement costs per barrel of oil. Using this approach, the decision maker would review his or her inventory of investment opportunities and choose those projects in which the costs of finding and developing petroleum were the lowest. This new profitability criterion was readily apparent in the "merger mania" of the 1980s. For many decision makers, replacement costs and risks were more readily minimized through the merger or acquisition process than through petroleum exploration and development.

Statistical studies (Herold, 1990) tend to suggest that it is still cheaper to buy reserves at the start of the 1990s than to drill for them. The differential between buying reserves and drilling for them, however, has narrowed in recent years in the United States due primarily to increased efficiency by U.S. companies in exploration and development during the period from 1985 to 1989. It could be argued that companies have tended to drill the better prospects (that is, less risk) during this period, thereby reducing exploration and development costs. This is in addition to the dramatic drop in drilling costs that has also occurred. Presumably if exploration activity were to increase dramatically, companies would be forced to drill those prospects in inventory that were riskier and would thus have the potential of driving up finding costs due to unsuccessful efforts. Drilling costs would also increase. This in turn would tend to shift reserve replacement in favor of reserves acquisitions as opposed to exploration and development. These observations coupled with the history of the oil and gas business over the last decade or so would tend to suggest that a company, large or small, is best served strategically by striking a balance in reserve replacement between acquiring reserves versus finding reserves with the drill bit.

CONCLUSIONS

The methods in oil and gas property evaluation presented in this chapter represent the basic methods most commonly used today by the industry in determining the value of oil and gas properties. As emphasized previously, the before tax and after tax cash flow models as shown in Tables 1 and 2 are complex. In effect, the cash flow models shown are nothing more than mathematical equations made up of numerous input variables. Property evaluators with engineering and/or geological backgrounds often tend to put emphasis on determination of reserves and reserve forecasts with little effort focused on price forecasts, costs analyses, taxes, or analyses of the economic interest(s) being evaluated. Reserves and reserve forecasts, however, represent only two variables in the cash flow equation. Such factors as prices, costs, and taxes can greatly impact the value of a property. Property evaluators must therefore be knowledgeable in the areas of legal, tax, accounting, and financial matters relating to property evaluation or at least be able to communicate with experts who are knowledgeable in those areas.

Even under the best conditions, the reserve evaluator must often resort to a "best guess" of some variables in the cash flow equation. The primary challenge in property evaluation, however, is to minimize the guess work by means of collecting and assimilating all available data. The final step in the process is to develop an understanding of the degree of uncertainty that exists in the calculated property value arising out of the application of best guess input data. Certainly all of these comments relating to the evaluation process need to be kept in perspective. It makes no sense to spend excessive time and money evaluating a property in which the cost to perform the evaluation is equal to or in excess of the estimated value of the property. In the final analysis, however, the reliability and confidence level associated with a property evaluation are strictly a function of the quality and quantity of the data that are input into the evaluation.

References Cited

Arps, J. J., 1956, Estimation of primary oil reserves: Trans. AIME, v. 207, p. 183-186.

Brealey, R., and S. Myers, 1984, Principles of corporate finance: New York, McGraw-Hill Book Company, p. 416-422.

Brown, G. G., D. L. Katz, G. B. Oberfell, and R. C. Alden, 1948, Natural gasoline and volatile hydrocarbons: Tulsa, OK, Natural Gasoline Association of America, p. 44.

Campbell, J. M., Jr., J. M. Campbell, Sr., and R. A. Campbell, 1978, Mineral property economics, v. 1: Economic principles and strategies: Norman, OK, Campbell Petroleum Series, p. 103-104.

Craze, R. C., and S. E. Buckley, 1945, A factual analysis of the effect of well spacing on oil recovery: Drilling and Production Practice, API, p. 144-155.

Diggle, F. J., and A. David, 1987, The perils of market value averages: Paper SPE 16841 presented at the 62nd Annual Technical Conference and Exhibition, Society of Petroleum Engineers, Dallas, TX, Sept. 27-30.

Eggleston, W. S., 1964, Methods and procedures for estimating fair-market value of petroleum properties: Journal of Petroleum Technology, v. 16, n. 5, p. 181-186.

Gallun, R. A., and J. W. Stevenson, 1986, Fundamentals of oil and gas accounting, 2nd edition, Supplement to Chapter 10: Tulsa, OK, Penn Well Publishing Company, p. 1-16.

Garb, F. A., 1985, Oil and gas reserves classification, estimation, and evaluation: Journal of Petroleum Technology, v. 37, p. 373-90.

Garb, F. A., 1990, Which fair market value method should you use?: Journal of Petroleum Technology, v. 42, n. 1, p. 8-17.

Gossett, R. E., 1984, An overview of well trades in Comprehensive land practices: Fort Worth, TX,

American Association of Petroleum Landmen, ch. IX, p. 5-13.

Gustavson, J. B., 1985, Comparison of methods for determining fair market values: Paper SPE 13769 presented at the 1985 SPE Hydrocarbon Economics and Evaluation Symposium, Dallas, TX, March 14-15.

Hammerlindl, D. J., 1971, Predicting gas reserves in abnormally pressured reservoirs: Paper SPE 3479 presented at the 46th Annual Fall Meeting, Society of Petroleum Engineers of AIME, New Orleans, LA, Oct. 3-6.

Herold, J. S., Inc., 1990, The John S. Herold, Inc., reserve replacement cost survey 1985–1989: Greenwich, Petroleum Outlook, v. XLIII, n. 8, Aug., p. 1-33.

Ikoku, C. U., 1985, Economic analysis and investment decisions (Chapter 4): New York, John Wiley & Sons, p. 95-122.

Katz, D. L., M. W. Legatski, M. R. Tek, L. Gorring, and R. L. Nielsen, 1966, How water displaces gas from porous media: Oil & Gas Journal, v. 64, n. 2, p. 55-60.

KPMG Peat Marwick, 1990, Energy industry profits from new tax law in Taxletter: New Orleans, KPMG Peat Marwick, Nov., p. 1-6.

Morgan, N. R., 1967, Criteria for determining type of drive in oil and gas reservoirs: Houston, Society of Independent Professional Earth Scientists, p. 1-9.

Newendorp, P. D., 1975, Decision analysis for petroleum exploration: Tulsa, OK, Penn Well Publishing Company, p. 9-57.

Russell, R. M., 1989, Environmental liability considerations in the valuation and appraisal of producing oil and gas properties: Journal of Petroleum Technology, v. 41, n. 1, p. 55-58.

Standing, M. B., 1952, Volumetric and phase behavior of oil field hydrocarbon systems: New York, Reinhold Publishing Corporation, pocket on back cover.

Standing, M. B., and D. L. Katz, 1942, Density of natural gases: Trans AIME, v. 146, p. 144.

Stermole, F. J., and J. M. Stermole, 1987, Economic evaluation and investment decision methods: Golden, CO, Investment Evaluation Corporation, p. 231-241.

Thompson, R. S., and J. D. Wright, 1985, Oil property evaluation, 2nd edition: Golden, CO, Thompson-Wright Associates, ch. 7, p. 1-22 and ch. 11, p. 1-25.

Van Sandt, B. K., 1989, Due diligence: Its importance, impact, and improbabilities: Paper SPE 18922 presented at the 1989 SPE Hydrocarbon and Evaluation Symposium, Dallas, TX, March 9-10.

Wiggins, G. B., and D. J. Tearpock, 1985, Methods in oil and gas reserves estimation: Prospects, newly discovered, and developed properties: AAPG Short Course, 1985 American Association of Petroleum Geologists Annual Convention, New Orleans, LA.

Wharton, Jr., J. B., 1948, Isopachous maps of sand reservoirs: AAPG Bulletin, v. 32, n. 7, p. 1331-1339.

Suggested Reading

Garb, F. A., 1988, Assessing risk in estimating hydrocarbon reserves and in evaluating hydrocarbon-producing properties: Journal of Petroleum Technology, v. 40, n. 6, p. 765-778.

Garb, F. A., and T. A. Larson, 1987, Valuation of oil and gas reserves: *in* Bradley, H. B., ed., Petroleum engineering handbook, Society of Petroleum Engineers, ch. 41, p. 1-25.

Garb, F. A., and G. L. Smith, 1987, Estimation of oil and gas reserves: *in* Bradley, H. B., ed., Petroleum engineering handbook, Society of Petroleum Engineers, ch. 40, p. 1-32.

Gustavson, J. B., and D. J. Murphy, 1989, Risk analysis in hydrocarbon appraisals: Paper SPE 18905 presented at the 1989 SPE Hydrocarbon and Evaluation Symposium, Dallas, TX, March 9-10.

Chapter Appendix

DEFINITIONS FOR OIL AND GAS RESERVES

(Approved by the Board of Directors, Society of Petroleum Engineers (SPE), Inc., February 27, 1987.)

Reserves

Reserves are estimated volumes of crude oil, condensate, natural gas, natural gas liquids, and associated substances anticipated to be commercially recoverable from known accumulations from a given date forward, under existing economic conditions, by established operating practices and under current government regulations. Reserve estimates are based on interpretation of geologic and/or engineering data available at the time of the estimate.

Reserve estimates generally will be revised as reservoirs are produced, as additional geologic and/or engineering data become available, or as economic conditions change.

Reserves do not include volumes of crude oil, condensate, natural gas, or natural gas liquids being held in inventory. If required for financial reporting or other special purposes, reserves may be reduced for on-site usage and/or processing losses.

The ownership status of reserves may change due to the expiration of a production license or contract; when relevant to reserve assignment such changes should be identified for each reserve classification.

Reserves may be attributed to either natural reservoir energy or improved recovery methods. Improved recovery includes all methods for supplementing natural reservoir energy to increase ultimate recovery from a reservoir. Such methods include (1) pressure maintenance, (2) cycling, (3) waterflooding, (4) thermal methods, (5) chemical flooding, and (6) the use of miscible and immiscible displacement fluids.

All reserve estimates involve some degree of uncertainty, depending chiefly on the amount and reliability of geologic and engineering data available at the time of the estimate and the interpretation of these data. The relative degree of uncertainty may be conveyed by placing reserves in one of two classifications, either proved or unproved. Unproved reserves are less certain to be recovered than proved reserves and may be subclassified as probable or possible to denote progressively increasing uncertainty.

Proved Reserves

Proved reserves can be estimated with reasonable certainty to be recoverable under current economic conditions. Current economic conditions include prices and costs prevailing at the time of the estimate. Proved reserves may be developed or undeveloped.

In general, reserves are considered proved if commercial producibility of the reservoir is supported by actual production or formation tests. The term proved refers to the estimated volume of reserves and not just to the productivity of the well or reservoir. In certain instances, proved reserves may be assigned on the basis of electrical and other type logs and/or core analysis that indicate subject reservoir is hydrocarbon bearing and is analogous to reservoirs in the same area that are producing, or have demonstrated the ability to produce on a formation test.

The area of a reservoir considered proved includes (1) the area delineated by drilling and defined by fluid contacts, if any, and (2) the undrilled areas that can be reasonably judged as commercially productive on the basis of available geological and engineering data. In the absence of data on fluid contacts, the lowest known structural occurrence of hydrocarbons controls the proved limit unless otherwise indicated by definitive engineering or performance data.

Proved reserves must have facilities to process and transport those reserves to market that are operational at the time of the estimate, or there is a commitment or reasonable expectation to install such facilities in the future.

In general, proved undeveloped reserves are assigned to undrilled locations that satisfy the following conditions: (1) the locations are direct offsets to wells that have indicated commercial production in the objective formation, (2) it is reasonably certain that the locations are within the known proved productive limits of the objective formation, (3) the locations conform to existing well spacing regulation, if any, and (4) it is reasonably certain that the locations will be developed. Reserves for other undrilled locations are classified as proved undeveloped only in those cases where interpretations of data from wells indicate that the objective formation is laterally continuous and contains commercially recoverable hydrocarbons at locations beyond direct offsets.

Reserves that can be produced through the application of established improved recovery methods are included in the proved classification when (1) successful testing by a pilot project or favorable production or pressure response of an installed program in that reservoir, or one in the immediate area with similar rock and fluid properties, provides support for the engineering analysis on which the project or program is based and (2) it is reasonably certain that the project will proceed.

Reserves to be recovered by improved recovery methods that have yet to be established through repeated commercially successful applications are included in the proved classification only (1) after a favorable production response from subject reservoir from either (a) a representative pilot or (b) an installed program, where the response provides support for the engineering analysis on which the project is based, and (2) it is reasonably certain the project will proceed.

Unproved Reserves

Unproved reserves are based on geologic and/or engineering data similar to that used in estimates of proved reserves; but technical, contractual, economic, or regulatory uncertainties preclude such reserves being classified as proved. They may be estimated assuming future economic conditions different from those prevailing at the time of the estimate.

Estimates of unproved reserves may be made for internal planning or special evaluations, but are not routinely compiled.

Unproved reserves are not to be added to proved reserves because of different levels of uncertainty.

Unproved reserves may be divided into two subclassifications: **probable** and **possible**.

Probable Reserves. Probable reserves are less certain than proved reserves and can be estimated with a degree of certainty sufficient to indicate they are more likely to be recovered than not.

In general, probable reserves can include (1) reserves

anticipated to be proved by normal stepout drilling where subsurface control is inadequate to classify these reserves as proved; (2) reserves in formations that appear to be productive based on log characteristics but that lack core data or definitive tests and which are not analogous to producing or proved reservoirs in the area; (3) incremental reserves attributable to infill drilling that otherwise could be classified as proved, but closer statutory spacing had not been approved at the time of the estimate; (4) reserves attributable to an improved recovery method that has been established by repeated commercially successful applications when a project or pilot is planned but not in operation, and rock, fluid, and reservoir characteristics appear favorable for commercial application; (5) reserves in an area of a formation that has been proved productive in other areas of the field, but subject area appears to be separated from the proved area by faulting and the geologic interpretation indicates subject area is structurally higher than the proved area; (6) reserves attributable to a successful workover, treatment, retreatment, change of equipment, or other mechanical procedure, where such procedure has not been proved successful in wells exhibiting similar behavior in analogous reservoirs; and (7) incremental reserves in a proved producing reservoir where an alternate interpretation of performance or volumetric data indicates significantly more reserves than can be classified as proved.

Possible Reserves. Possible reserves are less certain than probable reserves and can be estimated with a low degree of certainty, insufficient to indicate whether they are more likely to be recovered than not.

In general, possible reserves may include (1) reserves suggested by structural and/or stratigraphic extrapolation beyond areas classified as probable, based on geologic and/or geophysical interpretation; (2) reserves in formations that appear to be hydrocarbon bearing based on logs or cores but that may not be productive at commercial rates; (3) incremental reserves attributable to infill drilling that are subject to technical uncertainty; (4) reserves attributable to an improved recovery method when a project or pilot is planned but not in operation and rock, fluid, and reservoir characteristics are such that a reasonable doubt exists that the project will be commercial; and (5) reserves in an area of a formation that has been proved productive in other areas of the field but subject area appears to be separated from the proved area by faulting and geologic interpretation indicates subject area is structurally lower than the proved area.

Reserve Status Categories

Reserve status categories define the development and producing status of wells and/or reservoirs.

Developed. Developed reserves are expected to be recovered from existing wells (including reserves behind pipe). Improved recovery reserves are considered developed only after the necessary equipment has been installed, or when the costs to do so are relatively minor. Developed reserves can be subcategorized as producing or nonproducing.

Producing. Producing reserves are expected to be recovered from completion intervals open at the time of the estimate and producing. Improved recovery reserves are considered to be producing only after an improved recovery project is in operation.

Nonproducing. Nonproducing reserves include shut-in and behind-pipe reserves. Shut-in reserves are expected to be recovered from completion intervals open at the time of the estimate, but which had not started producing, or were shut in for market conditions or pipeline connection, or were not capable of production for mechanical reasons, and the time when sales will start is uncertain.

Behind-pipe reserves are expected to be recovered from zones behind casing in existing well, which will require additional completion work or a future recompletion prior to the start of production.

Undeveloped. Undeveloped reserves are expected to be recovered (1) from new wells on undrilled acreage, (2) from deepening existing wells to a different reservoir, or (3) where a relatively large expenditure is required to (a) recomplete an existing well or (b) install production or transportation facilities for primary or improved recovery projects.

GENE B. WIGGINS III, Vice-President and Director of Atwater Consultants, Ltd., in New Orleans, received a B.S. degree in Mechanical Engineering from the University of Houston in 1978 and an M.B.A. from Tulane University in New Orleans in 1985. He has worked for Zapata Technical Services, Hughes Tool Co., and Getty Oil Co. He joined Atwater Consultants in 1980 as a Petroleum Engineer and became Vice-President in 1985 and Director in 1991. Atwater Consultants is a geological and petroleum engineering firm that specializes in evaluation of oil and gas properties. Mr. Wiggins is a Registered Professional Engineer in four states and a member of SPE, SIPES, AAPG, API, the Society of Petroleum Evaluation Engineers, the New Orleans Geologic Society, and the American Society of Appraisers. He has taught various courses and seminars in oil and gas property evaluation for AAPG, GCAGS, SIPES, Tulane University, and the University of New Orleans.

Chapter 13

PC Software Considerations for Economic Evaluation of Investments in the Petroleum Industry

John M. Stermole
Investment Evaluations Corporation
Golden, Colorado, U.S.A.

Beware of a report that is data rich and information poor.
—Anonymous

OBJECTIVES

This chapter has three objectives. The first is to identify what potential users can expect to find in personal computer software for economic evaluation of investments in the petroleum industry. Second, it will illustrate many salient considerations that should be addressed in evaluating personal computer software for economic evaluation of oil and gas investments. Third, it will help to create a level of awareness of specific considerations that may require a more explicit explanation by a vendor of how a particular program handles various calculations. As an example, in reviewing software, it is easy for vendors to claim their product can provide a necessary computation but not to point out the additional manual entry that may be required to accomplish the task. That same task might be accomplished automatically in another program.

In addition to software considerations, the requirements of each user are very different and need to be thought out completely. As an example, firms involved in the business of oil and gas may be integrated producers who are immersed in every aspect of the petroleum industry, including exploration, development, transportation, refining, and marketing. These companies have evaluation needs that encompass a combination of income-producing alternatives that may involve selecting one or more alternative investments that will maximize profit to the firm. They may also have service investments that need to be analyzed to minimize cost for various considerations. In contrast, the producer might be a small independent or a royalty recipient who is interested in the current value of production from a single well. This range of participation in the industry can be translated into a vast array of economic evaluation considerations. With varying magnitudes of investments and risk, there are very different economic evaluation requirements to be considered. This can also be translated into a similar range of economic evaluation programs each of which tend to be directed to a specific user audience. Thus, each individual, partnership, or corporation must first consider their own unique needs and then consider programs that best fit those needs.

Some investors are looking for a program capable of tying together each of the varied aspects of their participation in the oil and gas business. They may want a program that will support a broad base of applications such as the ability to look at a range of values (due to geological, geophysical, or economic uncertainty), which may be necessary in exploration to determine which prospects hold the biggest economic potential. This might include advanced options such as Monte Carlo, or Latin hypercube simulation techniques or multiple branch decision tree analyses to build in expected value considerations. Furthermore, they may want the same program to assist reservoir and petroleum engineers in evaluating their production and reserve considerations.

On the other side is the small independent or royalty recipient who is interested in making a proper discounted cash flow analysis of his or her property, but who may not require a program offering a variety of tax and production data bases and engineering applications to determine the present value of a production stream from one or more wells. In the following pages, salient features to be considered in evaluating software are addressed in a general format.

Many comments here that are related to the current software industry are the results of a questionnaire sent out by the author to approximately 30 vendors concerning their available software packages. The Chapter Appendix presents an annotated list of these vendors. This list has been provided to assist interested readers with the question of where to start their own information search in determining the appropriate software package for their needs. The petroleum software industry is a very dynamic business consistent with the industry it helps support and, as such, no pricing information has been included. Finally, the intent here is not to critique or rank individual software programs but to provide a general understanding of what is available and a listing of where to look for more information.

PROGRAM DESIGN

Each software user has different preferences for user interface (the method by which data are entered then manipulated by the program and presented). Some users are dedicated to spreadsheets and template programs that can be installed into environments such as Lotus 1-2-3[1], Microsoft Excel[2], or Borland's Quattro Pro[3], and so on. Assuming one is literate in a particular spreadsheet program, these programs offer the advantage of a reduced learning curve for the evaluation program and the benefit of items such as graphics support that may already be provided by the spreadsheet. Depending on the program, some users can feel a little lost in template programs that present multilayered levels of data input and reports.

Other users prefer a *stand alone* system. One type of stand alone system includes those programs that are based on an interactive batch environment, which may involve the use of text editors to create a data file which is then *batched*, or sent, by the program. Some recent alternatives to the text editor allow the user to access a menu system for creation of the batch file, which is then batched to the program to be transformed into cash flows These stand alone software packages may require multiple programs for different routines, but they often offer a wide range of computational capabilities, including plant design, cost estimating, engineering, enhanced statistical considerations, and other features involved in the vertical market of oil and gas. One disadvantage of these systems is their complexity and the need to understand the use of text editors with each individual program.

Still other systems are completely *menu driven*, which means that the program accepts data, creates its own file data internally, and computes the appropriate criteria when called upon by the user. Menu-driven systems appear to be at the forefront of current software development due in part to a faster learning curve for users to become competent in the operation of these systems. A potential disadvantage to these programs is that users may feel locked into a perceived fixed environment.

One other consideration is the ability of the software and the personal computer system to interface with mainframe computers to access data bases that might be available either from the software company or from internal sources. These systems represent the most sophisticated level of program for personal computers and exist in one of the two stand alone environments discussed earlier.

HARDWARE REQUIREMENTS

Of the two primary personal computer hardware systems available today (IBM[4] compatibles and Macintosh[5]), most software systems available at the time of this publication are oriented toward IBM compatible units. In fact, of the companies interviewed for this article, only one (Palisade Corporation) indicated a program available for Macintosh users. Therefore, the balance of this section addresses hardware considerations for IBM compatible systems.

Although memory and disk storage space have become relatively inexpensive in recent years, users should be aware that depending on the software selected, storage requirements can range from less than 0.5 MB (megabytes) to more than 10.0 MB, which can represent a sizeable allocation of available memory. Other programs may actually require a hard disk to function properly due to internal program swapping that is built into the system. Furthermore, many programs require 100% of dedicated memory, which restricts the ability of users to install auxiliary pop-up window programs such as clocks, date books, and phone books. This problem is diminishing with advancements in PS/2[6] and Microsoft Windows related technology, allowing for several programs to operate concurrently within a single operating system. But to date, this technology has not been integrated into the architecture of many software programs.

[1]1-2-3 is a registered trademark of Lotus Corporation.
[2]Excel is a registered trademark of Microsoft Corporation.
[3]Quattro Pro is a registered trademark of Borland International Inc.
[4]IBM is a registered trademark of International Business Machines Corporation.
[5]Macintosh is a registered trademark of Apple Computer Inc.
[6]PS/2 is a registered trademark of International Business Machines.

Other considerations relate to available sources of output provided by the software. Almost all programs give the option of directing output to a monitor or printer, but some offer more options than others. For instance, if you are required to make numerous presentations, you would want a program with the ability to provide graphics and display those graphics on a screen or direct the results to a plotter for use in reports or to generate transparencies. Relevant questions in this area relate to the level of graphics display support provided on screen, such as video graphics array (VGA), enhanced graphics adapter (EGA), color graphics adapter (CGA), or monochrome displays. Furthermore, does the program support laser jet printer font escape sequences and plotter commands? As an example, all programs will most likely support output to a dot matrix printer, which supports graphics output but not with the same quality of resolution as a laser printer or the color capabilities of plotters.

Processing time is largely a function of the patience of each user. By necessity of the calculations involved, economic evaluation programs in general require significant computational capability from the hardware, which usually justifies the small additional expense of a math coprocessor. This will greatly accelerate computation time for all programs. This option is required for some programs. Other considerations not common to all software programs include mouse support, program protection devices that may need to be plugged into a parallel or serial port, modem support for interoffice transfer of data, and technical support and updates.

GENERAL SOFTWARE FEATURES

This topic addresses not only the software but the company or individual standing behind the software. Specifically, if you are interested in a program, many vendors have demonstration copies or will allow for a trial period to test and consider the product. Once the decision has been made to purchase a program, consideration must be given to how many people will be using the program and whether a purchase agreement, site license agreement, or lease makes the most economic sense. Some programs may be installed in a network environment, others may not. Software protection systems vary considerably from company to company. Unfortunately, there are no rules of thumb for determining a reasonable price or fee for software. For larger programs, the final figure tends to be a negotiated value based on the needs of each user. Smaller programs simply do not have the profit built-in to be very negotiable, although volume discounts might be possible.

Users should be aware that the same copyright laws that apply to printed matter work to protect the rights of software developers. As such, users should develop a plan that accurately reflects their needs. The cost associated with an additional copy of a program can usually be paid off with a single economic evaluation. In this author's opinion, bootleg copies are not worth the potential liabilities.

Maintenance is another area of concern. Some companies provide users with on-site installation and training, while other programs (templates and some stand alone environments) are much simpler and can be installed and operated by users simply by following instructions provided in the program manual, which should be a part of any program. A maintenance fee may be included in the purchase price of the software package. Updates might be included as part of an annual maintenance fee or available for upgrade fees when released by the vendor. The ability to protect sensitive areas of program files such as financial data or price forecasts is an additional concern. Internal file protection of this type is typically limited to larger scale programs.

In reporting data, most programs offer the variety of "canned" report formats that can be modified or tailored to individual needs. Depending on the program, this modification may be done either manually by the user or may be provided as a package option by the vendor. Most standard technology allows users some flexibility in designing their own cash flow report format.

A final consideration is the maximum evaluation life that a program can handle. The length of time over which a project may be established and the number of discrete or continuous interest compounding periods involved in an analysis are important investment analysis considerations.

TECHNICAL SOFTWARE FEATURES

This section addresses some of the more in-depth aspects of making an economic evaluation of hydrocarbon accumulations. Production, capital and operating expenditures, price forecasts, tax considerations, and the desired decision criteria are some of the primary areas of concern. Again, each individual will have different needs. The important consideration here is to purchase a package that addresses your investment situation.

Production Characteristics

Oil and gas production schedules can be developed in a variety of methods. Most programs provide several options including the simplest approach which is to manually enter anticipated production. While appearing to be the most archaic of the alternatives, this approach does allow users an added degree of flexibility depending on the input time required.

In some software packages, decline characteristics have been established that allow for more sophisticated production estimates based on exponential, hyperbolic, or harmonic declines. Gas deliverability calculations are also available in selected versions of the larger programs. Most often, production schedules are calculated on a monthly basis, but may vary from vendor to vendor. Decline curve considerations should also include the ability to change decline rates over time. The program should also have the ability to calculate secondary products separately as a unique production stream or to be considered as either a fixed ratio to a primary production stream or as a variable function of a primary stream.

With this information available, the real production questions then come to the surface: "How many wells can the program handle in a single evaluation?" "Can I complete more than one zone in a well?" For example, the price of some programs is a function of how many wells the individual wants to work with in a single program. One other consideration is the ability of a program to account for increasing production due to factors such as secondary or tertiary recovery methods.

Capital Expenditures

The degree to which capital expenditures need to be accounted for reflects the type of evaluation to be made. A proper after-tax analysis requires a certain level of detail to account properly for the tax effects of each cost. In addition, the financial position of each investor will impact how the benefits from those deductions are actually realized. This is discussed in the tax section of this chapter.

In general, capital expenditures can be broken down into several components. Those components might include acquisition or lease bonus costs, geologicalal and geophysical exploration costs, intangible drilling costs (IDCs), and tangible completion costs. Other components include land, plant costs, pipeline costs, and vehicles, all of which are handled differently for tax purposes.

Some programs offer the flexibility of entering as many costs of each type as previously described. In addition to the range of available costs, the manner of handling the timing of when costs can be entered into a program is another important consideration. Furthermore, some programs relate capital expenditures directly to production from a given well, while others require the user to tie the timing of capital expenditures to production in the data input process. The first approach more often relates to a database approach to economic evaluations, while in the latter situation, you would simply enter the appropriate capital costs and a separate production stream. The values in the second approach are not tied together internally to a specific well or specific well history.

Operating and Overhead Costs

Operating costs are expressed in a variety of methods, and most vendors allow users to specify costs accordingly. The most common method available in almost all programs is to specify costs as a monthly or period value, with the period value expressed on a per well basis or as a cost per unit produced. Within some programs, these values may escalate, de-escalate, vary in an arithmetic fashion, or remain constant over the producing life. Other programs allow these values to vary within specified maximum or minimum tolerance ranges after which the value may remain constant.

Oil and Gas Prices

Revenues are usually a function of production and a specified selling price per unit produced. Consistent with operating costs, most programs give users the flexibility to enter revenue dollar figures per period or to calculate period revenue from period production multiplied by selling prices. Selling prices can be expressed either as fixed, escalating, de-escalating, or arithmetically varying values within a maximum or minimum range of tolerance over the evaluation life.

Product Taxes (Excise and Severance Taxes)

Excise, severance, and ad valorem taxes along with compression and pipeline charges vary from state to state, both in the method of calculation and in the applicable rates. Most programs are designed to accommodate these taxes on a per period basis, as a dollar per unit produced, or as a percentage of a calculated net revenue value, which could range from a calculated gross value to net value (gross adjusted for royalties) adjusted for out-of-pocket expenditures.

Tax Considerations

Under the current tax law, there are three types of producers to be considered in economic evaluations: (1) integrated producers, (2) independents producing less than 1000 bbl of oil or oil equivalent per day, and (3) independents producing more than 1000 bbl of oil or oil equivalent per day. For each producer, the ability to make deductions is a function of their relative financial and producer status.

The deductibility of intangible drilling costs and the ability to use percentage depletion are affected by these different categories of producers, which can have a significant impact on project economics. Therefore, the program should be capable of differentiating these different producer categories. Most of the stand alone environment programs offering after-tax analyses address these differences, while many template programs do not.

Depreciation of capital equipment such as tangible completion costs, pipelines, plant costs, and so on is currently defined in the U.S. tax laws under the Modified Accelerated Cost Recovery System (MACRS). Most programs allow users to use the current depreciation method and alternative methods such as straight line or units of production. This can be an important consideration, especially if evaluations are being made outside the United States or if evaluations are performed subject to public financial reporting requirements.

The owner of an economic interest in a mineral deposit in the United States is entitled to applicable depletion deductions as the units contained in that property are depleted. Limitations on the type and extent of depletion do exist for the various types of producers previously mentioned. Cost depletion and statutory percentage depletion are the two approaches applied to determine the allowable deduction each year. A notable exception to this is that percentage depletion may only be taken on the first 1000 bbl per day of oil or oil equivalent produced. Consistent with the various producer classifications, if a program handles evaluations on an after-tax basis and addresses the producer issue, it most likely properly accounts for the above considerations.

Investment tax credits such as the Coal Methane or Devonian Shale credit should also be accounted for properly in the model. Tax credits may be calculated as a function of capital expenditures or selling price and represent a dollar for dollar reduction in tax liability.

Another potentially important subject is application of the Alternative Minimum Tax (AMT) to all corporate and individual producers. However, this tax relates to all corporate or investor revenues and deductions rather than to a specific project, so it is very difficult to approximate on a project by project basis. Simplified approximation routines for this tax may be developed in the next few years, but currently, very few programs are capable of allocating aggregate investor tax liability over individual projects. Therefore, the analysis of AMT does not appear in many of the programs listed in the Chapter Appendix since many do not address the overall financial position of the investor.

Federal and state taxes can be computed and are included with most programs. Some programs work with a single effective tax rate that combines both federal and state taxes, while others work with separate state and federal tax calculations. The ability to change tax rates over time maximizes program flexibility to the user and is available in many programs.

Project Participation (Joint Ventures)

This is an area of evaluation work where some significant segregation does exist. Oil field deals are structured in a variety of methods depending on the property, investors, and willingness of each party to assume a portion of the risk associated in its development. Accordingly, agreements may include leasehold or royalty interests, working interests, reversion interests that may include one or more reversion points, net revenue interests, carried interests, or production payments. Most programs can address the basics, but to view a deal from all perspectives quickly and easily, users may need to consider the larger, more complex programs.

Real and Nominal Dollars

Another consideration is the type of dollars that go into the economic evaluation model. Many investors today work in escalated (nominal) dollars that reflect the actual revenues and costs to be incurred in a project. These dollars include the effects of escalation, which includes inflation, over time. With the influx of foreign investors purchasing American companies or substantial interest in those companies, some firms must now justify project economics using constant (real) dollars, or dollars that have had the effects of inflation removed. Either approach (escalated or constant dollar) applied correctly will lead to the same economic conclusion, but many programs cannot properly execute a constant dollar analysis.

Related to this subject is the ability to account for currency exchange rates in international evaluations, which can have a significant impact on project economics. This does not necessarily have to relate to foreign investments by American companies. Companies that are owned by foreign investors might be interested in the impact of variations in exchange rates on their investments. American companies selling products in foreign markets need to be able to handle exchange rates in analyses to express all costs and revenues in the same currency.

Decision Criteria

There are a variety of ways to account for the time value of money in economic evaluations. These include the use of assumed discrete dollar values that can be realized at the beginning, middle, or end of a compounding period. These values can then be discounted using either discrete or continuous interest rates. One additional option is to assume that dollars are realized uniformly over a period and then discounted with a continuous interest rate. This is referred to as *continuous interest* for continuous flowing funds. With each of these options, you can work with dollars on a monthly, quarterly, semiannual, or annual basis or a combination of monthly and annual values, depending on the software program.

These considerations are important because it is necessary to be consistent in handling the timing of cost and revenues in the time value of money calculations. If users are looking for a small program to use in place of a company mainframe program, they may want the personal computer program to emulate the mainframe as closely as possible, which requires this type of flexibility. Furthermore, if users want to verify results with a financial calculator, this may require the use of discrete interest rates for discrete values, depending on the financial calculator.

Many calculators and software programs provide for each of these different options, but others are limited in application. As an example, spreadsheets utilizing built-in financial functions such as @IRR, for internal rate of return, or @NPV, for calculating net present value, are typically based on discrete end-of-period values and discrete interest rates.

The actual decision criteria calculated are fairly common, but some programs give you the option to create additional criteria. The most standard decision criteria results based on project cash flows include rate of return, net present value, ratios of present worth revenue to present worth cost or net present value to present worth cost, and payback criteria. Other breakeven calculations are also available, including the ability to calculate acquisition costs or breakeven product selling prices.

Some programs can identify anomalies such as "dual rates of return" that occur if a cash flow stream has costs following incomes (cost-income-cost). Other programs do not consider these cost-income-cost situations properly and require the user to address them by modifying the input file to correct the evaluation. The dual rates of return values have a combined rate of return and rate of reinvestment, which makes them useless for economic evaluations. It is very important, therefore, to handle this situation properly. Applications of these same criteria might also be applied to other project analysis situations such as incremental analysis when considering several projects from which only one may be selected.

Properly calculated ratios can also be used to rank projects when more than one may be selected, such as ranking drilling prospects in the order of decreasing economic desirability for a given budget. Some programs emphasize one well or field and do not address these later considerations, which again is a function of the needs of the user.

Other income-producing project criteria based on net income might include accounting rate of return, which is the rate of net income (not cash flow) divided by the average undepreciated asset value each year. The terms *return on assets, return on net assets, return on equity,* and *return on capital employed* are all interchangeable with *accounting rate of return.*

Some programs provide routines to evaluate cost of service investments and, as a result, calculate values such as present worth cost, annual cost, or monthly cost of service. Breakeven cost per unit produced is another calculation often used in these investment situations. Other criteria may consist of specific reports concerning reserves or other financial measures that may be required by law. The ability to generate such reports is unique to each program.

New Technology in Economic Software

Program enhancements in the area of sensitivity and risk analysis are on the leading edge of economic evaluation technology. These techniques are not new to the industry, but the ability to use these enhanced concepts in a personal computer format represents considerable cost and time savings over traditional mainframe programs. Specifically, in the area of sensitivity analysis, the application of probabilistic analysis techniques such as Monte Carlo or Latin hypercube simulation have recently appeared in several programs. This methodology associates probability of occurrence with project discounted cash flow criteria such as production rates, selling price, and capital or operating expenditures realized over a project life. Randomly selected values are placed in the model to compute cash flows and the desired criteria such as rate of return. This process is repeated 100 to 5000 times with the results typically summarized in a histogram. This provides the user with a probability distribution of likely results for the program.

A simplification of probabilistic analysis methods that gives the range over which project economic results are expected to vary without associating them with probability of occurrence is the "best/worst/most expected" analysis sometimes referred to as the *range approach.* This analysis methodology allows users to generate a best case, most expected case, and worst case scenario and does not associate probabilities of occurrence with each outcome. It is a simple but useful conventional sensitivity analysis.

Another project evaluation enhancement involves risk analysis and expected value or decision tree analysis of investments over time. This approach allows users to assign finite probabilities of occurrence to production streams from a single well to an entire producing field. Complex models allow for multiple reserve cases combined with multiple probabilities of success and failure and account for appropriate tax deductions and abandonment costs on dry holes and economic limit analyses. Results may include risk-adjusted net present value, rate of return, or ratios for a range of probabilities of success and failure.

CONCLUSIONS

In assessing economic evaluation software, a user needs to determine how complex he or she wants to be in the economic evaluation of prospects or projects. There is generally a positive correlation between price and the level of sophistication of evaluations that can be made. Each organization or evaluator must determine the program that best meets his or her needs considering the factors examined in this chapter.

Throughout this chapter, references have been made to larger, more sophisticated programs along with stand alone and spreadsheet template applications. Table 1 identifies the method of user interface for various economic evaluation programs.

Table 1. List of Programs by Type

Environment	Program
Spreadsheet/Template	
Palisade Corporation	@RISK
Sigma Energy	PETEC PLUS
Stand Alone	
Dwights Energy Data, Inc.	EnPAK
Garrett Computing Systems, Inc.	ARIES
Molli Computer Services, Inc.	MICA
Software Bisque	SEE
Strategic Systems Group	DISCOVERY
Texas Computer Works, Inc.	ECON
Integrated PC/Mainframe	
David P. Cook & Associates	OGRE
Garrett Computing Systems, Inc.	AIRES, POGO, & POGOPLUS
Michael Smith & Associates, Inc.	PROPHET
Scotia Systems, Ltd.	BASIL
Texas Computer Works, Inc.	ECON
Sigma Energy Consultants, Inc.	PETEC PLUS

Chapter Appendix

SOFTWARE VENDERS

This appendix lists alphabetically the vendors of software applicable to the petroleum industry that we surveyed. It also includes some brief comments from some of those vendors regarding their programs.

Artesia Data Systems, Inc.
5420 LBJ Freeway, Suite 1200
Dallas, TX 75240
Phone (214) 788-0400

Boffin, Inc.
6865 Main Street
P.O. Box 1619
Frisco, TX 75034
Phone (214) 377-9771
Fax (214) 377-3557

David P. Cook and Associates
6510 Abrams Road, Suite 410
Dallas, TX 75231
Phone (214) 349-6900
Fax (214) 343-9699

Program: OGRE®

From the vendor: DPC&A has been selling petroleum economics software and services since 1978. To date, the company has more than 600 clients, predominantly PC installations. Many of our clients look to us for a "total solution." DPC&A's economic system OGRE® is the oil and gas industry's most widely used product. More than 60 consulting firms nationwide use the software, as do most of the major energy lending institutions. Oil and gas companies, from one-person shops to majors, use OGRE for reserve studies. It is easy to use, flexible, and portable to many hardware platforms, ranging from PCs to major mainframes. DPC&A maintains the system to keep up with the ever-changing oil and gas industry. OGRE features Canadian tax scenarios as well as the standard U.S. tax scenarios. The program has also been adapted to work in other countries. DPC&A also offers several other software programs that interface with OGRE, providing the end user with a comprehensive suite of analysis tools.

There are three significant ways that a client can justify an investment in OGRE. First, the program increases the user's productivity. A petroleum engineer can analyze a property more quickly using our product than most other methods. In addition, data files are upwardly and downwardly compatible, so users can transfer data from one machine to another without tedious rekeying. Second, OGRE is extremely flexible and sophisticated, allowing the user to analyze properties more carefully. For example, a company can do complex pricing sensitivity analysis with OGRE, budget planning, etc. Third, OGRE is a precision tool. It produces accurate answers, and it allows the user to take into account the small details that can affect the bottom line. OGRE is available for Microsoft Windows 3.0 and will be compatible with subsequent versions of Windows.

Dwight's, A SoftSearch Company
1633 Firman Drive, Suite 100
Richardson, TX 75081
Phone (214) 783-8002
Fax (214) 783-0058

Program: EnPAK

From the vendor: Designed for reserve management and economic forecasting, there are six program areas that function as an integrated system. A powerful relational database is used to store data and interface with each of the program areas. A complete economic system computes present value cash flows from production, economic parameters, expenses, and investment data. The programs have been designed to facilitate modeling of the impact of changes to economic conditions through the use of sensitivity analysis and risk weighting of the information. There are over 20 different report options with a report writer for creating your own company standard report formats.

A more thorough analysis can be done with additional supporting programs that are included, such as Production Curve Analysis, Multiphase Production Curve Analysis, P/Z vs. Cumulative Production, and Volumetric Oil, Gas, and Waterflood. Commercial information (Dwight's production data) and company information such as historical production and economic information can be transferred both in and out of the EnPAK database.

Garrett Computing Systems, Inc.
2828 Routh Street, Suite 500
Dallas, TX 75201
Phone (214) 871-7117
(800) 825-8001
Fax (214) 969-0041

Programs: Aries, POGO, POGOPLUS

From the vendor: POGO is an oil and gas lease economic evaluation model that computes and reports the value of oil and gas leases. Although very comprehensive, POGO is exceptionally easy to use. Its flexibility allows the users to evaluate any type of interest owned in an oil and gas property to whatever degree of sophistication required. The program ensures accurate by detailed calculation of cash flows incorporating factors relating to production, prices, royalties, interests, capital investments, and income tax. Each is time-phased to provide reports that include revenue, expense, investment, profits, various profit indicators, and present value summaries. POGO is a powerful tool in the evaluation of oil- and gas-producing properties, exploration ventures, farm-out or farm-in proposals, royalty interests, reversionary (back-in) interests, foreign concessions, capital expenditure projects, and production payment/loan situations. It is also very useful in performing sensitivity analysis, incremental economics, forecasting gross, working interest, royalty interest and net reserves, determining a lease bid/sell price, and budget preparations. The program handles U.S., Canadian, U.K., and foreign tax systems.

POGOGPLUS is a windows-based, mouse-driven integrated system that allows end-to-end engineering evaluation on your PC—from tracking production to forecasting and projecting it, to corporate reporting and presentation graphics. The POGOPLUS suite of integrated systems include:

- POGOMART data retrieval software to access Canadian production history
- PRODUCTION HISTORY to store, maintain and report your production data
- DECLINE ANALYSIS to interactively analyze and project your production
- POGO to perform comprehensive evaluations with North America's leading economic analysis tool
- RESERVES system to manage and report your technical and economic reserves
- 10 PLANNER to customize and consolidate your corporate oil and gas planning applications
- THUMBPRINT engineering graphics to analyze and present complex data quickly and efficiently using a state-of-the-art graphics tool

Designed with a modular approach, POGOPLUS allows you to use any module of the system or the entire suite of software solutions. Whether you use one module or all of POGOPLUS you still get the windowing environment with pull down menus, point and shoot manipulation of data and the power of comparing multiple windows on one screen.

Intera Technologies, Inc.
6850 Austin Center Boulevard, Suite 300
Austin, TX 78731
Phone (512) 346-2000

Program: GO!

Metro Systems, Inc.
4010 South Yale Street
P.O. Box 4697
Tulsa, OK 74159
Phone (918) 665-6514

Program: SOGAS

Meyer and Associates, Inc.
RD 1, Box 458F, Route 908
Natrona Heights, PA 15065
Phone (412) 224-1440
Fax (412) 224-1442

Program: MNETPV-1

Michael Smith and Associates, Inc.
1200 Smith Street, Suite 1700
Houston, TX 77002
Phone (713) 650-6164
Fax (713) 650-3601
Telex 466239

Program: PROPHET

Molli Computer Services, Inc.
3630 Sinton Road, Suite 300A
Colorado Springs, CO 80907
Phone (719) 520-1790
(800) 869-7616

Program: MICA

From the vendor: MICA—software for Petroleum Engineers.—is available for personal computers and network systems. MICA offers decline curve analysis, economics, interactive curve matching, and gas well forecasting, all in one easy-to-use package. MICA's Decline module has the capability to plot up to 80 years of production data. The program can regress the production data, yielding a best fit line—either exponential, hyperbolic, or harmonic. In addition, the line can be manipulated interactively on the screen. MICA's Type Curve module allows the user to "slide" data points over a type on the video screen. As the data points are moved, the match point is continuously updated. MICA can present almost any type curve the user chooses. Special features are available when using M. J. Fetkovich's Advanced Type Curve for Decline Curve Analysis. Other features of MICA include the ability to perform the Texas Railroad Commission's Form G-1 calculations, to export production figures into a LOTUS 1-2-3 spreadsheet file, and to add production history using several different methods. In addition, production history can be imported from Dwight's PCD files, Petroleum Information's Format 97 files, ASCII files, and LOTUS 1-2-3 files. MICA supports several graphics standards including CGA, EGA, VGA, Toshiba T3100 (400 line), and Compaq III (400 line). Color graphs can be displayed on EGA and VGA screens. MICA supports several high-resolution printers including HP Laserjet, HP Deskjet, NEC 24 pin, Toshiba 24 pin, Epson LQ, Epson FX, Epson LX, Epson MX, and Epson RX. Any graph displayed on the screen can be printed using any selected resolution.

Palisade Corporation
31 Decker Road
Newfield, NY 14867
Phone (607) 277-8000
(800) 432-RISK
Fax (607) 277-8001

Program: @RISK

From the vendor: @RISK is a Lotus 1-2-3 add-in for risk analysis. Any work sheet built in 1-2-3 can be used with @RISK. The software uses Monte Carlo simulation to analyze uncertainty. Probability distributions are added to cells in the work sheet using 30 new probability @ functions, including normal, lognormal, beta, uniform, and triangular. Simulations are controlled from a Lotus-style menu that lets users choose Monte Carlo or Latin hypercube sampling, select output ranges, and start simulating. A separate output routine displays results graphically and provides detailed statistics.

The probability distribution @ functions can be used anywhere in a work sheet, either alone in a cell or as part of other expressions, and their parameters can be references to other cells. Dependencies can be specified. Up to 32,000 iterations can be run per simulation, and the seed of the random number generator can be specified. High-resolution graphics enhance manipulation of output results. Output graphics can be exported in .PIC format, and simulation data can be exported in ASCII format. Simulations can be paused and restarted. A version of @RISK for Microsoft Excel (PC and Macintosh) is available.

Petrocomp Systems, Inc.
1111 North Loop West, Suite 940
Houston, TX 77008
Phone (713) 880-0226
Fax (713) 861-7246

Program: PEEK

Petroconsultants, Inc.
6600 Sands Point Drive
P.O. Box 740619
Houston, TX 77274-0619
Phone (713) 995-1764
Fax (713) 995-8593
Telex 4620521

Program: GIANT

From the vendor: GIANT is a PC-based program that takes user-defined costs and technical parameters and combines them with the tax and contract terms in any country to generate a comprehensive after-tax economic evaluation. GIANT is the only international petroleum economics software that incorporates the following features:

- Evaluation of operations in countries worldwide and under terms of all contract types such as risk, concession, and production—sharing within a single integrated, user-friendly system.
- Ability to access on disk up-to-date fiscal terms, regularly monitored by Petroconsultants worldwide, and available to GIANT purchasers at a low cost.
- Ability to change input parameters and to run numerous sensitivities and scenarios easily with results generated across a wide range of evaluation criteria such as NPV, IRR, and Payout.
- Capability to evaluate the unique terms of any farm-in or farm-out throughout the world.

The flexibility of GIANT for international use eliminates the use need for constant reprogramming and modeling. Users may instead easily adjust the fiscal parameters for each country, however complicated the tax and contract regimes. GIANT is designed for IBM-PC/XT and compatible computers. It can be used with portable systems and thus can be taken to negotiations anywhere in the world, eliminating the need for reliance on head office facilities during time-sensitive discussions. A "Hotline" help service is available in Houston, London, Geneva, and Sydney.

Scientific Software–Intercomp, Inc.
1801 California, Suite 295
Denver, CO 80202
Phone (303) 292-1111,
(800) 526-5819
Fax (303) 295-2235
Telex 229836

Program: WPM

Scotia Systems, Ltd.
Two Energy Square, Suite 1150
4849 Greenville Avenue
Dallas, TX 75206
Phone (214) 987-1042
Fax (214) 987-1047
Telex 163186

Program: BASIL

From the vendor: BASIL is a database-driven cash flow and EMV risk analysis program designed to screen and analyze drilling prospects. Databases provide default values as a baseline and check against user input. County level drilling statistics, (1979–present, derived from over 700,000 wells nationwide) provide wildcat, stepout, and development well historical success ratios. Drilling costs, operating costs, and severance and ad valorem taxes are databased as well. BASIL runs three simultaneous cash flows at user-defined deal conditions to show comparatives. Expected value risk assessment, two-outcome or decision tree handling, and low case summation of entities using conditional probabilities are features of this program.

Sigma Energy Consultants, Inc.
1200 Travis Street, Suite 2330
Houston, TX 77002
Phone (713) 759-9011
Fax (713) 951-0079
Telex 705078 HOU UD C289

Program: PETEC1-PLUS

From the vendor: PETEC1-PLUS is an integrated work station that combines reservoir engineering calculations with a comprehensive oil and gas before and after federal income tax (BFIT and AFIT) net cash flow analysis. The program can be used independently or in conjunction with PETROCALC4-dPROD (Production History database), LOTUS 1-2-3, and, dJIB (Joint Interest Billing). Program features in the Economics Module include

- Production scheduling—exponential, hyperbolic, harmonic, constant, and direct
- Price and cost escalation
- Production, severance, ad valorem, and windfall profit tax calculations
- Five ownership interest reversions based on dollar amount, time, volume, or production rate
- Multilevel summaries
- Indexing and sorting
- Undiscounted and discounted net cashflow (ten user-defined PW rates)
- Profitability criteria calculations (ROR, ROI, Payout, etc.)
- After federal income taxes (1986 tax reform)—depreciation, depletion, intangible, and leasehold
- Loan calculations
- Unlimited number of wells to calculate (depending on disk memory)
- Monthly based calculations
- Double precision IEEE format math and import/export ASCII files.

Software Bisque
912 12th Street, Suite A
Golden, CO 80401
Phone (303) 278-4478
Fax (303) 279-1180

Program: Software for Economic Evaluation (SEE)

From the vendor: SEE is a comprehensive evaluation program for use in a wide range of investment scenarios, including petroleum ventures, chemical refining, service analysis, and mineral and general investments. As a fully self-standing program, SEE does not require other software in its operation, and its complete menu system makes the program simple to use.

Complicated analyses are easily handled with SEE's flexible and sophisticated evaluation features. Discounted cash flow criteria such as net present value, rate of return, growth rate of return and present value ratio are calculated with various user-specified rates from any of four investor tax positions. Sensitivity analysis and probabilistic risk analysis allow the user to gauge economic feasibility under changing conditions. With its integrated graphics capability and limitless variety of report configurations, SEE produces a tailored and detailed economic evaluation. In addition, the menu-oriented window environment and contact sensitive help screens provide efficiency and accessibility at all levels of use. SEE runs on IBM PCs, PS2s, and other IBM compatible hardware having at least 640K RAM and a hard disk drive. Graphics capability is recommended, but not required, with EGA or VGA display. SEE is available on both 3.5-and 5.25-inch diskettes.

Strategic Systems Group
2112 Malvern Hill Drive
P.O. Box 43348
Austin, TX 78745
Phone (512) 448-0949
Fax (512) 472-1558

Program: DISCOVERY

From the vendor: DISCOVERY, a petroleum exploration decision support system, provides the capability to measure, manage, and control the financial risks associated with petroleum exploration. Recent and proven theoretical developments in the area of risk analysis can point the way to practical improvements in decision making as it applies to oil and gas exploration. DISCOVERY implements economic and reserve evaluation, decision tree modeling, and risk preference or utility theory to enable the firm to consistently manage complex capital budgeting and investment issues.

The program enables the firm to assess its own particular risk preferences or risk aversion level and use that risk aversion level in the evaluation of all prospects or properties.

DISCOVERY employs an easy to use menu-driven format coupled with extensive on-line help screens. It has comprehensive modeling capabilities for production scheduling, ownership interests, economic parameters, etc. The program utilizes a risk-sharing module that enables the user to model specific working interest in a given prospect or group or prospects. Dazzling screen graphics takes place during the input stage to allow the user to react to "what-if" scenarios. To ensure increased accuracy, the program incorporates monthly production and cash flow computations, including all inflation, price change, and discount indexes. Dazzling screen graphics offer comparative analysis across an array of economic parameters. DISCOVERY's one-of-a-kind Prospect Ranking Report provides a ranked summary of all prospects evaluated and recommends the firm's optimal working interest level based on your firm's risk aversion level and capital constraints.

T&E Garland, Inc.
135 Classen Drive
Dallas,TX 75218
Phone (214) 348-7687

Program: WHATIF2

From the vendor: If you need a reserve and economic analysis program, you can't afford not to try WHATIF2. You may save thousands of dollars. A 90-day trial of the complete program is only $15.00. WHATIF2 Version 3.0 is a stand alone program that is ideally suited for rapid, accurate economic analysis and sensitivity studies of oil and gas production. It has most of the capabilities of programs costing many times more. The full screen data entry and menu-driven program is much easier to learn and use. You can create, revise, generate, and print a 30-year economic forecast on an individual or summary basis. Available standard decline curve techniques include exponential, hyperbolic, harmonic, or constant production for three product streams. Evaluation parameters and escalation schedules (taxes, prices, operating expenses, loans, etc.) can be entered rapidly for individual cases. For multiple cases, a "master" file containing the constant parameters can be used. Investments can be made during the first 5 years.

Gross and net income streams are generated for working and net revenue interests with one reversionary interest (payout or gross amount). Economic indicators include internal rate of return, net present value profile, net income to investment ratio, payout years, gross and net reserves, recovery per acre and acre-foot, and cost per BOE. A sensitivity analysis can be generated and a report printed showing five profitability indicators for the 16 cases resulting from four user-defined initial producing rates and four decline rates.

Texas Computer Works, Inc.
256 North Sam Houston Parkway E., Suite 202
Houston, TX 77096
Phone (713) 999-3175
Fax (713) 448-1104

Program: ECON

VTM Engineering
c/o Society of Petroleum Engineers (SPE)
222 Palisades Creek Drive
Richardson, TX 75080
Phone (214) 669-3377

Program: VTM ECONOMICS

From the vendor: VTM ECONOMICS is written for consultants, investors, and companies involved in oil and gas evaluations. It is designed to replace or supplement the timesharing systems currently in use by the industry. VTM ECONOMICS will save the time and storage costs of approximately $16–$25 per run of timesharing. Once the cost per run factor is eliminated, it is likely that many more what-if cases will be run and a better investment decision will be made. Also, VTM ECONOMICS can be run without special training. Some features of the program include the following:

- All calculations, escalations, and discounting are on a monthly basis, with a maximum project life of 25 years.
- Before-tax, after-tax, and summary reports (including incremental and risk factor analysis) can be generated, with no limit on the number of runs that can be summed. Several data disks can be scanned for the appropriate files.
- Internal rate of return, payout and banker's life calculations can be performed, and up to five rates can be specified for present value.
- Oil and gas price trends and operating cost trends can be scheduled.
- Decline calculations can be performed on four different production trends. Production can also be scheduled manually.
- Numerous error-trapping routines are included to aid in usage.
- The program is completely menu driven and runs on an IBM PC or compatible with 128K. A hard disk is compatible, but not required.
- The program is completely documented in a bound manual.

JOHN M. STERMOLE, Vice-President of Investment Evaluations Corporation in Golden, Colorado, earned a B.S. degree in Finance from the University of Denver and an M.S. in Mineral Economics from the Colorado School of Mines. He is involved in economic evaluation consulting and has developed personal computer software applications for the petroleum, chemical and mining industries. Mr. Stermole teaches a short course entitled "Economic Evaluation and Investment Decision Methods" for natural resource companies and general industries. He is also an adjunct professor at Colorado School of Mines where he teaches Economic Evaluation.

Chapter 14

Databases: Basic Considerations Using Well History Data

Robert W. Meader
Consultant
Littleton, Colorado, U.S.A.

. . . failure to use good and valuable information because it is not believed to exist is an unworthy excuse in present-day information management practices.
—Robert W. Meader

INTRODUCTION

To the geologist, the principle business of petroleum exploration is to find oil and gas, while to the engineer, it is to drill fewer dry holes. But to a manager, the goal is to make money. All would agree, however, that their best chance for success, by whatever measure, is through the careful use of all available information.

The recent decade has seen a significant increase in the amount of geoscience information. The variety of supporting information has come to have an ever more far-reaching impact, and to be competitive, the value of speed and access to analyzing this information has become a necessity. The word *database* is frequently used to identify these information sets, and the digital computer is the best tool to manage them.

Well History: An Example

In this chapter, a few of the more critical points of geoscience database development are emphasized. Well history data were chosen to illustrate these principles. The descriptions are directed to the explorationist, who is presumed to have minimal exposure to hardware and software. A drilled hole proves a reasonable model for a petroleum geologist or engineer, but it also serves as a good example for a geophysicist, geochemist, or mining engineer. All deal with a surface location and identification and an associated bank of subsurface information and resources data. Even remote sensing data are converted to a surface location.

Definitions

A database can be defined as an appropriate collection of geoscience information rendered into a predescribed scheme for computer storage and processing. It can also be defined more realistically by its reason for being, that is, by its use and expectations in the mind of its "owner." Particularly with technical considerations, the "why" *of* a database should precede the "what" *in* a database. Certain mind sets accept the digitizing of any and all data with little concern for its subsequent use, computer storage being seen as the panacea of information management.

A *database management system* (DBMS) is a library of computer programs developed to process this geoscience information. It is a series of integrated software instructions to capture the information, process and calculate it, print or plot it, and format it for subsequent technical or graphics applications. Such software systems give credence to computer handling because of the extensive selective processing and retrieval capabilities built into reputable database management systems.

The term *database* is used without connotation of size. It may have reference in the broadest dimension, such as a very large database made up of many smaller databases. A database for a basin analysis project, for example, might be made up of lesser databases containing stratigraphic, geochemical, and geothermal information. Equally appropriate, the term can also be used for a computer-serviced collection of 10 well locations with geological tops.

By now, you may have noticed that the word *database* might be seen as slang for all matter of "stuff." It could identify a universe of earth and sky information, or a single sample of rock or fluids. The important point is that a database should be seen as a collection of meaningful bits of information, logically arranged, with the expectation of being processed by computers under the direction of programmed control.

Row and Column Matrix

The predescribed scheme referred to earlier demands that these collected bits of information occupy precise and identifiable locations in the database and that they be organized into common sense associations. We can visualize this scheme by using a two-dimensional table, or *matrix*, made up of a number of rows and columns. Each column represents the position of one item of information or element of data, and each row is dedicated, for example, to an individual well. One such database structure of widespread use in the oil and gas industry that mimics that row and column matrix is the relational database scheme. A database might have 50 or more tables, each with as many as 40 columns of data for as many as 100 wells (rows). It is important to realize that the table and row/column definitions are critical in database design and development, but it is the presence of information—as digits and data within these row/column locations—that determine the worth and subsequent use of the database.

In the ever-challenging role for an explorationist, the search for the clue to discovery invites looking at the available data in new and novel ways. Yet such rules of thumb as "the more data, the better," or "when in doubt, digitize," may result in significant delay and lead to spurious trends. The justification for building or buying a database for exploration and/or production purposes turns out to be a compromise of many issues, not the least of which is trying to be clever in the use and display of its content, rather than trying to use the computer to randomly map geological data.

OWNERSHIP: TECHNICIAN OR PROPRIETOR

There was a time when geologists and engineers set about to assemble information necessary to put a deal together, solve a problem, and pursue a course of action. Whether or not they did all of the searching, correlating, and data handling themselves to build such a database, they had an awareness of the process and some sense of its strengths and shortcomings. That is, they knew what was in their various files, knew of the relationships among the contents of the files, and had some ideas of how to exploit them. They had, in other words, a sense of *ownership* of that database.

This is not to say that a closeness of a professional to his or her database cannot exist when not self-generated. It is apparent, however, that many databases are developed with one mind set, but used by another. A database can be used most effectively if the user has, or gains, a sense of dominion over the database—how it was put together and what can and cannot be done with its content.

The volume, complexity, and multiplicity of data, along with the capabilities of modern computer science, often place the geoscientist at the mercy of computer wizardry and program logic. All the same, the aforementioned success to be expected from such processing has a better chance to occur if the user knows what is going on. His or her primary area of expertise should be that of knowing the most about the relationships within the database and its content. The possibility of doing a calculation wrong is far less apt to happen than an incorrect assumption or an inept association of elements within the database. For example, the program logic to process zeros, blanks, and pseudovalues are routine matters for a DBMS, but to the unaware user or owner, the subsequent output could have major consequences for the geological analysis.

Technical know-how is every bit as important in the design of input screens and data capture instructions as is the analysis of output. Units of rates and measures, digitizing scales, and correct execution of quality control and logic algorithms have as much importance as the inputting of the data. Such things need to be owned at input if output credibility can be assumed. One has to ponder whether the so-called computer mistake, so commonly recited as an uncontrolled excuse, is in truth an example of insufficient ownership—a poor, if not disastrous, choice of priorities.

WELL IDENTIFICATION: UNIQUE NUMBER

It may seem strange that the surface location of a well is a more distinctive identifier than the operator name, well number, and lease name, yet database technology supports such a premise. Experience confirms that well numbers change; in some cases, there can be a different number for various departments within the same company. Corporate mergers and takeovers dilute the longevity of an operator name. In the vast majority of cases, however, once the well's surface location is surveyed, its permanence can be assumed.

A well from wildcat times may have several

"unique" numbers. In reservoir studies and especially in unitized fields, a well commonly has different segments of its information file tied to a variety of numerical and alphanumerical designations. Its permit number assigned by a regulatory agency, and the API Unique Number, has a dual definition. Five of the digits are location specific, whereas the remaining digits are intended to be one of a kind (within that location). Because the API number has no bearing on ownership and because early field development involves several operators and leases, a well numbering system, no matter how orderly, must be restructured if well identification is to become truly unique for reliable computer processing. Successful examples have demonstrated the value of arbitrarily assigning a sequential number (or numbers) or unique unit designations to all wells in a unit. Initial drilling sequence and original operator and lease labels are of lesser importance to well uniqueness. This reduced identification renders minimal congestion to posted information near the well spot, and the abbreviated designation is both logical and clear for sequence and processing requirements.

Database techniques make it relatively easy to build and retrieve from a master table in which all well identifiers are equated. Thereafter, the use or application may address a well with whatever number is familiar and retrieve whatever number is desired. As discussed later, adding a date to that identification could be helpful, especially in accounting or future legal applications.

The initial importance of a permanent unique well identifier may be set aside in favor of a design strategy in which the database accommodates all identifiers of both public and proprietary sources, over the life of the database. The most lasting significance of well number or identification in a database management system is the one-of-a-kind identification that is unique to computer sequencing and computer logic. Careful consideration of a unique well, point, sample, station, or observation designation is a must if reliability of data are to be maintained during processing and assumed for output.

LOCATION

Treatment of the location description can follow a pattern similar to well identification. That is, a number of description systems are used to locate a well (shot point, station, or spot) on the earth's surface. Several are numerical, while others are alphabetical and numerical, and like identification, different professions prefer certain forms. Meets and bounds; section, township, and range; and footages are only a few. A creditable database must include an appropriate transformation calculation if location is to be computer mapped because a digital representation must first be computed. This location's description is the familiar X-Y coordinate pair of graphic packages and Geographic Information Systems (GIS).

Latitude and Longitude

Several observations are noteworthy when considering information about location on the earth's surface. The latitude and longitude description is certainly the ultimate global location system, and Global Positioning System (GPS) technology is making it more conveniently available and accurate. However, latitude and longitude are not in common use, especially on larger scaled exploration maps, and are nearly absent on field maps. Although software systems are available to convert from any of several projections to digitized coordinates, these operations are usually transparent to the user. He or she must have access to local and familiar location descriptors, even though the latitude and longitude coordinates are an excellent and commonly used computer location designation that provide an accuracy well within the range of most applications.

Map Scales

Attention should be directed to the map scale if the location coordinates are taken (digitized) from a base map. The delightful ability to computer plot any map at any scale must be tempered by the reality that map accuracy can change drastically as original map scales are enlarged. Overlaying maps (for example, lease with structure) whose coordinates were sourced from different scaled bases and then computer plotted to match can create serious breaches of accuracy. Digitized coordinates for lease definition may result in well locations misplotted due to excessive magnifications. Thus, it is a good idea to add some identification of source or origin of location coordinates as well as scale, should there be any expectation of several magnitudes of enlargement.

Recent awareness stemming from the work of the National Geographic Association map projections raises questions about the future location definition. Thus, affixing a date code to any digitized coordinate pair would be worthwhile should reconciliation of variance arise from multiple projection sources.

Although directional drilling is a common practice in minerals exploration, the recent increased incidence of horizontal drilling has come to make three-dimensional location data extremely important. Companion databases of both dip and deviated hole data are excellent means by which this information can be stored and yet made directly and precisely accessible to the transformation calculations and processing software. Even so, the user must provide—or be assured of—the accuracy and sensitivity so as not to create an accuracy beyond the limits of the source. In this regard, it is uncommon to regard the depth (below surface) as a location parameter, especially in continuity with the latitude and longitude. Yet any depth can be tied in a program to the latitude and longitude, such as to define an X-Y-Z framework that is a unique identifier for any piece of geoscience information anywhere on or in the earth. Common practice with many GIS is to calculate the mappable coordinates every time a map is generated. A case can be made to calculate the X-Y coordinate earlier in the processing scheme of computed events and thus enable it to be used as a permanent and unique indexing parameter for processing and sequencing.

In summary, the frequently used identification of a point of data may not be unique for computer purposes. A case can be made that the identification can better be defined by its location. Yet our best location parameters are both unfamiliar and awkward to use, having to be transformed before plotting by program logic largely unfamiliar to most users. Thus, it could be argued that there is a degree of "bother" to both point identification and its location. Careful design of database structures, in harmony with the DBMS, will accommodate the requirements of uniqueness and the fixations for location. When properly done, the result is more accurate and has greater precision than the results of most previous handling techniques.

DATES AND THEIR FORMATS

Dates in a well history are vital to a database. They provide the order of events and set the framework in time to weave the fabric of the well history. In the scout report, a chronology can be deduced from the page on which the data were recorded. Earliest data are at the top of the page, with the most recent entry being the last information. Computer listing formats, in contrast, remove any implied sequence, and the date of computer entry has no relationship to drilling sequence. Particularly in secondary production and reservoir management work, the dates when engineering jobs were started (and finished) not only provide a sequence but can also confirm (or reject) a cause and effect.

A particularly annoying advantage to database management is the variety of formats by which the date parameter can appear. Fortunately, there is a style for nearly everyone; the accountant's month, day, year; the chronologist's year, month, day; two-digit years; fully written months; and others. This variety can be disconcerting when the printout of the date is in an unfamiliar style—for example, getting used to blank spaces that surround MAY because the maximum size of that format must accommodate SEPTEMBER, or the confusion of whether a six-digit number is in a month, day, year or a year, month, day format.

Another equally important value of the date in a database for professional use is its identification of quality or acceptability. The dynamics of computer-managed information systems encourage storing anything and keeping everything. Particularly valuable are computer files that contain multiple picks of formation tops from different sources (sample, log, core), from several geologists over time, or from successive committees in unit meetings. The date of file maintenance or systems update (an automatic result of computer processing) cannot be the date used for data reliability. Thus, the inclusion of a date, as part of the formation tops data, helps to identify their credibility and, along with other items in a database, can substantiate precision and quality, as well as author or source.

DEPTH: A LOCATION COORDINATE

Depth is the third coordinate of location for nearly any piece of geoscience information. Serving hand-in-hand with latitude and longitude (or their transform), depth is the third set of digits in an X-Y-Z location definition. Depth can be particularly helpful in the computer ordering and sequence control of information. Searching a well log library, containing any number of log runs or tool types, with the parameter of depth will catch every trace, irrespective of tool, date, repeat, or geological formation name.

Several concerns need to be noted. Every appropriate depth should be in a database, no matter what geological or engineering usage is envisioned. When in doubt, put in the depth. The label or identification of depth must be clear and forever associated with the numerical value of that depth. With a bit of ingenuity, a depth in meters can be discerned from the same depth in feet. Which total depth is meant in a well history of several, however, could defy all judgment in the absence of an identification or label. Equally problematic is the uncertainty of whether a depth is a true vertical depth (TVD) or a slant hole depth, and if a TVD, by which of several methods it was computed. Thus, both the identification and the source, as well as the derivation of the depth, is imperative for database stability and information reliability.

Depth Ranges

A second concern about depths is their use as a range—two depths, that is, from depth A to depth B, from an upper to a lower depth, or from top to bottom. Experience confirms that the process of capturing two depths works very well. It is common practice to input a single number twice to satisfy a depth range format when only one depth is known. However, should such a data element be subjected to a subsequent thickness calculation, the resultant difference may not necessarily be the number expected. In some modeling software, a single value for depth is needed, sometimes the top, but commonly a mid-point between the top and bottom. The concern is not about the software, but that in the use of a mid-point, the range (from whence it came) must not be lost. In well log analyses, such as porosity determinations, using ranges in some cases and mid-points in others is acceptable practice. Deriving mid-points from ranges is easily done, but the reverse calculation of defining a top and bottom from a single value is not. Thus, care must be taken to retain the original integrity of depth during downstream calculations.

Condition of No Depth

The X-Y-Z framework lends itself to an elegant data processing and information management control, yet certain engineering data may not be depth specific. So what can be done when there is no depth for downhole information? In a limited number of cases, a depth, or a range of depths, can be assumed with a measure of accuracy or point of reference. The data commonly relate to a second piece of data that is precise, for example, relating a recovered water sample to the top of the formation. In any case, professional prudence would opt for a companion data element to be included in the database to describe the no-depth relationship.

An alternative means of preserving computer access by depth when depth is not known, yet not compromising depth selectivity, is to indicate the availability of that information by tying it to a "meaningful" depth in the well history. This can be done by noting in some narrative form the surface (elevation) or top of pay. However it makes most sense, the possibility of a file search on the basis of depth (or travel time in a seismic application) is very useful, especially because it may catch all pertinent information, whereas a topic identification might not retrieve all appropriate data.

WELL STATUS AND CLASSIFICATION

Well status and well classification are well history data elements that are widely used and commonly available. The early practice of reporting these in alphanumerical strings or codes (D&A or wildcat) gave way to numerical codes and cumbersome mnemonics. More recent computer capability has brought back the use of the earlier styles, and database management systems provide reasonable access to the more than 100 different possible classes. Yet an understandable shortcoming of class and status is their frequency of change. Much to the chagrin of the database manager, the careful user may attach a high level of importance to the current correctness of these parameters.

The Lahee Classification is widely used and commonly understood. Regulatory agencies (state and federal) provide precise guidelines and support firm adherence to a well classification. There is a tendency to use status or classification for gross statistics: dry hole versus producers early in the life of a database; producers versus shut-ins (or injectors) in the later stages of a database. Yet, the importance of these two parameters gets lost in the routine of other data element update procedures and data maintenance. Consequently, a subfile generated out of a database when status or classification is a prime sort (or sequence) argument may produce results of uncertain reliability.

GEOLOGICAL TOPS AND SHOWS

Geological tops and shows are two categories of exploration and production information that may represent the very heart and soul of a database, certainly for the geoscientist. Furthermore, the development of computer graphics technology in the oil and gas industry may be attributed to mapping these geological tops. The early days of computer use in geology were characterized by immeasurable time being devoted to resolving the controversies of company tops and scout tops and of proprietary picks versus purchased picks.

An associated argument frequently heard was the reconciliation of basic data and interpreted data. Formation tops were mostly seen by the author as basic information. With the advent of computer access, these tops became available to a broader audience and to more multidisciplined users. Many of them saw these "basic" tops to be interpretive—as indeed many were—and thus subject to challenge or individual interpretation. Interestingly and more to the point of database matters, the energies devoted to computer contouring of these geological tops, and the subsequent defense (or discrediting) of the resulting maps, represented an interdisciplinary effort of colossal magnitude.

Both then and now, it is the geological identification, the name, that appears to be the source of greatest concern. From the point of view of database integrity, this concern is unfounded. That a computer-processed geological top, formation pick, or horizon marker is valid or not, right or wrong, basic or interpreted is beyond the boundaries of the integrity of a database system.

Hand-drawn contour maps have been made with identical data, but with much more acceptability than their computer-drawn counterparts. Without doubt, this acceptability mainly resulted from the contouring model, not the data values. In a similar setting, it is not uncommon to see posted (noncontoured) maps of the same numbers no more widely accepted. Apparently, the stigma of computer processing somehow taints the credibility of the data. It may be that an implied permanence is placed upon a geological top once it is entered into computer storage; a finality that was not so with traditional practices (by hand). In any event, the proponents of databases and their management systems must be continually vigilant in recognizing the proper assignment of quality control and information integrity. However far removed a geoscientist may be (or feel) from his or her data by the process of computer handling and program manipulation, a professional outcome requires professional awareness.

The advent of personal computers has dispelled much of that stigma, but the growth of large and diverse databases with multidisciplined users may raise a new sensitivity to biased and/or interpretive data, their integrity, and the reliability of the data processing and maintenance of them.

Subsea Depth and Thickness

Software capability within a DBMS can swiftly and accurately determine the subsea depth of a top or show. Once the resolution has been made of whether to use the surveyed or reported elevation (ground or rig floor datums), the computation is straightforward. Peculiar to areas with topographic relief, the calculation logic has to be designed to accept subsea depths that may be positive values. Rules of thumb for Kelly bushing or rig floor elevations when such are missing should not be built in to a universal processing program.

Formation Name

Accepting for the moment the credibility of the depth value of a geological pick, the worth of the identification of that pick takes on increasing significance as any user moves through the database. It is noteworthy that the allied geoscience users correlate and compose their geology by name (and geological age), not by depth and/or geodetic coordinates. The geochemist, the seismic modeler, and the paleontologist commonly go into a geology database making inquiries on the basis of geological nomenclature, not on the basis of depth and location.

Effective database design and development must provide for both variability and increase in stratigraphic nomenclature (more terms and subdivisions), as well as a degree of consistency and permanence in the use of geological names as a key index. Initial efforts of using codes and mnemonics for stratigraphic identities have been largely replaced by DBMS software tables to accept commonly used geological terms. This results in easier access and more reasonable reporting.

Geological Age

The numerical sequencing of depth and the labeling identification with a geological name are just two elements of a database with multidisciplinary geoscience import. Ongoing breakthroughs in age dating with microfossils, coupled with radioactive dating methods suggest that geological age may become an additional parameter for stratigraphic ordering. Millions of years before present or similar age designation for geoscience indexing will increase in worth both as a tool for computer sequencing and as a means to strengthen stratigraphic correlations.

Sequence

The mode of field operations in the drilling and coring of a well sets a sequence of happenings that predicts one event following another. In a database design sense, that sequence might well be chronological and ordered by date. However, as a well's history progresses through time, tests, logging runs, and engineering practices and reservoir procedures, the drilling chronology is lost by repeated operations, duplicate events, misruns, and paired exercises. Provision must be made, therefore, for an operation identification, tool run, and time and date of event or sample sequence so as to establish both an order and a unique "oneness." Such is done with a key identifier and/or data discriminator. Boolean operators, consisting of AND and IF statements, mediate this need for distinction and sequence, yet careful initial planning of the database design is warranted, particularly in consultation with a geologist or engineer.

Without the experience of dealing with database size, variety, and longevity, it is hard to know beforehand how detailed the indexing or definition of keys needs to be. Most DBMS provide means by which devices can be used to take care of initial oversights in sequence discrimination. The normalizing of geological and reservoir engineering data to ensure optimum control has not yet become an accepted practice. Furthermore, the emphasis upon standards and industry-wide practices currently in process will mitigate an alternative to trial and error, but that, too, takes time and acceptance.

CORES AND SAMPLES

The presence and position of rock data, especially cores, has been a standard item of interest from the earliest data file definitions. The core number, cored interval, and a few additional items were key ingredients of geoscience data files, but not altogether substantial pieces of geological information. Core numbers and intervals were noted, but greater detail about rock types and their descriptions were much less common. Sample logs and, later, mud logs may have been identified but were rarely if ever digitized.

A few commercial firms generated digital files of described lithologies, but they, too, were challenged—somewhat severely—for being too biased or open to question and individual interpretation. Private and company lithology files were, and no doubt still are, being generated. More than likely, however, such files are generated under special conditions or only in critical sections of the well bore.

One of the problems with the generation of rock data into a database is the volume of information and the tedious nature of generation. So-called bedsheets, some with as many as 70 fields (data elements) of recordable information for each foot of core, have been digitized for use in research projects. Thus, for a cored well with several hundred to a thousand or more feet of core, with an average of 35 data items per foot and some of these being 12–18 bytes per item, the resultant alphanumerical file would fill a good sized database. Codes and mnemonics were commonly used, and although they reduced the data volume, they required a translation and conversion setup for ease of readability upon output. The number of times the core description would have to be presented was generally infrequent. In many cases, the user would return to the bedsheet or the original source, finding it quicker and easier to understand than a printout from the database.

This provides a case in point that not all information about a well needs to be captured into a database. In the case of cores, their depths and identification, along with gross lithologic descriptions, serve the intended purpose. Unless a detailed rock analysis (porosity, fluids, etc.) is anticipated, the DBMS may provide only the identification (or "tickler") and perhaps a storage location of the core.

Two observations are germane to the generation of rock data (cores and samples) into computer-processable form. The greater number of lithologic columns generated by computers come as the result of log analyses derived from well logs. Lithologies are deduced from various logging tool responses to well bore wall rock. Nothing is meant here to reflect upon log analyses technology; from a database point of view, such lithologs are derived, and thus interpretive, results, not visual or direct lab observations of cored material.

A second observation of digitized core descriptions is their frequent use in statistics. There are those who would insist that the dependency upon statistics to justify digitizing lithologic data is insufficient. Feeding a computer families of numbers, without some foundation in use, is alien to applied technology. A more kindly appraisal of the analysis of rock properties using statistics may only reveal uncertainty as to the geological meaning of certain statistical variations in one or more parameters. That thought raises yet another question, namely, what are the anticipated geoscience questions that the database can be reasonably expected to answer? Basinal analyses and extensive geological modeling might warrant serious consideration of digitizing extensive lithologic descriptions, thin section work, geothermal analyses, and paleodata. This might also be considered in an exhaustive field study or in the study of an extensive field. One might be hard pressed, however, to design and develop a comprehensive database for a single exploration play.

All the same, the X-Y-Z concept of a location-in-space framework, upon which can be hung geoscience information of nearly every description, is a viable framework irrespective of the volume of data. Thanks to ever-improving database management systems, more data are accessible and available for ever-widening applications which will no doubt lead to a greater understandings of geology and the earth's resources.

WELL LOGS

Well log libraries (electric, sonic, or radioactive) are one of the few sources of database files in which the original data are captured in computer-compatible form. Consequently, reformatting the field tapes is invariably a necessary and early processing experience. This industry-wide practice has required the development of software capable of converting various logged runs to common in-house or general use tape formats. Subsequently, it has also given rise to moves toward standardization and equivalences across both the logging and computer industries.

Piecework Digitizing

Even with the on-going practice of field or site digitizing, there remains a concern for manually supervised digitization of paper copy logs. It is reasonable to assume that there will be an on-going need to carry out remedial and special analyses of logs on one or more sections of hole and on selected logging tools. Thus, a piecework approach to redigitizing certain traces or sections of certain logs can be expected. In such instances, the database manager will avoid confusion if he or she is able to sustain close and continuous contact with the user community and know what is happening to and expected from the well log database. It will be easier and more practical to digitize the critical sections from the paper copy in any number of cases rather than chase down the original tapes and reprocess them.

As an aside, a database designed to provide storage for the identification of these digitized tapes, their content and format, tape and processing parameters, and scales and intervals is an excellent adjunct to any reservoir or production field database. Particularly in the maturing stages of production, workovers, and EOR processing, a readily available identification database of the logging tapes can be a handy and direct source to a much larger, more remote resource of reservoir information.

Multiple Tool Types

One major challenge in well log database development and maintenance is the multitude of tool types. The large number of different logs and the additional complexity of contractor terminology is reason enough to consider a major database design effort so as to provide control of the well log library. Tables of codes with tool identities that reconcile redundancy, as well as differentiate basic and computed data, can be very helpful. These identification and processing variations must then be further subject to discernment because of the field practice of running many of the same tools several times during the life of the well, even over identical sections of the hole. The date and time sequencing of these redundant log runs is an important control, yet the date alone are not an exact discriminator of multiple logged results. The indexing and identification of logging information over the life of a reservoir is complex and challenging, yet vitally important.

Tapes and Headers

The database for a complete well log library exists in two parts: (1) the digitized traces of initial (field) and processed data on tape and (2) the "header" information and conditions associated with the logging run. In the first case, there are trace identities, mud conditions, logging constants, depths, dates, and tool and contractor specifications, while in the second case, there are computer specs, tape formats and densities, and processing parameters. These latter data make up a secondary but no less necessary subordinate database for the up-hole responses. Although the size of a single well's digitized log database can become large, the greater concern must be to keep it under computer-managed control so as to know exactly where and what is in any well log database subfile. The alternative of managing one's well log library off a shelf without dated listings can no longer be defended.

Stratigraphic Name

One add-on feature to a logging database that can provide a helpful downstream service is the access to a database-wide geological formation name, producing zone, or stratigraphic horizon. The depth control, by logged interval, fits nicely into an X-Y-Z information-spaced database. However, the additional access to a stratigraphic name provides a helpful retrieval argument and a handy way of selecting and comparing data from the log database with other databases, such as cores, perforations, and tests. In other words, the computer comparison may be made by depths (intervals), but the user's correlation "argument" will be made by formation or zone.

Longevity

The well log database may be one of the better sets of geoscience information that raises a question of database retention. How long do you keep the "original" data and how much of the subsequent computer-generated analyses do you keep? Archival storage, although not an early concern of database implementation, is neither trivial in pursuit nor inconsequential in storage. Unlike files in actual use, for which the changing and correcting of data is the issue, in long-term retention, it is the size, security, and accessibility of the logged files that is difficult to predict. The life of a project, reservoir, or field is a common measure to define retention of archival data. Increasing pressures from environmental concerns and/or atypical remedial work, however, may require archival files to be accessible for many years after abandonment. Retention of databases (not the source documents) for up to a decade may prove to be profitable if only for the litigation costs their absence might represent.

The analyses of well logs generate a large and varied number of subfiles and specialty databases. The user should know best what should be kept and for how long. A pride of accomplishment together with a fresh sense of labored achievement flavors the user's tendency to keep rather than scratch. All the same, a simple spreadsheet database built around the bookkeeping of log analysis computations can be of great value in the inevitable decision of retention. In many cases, that decision can be resolved by going back to the original and recomputing.

The generated data should not be scratched in all cases. In larger fields with longer production histories (and thus of greater corporate value), there is a strong tendency to keep everything forever. Personnel changes, committee hearings, unit meetings, and litigations may alone be sufficient cause to keep without discrimination. Common practice suggests that database maintenance gets downgraded and data integrity is sacrificed as use drops off. Well log computer databases and particularly the generated reservoir parameters and various formation characteristics may suffer the ultimate fate of becoming lost or perverted by people or practices unappreciative of their value or unaware of their worth. A well-designed exploration and production database can stand the test of time and change, but only as the user chooses to preserve its content, credibility, and access can it prove to be so worthy.

CORE ANALYSIS

Core analysis is geoscience information that has significant engineering value in the production phase of a field or reservoir, both routine and special. Core-derived porosities, permeabilities, and saturations provide critical data for reservoir evaluation. Relative permeabilities, capillary pressures, formations (resistivity) factors, and the like provide data to analyze well and reservoir behavior. By some arguments, these analyses may not be candidates for database storage. However, the number and complexity of reservoir engineering calculations fostered by these data preclude any thoughts to the contrary. Special core analyses generated into a database along with production history can find extensive use in later stages of a producing unit. Ordinarily not a consideration in the initial stages of database design, these special databases nevertheless must carry the identical controls and indexing used in the primary database. A solid well identification, precise dates, and meaningful stratigraphic names will give strength to the ease of use encountered by the replacement engineer or new development geologist who was not in on the original design.

The core analysis database has several design characteristics not commonly seen in other geoscience schemes. They are extraordinarily complex. The table definition of the electrical properties consists of as many as 40 data elements for a single analysis. Several parameters, such as the formation factor, are determined repeatedly at different overburden pressures. Laboratory conditions are key data elements that seem to defy standard input formats. Units of measurement and base conditions are a necessary part of the database, but they, too, can vary within a single analysis. Some data are given in parts per million in one instance, but as percentages in another.

This information database has its origin from a single or limited number of samples, yet the results are used over a much wider area—any number of adjacent wells. From an applications viewpoint, every nonsource well database should indicate where the analyses came from. As with other applications involving formations or stratigraphic identification, a retrieval logic relating depth to geological marker is extremely useful.

The database of a core analysis is but one good example of a specialized, limited database yet with applications of intense and far reaching consequence. Likewise, there are other sets of data in a geoscience information base that are not overwhelming in size, but still have significant worth in a project because of their intrinsic value. Thin section work, core photography, and other laboratory procedures likewise warrant consideration for database application. One thing that can be done is to put the availability—the fact that such data do exist—of such analyses into a database.

The X-Y-Z concepts of location and depth, together with a geological identifier as a minimum discriminator, to be included in an identification can save a good deal of time in a file search. More to the point, failure to use good and valuable information because it is not believed to exist is an unworthy excuse in present-day information management practices.

Annotated Selected Reading

Burrough, P. A., 1986, Principles of geographical information systems for land resources assessment: New York, Oxford University Press.
Contains extensive references and a glossary of GIS terms.

Geobyte, AAPG.
A bimonthly magazine published by the American Association of Petroleum Geologists. An excellent resource for current awareness of computer use, standards and earth sciences applications.

Pratt, P. J., and J. J. Adamski, 1991, Database systems, management and design: Boston, Boyd and Fraser Publishing.
A second-edition textbook. It includes three chapters on the relational model, an extensive glossary, and an appendix on file and data structures.

ROBERT W. MEADER received a Geological Engineering degree from the Colorado School of Mines in 1951 and served in the U.S. Army Corps of Engineers during the Korean conflict. He received an M.S. degree in Geology from the University of Minnesota in 1956. After completing graduate requirements toward a Ph.D. at Louisiana State University in 1959, he taught geology at Centenary College in Shreveport, Louisiana. Mr. Meader joined Marathon Oil Company in Findlay, Ohio, in 1961 and was transferred to the Research Center in Littleton, Colorado, in 1964. He worked there for 22 years on technical service with computer processing of exploration and reservoir data. He retired from Marathon in 1986 and became a consultant in Littleton, Colorado, specializing in database management systems for oil and gas reservoirs. He is a member of AAPG, AIPG, and the Rocky Mountain Association of Geologists.

PART IV

MANAGING THE BUSINESS

This part presents eight chapters covering a wide range of topics on managing an exploration business. Topics include managing an exploration program, people, building creative working environments, trend analysis, annual budgets, funding oil and gas ventures, putting a deal together, and enhancing prospect presentations. All these chapters can be read separately, but many can be related back to earlier chapters. For example, the business plan in Chapter 15 is a statement of what actions are appropriate in today's environment and how to manage after the business is established following the precepts in Part II. The next two people-oriented chapters discuss an important aspect of managing any program and are a logical follow-up to Chapter 15. They recognize that it all begins with people who add value to a company (Chapter 16) and that valuable, innovative ideas can come from anywhere in an organization (Chapter 17). Chapters 18 and 19, on trend analysis and annual budget, are a sequential pair that employ some of the same procedures as Chapters 6, 7, 10, and 11, but in a different context. This should be very helpful to those of us who do not fully comprehend a subject the first time we read about it. The last three chapters in this part are arranged in sequence from funding oil and gas ventures (Chapter 20), to putting a deal together (Chapter 21), to techniques for enhancing presentations (Chapter 22). A synopsis of the chapters in this part follows.

Chapter 15, "Managing an Exploration Program for Success," by Bill St. John, presents a management blueprint for a mid-sized independent oil and gas operator. It can be modified for a single consultant or a large corporation. The primary mission must be to make a profit, and an overview of the industry must be considered to determine how the organization can perform profitably. Current status of the organization must be critically examined. Then goals, strategies, and operational plans must be established. Finally, to survive and prosper, management must monitor results and continually relate back to specified goals. The author describes this blueprint in the text and in four useful summary tables.

Chapter 16, "It All Begins with People," by Marlan W. Downey, is a succinct statement of the author's reasoned convictions about exploration staff and management based on many years of experience in the industry. Here are some of the conclusions he reaches. Leveraging of risk capital by human intelligence and innovation yields extraordinary profitabilities from exploration success. Geologists and geophysicists are essential to exploration success. Well-managed companies with first-class staffs always outperform their competitors in the long run. Companies properly managed for the long term should hire staff carefully, prune judiciously, and avoid across-the-board terminations.

Chapter 17, "Building Creative Environments and Leading Creative People," by Kenneth F. Wantland, offers a variety of ideas, techniques, and examples to revitalize people and organizations in a petroleum industry undergoing large-scale change. Its driving concept is that everything starts with people. The author lists the critical roles of leadership in today's business environment and discusses the personal factors in developing people.

Chapter 18, "Trend Analysis," by Michael D. Smith and Donald R. Jones, describes a technique useful in mature exploration regions such as onshore Texas. The objective is to allocate resources into those areas that have the highest probability of achieving the target return on investment with a given size drilling program. By relating financial and exploration goals, the trend analysis technique serves as a reality check on investment decisions and can improve the quality and profitability of exploration. The authors guide readers through the technique using a series of data tables and graphs.

Chapter 19, "Annual Budget: A Short-Term Action Plan for Exploration," by Donald R. Jones, includes the following major topics: (1) evaluating prospects for inclusion in the drilling budget, (2) risk analysis methods, (3) prospect ranking and selection, and (4) budget performance reviews. The author considers the budgeting process the culmination of the explorationist's efforts and recommends that budgeting be done with diligence and on a sound technical basis. He considers four factors involved in the critical decision of whether to include a prospect in the drilling budget: risk cost, reward size, probability of success, and total available risk capital.

Chapter 20, "Funding Oil and Gas Ventures," by Lee T. Billingsley, is organized into three main parts: types of funding, sources of funds, and marketing. It is written for the independent geologist who needs to combine technical expertise with salesmanship to sell prospects, cause wells to be drilled, and achieve financial success. The author hopes this chapter will stimulate geologists to become students of salesmanship and thereby increase their fundraising effectiveness. He also recommends that geologists view acquisitions as an opportunity to develop low-risk prospects on producing acreage blocks.

Chapter 21, "Putting a Deal Together," by Albert H. Wadsworth, Jr., is offered as a guide for the beginning independent explorationist. It is brief, to the point, and very readable. The author states that printing stationery and hanging out a shingle does not constitute a viable business. An independent's success will be enhanced by the ability to judge what is salable, the timing of its presentation, and how to sell the concept.

Chapter 22, "Enhancement Techniques for Prospect Presentations," by Robert F. Ehinger, stems from a consulting position in which one of the author's primary assignments was to evaluate outside generated prospects for their economic and geological merit. Approximately 170 prospects were critiqued. It was found that a significant number of geologists had difficulty selling their prospects because of a lack of documentation, or because essential geological controls were not defined for the investing company. The author offers some observations and suggestions that may help prospect generators sell their prospect.

Chapter 15

Managing an Exploration Program for Success[1]

Bill St. John

Ethiopia Hunt Oil Company
Addis Ababa, Ethiopia

You have to continually press yourself to convert information into action.
—John Masters, in *The Hunters*

INTRODUCTION

Exploration, in the business sense, is only one segment of the many functions of a company, whether it be a major integrated international oil corporation or a one-person operation. Exploration must fit with the other functions of the organization, large or small, to operate efficiently. And even more important, exploration is not just the finding of oil and gas, but the finding of oil and gas *at a profit*.

The role of exploration within a company depends upon the mission, goals, and strategy of that firm. Therefore, to be a successful explorationist, one must understand the company's primary mission, goals, and strategy. Subsequent to determining these company objectives, one must establish goals, strategies, operational plans, and performance milestones at the district and functional levels. Doing so establishes the company business plan.

For the business plan to succeed, top management must see the necessity for it. The Chief Executive Officer (CEO) must be committed to the effort and must transmit his or her commitment to the line managers. A company that establishes reasonable and attainable guidelines and completes them in a timely manner should succeed and prosper.

COMPANY MISSION

Those involved with a company—the shareholders, employees, customers, service companies, host communities, and governments—all want different things from the company. However, the primary mission of a company is to increase its shareholder, or owner, value by increasing its asset base and profitability. Secondary missions of the company might include growth or enhancement of the company name. A company mission statement might include the following:

- Enhance company or shareholder value
- Be a top performer in industry
- Operate owned properties
- Show technical and professional leadership
- Be a good employer and good company neighbor

The company mission must be clearly defined and communicated to all subordinates for meaningful action to take place, and implementation of the mission must be organized and controlled.

[1]As this paper was substantially completed prior to the author having accepted his current position as General Manager and Vice President of Ethiopia Hunt Oil Company, nothing contained in this paper should be assumed to reflect the views of Ethiopia Hunt Oil Company or any affiliated company.

OIL AND GAS INDUSTRY OVERVIEW

Once the company mission is defined, the organization must consider an overview of the oil and gas industry and how the overview can be used for future planning. The U.S. energy industry has undergone dramatic changes in the 1980s. So wide ranging have been the changes, that industry participants remain uncertain about appropriate strategies to pursue in light of the consequences of this volatility. A study of the background, causes, and results of the changes provides an insight into possible future directions.

Late 1970s to Early 1980s

Starting in the late 1970s and continuing through the early 1980s, huge amounts of capital were being infused into the U.S. oil and gas industry in anticipation of $50per bbl oil prices and $9 per MCF gas prices. Early dominance by OPEC producers and a seemingly never-ending increase in demand for product and reserves caused price escalations.

The apparent economics drew attention from the financial and investment communities, which sought to capitalize on the situation. Oil and gas companies took advantage of availability of capital, despite high interest rates, to buy equipment and acreage/prospects and hire personnel at ever-increasing prices. Political guarantees and tax incentives from government, coupled with huge investment resources (mainly from financial institutions) promoted capital outlays in the industry far beyond its management and exploitation capability. This created enormous waste, abuses, and distortion of values in all sectors of the business from reserves to personnel, and most importantly, company equity values (Figure 1).

Despite skyrocketing inflation, interest rates, and costs of exploration and production, the energy producers—both established participants and inexperienced late entrants—sought high-cost reserves, believing prices would continue to rise in line with cost escalations. In fact, margins were maintained or increased for quite some time, encouraging even more outside investment and further increasing industry debt burdens. In this environment, mergers and acquisitions proliferated. The corporate raiders in the oil and gas industry caused large independents and even majors to restructure and reorganize, seek "white knights," assume large debt burdens, and even disappear. In effect, attention was taken away from the real business of the industry—to find efficient oil and gas reserves.

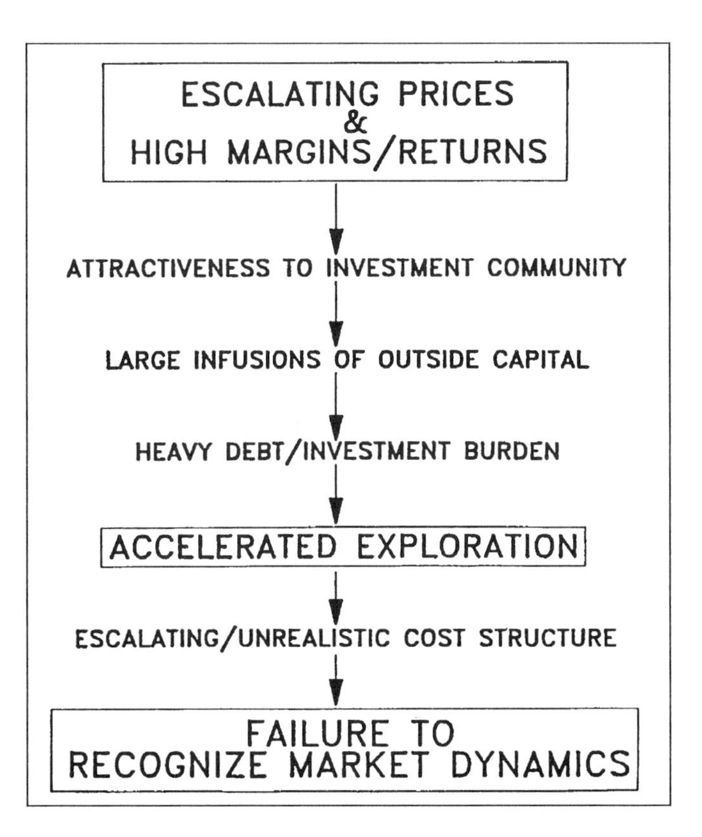

Figure 1. Oil and gas industry scenario for the late 1970s to early 1980s.

Mid-1980s to Present

Eventually, the Saudis and other Arabian Gulf producers lost market share to non-OPEC producers who were now the recipients of increased price benefits and other economic incentives, such as gas deregulation and tax breaks. But other factors were also coming into play that set the stage for the problems which were to surface for the industry in 1986.

The factors that resulted in the disaster scenario played out by the oil and gas industry in 1986–1987 and 1988 were part of a larger global shift, or reordering, of the whole economy. The value of the dollar has declined, as has the nominal interest rate and the rate of inflation. Oil prices, and gas prices in tandem, experienced a cataclysmic drop from the $30 per bbl range to below $10 per bbl in a very short period of time due to worldwide overproduction of oil and gas (Figure 2). Reasons for this change were numerous and diverse:

- Volatility and speed of change within industry yielding failure to attend to market considerations
- High prices indefensible in the market place
- Increased OPEC production as Saudi Arabia determined to discipline OPEC cheaters and regain its control in oil markets

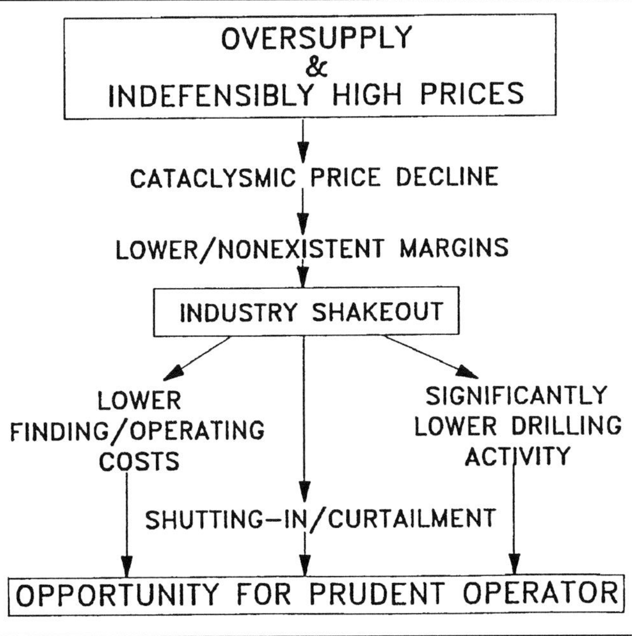

Figure 2. Oil and gas industry scenario for the mid-1980s to the present.

- Mature customers for product—client industries not growing
- More efficient utilization and cogeneration by industrial users
- Conservation in response to high prices and uncertain availability
- Environmental concerns
- Competition with other fuels—technology in place that facilitated switching to alternate fuels
- Deregulation in gas industry
- Large and persisting oil production surplus
- Creation of the "gas bubble," or surplus

Faced with the huge price decline, oil and gas companies saw margins narrowing and even disappearing on high cost-to-produce reserves. Realizing that a small percentage of their properties were producing the bulk of their profitability, companies concentrated on their best properties. Emphasis on finding, lifting, and transportation costs led to shutting in and curtailing production. Pressure was put on the federal government to repeal the Windfall Profits Tax and assist in other ways.

As the price drop persisted, companies were forced to even more drastic measures, as heavy debt and interest burdens threatened overall profitability. Cost cutting came in the form of slashed exploration budgets, personnel cuts, and budget cutbacks in every possible area of activity. Consequently, the service company industry foundered. Sales of equipment and supplies virtually ceased, as rigs became idled due to lowered exploration budgets.

Existing Opportunity

This brings us to the new oil and gas industry scenario, where the financially weak and inefficient no longer exist, be they operating companies or the equipment and service companies supplying them. Especially hard hit by inability to service huge debt loads, many independents (who historically have accounted for 85–90% of all exploratory drilling activity) went bankrupt. Squeezed by the incredibly low prices they received for their services and limping along with skeletal crews, the equipment and service companies remaining could scarcely handle demand if exploration activity were to increase significantly and rapidly.

The rich, relatively invulnerable industry giants who were able to weather this extreme downturn are turning this situation to their advantage by acquiring oil and gas reserves at bargain prices. They are operating leaner and more efficiently than in recent years. Supplier and service costs are likewise lower than ever experienced in recent history. These companies have recognized the existence of a strategic "window of opportunity," wherein costs are low and prospects are readily available to those willing to explore and acquire. Prudent operators realize that oil and gas will continue to be the major energy sources for the world for upcoming decades. Because supplies of oil and gas are finite and reserves decrease daily, these resources become more valuable every day. The question is not *if* prices will increase, but rather *when* and *how much* they will increase (Figure 3).

While estimates of future energy price movements vary, there is general consensus among industry experts that prices will increase. The price increase and production limits set by OPEC in June 1987 have improved the outlook for the oil industry. It appears that production is under control and that prices have not only bottomed out but are beginning a steady upward climb that should continue. It is generally believed, however, that exploration activity will remain limited until prices have reached a sustained minimum of $25 per bbl.

Domestic gas reserves, delivery, and price are a separate, but equally important factor. More exploratory wells in the United States, onshore and offshore, yield more gas than oil. A 10¢ increase per MCF is at least as meaningful as is a $1 per bbl oil price increase.

Although for the past several years the demise of the gas bubble has always been set at 18–24 months from the day of the projection, it appears this may soon be a correct estimate. That there have been only minimal

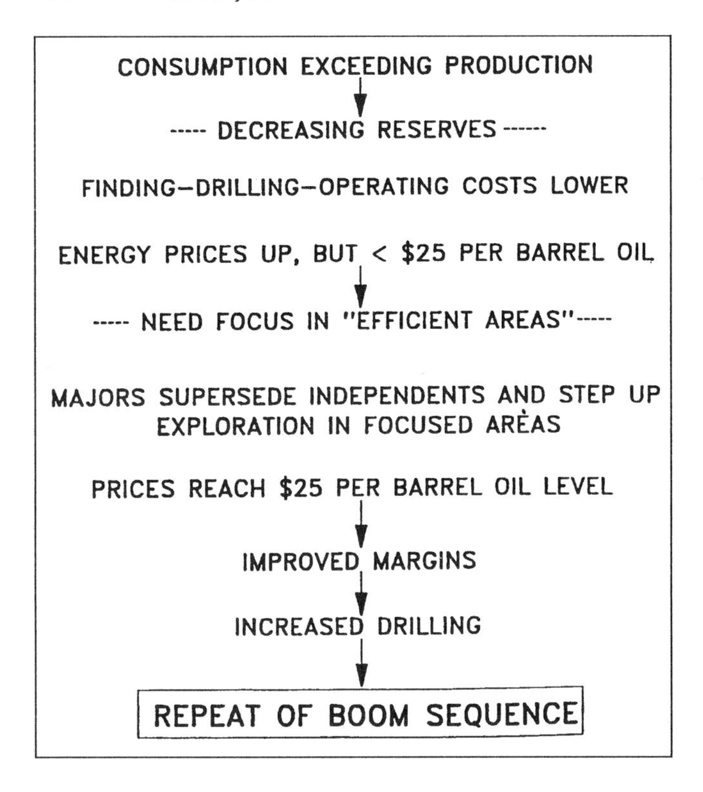

Figure 3. Dynamics of existing opportunity.

gas reserves added in the past few years is not questioned—more gas has been consumed than found since the end of 1984. There are the beginnings of doubt as to deliverability of adequate gas should an abnormally cold winter occur, hence the many published predictions of gas price increases.

With costs low and prospects for increasing prices good, the strategic window of opportunity remains open. As long as drilling remains dormant and industry reserve additions fail to replace oil consumption, the day that prices rebound draws ever closer. Investors who recognize that another major petroleum boom is in the cards in the years ahead, and act accordingly, stand to reap the highest investment returns when it takes place.

In general, low cost reserve replacement is the name of the game. Companies that consistently fail to replace production, either through exploration or acquisition, are in a liquidation mode. Low cost acquisition can be achieved by acting in a timely manner before a large increase in drilling, industry-wide taxes, and current equipment and service availability drive up their prices. It can also be effected through concentration of efforts in high potential and/or low risk areas.

Large operators in the oil and gas industry are concentrating their efforts on acquisitions and on exploration and development in Alaska, offshore California, offshore Gulf of Mexico, and in select international areas. This is true of non-U.S. companies also, as witnessed by the recent acquisition of Standard Oil by British Petroleum and the influx into the Gulf of Mexico of Elf and Total (France), Fina (Belgium), Wintershall (Germany), BP (England), Agip (Italy), Braspetro (Brazil), and various Japanese companies.

It is generally accepted that no sizable reserves remain to be found at shallow and intermediate depths onshore in the U.S. lower 48. Large gas reserves undoubtedly remain but will occur at greater depths—depths that are uneconomical at today's prices. Onshore prospects are increasingly smaller and riskier. Mid-sized and larger companies must seek large prospects or get out of the business. One-well fields will not support a viable exploration effort. For sizable reserves additions, companies must go to Alaska, offshore Gulf of Mexico, offshore California, or overseas.

Alaska has potentially very large reserves remaining to be discovered. However, the high costs of exploration, development, and transportation, plus the long delays involved, make this a major company venture. Offshore California is also a major company play because of the high lifting costs involved with producing heavy oil, coupled with the delays of federal and state permitting requirements and the active resistance of environmental groups.

A case for "bread-and-butter" activity can be made for the Permian basin of West Texas and other areas, depending largely upon lease positions and other special situations. To add significant reserves, however, majors and large independents are moving more and more to international areas and to offshore Gulf of Mexico. Observation of companies that are active overseas and in offshore Gulf of Mexico suggest those having the best financial results are those operating joint ventures and benefiting from the promoted interests of their nonoperating partners.

In summary, large independents will survive and prosper through select acquisitions and in exploration and development in offshore Gulf of Mexico and in select overseas areas plus bread-and-butter activity in lower 48 basins where special situations exist. The majors will follow the same strategy, adding Alaska and California to their long range plans.

CURRENT STATUS OF A COMPANY

Any successful company will periodically review its overall status and take a hard look at its financial condition, how its competitors are doing, and what its competitive position is regarding its strengths, weaknesses, opportunities, or threats. It will also consider its organizational structure and key personnel. Radical changes in the industry economy, such as what happened to oil and gas in 1986, will result in unscheduled and intensive reviews.

Financial Status

The industry price collapse beginning in late 1981 and climaxing in early 1986 brought harsh financial reality to many firms. Foreclosures, abandonments, Chapter 11 bankruptcies, and fire sales became commonplace. Many companies brought in outside consultants to appraise their situations, and the resulting evaluations were often devastating.

The options outlined were usually (1) divestiture, (2) caretaker, or (3) expansion. If debt and lack of assets were overwhelming, companies frequently had no other choice than to sell at the best price and as quickly as possible. Others, with conservative but viable assets and little or no debt, often elected to assume the caretaker mode, whereby activities were curtailed, staff reduced, and surviving management and staff settled down to wait for better times.

Those with strong cash positions and insignificant debt found themselves in a unique position: they could expand relatively inexpensively. Troubled companies were being forced to sell reserves of oil and gas at prices far below historical finding costs. The questions facing the healthy oil and gas firms were how low would prices fall and how long before the inevitable upswing would occur. When should they jump in? Notable acquisitions by some companies occurred during the downturn. Those acquiring companies will be much stronger and more prominent in the years to come.

Comparison with Competitors

Much can be gained by examining the success or failure of competitors. What are their strengths and weaknesses? Where are they successful that others are not? Data are available from annual reports, 10-K's, and published comparisons such as the *Oil & Gas Journal*'s annual "Top 400." These data sources reveal current reserves, changes in reserves for the year, developed and undeveloped reserves, and asset values. Also, annual expenditures, net income, operating costs, acquisitions, total acreage held, and even areas of activity and special developments are often revealed. There are usually enough data readily available to compare competitors' finding costs and profitability.

Competitive Position

In assessing the company's current status, great care should be taken to analyze strengths, weaknesses, opportunities, and threats. Some writers use the acronym SWOT to describe the combination (Hatten and Hatten, 1988). Strengths include the following:

- Financial strength—no foreseeable problems, no outstanding debt, strong and capable of expansion.
- Production and revenues—large reserves, abundant production and strong revenue stream.
- Mix—a healthy combination of both oil and gas reserves located domestically and internationally.
- Mineral and/or lease position—widely held minerals position in a producing basin, strong lease position in active trend or play area.
- Prospect inventory—good quality prospects, enough for a lengthy and active drilling program.
- Management team—experienced and successful individuals with a good record as a team.
- Qualified personnel—experienced and widely known and respected, no skill gaps.
- Contacts—oil and gas exploration is basically a partnership business, few wells are drilled 100% by the operator, and a wide network of contacts ensures opportunities.
- An outstanding reputation—of success or as a firm with which other companies would want to do business.

Weaknesses would include the following:

- Financially weak—can't afford to operate, reputation as being slow or a nonpayer, not a desirable participant.
- Inability to access financial data on specific wells or fields—don't know when they are operating at a loss.
- Minimal prospect inventory—problems with operating schedule, can't put together joint ventures, no leverage via farm-outs.
- Weak acreage position—can't drill if can't put acreage together.
- Weak data base—difficult to generate prospects.
- Image as nonoperator—failure to take advantage of leverage available.

Opportunities include the following:

- Improved economics—domestic leases are less expensive and terms are more reasonable overseas; also, seismic, drilling, and service companies' costs are significantly reduced.
- Reduced competition—fewer competing companies exist and those have reduced budgets; consequently, the quality of available prospects is much greater than in previous years.
- Acquisitions—domestic and foreign reserves and producing properties are available, many below historical finding costs, as are companies with prospective acreage and drillable prospects.
- Availability of quality exploration personnel—the demise of numerous exploration organizations and downsizing of others has made available a wide selection of exploration personnel and, although unfortunate, has reduced the compensation demanded by those personnel.

Although little can be done about external threats, much depends upon the proactive stance of management. Companies need to know they exist and make allowances for them:

- Oil price collapse—keep debt and obligations at a minimum, monitor expenses, retain efficient staff.
- Continuance of oversupply-driven soft gas prices and curtailed production—develop marketing skills.
- International risks—business interruptions and financial losses can be covered by insurance for (1) war, (2) expropriation, (3) inability to repatriate funds, and (4) inconvertibility of currency. Changes in tax laws or contract terms can usually be offset through negotiations.
- Imposition of punitive taxes—resist through organizations and support of positive representatives and enactments.
- Shortages—if activity were to increase suddenly, shortages of experienced personnel, equipment, and supplies could occur in the near future. Develop a contingency plan.
- Technology breakthrough—monitor developments.
- Cultural differences in management in foreign areas—train expatriate managers and hire experienced local managers.

Personnel and Organization

A periodic company review should include scrutiny of human resources and structure of the organization. The personnel and organizational structure might best be inspected by an outside consulting group. It is sometimes difficult for insiders to either see or face weaknesses in these areas. Are there management or skill gaps? Does the company have a weak manager in a key position? Is some key function simply not addressed?

Companies grow, and sometimes employees in key management positions do not grow at the same pace. For example, an exploration manager may have spent his or her formative technical years onshore in the Rocky Mountains, the Mid-continent, or the Permian basin and thus may find the offshore Gulf of Mexico bewildering. Consequently, he or she may tend to continue concentrating effort and exploration budget expenditures in areas of minimal returns.

How about gas sales? Has the firm added a gas sales manager or does it still rely on old contracts and assume gas buyers are honoring those contracts? Or a company might elect to go international in a big way and have several vast license areas requiring extensive reflection seismic surveys. Does the company have in its ranks an experienced seismic field quality control supervisor? If skill gaps occur in what will be temporary situations, consultants can be used for the period required.

What about the organizational structure? Sometimes a strong manager will assume responsibility for functions that would best serve the company in another department. For example, where does the responsibility lie for a company's electronic data processing (EDP)? Is EDP centralized or does each department and district have their own system? The use of computers has proliferated, and in many organizations, it has grown from a single personal computer (PC) in accounting and possibly another in the engineering section to a mainframe system with remote terminals plus numerous PCs. Ideally, there should be an EDP section in the organization having responsibility for a mainframe system, mainframe work stations, PCs, and specialty computers such as large map printers. A typical well-managed firm may have the EDP department reporting to the Chief Financial Officer or Controller because, as versatile as the modern computer is, its main contribution on a daily basis will be in the accounting and finance area. Or the company may have a full-time Chief Information Officer managing the EDP function. There are software programs available that can coordinate the data of accounting, land records, production, and sales (see Chapter 13).

Organizational restructuring may be necessary to make the organization more responsive to strategies, goals, and action plans. Once the organization and its key people have been evaluated it is upper management's responsibility to put the necessary changes into effect.

COMPANY GOALS AND STRATEGY

After the company mission has been defined, the overview of the industry completed, and an analytical appraisal of the company's current status made, attainable goals and associated performance measures must be established. Attaining the goals outlined in Table 1 via the listed performance measures should enable the company to fulfill its mission.

Company strategy, or strategic planning (Quick and Buck, 1983), must be designed to seize the opportunities provided by the evolving oil and gas industry scenario of the time. Strategic planning is dynamic, and annual reviews of the industry as a whole and the progress of the company will necessitate adjustments. For the company strategy to succeed, top management must believe in it and must convince the line managers of its necessity for survival and success. The line managers must be directly involved in the evolution of the company strategy so that they can provide the input of their knowledge about their specific areas of the business. It is also necessary that the line managers be involved for the goals and action plans to be realistic

Table 1. Company Goals and Performance Measures

Goals	Performance Measures
Become a top performer in oil and gas exploration and production, domestic and international	Replace and increase asset base annually. Asset base among top 20% of independents. Finding cost of barrel of oil 1/3 of market price. Finding cost of MCF of gas 2/3 of market price. Internal rate of return (IRR) ≥ 16% after federal income tax (AFIT) on individual projects and ≥ IRR of top 20% of independents.
Achieve financial objectives	Positive earnings. Market Value greater than debt. Positive shareholder equity. Return on investment of 16%.
Balance ongoing profitability with future growth	Increase asset base, revenues and skills base. Improve cash flow per employee. Improve return on assets.
Add low cost reserves of oil and gas	Incremental worth (IW) per equivalent barrel as compared to prior year adjusted for oil and gas pricing (see Chapter Appendix).
Improve efficiency of operations	Establish standard system of policies and procedures. Timely completion of project milestones.

and attainable; also, people support that which they help to create.

A company with a good strategy—well thought out, attainable, scheduled and controlled—will have a definite competitive advantage over those that do not. The general strategy must be outlined in detail with support and justification. Each district and functional area must play a role in the implementation for the strategy to succeed. Such a general strategy might revolve around these activities:

- Reduce costs
- Improve efficiency
- Leverage the company position
- Focus exploration efforts
- Acquire producing properties
- Form joint venture(s)

A specific and more detailed strategy for the overall organization would probably be broken down into individual functional areas, somewhat as shown in Table 2. After the company mission, goals, and strategy are defined, operational plans and performance milestones follow.

OPERATIONAL PLANS

Operational plans—goals, action plans, and performance milestones—flow from the basic strategy. The company and each subunit, or district, must consider and base the operational plans upon the following:

- Current status and overview
- Assessment of competitive position
- Key personnel and organization
- Financial projections (1-, 5-, or 10-year)
- Strategy

To illustrate, let's assume that analysis of our current status has shown that we should expand operations of one district, close a second district, and expand in the international area. Resultant goals and allied action plans might be as shown in Tables 3a, 3b, and 4.

International Exploration

Although there are some differences, the general goals of a company's international exploration efforts have much in common with domestic exploration goals. Both should assist in creating additional shareholder value, enhance the reputation of the company, be known as top performers in their individual focus areas, add low cost reserves, and meet all financial projections.

Few unexplored sedimentary provinces still remain in the domestic arena, and the overwhelming percentage of domestic exploration is done in relatively mature areas. Internationally, considerable exploration and development remains in known producing areas. However, many unexplored or lightly explored basins and sedimentary provinces remain. Even the major oil companies would find it economically and

Table 2. Detailed Strategy for Specific Company Areas

Company Area	Strategy
Exploration: Domestic	Focus on select basins. Generate prospects in-house. Emphasize development drilling and low cost reserves additions. Limit working interest to 50% of all exploratory prospects. Operate properties to maximize profitability and control budget expenditures. Leverage exploration position by selling promoted interest in in-house prospects and forming joint ventures. Accumulate additional leases in favorable areas at favorable costs.
Exploration: International	Expand exploration in areas currently profitable. Protect existing interests Monitor closely performance of nonoperated properties. Screen new areas on basin prospectiveness, profitablitity, and political stability, with view toward entering new ventures. Leverage exploration position by selling promoted interests in in-house prospects and by forming joint ventures.
Drilling and production	Operate to enhance return. Monitor/audit activities closely to reduce production costs on both operated and nonoperated properties. Maintain optimum production rates. Control drilling/completion costs.
Marketing crude oil and natural gas	Ensure best oil prices received by being aware of current market prices and avoiding extra transportation charges, gravity adjustments, etc. Ensure best gas prices received by being aware of current prices, being sure revenues received as per contracts, and placing shut-in gas on market at favorable prices. Improve gas curtailment situation via seeking alternate markets and seeking release from unsatisfactory contracts.
Acquisitions	Purchase price should allow meeting predetermined hurdle rate. Oil and gas reserves located in focus areas. Must fit with long-range plans. Consider production payments or stock exchange rather than cash purchase. Pursue in coordination with district and marketing functions.
Asset redeployment	Sell low revenue properties to ease administrative burden. Sell high depreciation, depletion, and amortization (DD&A) burdened properties. Consider selling any asset at price substantially higher than fair market and/or book value. Coordinate any asset sale with district to factor in staffing and budget implications of the sale.
Financing	Anticipate capital requirements over 1-, 5-, and 10-year periods.
Revenue and receipt control	Continually monitor contracts to verify gas revenues. Establish appropriate compensation for owned mineral leases by comparing state drilling reports against lease position.

Table 3a. Ongoing Domestic District (Company Subunit) Goals and Actions for Operational Plans

Goals	Action Plans
Become top performer in focus area	Increase production to be one of the top 20% of producing companies. Achieve targeted hurdle rates and financial performance measures. Maintain district staff experienced in focus area and capable of handling multiple activities and large volume of work. Support professional organizations and cultivate industry contacts.
Focus exploration efforts	Identify and define key exploration plays in focus area. Complete focus area regional framework maps. Build and maintain data base requisite to achieving exploration objectives.
Optimize expenditure of exploration dollars by lowering exploration risk	Concentrate effort on known producing areas and trends. Form joint venture with company as general partner. Seek partners in select areas where potential partner has superior information or better lease position. Confirm that potential partners are financially able, technically capable, and trustworthy. Secure partners for front-end acreage and seismic costs. Promote exploratory wells. Explore low-risk objectives underlying existing shallower production. Inventory leads and prospects for short- and long-term to be used as dictated by economic and/or technological developments. Consider tactics to sever deep rights from shallow leases held by production. Limit time and effort applied to screening outside submittals.
Operate working-interest properties when feasible	Insist on operating all internally generated prospects. Assume operations of properties where company owns major working interest or where current operator is unsatisfactory and can be voted out. Acquire operated properties that meet economic guidelines and fit into district strategies.
Improve profits from nonoperated properties through closer supervision	Designate one person to spend up to 100% of his or her time to review nonoperated properties. Ensure that field maps and production reports are correct and up to date. Review monthly joint interest billings to ensure that company is not being overcharged for overhead, labor, or materials. Check each property to verify proper charges based on gross reported oil and gas production and company's net revenue interest. Evaluate uneconomic properties and either improve, sell, or abandon them. Become operator where company is major working interest owner.
Exploit mineral ownership position	Take full advantage of owned mineral acreage with regard to exposure, leverage, selectivity, access to information, and budget flexibility. Contribute proration unit lease, but reserve option to participate via working interest in offset development wells. Initiate joint venture for purchase of mineral interests with company responsible for ongoing stewardship. Systematically review underdeveloped acreage with view to development.
Acquire good value producing properties	Identify and evaluate all acquisition opportunities within district or focus area. Make special effort to identify acquisition properties where company can become operator. Ensure all acquisition prospects fit district strategy and meet economic guidelines.

Table 3b. Discontinued Domestic District (Company Subunit) Goals and Actions for Operational Plans

Goals	Action Plans
Close operations at least cost while maintaining company reputation	Transfer, early retirement, or acceptable termination payment for all employees. Sublease or negotiate settlement for all office, storage, and work space leases. Transfer or sell vehicles, office furniture, supplies, and equipment.
Protect ongoing company interests	Farm-out lease inventory at best terms available. Consider arrangements with independents to use data and acreage to generate prospects and retain carried interest for company. Have one knowledgeable employee or local consultant continue to monitor producing properties, supervise subsequent development drilling, and oversee interest in any exploratory efforts.

operationally prohibitive to attempt exploration in each basin or province worldwide.

The problem then becomes one of basin or province selection and how the firm best utilizes its personnel and allocated budget. Basin analysis is a subject in itself, so let it suffice to state here that the number and location of basins or sedimentary provinces can be screened, according to selected parameters, and reduced to a manageable number. For discussion purposes, we address the operational plans, that is, goals, action plans, and milestones, that a mid-sized independent oil and gas company might follow. A major corporation or a small independent would follow a slightly different approach.

Performance Milestones

Specific performance milestones should be defined and set at the company and district level and should also be applicable in functional areas. The milestones are used by management to monitor the timely completion of planned activities and should relate back to company goals to ensure relevance. An incentive system should be tied to the performance milestones to ensure that the desired results are achieved. The incentive system should provide a method of rewarding the managers in positions of implementing the strategies and action plans. The incentive system, which provides motivation, needs to be part of the overall business plan. Although there are nonmonetary motivators, stock options or cash rewards for meeting the performance milestones are probably most effective.

Examples of performance milestones that would fit with the sample goals and action plans previously given for an ongoing district are as follows:

- Fill in the management and/or technical skills gaps and complete hiring of exploration staff within 6 months.
- Secure and organize exploration data base within 9 months.
- Develop a district regional framework within 18 months. Include cross sections, maps, and geological and economic models; address capital requirements, anticipated risks, chances of success, and short- and long-term budgets.
- Incorporate regional framework into focused exploration area(s) within 6 months of completion of regional synthesis.
- Participate in upcoming federal and state lease sales on a scheduled basis.
- Identify and bid on ten offshore Gulf Coast blocks within 12 months.
- Drill and test eight internally generated prospects within 12 months.
- Farm-in and drill two exploration prospects within 6 months.
- Participate in four outside generated exploration prospects within 12 months.
- Form an exploration joint venture with company as operator within 12 months.
- Update geological reviews of all major properties within 6 months.
- Develop standard operating procedures (SOPs) for use by exploration personnel and contract service groups within 3 months.
- Organize and staff a district exploitation group within 6 months.
- Drill and test at least eight internally generated development wells within 12 months.
- Participate in at least four externally submitted new development projects within 12 months.
- Prepare plan of action to take over operation of some nonoperated properties within 6 months.
- Become operator of at least ten new properties within 12 months.
- Prepare a plan for systematic review of significant nonoperated producing properties within 3 months.
- Develop within 3 months a checklist and schedule for reviewing underdeveloped and held-by-production (HBP) acreage.

Table 4. Goals and Actions for International Operational Plans

Goals	Action Plans
Enhance reputation as successful participant in international oil and gas business	Build experienced international staff capable of basin selection, play analysis, risk assessment, economic analysis, and negotiations. Honor existing contracts and develop discoveries expeditiously. Maintain positive relationship with representatives of host government. Be active in international technical organizations. Cultivate contacts among industry associates.
Meet company financial performance measures	Seek opportunities for early payout. Seek opportunities for high return on investment.
Increase asset base and future revenues	Expand activities in countries where successful and profitable operations of company already exist. Evaluate acquisitions of small companies with producing properties in select target countries with view to achieving "producer" status in that country quickly. Use leveraged partners to limit exposure and reduce risk. Obtain participation in new "company maker" type prospects or licenses.
Achieve lower finding costs, more barrels of oil equivalent per discovery, and higher success rate than attainable domestically	Prudent participation in international exploration should accomplish these goals. Recent studies (K & A Energy Consultants, 1987) have shown that finding costs are markedly lower internationally than in the United States; that finding rate, i.e., barrels of oil equivalent per foot of hole drilled, is approximately 30 times greater internationally than domestically; that finding and development costs are approximately 25% of that in the United States; and that lifting costs overseas may be considerably less than they are in the United States. Be operator and take advantage of leverage, use of in-house talent, control of operations and budget, and access to new opportunities. Use proven consultants in their areas of expertise.
Pursue high risk and high reward search for "company maker" discovery	Improve screening abilities. Reduce political and economic risks. Participate in testing major size prospects.

- Identify and nominate within 12 months uneconomic and marginal properties for sale.

For the district deemed uneconomical and having a negative long-range outlook, performance milestones are useful to ensure minimum loss and prudent care of company assets. Typical milestones are listed here.

- Following disposition of district staff, transfer or sell all vehicles, equipment, furniture and supplies within 2 months.
- Sublet all office and warehouse space within 4 months. Negotiate settlement of leases if subletting not feasible.
- Drill via farm-outs a minimum of four prospects on company acreage within 12 months.
- Maintain cost of overhead and lease maintenance at $300,000 or less for 12 months.

Many of the domestic district performance milestones are applicable to the international exploration effort, such as staffing and data base acquisition. However, because of some basic differences (for example, vast number of potentially prospective areas, long lead times, and size of exploration tracts), some performance milestones in the international area will be different. The following would be reasonable for a start-up international effort:

- Screen the world's approximately 700 sedimentary provinces and eliminate those not considered viable for the company. Rejections can be because of poor hydrocarbon potential, politically unacceptable, unsafe for personnel, harsh environment, expensive operations, too much competition, or unacceptable host country contracts. Knowledgeable consultants can be used to aid in the screening process. The screening should be finalized within 6 months.

- At the company level, determine the scope and budget of desired international activity. Decide the number of exploration projects the company will support and select the most promising from those surviving the screening process within 2 months of completing the screening.
- Complete international staffing adequate to handle the projects selected within 6 months of the project's selection.
- Acquire at least two of the exploration permits within 18 months.
- Complete the preliminary and relatively inexpensive exploration, including photo interpretation, satellite imagery, magnetic and/or gravity surveys, surface geology, geochemistry, and reflection seismic, within 2 years of acquiring permits.
- Drill one or two exploratory tests, with leveraged partners, during the third year.
- Complete feasibility, economic, and engineering plans with firm development schedules within 6–12 months of completing a discovery.
- Complete development of field, transportation, and sale of product on schedule.
- Participate with third party in testing at least one "company maker" concession or prospect under favorable terms each 12 months.

CONCLUSIONS

The management blueprint presented is envisaged as appropriate for the scale and activities of a mid-sized independent oil and gas operator. However, with modifications, the same approach could be applicable to a major corporation or small independent—even to a one-person consultant or prospect generator.

Whatever the size of the organization, it must have a plan. The primary mission must be to make a profit, thereby enhancing the value of the investment of the owner or owners. The overview of the industry must be considered and how the organization can perform profitably within the defined limits of the overview. The current status of the organization or company must be critically analyzed, covering financial status, comparison with competitors, competitive position—strengths, weaknesses, opportunities, and threats (SWOT)—plus a review of personnel and organization. Subsequent to the definition of mission, analysis of industry overview and current status of the organization, goals, strategy, operational plans, action plans and performance milestones must be established. To succeed, survive, and prosper, management must monitor the timely completion of the planned activities and should continually relate back to specified goals.

References Cited

Hatten, K. J., and M. L. Hatten, 1988, Effective strategic management: Analysis and action: Englewood Cliffs, NJ, Prentice Hall, 338 p.

K & A Energy Consultants, Inc., 1987, United States versus foreign finding rates and finding and development costs. Report prepared for Primary Fuels, Inc., Sept.

Oil & Gas Journal, 1991, Top 400 oil and gas companies, Sept.

Quick, A. N., and N. A. Buck, 1983, Strategic planning for exploration management: Boston, International Human Resources Development Corporation, 161 p.

Chapter Appendix

Incremental worth (IW) per equivalent barrel as compared to the prior year as adjusted for oil and gas pricing is measured by taking the future value (FV) minus the capital expenditures (CE). FV is determined by multiplying reserve additions (RA) in barrels of oil equivalent (BOE) times net profit per barrel (NPPB). NPPB is measured by subtracting lease operating expense (LOE) from oil and gas revenues and dividing by the current year production. Reserve revisions are treated as prior period adjustments in determining the prior period's IW/BOE. In arriving at the prior years' IW/BOE, current year oil and gas pricing is used. Here are the equations:

$$\text{NPPB} = \frac{\text{Revenue} - \text{LOE}}{\text{Current year production}}$$

$$\text{FV} = \text{RA} \times \text{NPPB}$$

$$\text{IW} = \text{FV} - \text{Current year CE}$$

$$\text{IW/BOE} = \text{IW/RA}$$

BILL ST. JOHN, Vice President and General Manager of Ethiopia Hunt Oil Company in Addis Ababa, received his B.S., M.A., and Ph.D. degrees, all in Geology, from the University of Texas at Austin. He served in the U.S. Marine Corps from 1951 to 1954, including Korea in 1953. He worked in Libya, Mauritania, and Senegal from 1960 to 1963. Dr. St. John worked for Exxon from 1965 through 1972 in Norway, England, Morocco, and Houston. In 1973 he joined an independent in Tulsa, Oklahoma, as International Manager, later becoming President of Agri-Petco International. From 1981 to 1989 he was President and Director of Primary Fuels, Inc., a wholly owned subsidiary of Houston Industries. He was an international consultant in 1989 and 1990, spending one year in Addis Ababa on a World Bank–U.N. funded project as Technical Advisor to the Ethiopian Institute of Geological Surveys, Ministry of Mines and Energy.

Chapter 16

It All Begins with People

Marlan W. Downey
ARCO International Oil and Gas Company
Plano, Texas, U.S.A.

Profitable exploration requires wise investment of risk capital in people's ideas.
—Marlan W. Downey

INTRODUCTION

Discussions about the business side of geology sometimes presume that attention can be restricted only to financial matters. Indeed, most of the other chapters in this Handbook provide guidance for the proper management of investors' risk capital. Finding an oil field with a prudent allocation of risk capital is the task of successful exploration managers. *It is the leveraging of risk capital by human intelligence and innovation that yields extraordinary profits from exploration successes.*

How important is the role of *people* in exploration success? Can one quantify the worth of staff contributions versus the contribution of risk capital? Are geologists and geophysicists really necessary for exploration success? Can investments in research and in training and motivating staff be justified on an economic basis? Is "intellectual capital" a real and measurable benefit to a company? (*Intellectual capital* is defined as the *net* value provided an organization by its investment in nonoperating technical staff.) Can companies see the worth of their intellectual capital investment reflected in their earnings? Can a corporation afford exploration staff in difficult economic times?

ROLE OF PEOPLE IN EXPLORATION SUCCESS

Exploration success has a number of similarities to successful treasure hunting. Each exploration success is a triumph of originality and risk taking—finding "treasure" where no one else has found it.

Because of the extraordinary need for innovation and risk taking in successful exploration, a management structure that curbs original thinking will harm successful idea generation. Able exploration groups are hot house flowers requiring continuous encouragement and challenges and are blighted by dogmatic and hierarchical management styles. Many of the most successful exploration teams flourish in entrepreneurial settings, and exploration staff in successful companies are often seen to be mavericks in conventional corporate culture.

Proper management of exploration people requires attention to staff morale, to creating mutually shared goals for achievement, to prompt and public tributes to excellence, and to private and thorough analysis of failures. When people in an organization are encouraged to do more work, better work, and work that is more useful to the company, then the organization can be said to be well managed and productive.

IS WISE INVESTMENT OF RISK CAPITAL ENOUGH?

It is sometimes argued that exploration success can be ensured by merely the wise investment of risk capital; that it is the wise investment of *money*, not the management of money *and people* that is important. Successful exploration belongs to a unique class of investments; it requires constant creation of high quality new venture opportunities. The creation and assessment of new ventures is done by people, and their innovative thinking gives value to their creations.

All the wisdom of Solomon can be applied to the management of risk capital invested in a poor quality exploration venture, and it will not raise the probability of success one iota. Profitable exploration requires wise investment of risk capital in people's ideas.

ORGANIZATIONAL STRUCTURE FOR SUCCESSFUL EXPLORATION

The proper organizational structure to carry out successful exploration varies with the character of the problems faced by individual companies. It depends on the relative importance that is attached to managing risk capital versus managing people *and* risk capital. It depends on whether the company can rely on off-the-shelf technology to solve its important problems. It depends on whether the company has the financial resources to look at long-term benefits versus short-term constraints.

At first glance, it may appear that the most cost-effective successful exploration strategy requires only a few people on a "technical loan committee," wisely allocating money to submitted ventures. A company established to simply select and fund the most attractive ventures provided to it by others would seem to have extraordinary cost advantages. Staff, overhead, and research charges would be minimal. Indeed, in difficult times, many companies behave as if their optimal organization is composed only of financial management, production staff, and a pool of investment capital. Such a "loan committee" company provides an excellent way of maximizing return from a provided selection of investment opportunities. However, a loan committee company, lacking staff and research, has no guaranteed access to highest quality exploration ventures. Such a company is able to make efficient use of the ventures supplied to it, but it cannot create its own agenda for action or see deeply into the technology and risks of the offered ventures. A loan committee company minimizes front-end expenses, but self-restricts the company's likelihood of participating in unique highest quality exploration ventures.

In contrast, a "major" company that supports its own staff and research chooses to accept the burden of their costs. It accepts the cost burden only because the company perceives benefits. Internal (captive) staff provide their companies the very best ventures they create (as well as a review of outside offerings). High quality technical staff, supported by an active research group, allows a company to broaden the choices for exploration ventures. Such a company does not need to restrict its exploration search to, say, onshore shallow drilling objectives. Major companies who support their own exploration staff do so because they expect that technical staff will provide worth to the company substantially in excess of their cost. The intellect and experience of a company's captive exploration staff should provide their company with highest quality exploration ventures, eventually providing lowered finding costs and greater unit profits.

ARE GEOLOGISTS AND GEOPHYSICISTS NECESSARY FOR EXPLORATION SUCCESS?

Many people can make money from exploration successes—those who sell services, those who buy and sell leases, those who loan money. These people *profit* from exploration successes, but are not *necessary* for success.

Geologists and geophysicists are *essential* to exploration success. Their value arises because their earth knowledge and technical skills provide a threefold benefit to investors: (1) *location* of exploration prospects, (2) studies *narrowing the range of uncertainties* in parameters necessary for success, and (3) providing a more *accurate assessment of risk and possible reward.*

ECONOMIC JUSTIFICATION FOR RESEARCH COSTS

Major oil companies do not support large research expenditures for altruism or for public image. They support staff and research expenditures because major companies find that technical staff expenditures provide a proprietary advantage. Major companies have created a captive pool of staff intelligence and experience that they value because it returns profit as a multiple of its direct cost. Major companies consider that staff and research expenditures justify their own costs by providing an enhanced return on the company's exploration investments.

Smaller companies are accustomed to pay a "premium" when joining with a major in its risk investments. When smaller companies pay a premium to join with a major, *it is a tacit recognition of the value added by the larger company's investment in captive staff and research.*

MEASUREMENT OF INTELLECTUAL CAPITAL

In assessing the value of the intellectual capital amassed by companies, it is instructive to review how costs vary among competing companies. Data amassed on 30 large energy companies (Salomon Brothers, 1989) illustrate the spread of cost efficiencies between companies in finding and developing hydrocarbons. These differences among companies, over numerous trials and during an 8-year period, are probably an accurate reflection of the differences between, say, the "most efficient" 15 companies and the "less efficient" 15 companies in the 30-company group.

On a weighted dollar per barrel basis, the most efficient group of companies found and developed reserves for $6.40 BOE versus $9.10 BOE for the less efficient group. Very large differences in cost efficiencies are apparent between the two groups. Volumes found by the first group should yield profits; volumes found by the second group probably will not. All these companies have the ability to invest anywhere in the world. What makes the difference in performance over the time period? Is the difference the benefit of luck, or the value of the intellectual capital of the organization? Luck can certainly play a part (in the sense of a beneficial outcome earlier in the chance sequence than the statistical average), but luck is not a consideration when a large number of wildcats are drilled and many investments are made over a long period of time. *The differences in economic outcome are the benefits from having high quality staff and management.*

Simply stated, well-managed companies with first-class staff always outperform their competitors in the long run.

CAN A COMPANY AFFORD EXPLORATION STAFF IN DIFFICULT ECONOMIC TIMES?

During good times, it is an extraordinary temptation for entrepreneurs to expand companies vastly using cheap debt and additional staff. A rampant increase in staff numbers often indicates a company out of control. If ever cash flow cannot cover bank debt and overhead costs, such companies are found to be paper tigers and falter immediately. Staff expenses are only a very modest portion of the costs of running an oil company, but staff costs are often seen as the only *controllable* part of corporate expenses.

It is one of the paradoxes of modern business that staff are always described as the core and heart of a business in good times, but all too often, are the first thing cut out of a company fallen on bad times. In difficult economic times, costs must be decreased to allow survival of the company. Given ensured survival of the company, further cost cutting by staff and research reductions must be carefully examined by officers of the company to see if these reductions may cause a deterioration in the company's long-term ability to make money.

Successful exploration and production is a long-term business, and companies that are not financially strong and well managed offer shaky platforms for technical careers. Companies properly managed for the long term should *hire staff carefully, prune regularly and judiciously, and avoid across the board terminations.*

Reference Cited

Salomon Brothers, 1989, Proved petroleum reserves of 30 large energy companies, 1981–88. Oct.

MARLAN W. DOWNEY, Senior Vice-President of Exploration for ARCO International Oil and Gas Company in Plano, Texas, was educated in Nebraska and has degrees in Chemistry and Geology. He worked for Shell Oil for 30 years in operations, research, and management. He also served as Vice-President of Shell Oil Company and President of Pecten International. His publications include operational applications of petroleum geochemistry, criteria for Miocene reef production, seals for hydrocarbon accumulations, and fault control of hydrocarbon fields. Mr. Downey was a member of the Gordon Research Conferences, chairman of the AAPG Symposium on Seals, and AAPG Distinguished Lecturer. He has been an invited speaker for the World Bank, SEG, International Association of Petroleum Negotiators, and the OTC.

Chapter 17

Building Creative Environments and Leading Creative People

Kenneth F. Wantland

Wantland and Associates
Tulsa, Oklahoma, U.S.A.

It is essential to see change as normal, time as an ally, and information as the most significant variable in the equation for success.
—Kenneth F. Wantland

INTRODUCTION

Driving Concept

Everything starts with people. The key strategy for the renewal of the petroleum industry is "to discover and implement the innate creativity of its people" (Fabun, 1972). This is the driving concept upon which companies in the petroleum industry can build programs of renewal and environments that foster creative thinking and creative leadership. The corporation is the vehicle through which people express their creative spirit and abilities, and leadership is the catalyst of change, renewal, and transformation required.

Leaders must build environments to foster the increased need for innovation, teamwork, technical competitiveness, and improved flow of information. To do this, they need to create the vision of success for their groups in a business climate characterized by faster pace, increased complexity, and higher levels of applied technology. Leaders must involve people in the vision, turn the vision into a plan, and empower people to execute. The leader must create, manage, and defend an environment that supports the creative capacity of the workforce and realizes the mission of the group. He or she must have the insight to develop a compelling vision instead of a meeting agenda, dynamic teams instead of task forces and committees, and to share that vision in a compelling way. In turn the individual contributors must take a larger responsibility for planning and managing their careers and must continuously seek ways to increase their value and influence in their organizations.

Purpose and Scope

This chapter offers a variety of ideas, techniques, and examples to renew and revitalize people and organizations in the petroleum industry. Also, it integrates them into a practical design for building creative environments and leading technical people. The emerging business climate requires rapid response coupled with long-term planning, higher and faster levels of applied technology, and the integration of diverse technical disciplines through teamwork. Above all, it requires that companies rethink their approach to leading and managing the professional workforce.

Both leaders and followers have new roles, new responsibilities, and new challenges in this changing business climate. Leaders and followers will be increasingly clustered in a narrow range of age and experience. People selected for leadership must have the ability to lead peers and colleagues, many of whom will have specialized skills, knowledge, and experience. They must be able to extract all the potential from their people and focus it on mutually satisfying goals. This requires vision and an in-depth knowledge of the talents and aspirations of people. The leader must establish a climate of trust in which people can be themselves and express themselves freely. This is the foundation for creativity and innovation.

Individual Contributors

Individual contributors must increasingly take control of the planning and management of their own careers. Only the individual has the capacity to clarify the personal values and articulate career needs that inspire and motivate themselves. Only the individual is capable of analyzing the attributes of a job that best utilizes their natural abilities and meets their unique conditions for high job satisfaction. Each individual must become knowledgeable about the principles and processes of career development just as they are knowledgeable about their technical disciplines.

They need the understanding of career dynamics in organizations in order to improve "job fit" by personal growth and change, changing the job itself, changing jobs, or creating new jobs to have satisfying career paths and successful career aspirations. As individuals mature, they need to understand the concept of "career stages," their need to grow throughout their careers and increase their value and influence within their organization. More and more, they must be able to redefine success on their own terms. Together leaders and individual contributors must build a creative climate based on mutual trust and respect, personal values, and constructive goals.

The forces of change in the industry magnify inherent paradoxes in many organizational systems. Companies welcome new ideas, but most do little to enhance or understand the environment, people, and processes that produce them. They value high technology, but do not necessarily demonstrate that they value the technical people that produce it. They welcome the results of teamwork, but foster individual performance and personal competition for rewards and recognition. They want a highly motivated workforce, but do not examine and align all their systems to eliminate those that defeat their motivational intent. These paradoxes need to be resolved.

Some people have thought that the ideas and methods assembled in this chapter are applicable only in a "research" environment. They are, in fact, universally applicable ideas and concepts for leading intelligent, creative people under any circumstances. The factor that makes this universal is that everyone has a unique role in the analysis and dissemination of information. Higher levels of technology were at one time the focus of applied research or special staff groups attached to headquarters. The need for this level of applied technology is swiftly moving toward operations. The organization must be tailored to the informational needs of the group, and for this reason, the command and control structure is being gradually replaced by organizations that resemble the informal informational network.

As described by Davis (1987), it is like comparing old and new television sets. To change channels from two to seven on an old set, you clicked through three, four, five, and six on the way to seven. On a new set, the change is instantaneous, two to seven, with no clicks in between. The new business climate requires that kind of speed, and the accuracy that goes with it. The ability to define personal informational needs is a requirement of the new organization. People have always known who they had to supply information to. Now they need to define what information needs to come to them and where to get it and in what form. In addition, the level of technological expertise and need for integration of disciplines that was true of applied research a few years ago is very comparable to the needs of all parts of petroleum companies now, especially operations. The principles outlined in this chapter may have been tested and refined in research groups 20 years ago, but their relevance is broadly applicable today.

Leadership

Enterprises in the petroleum industry are undergoing significant transformation. The leadership is being transformed gradually, being driven by the changes reshaping the industry. We are in the midst of this transformation. These ideas, concepts, and examples form the framework for an emerging philosophy of leadership that is more appropriate to the times than the command and control philosophy rooted in military and industrial manufacturing experience.

The core of the emerging style of leadership is the need to lead intelligent, creative people and to manage the flow of information and ideas. To compete in a faster and more complex business environment, new models of leadership and organization need to be implemented. The informational network is superseding the formal hierarchy as the dominant organizational structure. It is more important to the success of the business to make the right decision than to identify the individual making the decision. Ideas can come from anywhere; therefore, the hierarchy and the organizational boundaries, vertical and horizontal, need to be porous and permeable. The crucial organizational model/element/style is the informational network. New models are premised on new visions of what an effective and creative enterprise would be like.

The critical roles of leadership are to do the following:

1. Create the new vision and present it in a compelling way.
2. Translate the vision into an action plan for change,

renewal, and transformation of the enterprise.
3. Establish, manage, and defend a creative environment in which information and ideas flow effectively.
4. Arouse the curiosity and commitment of the people.
5. Launch an unending war against incompetence, indifference, apathy, and deeply buried assumptions and viewpoints in the organization that can sabotage positive change.
6. Empower the people to share in the vision and to be involved in the plan and its execution.
7. Reshape the organization and its systems to support and be consistent with the vision.
8. Guide the process, develop people, and lead so that everyone shares in the realization of the vision.
9. Understand and enhance information flow in their organization. Where do ideas come from? How does information flow? Are there unnecessary barriers to information flow? How can it be improved?

Genealogy of Ideas

This is not an exhaustive review of the literature in organizational development. What follows is the "genealogy of ideas" that have shaped the philosophy expressed in this chapter. The practical application of these principles by the writer was developed over 10 years of building and managing an applied research and technology group for Cities Service Oil Company and was refined over several years of consulting experience.

The driving concept was articulated in a Kaiser Aluminum article on the corporation as a creative environment (Fabun, 1972). In it, the fundamental purpose of any business enterprise was defined as, "the discovery and implementation of the innate creativity of people." Barron (1977) described the essential characteristics of a creative environment: "Freedom of expression and movement, lack of fear of dissent and contradiction, willingness to break with custom, spirit of fun coupled with dedication to work, and purpose on a grand scale."

Vance (1980) proposed the following definition of *creative leadership* based on experience at Disney Productions: ". . . the ability to establish and manage a creative climate in which people are self-motivated toward the achievement of long-term, constructive goals in an environment of mutual trust consistent with personal values." Montessori (1967) outlined the requirements of being an effective teacher that are the same as those needed to lead creative people. She instructed her teachers to "create the environment, arouse curiosity, and get out of the way."

The nature of creativity itself has been analyzed by Vance (1980), Perkins (1981), May (1975), and Barron (1977), with special attention to the role of creativity in organizations. Vance defined *creativity* as the "invention of the new and the rearrangement of the old in new ways." Perkins defined things as being creative when they meet the criteria of being "original, useful and high quality."

Hall published a significant study of management practice (Hall, 1980) and wrote a series of probing essays on the subject (Hall, 1982). Ray and Myers (1986) have offered a different and valuable view of creativity in organizations. Peters (1987), Peters and Austin (1985), and Peters and Waterman (1982) have written extensively on management practice. In his handbook *Thriving on Chaos*, Peters (1987) outlined the basic process of change: "create the vision, turn the vision into a plan, empower the people and shape the systems to support the vision."

Dahl and Sykes (1988) outlined a process for value-based, goal-oriented personal career planning. It links values and needs to set goals that engender commitment, and it identifies our viewpoints or attitudes as major, often underestimated, factors in the success or failure of goal achievement. Dalton and Thompson (1986) have crystallized the concept of nearly universal career stages in which people go through a predictable sequence of developmental stages and transitions during their careers in organizations.

Drucker (1980) has written a milestone book on entrepreneurship and innovation in organizations in which he identifies the major sources of new ideas. D'Aprix (1976), Mitchell (1983), and Yankelovitch (1981) have addressed the impact of values on people and the workplace, and Waterman (1987) has written about renewal of organizations. Davis (1987) introduced original concepts about time, space, and materials and their impact on organization systems of the future, much of it based on the premise that "the present is the past of the future." The writer has commented on the application of many of these principles in the petroleum industry in a series of columns for the AAPG *Explorer* (see Wantland 1988–1991 in the Further Reading list at the end of this chapter.) Masters (1990) has presented a provocative challenge to existing management practices and organizational structures in the industry.

In a recent article in *Fortune* (Dec., 1990) the most crucial ideas for business in the 1990s were outlined. Among them was the concept that corporations are becoming more and more the source of education for ethics and values in the workforce and that it is good business, not just altruism, that drives the need for this effort.

BUILDING CREATIVE ENVIRONMENTS

Attributes of a Creative Environment

Every leader, by his or her actions, establishes a particular kind of climate in the workplace. It may be regimented or loose, directive or participative, formal or informal. The premise of this chapter is that, to utilize fully the creative capacity of the workforce, leaders need to be fully aware of the environment they establish and attempt to make it the most conducive climate possible for people to both discover and implement their abilities.

The framework for building a working environment that fosters creativity and teamwork is based on open communications, risk taking, and high purpose. Barron (1977) described this framework from his broadly based research on creative people and the creative process as having these attributes:

- freedom of expression and movement,
- lack of fear of dissent and contradiction,
- willingness to break with custom,
- spirit of fun coupled with dedication to work, and
- purpose on a grand scale.

These words reinforce and strengthen what people commonly express as their needs today. These parameters outline a philosophy that can be a significant guide to day-to-day decision making by leaders and followers alike. If these broad tenets are adhered to, there is no need for detailed policy to govern every eventuality in the organization. It forms an umbrella of preferred behavior, performance, and expectations. People have great freedom to act so long as these basic principles are not violated. The philosophy also provides a powerful recruiting tool; it gives potential employees a taste of the reality in the workplace.

Freedom of Expression and Movement

Creative people need freedom to talk to anyone up, down, and sideways without fear of unjustified reprisal. Organizations are being permanently altered by the need for faster and more accurate flow of information. Filtering information up and down the standard hierarchy is too slow and too filled with distortion to be competitive in the emerging business climate. It also inhibits creative expression. People need to try out ideas in an environment of mutual trust.

Lack of Fear of Dissent and Contradiction

Creative people and creative ideas need positive, constructive critique. Negative, destructive criticism inhibits the free flow of ideas, the useful examination of assumptions, and the illumination of incongruities that lead to innovation. Leaders can and must play a significant role in suppressing negative criticism and promoting healthy, constructive debate.

Willingness to Break with Custom

As individuals and groups set new goals, it is inevitable that new and better ways to accomplish them will be discovered. As the goals change, the processes must change as well. Periodic examination of basic assumptions and established processes is essential to the innovative organization.

Spirit of Fun Coupled with Dedication to Work

Creativity has always had a childlike quality. It reflects the openness and acceptance that fosters new ideas. A spirit of fun and spontaneity helps to keep people out of the ruts of repetitiveness that are the hallmark of organizations that are "running smoothly" and "businesslike." Creative organizations commonly have an edge of near chaos about them from time to time. But this is not to be confused with a lack of goal orientation, dedication, and commitment. Creative people do not have to be driven to work; they drive themselves and the people around them. And they do not tolerate incompetence.

Purpose on a Grand Scale

People respond to high purpose with exceptional performance. People want to be part of a larger mission and purpose. This is the critical role of vision in the life of the organization.

Creating the Vision

Effective leaders are both catalysts and liaisons. This is true in technology, it is true in creating teamwork, and perhaps most importantly, it is true with respect to time. Effective leaders are liaisons between the present and the future. They are the bridges between what is and what will be. The vehicle for this activity is their vision of the future. Exceptional leaders live in the future with frequent excursions to the present. They act on the premise that "the present is the past of the future" (Davis, 1987). They create a multidimensional vision of the future, and from that vantage point, they let the present come rushing toward them.

Vision springs from clear knowledge of what is important to the enterprise and its people. Each leader needs to discover what is important about significant planning dimensions, including technology, communications, people development, image and reputation of the group, and management of resources. These are fundamental dimensions around which a vision and a plan can develop.

The vision is the leader's excursion into the future. The vision needs to be multidimensional, an entity that one not only sees in the mind's eye but walks around in. A "vision" that one can feel, smell, taste, and hear as well as see. When the vision becomes this real, two things commonly happen: success becomes inevitable and the timetable, or agenda, for success is shortened.

Turning the Vision into a Plan

The fundamental elements in a creative plan are the clarification of values (knowing what is important) and articulation of what is needed to satisfy those values in specific planning areas such as technology, developing people, and communications. Values and needs lay the groundwork for setting goals that engender commitment. It is crucial that the values and needs represent truth. This requires an inordinate attention to underlying assumptions and viewpoints in the corporate culture that bias the establishment of values and needs in a truthful way. The values and needs are the basic elements of the foundation that underlies creative planning. Since creativity is the "invention of the new and the rearrangement of the old in new ways" (Vance, 1980), it would follow that if any of the "old" is not true, then the rearrangement of the old in new ways, no matter how creative the process, may not be true. If the values and needs a group defines are not true, then the goals, strategies, and tactics that result from them cannot be true, useful, or valid.

It is essential that the group create a set of goals for technology, communications, image and reputation, developing people, and resource management. It is also essential that the goals be consistent with and spring from the needs, values, and vision the group has established.

For example, when the decision was made to determine if applied geological research was worthwhile at Cities Service Oil Company in 1969, the opportunity was opened to create a new entity. The existing group was too small and limited in scope to carry out the mission of teaching, technical service, and applied research expected of it. It was possible to envision the diversity and capabilities of technology needed, the framework of a creative climate to attract and develop technical people, the network of contacts and communications required to transfer technology more effectively, the image and reputation necessary to command respect both inside and outside the company, and the management style needed to bring people, equipment, and technology together to solve complex problems and support the oil-finding efforts of operations. Need-driven goals and strategies were put in place in each of these areas.

Viewpoints

Values and needs define goals. Whether or not goals are achieved depends on a variety of factors. Key among them, and often unappreciated, are the *viewpoints* that people and companies hold about crucial issues. Viewpoints are attitudes. Unlike values that are deeply rooted and difficult to change, viewpoints are somewhat transitory. They are like assumptions that are held to be true until a new perspective is attained that demonstrates them to be inappropriate. Then they can be examined and, if necessary, changed.

In the previous example, some viewpoints had to be altered. One such corporate viewpoint held that mediocre equipment was adequate to meet the needs of the technical support staff. It became clear that to meet the goals of becoming competitive and to be small and flexible, we needed top-of-the-line equipment. Another viewpoint held that publication and participation in professional activities did little to advance the interests of the company. This viewpoint ran counter to the goal of establishing a strong technical reputation and attracting high quality people. The serendipity of publication and involvement was an increased flow of information, simply by being closer to the mainstream of the industry. Much more was gained than was ever divulged in the program of publication and presentation of activities. Thus, the ability to examine and alter inappropriate viewpoints enhanced the likelihood of achieving significant new goals.

It is important for leaders to have a clear idea of what they and the organization hold to be true about such topics as failure, success, power, time, and process. The viewpoints about these topics have a major impact on the ability of leaders and individuals to set and achieve long-term, constructive goals.

For example, one organization was having difficulty obtaining a consensus about what it took to be respected and rewarded as a leader, especially among exploration managers. This was true until senior management made it clear by their actions that they expected a significant shift toward a team concept throughout the company. Once it was clear that all performance was being seen through the lens of a team-building process, expectations and success came into focus. Understanding the newly defined viewpoint about *process* clarified what was true about *success* and significantly improved both performance and results.

It is commonly true that the way something is done, that is, "the process," is part of the culture of the company and not explicitly articulated. It is among those things that people learn as part of their initiation to the organization, "learning the ropes." Since there are acceptable and unacceptable ways of doing things, the way things are done can have enormous influence on the acceptance and judgment of the results. Couple this with the fact that in times of change, some of those cultural, deeply buried ways of doing things may no longer be the best way to achieve desired results and the stage is set for considerable misunderstanding. If, as in the previous case, there is a significant lag between a change in viewpoint and the understanding of that change throughout the organization, even more misunderstanding is possible. It is necessary to reexamine and articulate the process clearly so that everyone can factor it into their strategies for success.

The perception of success is one of the most critical factors in the development of professional people and their careers in the petroleum industry at this point in history. The underlying assumption in most companies is that success is attaining a position in senior management. Despite organizational systems such as dual ladders that offer the opportunity for progression parallel to the management ladder, popular perception is that real rewards and real status are found in management. Efforts to change this perception have generally not been persuasive.

With reduced numbers of people in the industry and prospects for slow growth in terms of creation of new management jobs, the opportunity for traditional, management-driven pathways to success are limited. In addition, the mean and median age of professionals in the industry is 38–40 years of age (L. Nation, pers. comm., 1990). This means that a large number of people are reaching peak career periods at the same time. This is creating pressure to invent new ways to develop multiple pathways to satisfying successful careers. Part of the answer lies in the viewpoint of success. Success must be redefined in terms of *influence* rather than position. An individual who significantly influences the direction of an enterprise is a success whether he or she holds a management title or not. Recognizing, rewarding, and encouraging continual growth in personal influence and increased personal value among technical professionals is a major challenge to organizational systems.

LEADING CREATIVE PEOPLE

Leaders need a wide array of interpersonal skills and abilities to carry out their role in the development of creative people. They need to be aware of the influence of the overall environment on the productivity and innovation of the people. They need to understand the role of values in setting goals, and they must be aware of all the factors, both personal and corporate, that go into the establishment of an effective and creative career plan. They also need to know that planning is only the first step in producing results. Effective leaders must invent creative ways to keep people at the top of their agenda, to be accessible, to listen, and to seek continuously to better understand and learn more about the people around them.

Arousing Curiosity and Commitment

Much is said about the role of the manager or leader as a motivator of people. No one hires people who are not motivated. The role of the leader is not to motivate people as much as it is to challenge them. The role is to present a vision of high purpose and to invite people to find a place in realizing that vision. The workforce today is a volunteer army less subject to command and control and more responsive to shared goals with shared recognition and reward. The job of the leader is to have a supply of ideas, intriguing problems, and opportunities to challenge people. It is also the job of the leader to know the unique abilities each person is motivated to use. People bring a broad array of talents to the enterprises they join. Their talents are only partially defined by their education or their technical specialties. Leaders need to take the time to know more than one dimension of their people. If leaders are expected to accomplish more with fewer people, it is imperative that they know their assets, their resources, and all their complexity and depth. One common measure of the importance of a manager's position is the number of people under his or her control. What if the rules changed and the manager's job was instead judged by his or her ability to get the job done with the fewest number of people? Perhaps then it would become crucial to know all the abilities people possess and are motivated to use.

Let's look at an example. A young geologist was hired to fill a vacancy in computer applications to enhance exploration. Although he came with a good academic record, his performance on the job was less than outstanding. In searching for the cause, a simple truth was discovered. He was capable of doing the job, but he did not want to do it. He wanted to do petrology; he had a clear record of accomplishment in petrology buried in his credentials. Computer applications had been a means to get a job. When he was turned loose on the rocks and allowed to apply the skills and abilities that he was motivated to use, he "blossomed" into an outstanding performer who made significant contributions to both the company and to science. The loss in computer applications was not missed.

We hire whole, complete human beings with their own agendas, ambitions, aspirations, and dreams. If we hire talented people who share the vision and values of the group, they will make important contributions to the success of the enterprise. It just may not always be the contribution that we predict. People need the opportunity to find their own pathways to increase their value and influence. They need the option to change themselves, to change the job they are in to express their abilities more fully, to change jobs as they grow and develop, and occasionally, to create jobs uniquely suited to their talents that did not exist on the organization chart. Some people do not fit into boxes at all; some are more suited to occupy the spaces between them. This requires trust and commitment by everyone involved.

People are driven by different key motivational factors in their accomplishments at work and in life. Hanson (1985) outlined some of these key motivational payoffs for individuals. Some people are motivated to stand out, to be key, to be unique, to excel. Others are motivated to exercise power, to be in charge, to overcome obstacles, and to have impact and influence on people, ideas, and organizations. Still others are motivated to reach objectives, to win. Some are motivated by the process, to delight in the journey, to explore, to realize concepts. And there are those who are motivated by challenge, by meeting the test.

Each individual needs to know the source of his or her own motivation. It is a major factor in realizing satisfaction in a job. If someone gets great satisfaction from winning, he or she needs to be in a job where the opportunity exists to keep score. Commitment comes from the ability to link satisfying goals in work to the personal values and needs of individuals.

Developing People

The key to developing people is the process of creating long-term constructive goals consistent with personal values. This is a joint effort between leader and follower in which both need to understand the factors that influence long-term career planning and career management as well as the dynamics of goal achievement in organizations. The factors that influence career planning and management are both personal and corporate. That is, some factors must be addressed independently by the individual, while others are the result of being part of a particular corporate organization. The personal factors fall in three basic categories:

1. values, needs, and wants;
2. goals, strategies, and tactics; and
3. viewpoints and obstacles.

All individuals must be clear about what is important to them about their careers independent of the opinion of others. People who primarily value professionalism and technical excellence will have sets of needs and goals in their careers that are different from those who primarily value authority and winning. We are all unique in our suite of values, needs, and wants. The ability to articulate a set of values is a significant foundation to career planning. As Roy Disney is quoted as saying, "When my values are clear, decisions are easier." Without a foundation of personal values, individuals are likely to accept someone else's values as their own. An understanding of values makes it possible to define career needs. If, for example, you value technical excellence, you may have the need to be constantly on the leading edge of your discipline. That need translates easily into a career goal: to establish and maintain a position on the leading edge of your profession throughout your career. A goal like this is likely to create commitment because it is rooted in your personal needs and values. The question is whether or not the goal can be realized in the particular corporate environment in which you work.

The corporate factors that influence career planning and management are the values, vision, mission, purpose, and goals of the company, as well as its unique culture, structure, demographics, and viewpoints. Some companies have made the effort to clarify corporate values and use them as a motivating, unifying force around which individuals can find focus and purpose. Other companies do not find this to be a useful activity and avoid articulating corporate values. For the individual, it is vital to have some idea of what the company holds to be important about technology, developing people, image and reputation, resource management, and communications because these will strongly influence the individual's ability to set and achieve long-term constructive career goals. The individual can obtain insight into the corporate values by simply observing and listening to corporate leaders. The way the leaders spend their time and allocate their resources reflects their values. If they value the merits of "management by walking around," they will in fact spend some of their time walking around. If they value technical excellence, they will expend the time and resources to set standards, establish role models, fund training, outline incentives, and provide the materials essential to achieving that goal. Leaders can obtain insight into the perspective their people have on corporate values by asking them why they stay with the company. Any group of employees can generate a rapid list of reasons why they stay. Among these are the positive values they see expressed in the company.

In addition, being part of an organization introduces the need to understand the concept of *career stages* (Dalton and Thompson, 1986). The career stage model defines a series of stages and transitions that people pass through in the course of developing their careers. People enter organizational life as apprentices (Stage I), then develop credibility in their profession and emerge as independent contributors (Stage II). With continued growth and maturity, people enlarge their sphere of influence and become leaders or mentors (Stage III), and some pass through a transition in which they assume broad powers of authority and become corporate directors (Stage IV) with the ability to influence significantly the direction of the corporation.

Each stage contains challenges, opportunities, and problems. The value of the model is that it provides experience-based guidance that can help people anticipate problems and establish goals and strategies consistent with their stage of development. For example, people in Stage II (independent contributor) need to focus their attention on securing their reputation and deepening their abilities in their chosen field before they make the transition to Stage III. This ensures that they have the credibility and respect needed to expand and broaden their influence significantly with and through other people.

Perhaps the most pressing issue in career development today is the need to have multiple pathways for increasing influence and value, that is, a variety of ways to succeed as a mentor or leader in Stage III. A large number of people are maturing together in the petroleum industry at the same time that only limited new opportunities exist for traditional leadership roles. This means that a large number of people must find alternative ways to express their ability to have influence in the organization.

The solution to this problem requires imagination in the concept of professional, technical ladders, the organizational alternatives to the managerial hierarchy. Experienced professionals can play a number of extremely valuable roles that will increase their value and influence in the company and that are not managerial or supervisory in nature. They have to do largely with the analysis and dissemination of information. The need to analyze data is greater than the need to relay it up and down the management chain of command. This creates the need for people who are information hubs, informational liaisons among various disciplines, gatekeepers to new information from outside sources, teachers and mentors who are role models for the analysis and application of information, and idea generators who integrate information. Companies need to rethink the value of new roles in the organization in the light of the growing demands on the system to analyze data and refine information. Being a key person in the analysis of technical information may be as important to the success of the company as any number of key managers. New ways to measure the importance of jobs are needed in the new realities of today's business climate.

Managing Values and Vision

It is not enough to envision and establish a creative climate. Such an environment must be managed and maintained. Values must be reaffirmed and reinforced by action. Freedom of expression must be encouraged. Destructive dissent must be dealt with. Leaders must ensure that individuals and the group do not get into a rut of doing things over and over the same way. Underlying assumptions need to be brought to the surface and discussed. The group needs to explore any topics that are considered "undiscussable." Periodically, it is a good idea for the group to meet to review assumptions in one or more of the major planning dimensions. Are there assumptions about the technology that are no longer valid? What new ideas have been learned that need to be incorporated into the mainstream of the thinking and action of the group? What incongruities and inconveniences are being tolerated that could be done in a better way? Leaders need to be tireless in their pursuit of better ways to do things.

Issues in Leadership

There is an all-encompassing need to rebuild trust as a basis for creativity and commitment within the workforce of the petroleum industry. Rebuilding trust requires that developing people go to the top of the list of leadership priorities. Leaders must establish a clear set of goals that are focused on developing a productive, creative, and motivated workforce. They must reorder all systems (and the departments that administer those systems) to be aligned with the goals and the vision and intent that the goals reflect.

Commitment comes from value-based, goal-oriented career planning and management. Equally important, it comes from long-term, life-long thinking. Although neither the individual nor the company can make lifelong commitments, there is a mutual responsibility to support and further the long-term needs and goals of both.

To enhance creativity, the petroleum industry needs to seek the widest possible diversity of input to the workforce. Creativity thrives on diverse ideas and perspectives. Over the next decade, women will assume more leadership roles in many industries and white males will become a minority in the workforce for the first time.

The industry has experienced significant change over the past few years. It will experience even more change over the next decade. The challenge to everyone, leaders and followers alike, is to see the opportunities rather than the threats in this change. In fact, the definition of an *entrepreneur* is one who sees opportunity in change (Drucker, 1986). It is essential to see change as normal, time as an ally, and information as the most significant variable in the equation for success.

CONCLUSIONS

The petroleum industry is undergoing large-scale changes. The corporate mergers and cutbacks driven by the collapse of oil prices have overshadowed equally significant personal and organizational changes driven by the need for enhanced technology, integration of disciplines, and innovation. The rapid change has created and exposed incongruities and paradoxes. Companies need to shift from focus on the individual to teams, from an environment of internal competition to one of cooperation and harmony, from a formal hierarchy to informal networks. Long-term planning must be balanced against short-term performance. Five-year plans must be superseded by compelling vision, and committees must be replaced by problem-solving teams. Strong, clear goals and strategies for better communication and information flow must take their place alongside goals for replacement of reserves and the finding cost per barrel of oil.

Most of all, the long-term development of people

must come to the top of the corporate agenda. The responsibility for this effort is shared by individual contributors and all levels of organizational leadership. Leaders have the responsibility to create a vision of the future, establish a creative environment, and reshape organizational systems to reinforce the creative development of the workforce. The fundamental tools are value-based, goal-oriented planning for individuals and the group as a whole, and an inventory of challenging problems. Leaders cannot be afraid to get to know their people in order to build long-term commitment.

Individual contributors must take the responsibility for planning and managing their own careers. Successful people in this industry consistently are those who possess clear personal values, well-defined goals, and the initiative to get started and persevere. Individuals must learn about the factors that affect their professional development and incorporate this knowledge into their planning just as they would incorporate new technical knowledge into their day-to-day work. Above all, they must continuously develop creative strategies to increase their influence and value in their profession, organization, and industry. Accomplishing this requires honest self-appraisal and a commitment to articulate personal values, needs, and goals.

"The present is the past of the future" (Davis, 1987). The future is upon us; change is the norm. Finding opportunity in change is the key to transformation and success.

References Cited

Barron, F. X., 1977, *in* J. Douglas, The genius in everyman (2): Science News, v. 111, p. 285.

Dahl, D., and R. Sykes, 1988, Charting your goals: New York, Harper and Row, 160 pp.

Dalton, G. W., and P. H. Thompson, 1986, Novations: Strategies for career management: Glenview, IL, Scott, Foresman and Co., 280 pp.

D'Aprix, R. M., 1976, In search of a corporate soul: New York, AMACOM, 218 pp.

Davis, S. M., 1987, Future perfect, New York, Addison-Wesley, 243 pp.

Deal, T. E., and Kennedy, A. A., 1982, Corporate cultures: The rites and rituals of corporate life: New York, Addison-Wesley, 232 pp.

Drucker, P. F., 1980, Managing in turbulent times: New York, Harper and Row, 239 pp.

Drucker, P. F., 1986, Innovation and entrepreneurship: New York, Harper and Row,277 pp.

Fabun, D., 1972, The corporation as a creative environment: Raiser News, Oakland, CA, Kaiser Corporation, n. 1, 34 pp.

Hall, J., 1980, The competence process: The Woodlands, TX, Teleometrics International, 275 pp.

Hall, J., 1982, Ponderables: Essays on managerial choice: Past and future: The Woodlands, TX, Teleometrics International, 93 pp.

Hanson, M., 1985, Job fit and creativity: IEEE Potentials Magazine, Oct., p. 13-15.

Masters, J. A., 1990, Exploration de-organization: Houston Geological Society Bulletin, May, p. 24-31.

May, R., 1975, The courage to create: New York, Bantam Books, 173 pp.

Mitchell, A., 1983, The nine American lifestyles: New York, Warner Books, 302 pp.

Montessori, M., 1967, The absorbent mind: New York, Dell Publishing Co. (English translation), 304 pp.

Perkins, D. N., 1981, The mind's best work: Cambridge, MA, Harvard University, 314 pp.

Peters, T., 1987, Thriving on chaos: New York, Alfred A. Knopf, 561 pp.

Peters, T., and N. Austin, 1985, A passion for excellence: New York, Random House, 437 pp.

Peters, T., and R. Waterman, 1982, In search of excellence: New York, Harper and Row, 360 pp.

Ray, M., and R. Myers, 1986, Creativity in business: New York, Doubleday and Co., 222 pp.

Vance, M., 1980, Management by values: Guidebook with cassette tape set: Phoenix, AZ, General Cassette Corp., 10 pp.

Waterman, R. H., Jr., 1987, The renewal factor: How the best get and keep the competitive edge: New York, Bantam Books, 338 pp.

Yankelovitch, D., 1981, New rules: Searching for self-fulfillment in a world turned upside down: New York, Random House, 278 pp.

Further Reading

Allen, T. J., 1977, Managing the flow of technology: Information flow in R & D organizations: Boston, MIT Press, 320 pp.

Beer, M., R. A. Eisenstat, and B. Spector, 1990, Why change programs don't produce change: Harvard Business Review, v. 68, n. 6, p. 158-166.

Drucker, P. F., 1986, The frontiers of management: New York, Harper and Row, 368 pp.

Drucker, P. F., 1989, The new realities: New York, Harper and Row, 276 pp.

Hawken, P., 1983, The next economy: New York, Ballantine Books, 242 pp.

Wantland, K. F., 1988, Survivors feel intangible losses: AAPG Explorer, Dec., p. 10-11.

Wantland, K. F., 1989, Austere climate stifles creativity: AAPG Explorer, Jan., p. 16.

Wantland, K. F., 1989, Survivors need stable leadership: AAPG Explorer, Feb., p. 27.

Wantland, K. F., 1989, Survivors require "intact" fuel lines: AAPG Explorer, April, p. 34.

Wantland, K. F., 1989, Personal goals generate results: AAPG Explorer, May, p. 9.

Wantland, K. F., 1989, Geologists must beat fear factor: AAPG Explorer, June, p. 26.
Wantland, K. F., 1989, "Old" career paths not always best: AAPG Explorer, Aug., p. 4.
Wantland, K. F., 1989, Careers thrive after self-discovery: AAPG Explorer, Sep., p. 10-11.
Wantland, K. F., 1989, Improvement? Take creative steps: AAPG Explorer, Oct., p. 24.
Wantland, K. F., 1989, You hold the key to improvement: AAPG Explorer, Nov., p. 10.
Wantland, K. F., 1989, Survivors plan to manage crisis: AAPG Explorer, Dec., p. 18.
Wantland, K. F., 1990, Individual initiative needed in the 1990's: AAPG Explorer, Jan., p. 4.
Wantland, K. F., 1990, Empowered people are industry MVPs: AAPG Explorer, Feb., p. 18.
Wantland, K. F., 1990, Company changes don't end careers: AAPG Explorer, March, p. 34-35.
Wantland, K. F., 1990, Employees redefine meaning of success: AAPG Explorer, May, p. 45, 49.
Wantland, K. F., 1990, Intangibles can help career moves: AAPG Explorer, June, p. 18, 19.
Wantland, K. F., 1990,Three factors critical to success: AAPG Explorer, July, p. 8, 9..
Wantland, K. F., 1990, Before you talk, you have to trust: AAPG Explorer, Aug., p. 8.
Wantland, K. F., 1990, Will we meet the challenges?: AAPG Explorer, Sept., p. 40.
Wantland, K. F., 1990, Just what IS this "vision thing"?: AAPG Explorer, Nov., p. 28.
Wantland, K. F., 1990, Street talk indicates attitude shift: AAPG Explorer, Dec., p. 10.
Wantland, K. F., 1991, Command and control won't work: AAPG Explorer, March, p. 50.
Wantland, K. F., 1991, Fresh viewpoint aids creative thinking: AAPG Explorer, July, p. 10.
Wantland, K. F., 1991, Balance of present, future needed: AAPG Explorer, August, p. 8-9.
Wantland, K. F., 1991, How does your future look today?: AAPG Explorer, Sept. , p. 12-13.
Wantland, K. F., 1991, Information is the name of the game: AAPG Explorer, Oct., p. 11.
Wantland, K. F., 1991, May all the forces be with you: AAPG Explorer, Nov., p. 32.
Wantland, K. F., 1991, Future requires anticipation, action: AAPG Explorer, Dec., p. 14.
Wantland, K. F., 1992, Make career strategy top priority in 1992: AAPG Explorer, Mar., p. 31.

KENNETH F. "FRANK" WANTLAND, consultant and owner of Wantland & Associates in San Francisco, received a B.S. degree in Geology from the University of Cincinnati and M.S. and Ph.D. degrees in Geology from Rice University. Prior to entering the consulting business, he held a variety of positions in the petroleum industry, including Research Scientist and Manager of Geologic Research for Cities Service Oil Co., Director of Technical Human Resources and Director of Training for Cities Service/Occidental, and Exploration Training Advisor for Exxon, USA. Dr. Wantland also was Professor of Geology and Director of the School of Geology & Geophysics at the University of Oklahoma. He currently conducts seminars and workshops on career planning and development, leadership of creative people, building creative environments, and renewal of organizations after restructuring. He is a member of AAPG, writing a monthly column on career issues for The Explorer. He is also a member of Sigma Xi and the National Speakers Association.

Chapter 18

Trend Analysis

Michael D. Smith
ENSYTE Energy Software International, Inc..
Houston, Texas, U.S.A.

Donald R. Jones
Consultant
Houston, Texas, U.S.A.

The near future can be modeled by the recent past.
—Michael D. Smith and Donald R. Jones

INTRODUCTION

The decision to commit capital to exploration is made after analyzing a large quantity of geological and financial data. Interpretation of these data is largely subjective, with varying levels of uncertainty that, in aggregate, establish venture risk. Exploration companies spend large sums of money to reduce data uncertainty. For instance, increasing the application of costly technology to processing seismic data and interpreting the results has improved the ability to find and delineate traps, thus increasing the probability of success. Companies can justify spending money on technology to improve data quality subject to the expectation that the risk of exploration will be reduced and the pay-off increased.

In expensive frontier ventures with high pay-off potential, the cost of improving data quality with technology is insignificant compared to the cost of the drilling component of the proposed exploration program. Thus, with only a few such ventures per year, the cost to improve data quality can be considered reasonable. In such ventures, the optimum place to explore is the venture in which risk can be most reduced by the application of exploration technology.

Contrast this high cost, high risk, high return exploration option to lower cost, lower return exploration activity in mature onshore and offshore petroleum provinces. Finding the most profitable places to invest exploration dollars in the latter instances is a different question. The risk reduction achieved through technology is comparable, but the cost is much higher relative to the cost of drilling, and smaller returns are anticipated. For example, it is unlikely that the cost of prospect generation could be justified if a risk-reduction technique such as extensive 3-D seismic were used to support exploration in these areas.

In mature areas, to limit or cull the active areas and focus on those better suited for spending exploration budgets, the explorationist can use the technique of *trend analysis*. The objective is to allocate resources into those areas that have the highest probability of achieving the target return on investment with a given size of drilling program. Target rates of return and expectations from investing in exploration drilling result from goal setting during strategic planning. Establishing exploration goals is a prerequisite to conducting trend analysis.

EXPLORATION GOALS

The objective of exploration is to find oil and gas reserves that can be developed, produced, and marketed with an after-tax return on investment commensurate with the risk taken. Exploration must be related to a company's overall financial goals. (The term *company* is used here to describe any exploration organization, including partnerships, individuals, etc.) Relating exploration to financial goals is best done through long-term planning, a process of developing, monitoring, and adjusting action plans. Simply put, long-term planning is the time-phased allocation of a company's resources to achieving goals. It follows, then, that a company of any size must have a clearly stated goal for exploration before it can build an effective plan. The goals can be stated in various formats, such as return on investment or reserve addition criteria. For purposes of this chapter, the goal is stated in financial terms using probability estimates. For example, a goal could be stated as such: "First, we want a 60% or higher probability of achieving at least a 15% DCFROR (discounted cash flow rate of return), and second, we want no more than a 15% chance of less than an 8% DCFROR." (The 8% figure represents the alternate "safe" return from, say, long-term U.S. Treasury Bonds.)

If a company's resources are limited relative to the size of the exploration investment, we could state the goals differently: "First, we want a 60% or higher probability of achieving at least a 15% DCFROR, and second, we want no more than a 5% probability of losing $20 million." (If $20 million represents bankruptcy, then we are willing to accept no more than a 1 in 20 chance of this happening.) Goals stated in this fashion clearly merge financial considerations with the risk inherent in oil and gas exploration.

A properly designed exploration goal will have the following characteristics. It will

1. represent the goals of the stockholders as articulated by decision makers;
2. express plans and operations quantitatively so that they can be measured by objective standards;
3. relate potential returns to investments;
4. incorporate probability and risk-aversion concepts; and
5. be reality based with a reasonable chance of success.

Trend analysis helps relate financial and exploration goals. It is a technique for performing a series of reality checks to determine which trends have the desired probability of achieving the company's goals.

WHY DO TREND ANALYSIS?

The trend analysis process is consistent with a fundamental concept of business. That is, in any economic environment, *intelligently presented information* is the backbone of investment decision making and is often the key to attaining a competitive advantage.

Given that information plays such an important role in resource allocation, the question is often asked, "Why aren't more exploration decision makers using formal trend analysis techniques?" Some possible explanations include the following:

1. It is a data-intensive process with considerable start-up costs.
2. Many companies simply explore in the trends where their prospect-generating geologists have been successful, without formally considering the decision.
3. New techniques catch on slowly, especially when they appear to infringe on the domain of the prospect generator.
4. It is not an easily reached intuitive conclusion that an investment in one geological trend is superior to an investment in another.

The primary reason for establishing a formal trend analysis process is to capitalize on a wealth of empirical and historical facts to help predict the future. A statistical projection of historically factual data supports intuitive judgment about how the future *could* unfold. The choice of which trends to commit exploration funds could be a "make or break" decision for a company. It is even more important than the choice of individual prospects because the correct trend selection permits the concentration of resources at a level sufficient to make a difference. Choosing the wrong trends can force a company to depend on random, disjointed prospect selection and luck to achieve its exploration goals.

Trend analysis is a reality check on investment decisions. It helps answer the hypothetical question, "What chance does the exploration budget have of meeting our company's corporate financial goal for adding reserves?" Trend analysis provides a valuable methodology for improving a company's exploration performance. The trend analysis process yields explicit economic statements about geological plays. It brings discipline to the process of allocating exploration budgets by addressing the question, "Where should a company explore to have the highest probability of achieving its corporate goals?"

Note that trend analysis is intended to answer a different question than that posed by the related field of "resource estimation." Resource estimation seeks to answer the question, usually with probabilistic statements, of how much oil and gas remain to be found in a trend, basin, or region. Resource estimation is usually carried out by governmental bodies for economic planning purposes; trend analysis is performed by exploration companies to improve their chances of economic success.

Trend analysis does not attempt to substitute statistics for the real work of exploration. Instead, it provides the exploration manager with a tool that helps solve the practical strategic problem of deciding how to allocate a company's (or division's) exploration effort, over the many plays available, so as to maximize its chance of success. The crucial decision of where to explore is made after explorationists and managers identify the potential plays available for exploration. A trend analysis is then conducted that blends intuition (experience) with concrete historical data. More specifically, trend analysis organizes and uses the available empirical facts to provide the information necessary for managers to make strategic exploration decisions in an objective manner.

WHAT IS A TREND?

A *trend* (*play* or *fairway*) is a geographically contiguous area where a formation, zone, or grouping of vertically contiguous zones produces oil or gas from similar types of traps. Trends are geological formations, states of nature, that cannot be defined by political or survey subdivisions nor by fixed depth brackets.

For practical use in trend analysis, the database for a trend must be large enough to allow meaningful probability estimates from the basic statistics. As a rule of thumb, a trend should have a record of 30 exploratory wells and 8 discovery reserve estimates to provide meaningful statistical results. If there are fewer data, as in a new or emerging play, the analyst should use analogies to geologically similar trends. This situation (insufficient data) also often arises from the tendency of geologists to be "splitters" rather than "lumpers," that is, they define trends too narrowly. In this case, the analyst should combine the trend with vertically and/or horizontally contiguous trends that are reasonably similar.

Generally, the time span for a trend study should be restricted, starting from the most recently available data and going back no more than about 10 years. This should be done for two reasons:

1. In relatively mature petroleum provinces, the large discoveries that were found 30 or 40 years ago can no longer be expected. This has been demonstrated in many areas by discovery function studies. Larger discoveries tend to be made early in the exploration history of a trend.
2. The technology of exploration, drilling, completion, and production methods in the trend for the time span studied should not be radically different from what is expected for the trend during the projected future exploration period. As an example, the change to digital seismic recording and analysis (early 1970s in most areas) means that success ratios from the 1950s and 1960s are not useful for estimating the risk of exploration in the 1990s.

For trend analysis to be successful, the information database must contain statistically relevant data so as to provide the best estimates of the probabilities of success and discovery sizes that will apply in the near future.

COMPUTER SOFTWARE TOOLS

Trend analysis cannot be accomplished in a reasonable time, nor to acceptable quality, without the use of the computer and the proper software tools. Fundamentally, trend analysis is a process of drawing statistical inferences from a large information center. The software tools necessary to accomplish this process fall into three categories:

1. *Information Database:* The data for a trend analysis may occupy from 10 to 300 megabytes of data storage. This volume of data requires a user friendly database system for query and data manipulation and correction. This database system must be compatible with commercial data sources such as those offered by Petroleum Information and Dwight's EnergyData.
2. *Data Conditioning Programs:* These programs are used to calculate data parameters from raw input. A typical example is automatic calculation of reserves by decline curve analysis of monthly production data.
3. *Statistical and Analytical Programs:* These programs use data from the database to calculate the results of a trend analysis. Typical examples are programs to perform economic analysis (deterministic and Monte Carlo simulation) and programs to perform statistical sampling and curve fitting.

TREND INFORMATION DATABASE CENTER

The database for individual trends is the critical link in the trend analysis process because it is

1. the repository of historical drilling and production data for the trends subject to analysis,
2. the repository of intermediate and final results from the analysis process, and
3. a data source for generating prospects (e.g., by locating "sweet spots" in the trends).

Once created, the database becomes the source for long-term development and evaluation of exploration strategy. Substantial data verification and modification are required before the database can be considered a reliable source of information.

Data Sources

Typically, commercial sources are used to provide an initial starting point for the trend analysis database. There are four primary commercial data types:

1. *Well History Data:* Data in this category are obtainable from Petroleum Information's Well History Control System (WHCS), AAPG/API well data, or other commercial or regulatory bodies. These sources can provide the key well statistics. These databases are compiled from data submitted by operators to regulatory agencies. Data errors occur, particularly with formation identification, and must be detected and corrected in the data conditioning process.
2. *Production and Well Test Data:* Data in this category provide the basic information for computing reserves and well performance statistics and can be obtained from Petroleum Information's National Production System (NPS), Dwight's EnergyData, or other similar sources. The data are compiled from state and federal regulatory reports and are generally reliable.
3. *Oil and Gas Field Sizes:* Databases compiled by university, government, trade, and commercial organizations present reserve projections, production statistics, geological interpretation, and other data parameters for discovered oil and gas fields. An example is "The Significant Oil and Gas Fields of the United States" available from NRG Associates, Colorado Springs, Colorado, and in particular, their "Field/Reservoir Clusters of the United States." Unfortunately, many other data sources are of limited use because they provide only total field data on multiple reservoir fields, or because they do not include the smaller fields.
4. *Land Data:* Land ownership is a very important practical aspect of trend analysis. Terra-Image, a combined effort of Geomap Company and Stewart Title Company, provides a comprehensive digital mapping package. Along with other data, the database contains property ownership and lease information that is updated monthly. As of 1991, the system is available for parts of Louisiana, Oklahoma, and Texas. In areas without such database coverage, the required information is commercially available in traditional map format. It must be analyzed the old-fashioned way—with color/pattern maps for open acreage, held by production (HBP) acreage, yearly expirations, etc.—to gain an insight into a company's position, or the potential for acquiring a position, relative to its competitors in the trend.

In this chapter, we demonstrate how to build a database for probability of success, reserve size distribution, and production performance from raw well and production data. Mere transfer of the data from a commercial source into the trend database is only the first step. To be useful, the data must be conditioned; that is, they must be corrected for errors and then converted into usable form (production records must be converted into reserve estimates).

Well History Data Conditioning

In its raw form, commercial well history data are of limited use. There are three primary objectives of the conditioning process:

1. *Correct Well Type Classifications:* Each well is classified in one of three ways: new field wildcat, other exploratory (step-outs, deeper tests, etc.), or development. Assignments should be made in accordance with AAPG definitions; however, many wells are incorrectly classified in the commercial data. The most common error is the misclassification of other exploratory wells as development wells, or vice versa. It is essential to review all well classifications, using maps and cross sections, and, taking into account the well control at the time the well was spudded, to estimate the operator's original intent.
2. *Correct Producing Formation Code:* Producing formations are often misclassified because the formation code is too broad for meaningful trend analysis. For example, the FRIO code will be used as the producing formation code for Lower, Middle, and Upper Frio over a vertical interval of thousands of feet and varying depositional facies. It is important that producing formation codes be narrowly assigned, at least at first. For example, if later analysis indicated that there were no significant differences between the Middle and Upper Frio, then these could be lumped together. Conducting an analysis with formations lumped together in the original data may provide meaningless statistics. Unfortunately, one cannot "unscramble the eggs" if the lumped data prove misleading or incomplete.
3. *Correct Objective Formations:* The data for formation at total depth provides important insight into the objective of the operator when the well was spudded. In the absence of any other information, we must assume that the well's primary target was the deepest potential producing zone penetrated. We thus assign the well to the deepest trend (formation at total depth) penetrated.

These three data verification steps are the backbone of the trend analysis process. They contribute directly to the calculation of success probabilities and to the assignment of production and reserves to the correct trend. Reclassification of data, while it requires a significant commitment of time by knowledgeable geologists, is an important step in trend analysis, and one that is vitally necessary for the process to yield meaningful results.

Production Data Conditioning

The major issues in this phase of building the database are to ensure the following:

1. Well identification records from the production data source must match those from the well history data source. During the developmental stages of trend analysis as a statistical tool, considerable effort went into this process. Petroleum Information, one vendor of data, has eliminated much of this uncertainty by internally cross-referencing well history and production data.
2. Completion depth intervals must be correct, so that production related data can be cross-referenced by depth to trend. This is especially critical when working with multiple completions and/or recompletions.
3. Production records must account for all production including cumulatives prior to the recording of monthly data. Some production databases do not contain production before a "system start" date or do not contain this presystem data for all wells.

Production Data Conditioning Programs

After the production and well history data are merged and verified, production statistics are created by special purpose data conditioning computer programs. These programs sift through the data to create and insert into the database additional data parameters such as the following:

1. *Production Parameters:* Actual production data streams for each completion are scanned. For the primary stream, significant data such as the initial production rate, peak production rate, and production delay time are captured. The initial, final, and average secondary production ratio (condensate yield or gas to oil ratio) are also calculated. All depleted wells are scanned to calculate for each completion the production life, final production rate, cumulative production, and average decline rate.
2. *Estimated Completion Reserves:* Historical production and pressure data from a currently producing completion must be analyzed and projected to calculate estimated ultimate recovery. The model is a standard decline curve (rate versus time, rate versus cumulative, or P/Z versus cumulative) program with software rules added to process large quantities of data with a reasonable accuracy expectation. Rules should be installed to handle completions that

   a. are not declining or are declining at a low rate,
   b. are exhibiting inclining rates, or
   c. have not exhibited enough history.

   The computer program should automatically calculate estimated ultimate recoverable reserves and insert the data into the database along with a confidence indicator such as the goodness of fit or reserve life index. This process should be attempted for all completions in order to obtain the largest data sample possible. The program should also compute the appropriate decline parameters for each completion. For example, the program should convert observed cumulative production, initial rate, final rate, and production life data to hyperbolic or exponential decline curve parameters.

At the conclusion of this type of process, the database is conditioned for supporting a trend analysis project. Two things should be kept in mind about database conditioning:

1. The better the source data, the more reliable the analytical results.
2. The major variable cost (computer software is considered a fixed cost) is the personnel time required for data verification and conditioning.

Can We Take Shortcuts?

When faced with this rigorous process for obtaining a reliable information base, many people will ask if meaningful results can be obtained with less data work. If the objective is to make a cursory pass through an incredibly voluminous database to identify areas for further study, or to pick out geographic areas for accepting prospects for review, the answer is yes. If the objective is to build an information center for guiding exploration in mature producing provinces, the answer is no. The information in the trend analysis database will never be 100% reliable. However, unless the hard work of data verification and conditioning takes place, the analyst will be unable to recognize that he or she may be facing the Achilles' heel of computer applications to any business problem: "garbage in, garbage out." The desired result of reliable, valuable statistical analysis can be achieved only through complete data verification and conditioning.

One can save oneself a great deal of work by eliminating some trends early in the process. For example, always check the land situation first. Is land available, held by production (HBP), or under long-term lease? Are there wells being drilled by operators other than those who hold the acreage? That is, can one get into the play?

Similarly, an early look at well costs, maximum discovery sizes, and so on can eliminate trends that clearly do not fit a company's goals. For example, it is unlikely that a major company would be interested in a trend, no matter how high the success probability, where the maximum discovery size is less than 1 million bbl. Other companies may eliminate deep, overpressured trends due to cost and risk.

STATISTICAL PROCESSING

The information about wells and completions required in a database for even a modestly sized trend (one comprising several hundred wells) will be massive. This raw information must be reduced to statistics to be understood. These statistics form the basis for the projections of probabilities for various economic returns from exploration in the trend. In this section, we discuss the information needed for this process and how it can be extracted from the database.

What Information Do We Need?

The two most important types of probability distributions that can be developed from the available data are

1. an estimate of the probability of success for a typical prospect, and
2. the probabilities for the range of successful outcomes (discovery size distribution expressed in reserves of barrels and/or MCFs).

If a company's goals are simply barrels or MCFs, for example, increasing reserves by a stated amount or maintaining the reserves to production ratio at a desired level, these data are all that we would need to analyze the trend. If the goal is stated in terms of finding cost, then estimates of exploration cost per prospect would also be needed. In the usual case, where a company's goals have been stated in terms of cash flow, we must develop the data necessary to convert these probability and reserve numbers to dollars. We will need development costs, production rates, product prices, etc. Most of these input data variables are probabilistic in nature (exhibiting uncertainty) even within a given trend. However, the ranges for many input variables within a given trend, or the differences between trends, may be sufficiently small enough to treat them as fixed. Again, always try to simplify as much as possible by using fixed values for variables for all trends (for example, gas price projection) or within a given trend (for example, production decline rate).

PROBABILITY OF SUCCESS

Obtaining an estimate of the probability of success, P_S, is difficult, particularly in a geographic area (or basin) where there are multiple exploration targets. In an area with multiple trends, the analyst must develop a consistent method for deciding whether or not to include a given exploratory well in the P_S statistic for a given trend. As stated before, in the absence of any other information, the best method is to assign the well to the deepest potential producing zone penetrated. Also, P_S will generally be different for new field wildcats (NFWC), other exploratory (OEXP), and development (DEV) wells. Each statistic is important since the trend analysis model will include all three types of wells.

In a mature producing area, most of the exploration occurs in or near producing fields; the true new field wildcat is rare. Thus, other exploratory wells (extensions, new fault blocks, deeper pay zones, etc.) account for the bulk of the activity and the successes. It is advisable to take the OEXP category of wells into account, unless the objective of exploration is a relatively unexplored area or company policy is to explore for new fields only.

There are two ways to estimate the probability of a discovery:

1. The first method combines estimates by a knowledgeable geologist of the probabilities that:
   a. a structural or stratigraphic trap exists (P_T),
   b. reservoir rock is present (P_R), and
   c. hydrocarbons have accumulated in the closed trap (P_H).

   Assuming that these variables are independent, the probability of success (P_S) is equal to

   $$P_S = P_T \times P_R \times P_H$$

   This equation provides a single estimate for P_S. The geologist should be encouraged to establish a minimum, a most likely, and a maximum estimate for each of these probabilities and compute, with a Monte Carlo simulator, a distribution for P_S for the trend. Generally, this method will provide at best an educated guess. If enough data are available (for instance, regional seismic structural interpretation and well logs), then the presence or absence of each of these necessary conditions (trap, reservoir, and hydrocarbons) can be determined for each exploratory well. (Note that all three must be present for a discovery and one or more will be absent for a dry hole.)

   Frequency statistics can be developed from these determinations (for example, reservoir rocks occurred in 70% of the NFWCs). These frequen-

Table 1. Success Ratio Statistics for Trend 6-A NFWCs

	Base Data						
Year	Oil	Gas	Total Productive	Dry	Total Wells	Success Ratio	Three-Year Running Average
79	1	2	3	7	10	.300	
80	0	3	3	8	11	.273	.294
81	1	3	4	9	13	.308	.300
82	2	6	8	17	25	.320	.293
83	1	7	8	24	32	.250	.277
84	2	9	11	31	42	.262	.228
85	0	6	6	29	35	.171	.216
86	1	3	4	15	19	.215	.176
87	0	2	2	12	14	.143	.203
88	0	3	3	9	12	.250	
10-year totals	8	44	52	161	213	.244	

Oil discoveries = 8/52 = .154
Gas discoveries = 44/52 = .846

cies are then used as probabilities for the trend (P_R = 0.70). This approach, although time consuming, provides a quantitative measure of the components of P_S. In some trends, geological characteristics more specific than the generalized trap, reservoir, or hydrocarbon parameters used above may be directly related to P_S. A simple example would be a trend in which 75% of the prospects with seismic "bright spots" have been successful. If we use the presence of bright spots as a requirement for drilling, then we should use $P_S = 0.75$. (Usually more than one variable would be involved, resulting in an equation for P_S similar to the previous equation, but using different variables.)

2. The second method provides estimates calculated empirically as a time series data plot of the exploration activity. Success ratios for NFWC, OEXP, and DEV wells are plotted for each year of the study period. A curve is then fitted to the points to project the data into future exploration periods.

We recommend the use of method 2. The annual success ratios should be plotted as a time series to project the trend into the future; use 3-year running averages as a smoothing technique if necessary. Table 1 and Figure 1 provide examples of the data calculations and plots. Note that these examples clearly present a decreasing success ratio. Sometimes no trend is apparent; on these occasions, the estimated P_S should be described as a range using all the data. Generally, it is more accurate to project the trend rather than use a simple average over the entire time span. Similarly, the difference between new field wildcats and other exploratory wells often is significant enough to justify treating them separately. Table 1 demonstrates in full detail the derivation of the points plotted in Figure 1. Fitting the average, minimum, and maximum curves to the data points can be done by statistical techniques such as least-squares methods, but often a simple visual fit will be adequate.

One other factor needs to be discussed. Generally, it is better to classify all productive wells, no matter how small, as successful, rather than to make an arbitrary decision as to what size discovery is "commercial" and classifying discoveries below this size as dry holes. This way, noncommercial discoveries will occur in the trend analysis model, just as they do in real life. As we all have experienced, noncommercial "discoveries" cost a lot more than pure dry holes. That factor needs to be included in the analysis.

DISCOVERY SIZE DISTRIBUTIONS

Within a given trend, creating a probability distribution for discovery size in the time frame of interest is of utmost importance. Generally, a distribution of reserve size for reservoirs (pools) discovered during the study period represents a part of the underlying discovery size distribution in the trend since the first discovery. The assumption is made that the distribution of reserves discovered in the near past is a reliable indicator of the discovery size in the near future. The emphasis is on reservoirs (or pools) since *field size distribution*, a common term, is often a misnomer in trend analysis because a multireservoir field may represent several trends. This distinction is one of the limitations on using commercial data on field sizes as a basis for

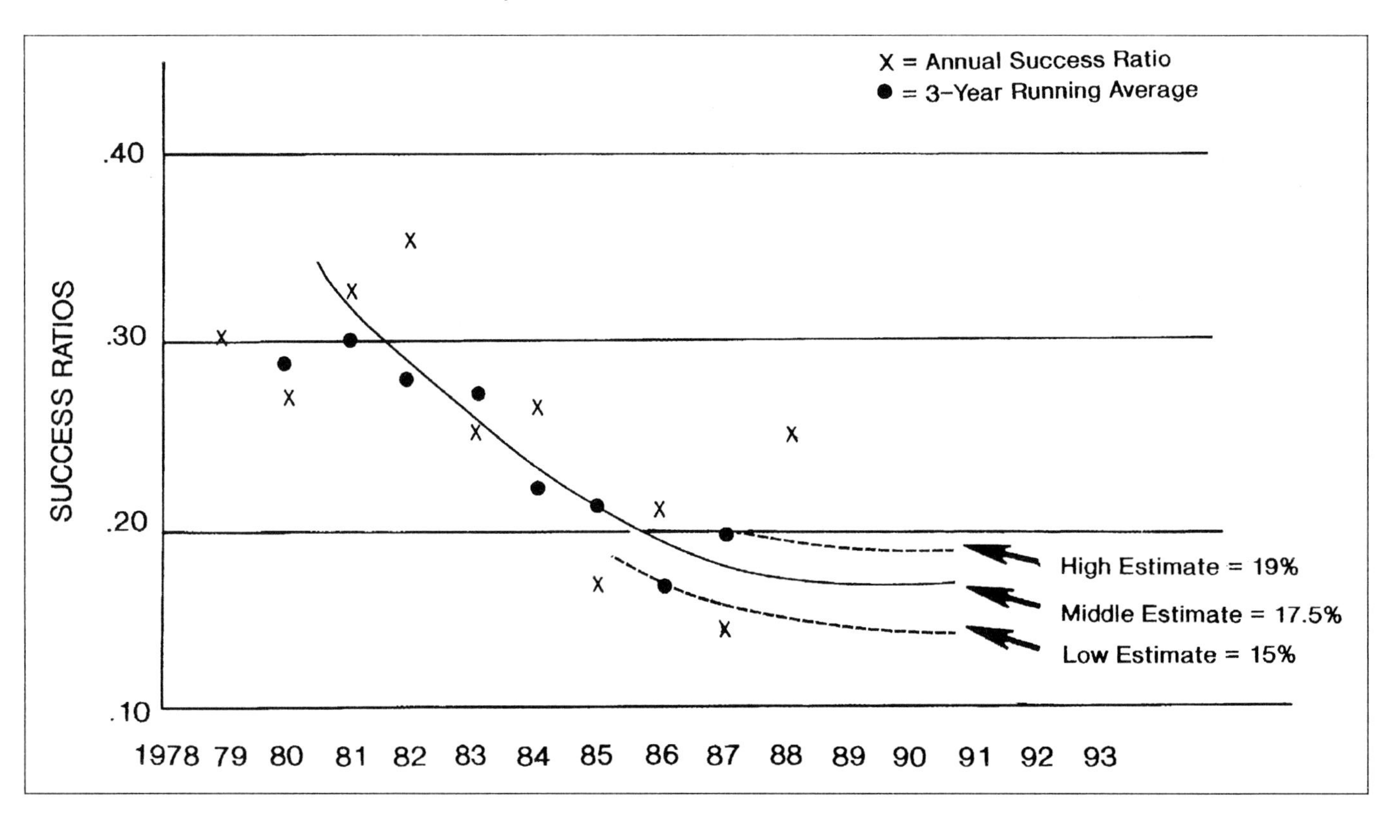

Figure 1. Success ratios.

trend analysis. Note that reservoir size distributions may also overstate the potential to the discoverer due to the inability to control the entire land area of the discovery.

The key to obtaining the distribution of reservoir reserves found during the study period is to calculate the reserves attributable to each discovery well. The method used to calculate the total ultimate recoverable reserves for each discovery in the trend during the study period is to group trend completions that can be logically associated with the discovery well. The general methodology is as follows:

1. For each discovery well, identify the subsequent other exploratory wells, development wells, and recompletions from other zones to be credited to that discovery. Deciding which wells to include with a specific discovery requires judgment supported by empirical data. The standard approach is to sort the database for all wells drilled in the area after the discovery and producing from the trend. These wells are inspected on geological maps, and the relevant wells are then selected. Note that because of multiple pools (e.g., fault blocks) and multiple producing zones within a trend's stratigraphic interval, it may be logical to associate a subsequent "other exploratory" well that finds a new reservoir with a previous discovery well rather than setting it up as a separate discovery. In other words, the well is a development well in the broader sense of the "trend," rather than from the stricter reservoir engineer's or regulator's viewpoint.
2. List the failures for development wells and/or recompletions (these statistics will be needed later). Also, if a development well finds reserves in a new reservoir in the trend, credit those reserves and any reserves from further development to the original discovery well. Remember, we are trying to model the return from exploration. The model should include the pleasant surprises experienced during development as well as the disappointments.
3. Allowance needs to be made for the fact that the discoverer often does not control all the acreage on the discovery and thus cannot be economically credited with all the development wells and recompletions (even if he deserves the professional credit). Statistics should be developed for the number of development wells and recompletions the discoverer drills in comparison to the total number of wells attributed to the discovery. Care should be taken in considering possible company name changes, property sales, or other nomenclature changes that may affect the

Table 2. Tabulation of Gas Reserve Discovery Sizes for New Field Wildcats in Trend Y

Discovery	Reserves MMCFE	Log_e Reserves	Rank	Fraction ≥	% ≥	% <
A	250	5.5215	15	15/15	100.0	0
G	350	5.8579	14	14/15	93.3	6.7
O	620	6.4297	13	13/15	86.7	13.3
D	700	6.5511	12	12/15	80.0	20.0
B	1,300	7.1701	11	11/15	73.3	26.7
L	1,750	7.4674	10	10/15	66.7	33.3
M	2,400	7.7832	9	9/15	60.0	40.0
K	3,700	8.2161	8	8/15	53.3	46.7
E	4,700	8.4553	7	7/15	46.7	53.3
N	6,700	8.8099	6	6/15	40.0	60.0
I	10,000	9.2103	5	5/15	33.3	66.7
H	17,000	9.7410	4	4/15	26.7	73.3
J	23,000	10.0432	3	3/15	20.0	80.0
C	30,000	10.3090	2	2/15	13.3	86.7
F	48,000	10.7790	1	1/15	6.7	93.3
Totals	150,470	122.3447				

Estimate of mean = Average = 150,470/15 = 10,031 MMCFE
Estimate of median = (3700 + 4700)/2 = 4200 MMCFE

Logarithmic mean = Estimated median of lognormal distribution = 122.3447/15 = 8.1563
Antilog = $e^{8.1563}$ = 3485 MMCFE

development of this statistical information.

4. If a discovery well finds reserves in a zone in a different trend (by stratigraphic interval) from the objective trend, it is improper to credit the reserves to the objective trend. Reserves should be assigned to the set of data (trend) to which they belong. Such serendipity (or good geological planning, if one prefers) in areas where trends overlap vertically should be accounted for in the analysis process.

After calculating reserves for each discovery, the next step is to determine the probability distribution for discovery size within the trend. It is generally the case that the distribution of discovery sizes will be lognormal. If the data for a trend are plotted on lognormal probability paper, a straight line will usually represent a good fit of the data between the 10% and 90% points. If the data time period is extended to include past discoveries, the upper and lower "tails" will tend to approach the lognormal line. An analytical program can be designed to accept any form of probability distribution, but as a rule, the lognormal approximation is best for presenting and understanding the data. The important thing is to *model the actual data*.

In the analysis, the key assumption is that future exploration will yield results similar to those experienced in the study period. This assumption can be modified slightly if there are enough data to support better or worse performance. The procedure for creating the distributions is given here.

1. Make lists of discoveries and their corresponding total reserves. Sort the data by new field wildcat and other exploratory categories, with each category broken down into oil discoveries and gas discoveries, if necessary. Secondary products can best be accounted for later in the analysis by means of condensate yields and gas to oil ratios in the projected cash flow streams. However, if their value is small, they can be converted to barrel or MCF equivalents at this point. (Convert condensate in a gas discovery into its MCF equivalent and add it to the gas reserves. Similarly, convert associated gas in an oil discovery to its barrel equivalent and add it to the oil reserves. The equivalency ratio used can be an energy content ratio, where 6 MCF = 1 bbl, or more conservatively, an economic value ratio, which has historically been about 10 MCF = 1 bbl.)
2. List the discoveries in increasing order. An example of a list using MMCFE (MMCF equivalents) is shown in Table 2.
3. Plot the points for each list on log probability graph paper and fit a straight line through the data emphasizing the middle of the data range. Figure 2 provides a plot of the points listed in Table 2.
4. Assess the quality of the fit either by visual analysis or by more sophisticated statistical goodness-of-fit tests. If the fit is poor, reexamine the trend criteria. Is more than one trend (trap type, reservoir type, etc.) being mixed? Are there

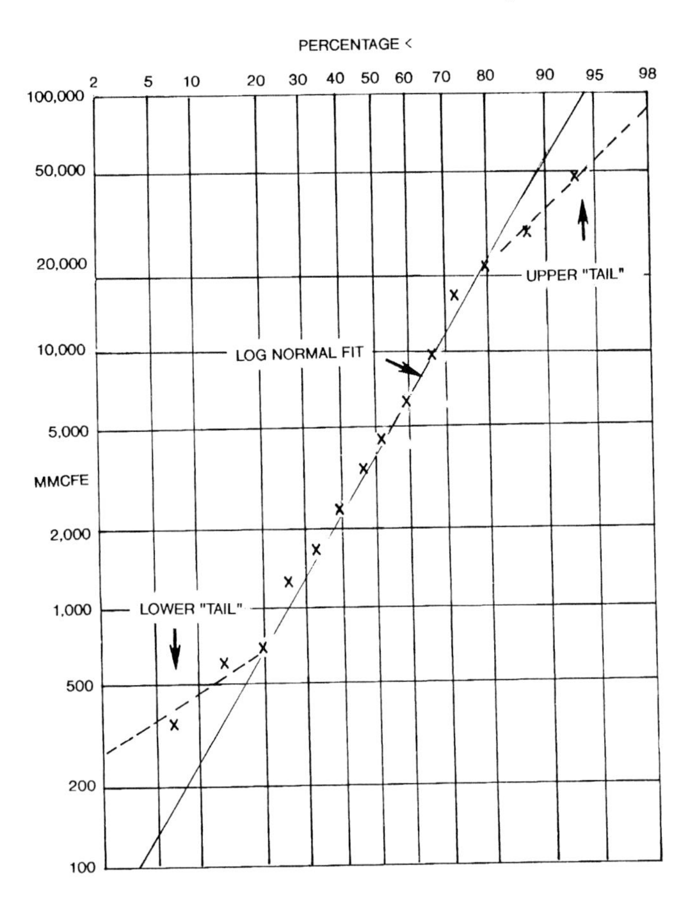

Figure 2. Discovery size distribution.

enough data in the trend as defined? Consider combining the data with data from other vertically or horizontally contiguous trends to create a new trend with adequate data. Or consider using data from an analogous trend.

MONTE CARLO SIMULATOR CALCULATIONS

Economic analysis can be conducted with a deterministic or probabilistic computer program. A deterministic program suffers the inherent disadvantage of assuming that there is no data uncertainty. To appreciate the variability of economic returns available in a trend, an economic Monte Carlo simulation program is recommended so that the distribution of possible economic returns from exploration can be presented. The program should be an acceptably accurate model of the real world to provide for meaningful consolidation and interpretation of data. The Chapter Appendix contains a step-by-step description of how to develop data for input to the Monte Carlo economic simulator.

The *Monte Carlo simulator* is simply an economic evaluation program that uses random samples from input probability distributions to calculate the economic results of exploration. When this is done over a large number of trials, the resulting probability distributions for various economic parameters (e.g., net present value) will represent the combined effects of all the input variables.

A simple example of this procedure involves the toss of a coin. What is the probability that a coin toss will result in heads? Using the Monte Carlo approach, one would simply flip the coin a large number of times, examine the cumulative results of these trials, and calculate an estimate of probability. If the coin is tossed 1000 times, then approximately 500 trials will result in heads. The estimate of the probability of achieving heads on any single toss of the coin would be 0.50.

Calculating the economic results of exploration is, of course, more complicated. A realistic economic calculation is performed relative to each pass of the simulator. To be meaningful, the simulator should experience at least 500 successes to calculate the desired distributions for economic indicators. Each pass (trial) of the simulator begins with the drilling of a prospect. In proportion to the probability of success, the well will be either dry or successful. If dry, the full dry hole cost (including nondrilling exploratory costs) is recorded as exploration expense (dollars at risk), and the trial is recorded as a failure costing that amount. If successful, the program randomly selects a discovery size from the reserve distribution, randomly selects values for all of the other data variables (reserves per completion, initial production rate, oil prices, etc.), calculates the economic results, and records those results as a successful trial. After all the trials are completed, the frequency distribution of the results represents the range of what could happen if a large number of exploratory wells were drilled in the trend.

It is important to note that if there are secondary vertically overlapping trends to be analyzed, each trend should be treated separately. Combining the data from trends with significantly different probability distributions will result in input probability distributions that have no real meaning.

The correct methodology for handling potential reserves in a secondary or tertiary trend is to permit the simulator to develop the reserves as

- completion in the primary with recompletions,
- multiple completions, or
- produce with separate wells.

The selection of development method can be based on industry experience in the trend or on the company's preference.

After the random selection of values for the input variables, the Monte Carlo simulator follows the same basic steps as any other petroleum economics program. The development scheme (single, dual, recompletions) and the resulting overall production schedule may have a significant effect on the economic results. For example, if production is stretched out by recompletions, the economic value of the recompletion reserves will be minimal due to the deferred receipt of cash flow. However, rapid production of reserves will improve rates of return due to the early receipt of cash. For long development schedules, high investment "hurdle rates" may not be the best indicator of value to the company. For example, would the company prefer short-term, high cash flow, high rate of return (ROR) projects or long-term, low cash flow, low ROR projects?

For each successful economic pass, the reserve and production section of the simulator must calculate the completions per trend required to recover the reserves and, based on the development scheme, the number of successful development wells required.

Another key component of the simulator is an automatic well scheduler that determines the time and cost schedule for adding successful and dry development wells, completing the wells (and scheduling recompletions), adding production facilities, and connecting to pipelines. The simulator should lay out an investment schedule that approximates the real life situation as closely as possible. Most economic analysis programs require manual investment scheduling, which is impossible for simulation programs where 500 or more successful passes are required to achieve acceptable stability in the output.

Scheduling wells for offshore trends presents another degree of difficulty. The wells must be scheduled to platforms of various sizes, and multiple platforms must be considered for large offshore developments. Also, the model must reflect the extreme time sensitivity of offshore operations. This means that exploration, platform setting, development drilling, production facilities, and pipelines must be keyed to realistic time delays.

The output of the well scheduler subprogram is an investment and production schedule. Using the price and cost input data, the next stage of the program converts this schedule to cash flow and calculates the economic results for that pass.

TREND SIMULATION OUTPUT

The simulation program should allow the output of any time series data stream (calculated or input) for data checking. After data verification, the simulator should present reserve or cash flow parameters plus economic profitability indicators as tables and graphs of cumulative probability distributions. Each output parameter can be presented either as unrisked (that is, the probability distribution of successful results only), or risk adjusted (the distribution of all outcomes, dry or successful). The data distributions illustrated in Figures 3a and b and discussed here are representative.

1. *Reserves*: The cumulative probability reserve distribution (oil or gas inMMbbl, BCF, BOE, or MMCFE,) represents either the primary trend or the primary trend plus any secondary trends (Figure 3a). Read the figure by picking a probability such as 70% and identifying the risked and unrisked reserves discovered. In the example, a 70% probability exists that risked reserves will be less than 2 BCF and unrisked reserves will be less than 12.5 BCF. In the case of an analysis of a single trend, the output distribution (unrisked) should directly match the input reserve distribution.
2. *Economic Indicators:* Economic indicators bring all data uncertainty into a single distribution representing the potential outcome that can be expected from exploration in the trend. If primary and secondary trend reserve discoveries are possible, the economic indicator represents the statistical combination of the reserve distributions and the subsequent calculation of economics. The choice of economic indicators should fit the company's preference for presenting economic value data. On a risk-adjusted or unrisked basis, and before or after income tax, typical choices include the following:
   a. Discounted net present value
   b. Discounted profit to risk investment ratio
   c. Discounted cash flow rate of return (Figure 3b)
   d. Discounted growth rate of return (assume reinvestment)

The graphs can be used to read the probability of achieving a desired value for an economic parameter. In Figure 3b, the probability of achieving a risk-adjusted DCFROR of 20% or better is 40%.

Again, we stress that these distributions represent the results that can be expected from drilling a very large sample (hundreds) of prospects in the trend. Similarly, the expected values of these distributions represent the average results one would achieve over that very large exposure. In the real world, exploration capital (or management's willingness to continue to expose it) is limited. A company must evaluate the probabilities for the short run over a limited number of prospects. This problem is addressed in the following section.

SHORT-RUN OR LIMITED CAPITAL PROBABILITIES

A helpful aid to understanding short-run economic probabilities over a relatively small number of prospects is the concept of "gambler's ruin." Suppose the probability of success in a trend is 20%. If five wells

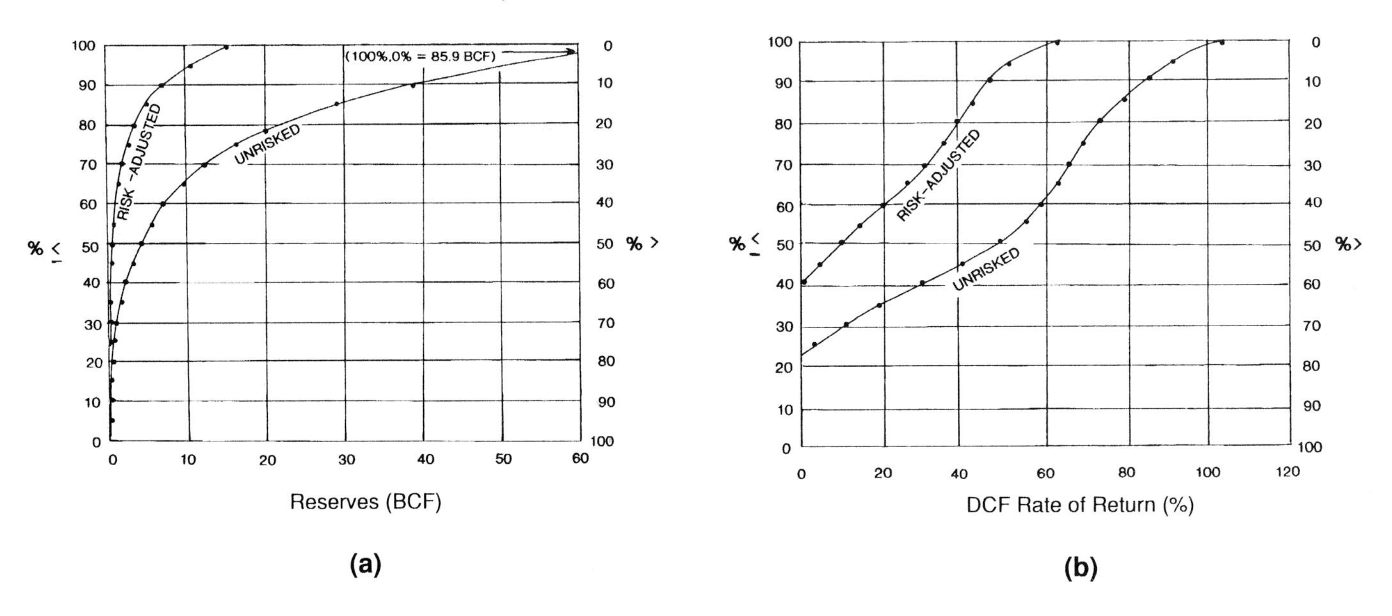

Figure 3. (a) Risked versus unrisked reserves from Monte Carlo simulation. (b) DCF rate of return—distribution from Monte Carlo simulation.

are drilled, the expected number of successes is 5 x 0.20 = 1. One would logically assume that one of the five wells is successful. However, as experience dictates, this is not necessarily true. Expected value actually means that if a large number of five-well programs are drilled, some programs would result in five successes, some in four, some in three, and so on to none. The *average* result of this large number of five-well programs would be one success per program. But, what if there is only enough money for a single five-well program? What is the probability that we will have a run of bad luck (gambler's ruin) and drill five dry holes? Or conversely, what is the probability that we will have at least one success?

That one success could be a small discovery. What we are really concerned about is not the probability of at least one success, but the probability of *economic success*. In other words, the desired result is an estimate of the probability of discovering enough reserves in the success(es) to yield a satisfactory profit on the total money invested, *including the dry holes*, when the number of wells we can drill is limited. There is no rigorously correct short-cut method to calculate this probability. The only way to derive this information is to use a Monte Carlo simulator to calculate, for programs of various sizes, the probability of achieving a profit objective. One then plots the output as in Figure 4 (from Jones and Smith, 1983). Note that the various program sizes (number of wells) shown on the horizontal axis for three different trends have been converted to the expected value, in dollars, of the risk investment, to better compare different trends.

In the example of Figure 4, the goal is at least a 3:1 present value return on the risk investment. Note that for each trend, the probability of achieving the goal initially rises with increased investments. This example illustrates that the probability of achieving the goal is higher for a five-well program than for a single well. However, the curve for one of the trends indicates a maximum value beyond which increasing investments results in a lowered probability of achieving the goal. In such a case, the downturn can be explained by the combined effects of an upper limit on discovery size and the binomial probability distribution. For good trends, this maximum occurs at a large program size; that is, statistically, the data for the study period indicates that the projected average discovery size is high relative to the finding cost. For any given trend, the curve is ultimately shaped by the interacting effects of the investment cost for a prospect, the probability of success, the discovery size distribution, and all of the other probabilistic variables.

APPROXIMATION METHODS

Trend analysis, as previously described, involves a significant commitment of time and money to gather the necessary data and analyze the results. However, it should be clear that a limited, rudimentary effort to look at success ratios and discovery size distributions will provide valuable information in determining where to explore. Thus, no matter how limited a company's resources may be, trend analysis can provide useful information.

In the absence of a complete database and a Monte Carlo simulator, one can still use gambler's ruin theory to provide some insights into exploration in a trend.

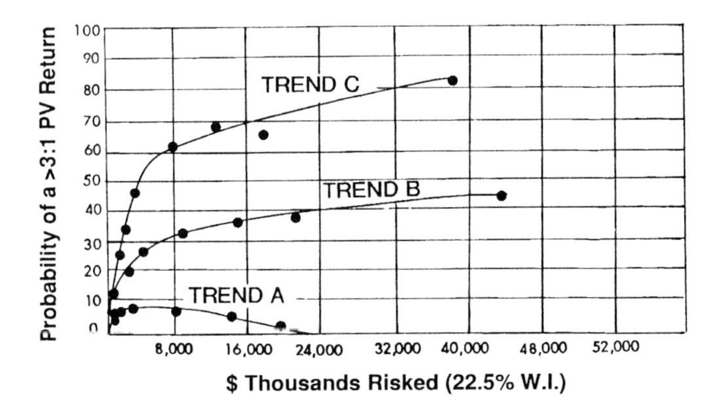

Figure 4. Probability of achieving a goal versus dollars risked.

Let us again pose the question asked earlier. If the probability of success (P_s) for a trend is 20%, what is the chance that, in a five-well program, all five wells could be dry holes? The probability of failure is $(1 - P_s)$ for a single try. For two tries that are independent, the probability of both being failures is $(1 - P_s) \times (1 - P_s)$, and the probability V of n failures is

$$V = P(n \text{ failures out of } n \text{ tries}) = (1 - P_s)^n$$

So in our example,

$$V = P(5 \text{ failures out of 5 tries}) = (1 - 0.20)^5$$

$$= (0.80)^5 = 0.328 = 32.8\%$$

In other words, with a small (five-well) program, there is still a 33% chance that all five wells will be dry, even if the expected value is one success and four failures. The complement of five failures is one or more successes. Thus, there is a $(1 - V) = (100 - 33) = 67\%$ chance of at least one success.

If one success is enough to carry the program (that is, the expected net present value from one success is greater than the exploration cost of drilling the one success and four failures), then we can estimate a 67% confidence that we will avoid gambler's ruin. It is important to note that we used the expected values of the return and the risk investment to determine that one success could carry the program. The confidence estimate is thus only valid if there is a small variance in these dollar amounts.

Suppose that 67% confidence is not satisfactory. We want a higher confidence, say 90%, of avoiding gambler's ruin. The need exists, then, to drill more wells, but how many more? The answer is provided by rearranging the previous equation to solve for n:

$$V = (1 - P_s)^n$$

$$\text{Log } V = n \text{ Log } (1 - P_s)$$

$$n = \frac{\text{Log } V}{\text{Log } (1 - P_s)}$$

Let's define C as the probability of at least one success. Then V = probability of all failures = $(1 - C)$, so

$$n = \frac{\text{Log } (1 - C)}{\text{Log } (1 - P_s)}$$

Thus, for a confidence of 90% of at least one success,

$$n = \frac{\text{Log } (1 - 0.90)}{\text{Log } (1 - 0.20)} = \frac{\text{Log } (0.1)}{\text{Log } (0.8)} = \frac{-2.3026}{-0.2231} = 10.32$$

or 11 wells since n has only integer values.

This simple method of determining how to avoid gambler's ruin depends upon the assumptions that the tries (wells) are independent of each other and that there is an unlimited number of prospects. In reality, drilling exploration wells in the same trend are not independent events. The sampling is without replacement (in other words, each time a prospect is drilled, there is one less prospect left). Thus, the proper distribution for use is the hypergeometric. In practice, the binomial distribution, which we have used, is a close approximation of the hypergeometric as long as the total number of undrilled prospects remaining in the trend is at least ten times larger than the number of tries under consideration. (In the example with five tries, for instance, there would have to be at least 50 prospects left.) Since this is usually the case, this method provides a quick way of finding the probability of all failures and the corresponding probability of at least one success.

MULTIPLE TREND ANALYSIS

After all trends in a study are analyzed, the confidence of success versus dollars expended for each trend should be plotted together on a common scale and compared. This comparison will clearly eliminate some trends where the maximum confidence of success is unacceptably low. One can also compare the confidence levels for an exploration level of effort corresponding to the current or proposed effort in the different trends. Examine Figure 4 again. It presents a comparison of three trends for the goal of a 3:1 12% present value (PV) return rate. At the $16 million expenditure level, the confidence of achieving at least a 3:1 PV return ratio is at 68% in Trend C versus 6% in Trend A.

Suppose, however, that the company has a budget of only $3 million. Then for Trend C, and for the 22.5% working interest displayed, the probability of achieving the goal is only about 40%, even though the trend's probability is much better at higher levels of risk investment. Suppose the company wants at least a 50% probability of achieving the goal. This can be accomplished by *reducing the working interest*.

The risk investment (horizontal scale in Figure 4) represents the dollar investment in a certain number of wells (say, ten) for the 22.5% working interest. If the working interest is reduced by a factor of two to 11.25%, then the same $3 million can be invested in twice as many (20) wells. A $3 million investment at 11.25% working interest is equivalent in risk to a $6 million investment at 22.5% working interest. At $6 million and 22.5% working interest on Figure 4, the probability of achieving the goal in Trend C is now an acceptable 60%.

Even after eliminating those trends with an unacceptably low probability of meeting the company's goals (e.g., Trend A in Figure 4), there are often several trends in the study area that theoretically may meet the company's goals (e.g., Trends B and C). How is a company's exploration effort best divided among these acceptable trends?

At first glance the answer to this question appears simple—concentrate an adequate effort in the trend with the highest probability of achieving the company's goal. This straightforward approach is often impractical for a variety of reasons:

1. If the best trend has a high per-well risk investment, the company may not have enough exploration funds to achieve a satisfactory probability of reaching its goals in this trend alone. The company may not have the desire or the ability to raise the required funds.
2. Increasing the probability of success by participating in more prospects at a reduced working interest may not be feasible if the company cannot generate enough prospects (or participate at favorable terms in prospects generated by others). There are real-world constraints on the number of prospects in each trend that can be generated or acquired during the time span for the program. These constraints arise from competition, lease sale schedules, staff size, and time limitations.
3. The decision makers may be reluctant to commit all available exploration resources to a single trend. In other words, decision makers often show a preference for diversification of effort.
4. The decision makers may consider it prudent to have a mixed strategy of risk and reward (that is, some high risk, high return plays, and some lower risk, lower return plays).

Ultimately, any decision on exploration involves a decision on the allocation of resources. What is the best mix of a company's total exploration effort among the available acceptable trends? To define *best mix* more accurately, what combination of numbers of prospects in each of the trends will have the highest probability of achieving the company's goals?

Multiple Trend Program

This allocation problem is often discussed under the general category of *portfolio analysis* theory. Derived from financial investment studies, portfolio analysis attempts to identify, from a set of potential investments with varying returns, risks (probabilities) and investments (risk costs), the investments that constitute the "efficient frontier"—the set of investments that provides the maximum expected return for a given risk.

This efficient frontier concept can be used to construct an optimization model (another Monte Carlo simulator since probabilities are involved). Part of this model will determine the optimum working interest to be taken in prospects in each trend.

Along with the previous probability and economic data for each trend, constraints are input to the optimization program. Some examples of input constraints include

1. total exploration effort (in dollars),
2. maximum working interest in each trend,
3. maximum number of prospects in each trend, and
4. minimum number of prospects for low risk trends.

This Monte Carlo program is used to calculate the marginal return (either the risk-adjusted expected value or the probability of securing a desired economic return) achieved from the possible candidates for the next prospect. It then allocates budgetary resources to the prospect with the highest marginal return. The result is a table of n well programs for each trend, representing the optimum mix of prospects. If constraints are set on the number of prospects in the trends (e.g., "no more than ten prospects in Trend X"), then some trends in the optimum mixture may be fully invested to the maximum allowed. However, often the outcome is a mixture of investment in all trends. For example, in Figure 4, Trend C is clearly the best trend. Note, however, that the probability of achieving a 3:1 PV return increases dramatically in the early stages of investment. The rate of change then decreases with larger investments. At what investment level would funds be allocated to Trend B? At some point, the marginal return (incremental increase in economic rewards) for investment in Trend B will exceed the marginal return for another increment of investment in Trend C. This cannot be determined directly from Figure 4; the Monte Carlo multitrend allocation program is required. This analysis of marginal returns obtained from incremental investments in different trends will ultimately determine the optimum allocation of the total exploration budget.

Risk Profiles

After completing the multitrend analysis, the allocation of investments—the cost of the prospects to be drilled in each trend—can be plotted as a function of risk (probability of failure). On a *risk profile* curve such as Figure 5 (from Quick and Buck, 1984, figure 6), the risk expenditures (as a cumulative percentage of the total) are plotted as a function of increasing risk.

The shape of the resulting curve graphically reveals the risk aversion strategy of a given program. A diagonal straight line represents a balanced strategy, one in which the investment is spread evenly over all risk levels. A curve entirely above the diagonal represents a risk-averse strategy, where the largest proportion of money is invested in low risk trends. Conversely, a curve that stays below the diagonal represents a program with a risk-seeking strategy where more of the investment is in high risk projects. A medium risk strategy, with most investments in the mid-risk range, would rise steeply and cross the diagonal near the middle of the risk range. Note that the concept of risk aversion is subjective and should be consistent with the corporate goal. Suppose that the total program is $40 million and our goal is a 3:1 PV return on the risk investment. Suppose also that the multitrend allocation program results included an $8 million risk investment (at 22.5% working interest) in Trend C of Figure 4. Then this 20% ($8 million/$40 million) increment of cumulative expenditure would, from Figure 4, have a 60% probability of achievement and a corresponding 40% probability of failure to meet the company's goal. On the risk profile, this 20% increment would be plotted at the 40% risk level.

The resulting risk profile can be used as a check against management's risk aversion strategy for a company's planned exploration program. The two should agree. If they do not, then the goals for the exploration strategy were probably not defined properly. A risk profile can also be used to help define the constraints for a multitrend optimization program. In the absence of a formal multitrend analysis, such a profile can be used to construct a recommended program to fit a desired risk aversion strategy.

CONCLUSIONS

Trend analysis works to make the statistical history of established trends complement the judgment of the explorationist. In addition, trend analysis provides valuable analogy information for the economic assessment of new plays or ideas. The development and analysis of a database of established trends can substantially improve the vision of explorationists and the quality and profitability of exploration. Trend analysis, in summary, is a reality check for long-range exploration planning.

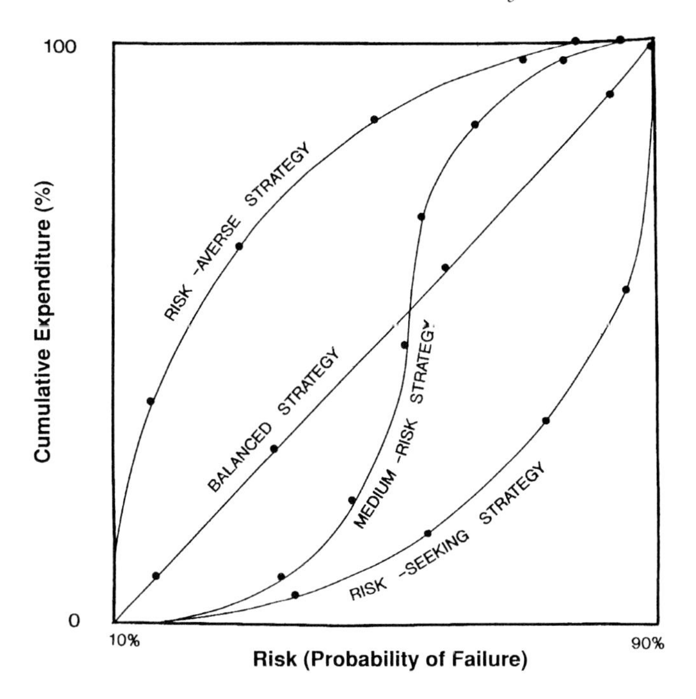

Figure 5. Risk profiles.

Annotated Selected Bibliography

The following list is a representative sample of key articles on trend analysis subjects selected from publications readily available to most explorationists. The authors' comments relate the articles to topics discussed in this chapter.

Arps, J. J., and T. G. Roberts, 1953, Economics of drilling for Cretaceous oil on east flank of Denver–Julesburg basin: AAPG Bulletin, v. 42, p. 2549-2566.
First demonstration of lognormal discovery size distribution. Early effort at trend analysis, but is actually a determinate resource estimation model.

Caldwell, R. H., 1986, The use of computerized statistical data in risk-appraised economic analysis of oil and gas prospects: Journal of Petroleum Technology, v. 38, p. 447-452.
Good discussion of use of API drilling statistics to calculate success probabilities. Lists nine problems to watch for in data.

Capen, E. C., 1976, The difficulty of assessing uncertainty: Journal of Petroleum Technology, v. 28, p. 843-850.
Thought-provoking discussion of problems with "experts" estimating uncertainty.

Davidson, L. B., and D. O. Cooper, 1976, A simple way of developing a probability distribution of present value: Journal of Petroleum Technology, v. 28, p. 1069-1078.
Alternative to full Monte Carlo simulation. Equations for calculating mean and standard deviation of any type distribution given 10th percentile, 90th percentile, and most likely values.

Greenwalt, W. A., 1981, Determining venture participation: Journal of Petroleum Technology, v. 33, p. 2189-2195.
For trend analysis, use venture participation percent to maximize expected profitability and minimize financial failure ("gambler's ruin"). Includes risk aversion level of management.

Jones, D. R., and M. D. Smith, 1983, Ranking South Louisiana trends by probability of economic success: Oil & Gas Journal, Oct. 24, p. 149-156.

Ranking of trends by probability of achieving goal as a function of size of risk investment. Hackberry trend used as example. Gambler's ruin effect discussed.

Kinchen, A. L., 1986, Projected outcomes of exploration programs based on current program status and the impact of prospects under consideration: Journal of Petroleum Technology, v. 38, p. 461-466.
Excellent paper. Probability that a program will meet a goal is discussed along with the effects of dilution of interest and partial completion of program.

Kumar, N., 1985, Estimating exploration potential in mature producing area, Northwest Shelf of Delaware Basin, New Mexico: AAPG Bulletin, v. 69, p. 1903-1916.
Demonstration of geological analysis necessary for trend analysis. Analyzed two plays (San Andres and Siluro-Devonian) and discussed selected areas of interest. Follow-up study 4 years later confirmed predictions.

McCray, A. W., 1969, Evaluation of exploratory drilling ventures by statistical decision methods: Journal of Petroleum Technology, v. 21, p. 1199-1209.
This is the earliest trend analysis paper. Discusses triangular distribution for net profit outcomes, Monte Carlo simulator use, and gambler's ruin degree of participation.

Newendorp, P. D., 1976, A method for treating dependencies between variables in simulation risk-analysis models: Journal of Petroleum Technology, v. 28, p. 1145-1150.
Cross plots of variables to determine independence, complete dependence, or partial dependence. Calculation technique to account for partial dependence.

Quick, A. N., and N. A. Buck, 1984, Viability and portfolio analysis as tools for exploration strategy development: Journal of Petroleum Technology, v. 36, p. 619-627.
Monte Carlo simulator. Viability equals ability of program to sustain itself. Effect of risk-sharing and targeting of acceptable probability of ruin. Portfolio analysis and "efficient frontier" concept. Risk profiles and characterization of risk nature of programs (risk-balanced, risk-seeking, risk-averse, etc.).

Rose, P. C., 1987, Dealing with risk and uncertainty in exploration: How can we improve?: AAPG Bulletin, v. 71, p. 1-16.
"Risk" versus "uncertainty." Methods to improve consistency in risk decisions—gambler's ruin, risk aversion factors. Methods to reduce uncertainties (improve forecasts by removing biases)—cross plots, year-to-year trends, etc.

Smith, M. B., 1970, Probability models for petroleum investment decisions: Journal of Petroleum Technology, v. 22, p. 543-550.
Trend model included. Detailed equations for uniform and triangular distributions. How to approximate normal distribution with triangular. Risked versus nonrisked results. Trend analysis by Monte Carlo simulator. Good appendix of definitions of probability and statistical terms.

Steinmetz, R., 1978, Statistical summary of wells drilled below 18,000 feet (5,500 meters) in West Texas and Anadarko basin: AAPG Bulletin, v. 62, p. 853-863.
Good example of initial "quick-look" trend analysis of areas dominated by a major deep trend. Statistical data developed from well history and production databases only.

Woods, T. J., 1985, Long-term trends in oil and gas discovery rates in lower 48 United States: AAPG Bulletin, v. 69, p. 1321-1326.
Discussion of AAPG statistics. Discovery size statistics are pessimistic (only revised once 6 years after discovery—most discovery ultimate reserve continue to grow significantly beyond 6 years after discovery).

Wright, J. D., and R. A. Fields, Jr., 1988, Production characteristics and economics of the Denver–Julesburg basin Codell–Niobrara play: Journal of Petroleum Technology, v. 40, p. 1457-1468.
Good example of production data analysis needed for trend analysis. Mapping of production quality to locate "sweet spots."

Chapter Appendix

This Appendix describes how to develop the data needed for input to the Monte Carlo simulator. The discussion suggests simplifications, where possible, as well as the more complex cases in which modeling becomes difficult.

I. Single-Valued (Fixed) Input Variables
   An example of this type of variable is the discount rate used for calculating present value. Many other variables (e.g., well costs) can also be treated as single-valued if there is little uncertainty in the estimate.

II. Probabilistic Variables
   A. Use of Triangular Distributions
      1. Input minimum, most likely, and maximum values.
      2. Triangular distributions are sufficiently rigorous for most trend analysis variables and approximate data language commonly used for petroleum economic sensitivity analysis. They simplify input and eliminate the necessity to describe probability distributions with parameters that are difficult to calculate.
   B. Probability of Success (P_S) for Exploratory Wells
      1. Extract from database the P_S values for each type of exploratory well (NFWC and OEXP). It is also necessary to extract or specify the proportion of each type of well in the exploration program. For example, 70% of all exploratory wells will be NFWCs and 30% will be OEXPs.
      2. For each trend and exploratory well type, set up a triangular distribution from the time series trend lines.
         a. Single trend example from Figure 1 (NFWC success ratios, Trend 6A). The time series trend defines low, average, and high values for P_S as 15%, 17.5%, and 19%, respectively. These are entered as the minimum, most likely, and average values for a triangular distribution (Figure 6).
         b. General properties of triangular distribution
            i. Probability of selecting an end-point is zero; therefore, minimum and maximum values will not occur.
            ii. Unless the triangle is symmetric, the most likely value (mode = 17.5% on Figure 6) will be different from both the median (50th percentile) and mean (expected) values.
         c. Multiple (overlapping) trend areas
            i. Definition—Area where a well drilled to the primary trend will evaluate secondary trend(s).
            ii. Trends with independent probabilities of success, such as pure stratigraphic trap trend overlying a structurally controlled trend.
               (a) Success in secondary trend is in no way related to success in primary trend.
               (b) Often a shallow "bail-out" zone (lower risk, but uneconomic on its own) above deeper, higher risk primary trend may be exploited if primary trend is unsuccessful.
            iii. Dependent secondary trends where success in the secondary trend is related to success in primary trend (e.g., directly overlying structural traps).
               (a) Analyze the data to determine *conditional* probabilities of success for secondary trend given either success or failure in the primary trend.

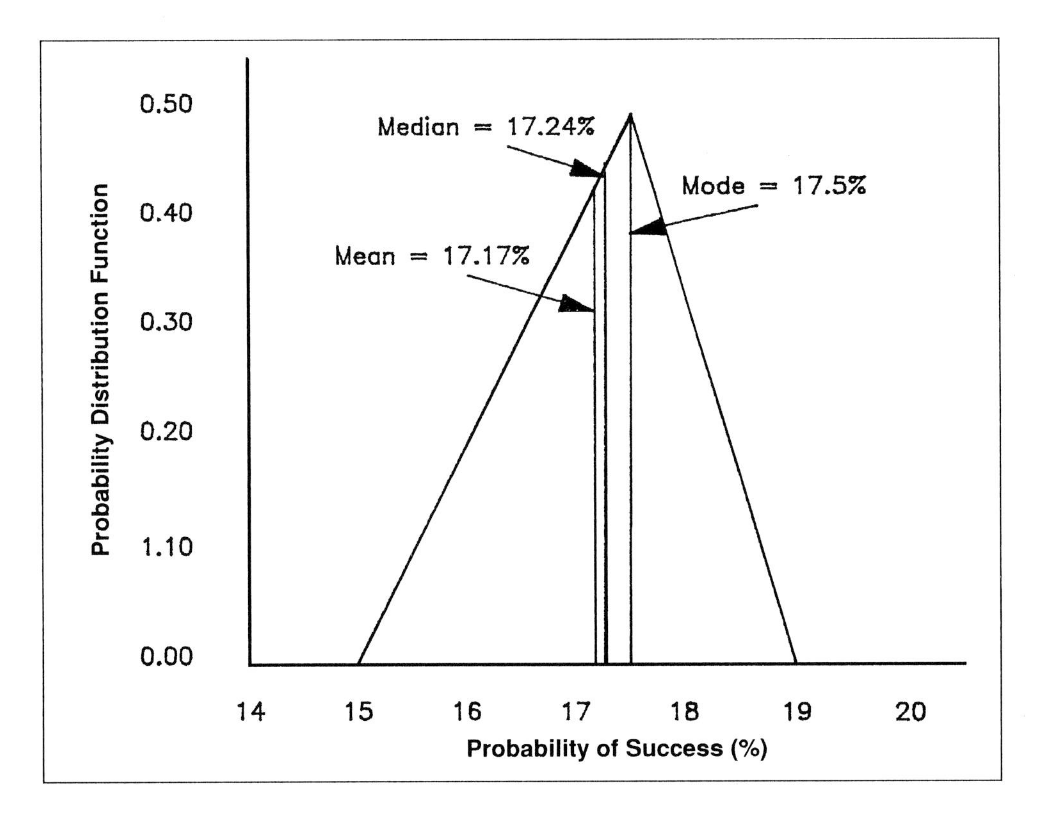

Figure 6. Triangular distributions.

(b) If more than two secondary trends are present, the analysis becomes complicated. An approximation may be achieved by lumping multiple producing sands as the tertiary target.

C. Probability of Success for Development Wells and/or Recompletions
   1. Extract from database time series trend lines as done for exploratory wells.
   2. If multiple trends involved, extract for each trend. (For example, if a development well in the primary trend is dry, the recompletion attempt in the secondary trend will usually have a different probability of success.)
   3. Use triangular distributions, or simplify to single valued variables if possible

D. Probability of Primary Product for Discovery
   1. Determine the percentage of discoveries that are oil or gas discoveries
   2. Assign these percentages as the conditional probabilities, that given a discovery, the discovery will be oil or gas
   3. These conditional probabilities are complementary, that is, they must add up to 100%
   4. The occurrence of oil and gas discoveries in a trend should be studied to see if it can be divided geographically or by depth into separate oil and gas plays. If so, it should be treated as two different trends.

E. Discovery Reserve Size Distributions
   1. This is the exception to the rule to use triangular distributions. Use the actual distribution (usually lognormal). Otherwise, use triangular distributions.
   2. Input separately for each type of exploratory well (NFWC or OEXP) and for oil and gas if both are present.

3. Example from Figure 2 (distribution of gas discovery sizes for NFWCs, Trend Y). Input the lognormal cumulative probability data. The input should follow the actual data in the upper and lower "tails" where the data points depart from the lognormal approximation.
4. Volumetric reserve data (alternate method)
   a. Data are collected for volumetric input variables (acres, net pay, saturation, etc.) *rather than reserves.*
   b. Input as triangular distributions to approximate data variance (e.g., net pay as 10, 24, and 32 ft).
   c. Potential short cut (avoids development of reserve size distribution from production data).
   d. Usually do not have enough data, and if volumetric input variables are merely estimated, resulting reserve size distributions may not match actual reserve data very well

F. Reserves Per Completion
1. Extract reserve data for all completions in the trend (separately for oil and gas).
2. Construct distribution in one of three ways:
   a. Cumulative frequency distribution.
   b. Normal distribution (compute mean and standard deviation).
   c. Approximate with triangular distribution.
3. Conditional probability distribution
   a. Often a relationship exists between discovery size and reserves per completion. Large discoveries result in large per-completion reserves, and small discoveries have small per-completion reserves.
   b. Confirm existence of relationship by cross plot of reserves per completion versus discovery size. An example for generalized random variables X and Y is shown in Figure 7 (modified from Newendorp, 1976, Figure 7).
   c. Fit equations for the boundaries Y_{max} and Y_{min} as functions of X.
   d. Normalize all Y data points to Y_{norm}:

$$Y_{norm} = \frac{Y - Y_{min}}{Y_{max} - Y_{min}}$$

   e. Determine the frequency distribution of Y_{norm}. It may be rectangular (uniform), normal, triangular, or any other distribution function. The triangular distribution will often be adequate.

G. Production Profile
1. Extract production profile data from the database for all completions in each producing zone for vertically overlapping trends. These will provide the necessary data to approximate the expected production profile over the life of a completion. The data needed include profile type (exponential, hyperbolic, etc.), initial production rate, reserves per completion (already obtained), and any other parameters necessary to define the profile (e.g., annual percent decline for an exponential decline).
2. Determine the most common profile type or a distribution for types (for example, 70% exponential, 30% constant rate).
3. Specify a triangular distribution for each parameter.
4. Initial production rate data
   a. *Do not use* initial potential tests or absolute open flow calculated rates. A good approximation is the average rate for the first three months of production.

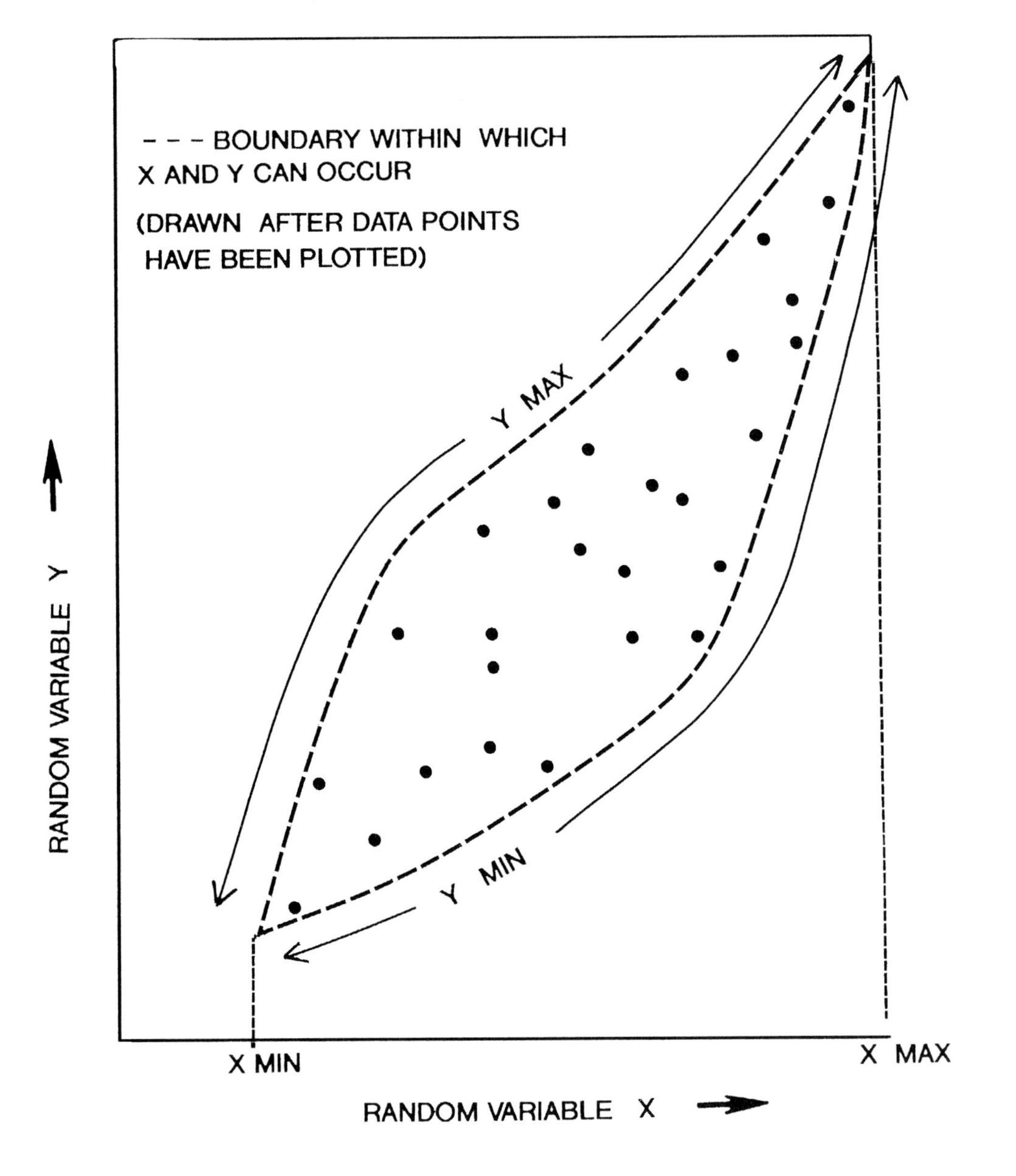

Figure 7. Partially dependent variables.

b. Dependency (conditional probability) often exists for initial production rate as a function of net pay thickness (productivity index) for producing zones with similar rock, fluid, and pressure properties. Thus, it should be expected for a trend. However, we rarely have adequate data for net pay thickness (completion interval thickness may be used if available). Use the same methods discussed previously for reserves per completion.

c. When net pay information is not available, often a simple correlation can be made to relate initial production rate to reserves per completion. (In general, big initial rates go with big reserves per well.) This correlation is generally preferable to no recognition of dependency. Again, use the previously discussed conditional probability methods.

H. Exploration (Finding) Costs

1. Exploration drilling and testing costs. Well costs are usually the largest component of exploration costs, so the data needs careful

treatment. Time series projections for well costs in various depth brackets are desirable. Extract the data for all exploratory dry holes in the trend from the database, analyze it, and enter the data as a triangular distribution.

2. Front-end costs (acreage, seismic, G & A costs). Expected prospect development costs are estimated including an allocated share of the costs of nondrillable leads, regional exploration, etc. Usually, these data are not available in commercial databases and must be developed from estimates by knowledgeable experts and entered as triangular distributions. The estimates should be checked against recent experience in the trend.

I. Development Costs

1. Development drilling costs (successful and dry hole). Analyze in the same manner as for exploratory well costs and enter as triangular distributions.
2. Completion cost of discovery well. Generally these data are not directly available but can be generated from the difference between successful and dry exploratory well costs to the same depth bracket.
3. Production facility costs. These are not generally available and need to be developed from estimates by knowledgeable experts. Usually they can be expressed as a simple function of discovery size or maximum production rate.
4. Recompletion costs. These data are necessary if secondary trends will not be developed separately. Data on such costs are not generally available from databases and must be developed from estimates by experts.
5. Development scheme. To analyze trends in areas of multiple overlapping trends, or in trends with more than one producing zone, it is necessary to determine the development scheme to be used. Will one use dual completions, later recompletions in the secondary zones, or separate development drilling to each zone? The scheme is specified as input so that the Monte Carlo program can correctly simulate development costs.
6. Discoverer's share of development. In many active trends, it will not be possible to control all the acreage for a discovery. So we need to develop data on the fraction of total development that the discoverer earns. An example would be a trend where the discoverer has drilled from 35% to 80% of the subsequent development wells in a discovery, with an average of 60%. One would then input these percentages as a triangular distribution.

J. Operating Costs

1. Generally not available in commercial databases—must be derived from information in company records or estimated by experts.
2. Types of costs
   a. Direct operating costs—Can be expressed as unit costs ($/bbl or $/MCF). Usually best to express as $/year.
   b. Remedial workover costs—Cost and frequency.
   c. Abandonment costs—Usually significant only in offshore trends where platform abandonment costs are high.
   d. Production taxes—Usually fixed in each state (none in Federal offshore waters). Severance taxes are expressed as unit costs (per unit of production). Ad Valorem taxes can be approximated as unit costs or related to remaining reserves each year.
   e. Royalties—These are not operating costs per se, but are included here for completeness. These should be estimated by knowledgeable experts.

K. Product Price Projections. Oil and gas price projections are extremely critical in assessing the value of exploration. It is generally preferable to run trend sensitivity analyses on fixed price estimates instead of incorporating uncertainty in prices or inflation estimates. Use a realistic price estimate as a base to focus the analysis on geological and development potential.

L. Price and Cost Escalations
   1. Inflationary changes. Inflation, as used here, represents price and cost increases due solely to decreases in the purchasing power of the currency. Inflationary price and cost scenarios are generally related; i.e., if oil and gas prices go up, then costs go up also (although not necessarily simultaneously or at the same rate). This offsetting effect means that for the purpose of creating trend economics and comparing and ranking trends, inflation adjustments *usually are not necessary* (providing that the same *real* price and costs adjustment scenarios are used for all trends).
   2. Real changes. Real price and cost changes, as used here, refer to expected changes in the real values (that is, with inflationary effects taken out). For example, prices can be adjusted to a base year using an inflationary index such as the Consumer Price Index. A typical real price expectation might be that over the next 20 years, as oil supply declines, the real price of oil will go up.
   3. These price and cost scenarios are typically expressed as a percent change per year for a number of years. Given the usual discounting rates, the first 10 years of the forecast is most important.
   4. Price sensitivity analysis. These are typically done to determine at what price a given trend will become economically attractive. If a deep gas play is uneconomic at $1.65/MMBTU of gas, will it meet the investment criteria at $2.00/MMBTU? Such analyses can be used to reject trends with a small probability of economic success.

MICHAEL D. SMITH is President and CEO of ENSYTE, Energy Software International, Inc. (formerly Michael Smith & Associates, Inc.) in Houston. He received a B.S. degree in Engineering Science from the U.S. Air Force Academy, an M.S. in Management Science from the University of Southern California, and did Ph.D. graduate work in Mineral Economics at the Colorado School of Mines. Mr. Smith was President of Resource Analysts, Inc., from 1976 to 1981 and Vice-President of Intercomp, Ltd., from 1979 to 1981. In 1982, he formed ENSYTE, an elite corps of petroleum software specialists in the application of computer software technology to the international petroleum industry. Mr. Smith has managed projects in the United States, Canada, North Sea, Middle East, and Malaysia. Projects include database and application program development, such as reservoir simulation, production operations, material inventory and purchasing, and platform and gas plant maintenance.

DONALD R. JONES, an independent consultant in petroleum evaluation in Houston, received a B.A. degree in Geology from UCLA in 1953 and M.S. in Operations Research from the University of Houston in 1970. After serving 2 years in the U.S. Air Force as an intelligence officer, he was employed in 1955 by Marathon Oil Co. as an exploration geologist in Midland, Texas, and subsequently in Corpus Christi. In 1961 he became a log analyst at Marathon's Research Center in Littleton, Colorado. Mr. Jones was employed in 1966 by Offshore Operators Inc. (a predecessor of TransOcean Oil Inc.) as an evaluation engineer in Houston and later served as a reservoir engineer and Manager of Reservoir Engineering. In 1980 he became Vice-President of Exploration and Production of American Exploration Co. in Houston, and from 1987 to 1988 was Vice-President of Engineering for Oxoco Inc. He is a member of AAPG, SPE, SPWLA, and the Houston Geological Society.

Chapter 19

Annual Budget: A Short-Term Action Plan for Exploration

Donald R. Jones
Consultant
Houston, Texas, U.S.A.

A pinch of probability is worth a pound of perhaps.
—James Thurber

INTRODUCTION

All businesses, including individuals, should have a budget—that is, an estimation of income and expense over an annual cycle. For companies, the budget is generally prepared and approved about one quarter before the start of the company's fiscal year and is updated and revised each quarter during the year. The budget is prepared on a cash flow basis and is then converted to *pro forma* projections of next year's accounting reports (income statements, balance sheets, etc.). Budgeting can be most easily accomplished using a spreadsheet computer program to organize and manipulate the estimates.

Although budgeting is a task dreaded by most exploration managers, it is usually the vehicle by which drilling prospects, the heart of any exploration program, are "sold" to the final decision makers. The budgeting process should be viewed as an opportunity rather than as a chore to be completed as quickly as possible.

PREPARING BUDGET ESTIMATES

Income Projections

The income projection side of the budget is usually prepared by the production and accounting departments of the company. A one-person consulting shop will do this himself or use consultants and accountants. However, even in a larger company, explorationists are involved in preparation of the income budget, particularly in the projection of start-up production rates for successful exploratory wells and provisions for development wells resulting from recent discoveries.

In many companies, the explorationists' bonus compensation is related to the company's or division's performance compared to the budget. It thus behooves the explorationist to review the income projections in the budget carefully. It is important to avoid the "Five Year Plan" syndrome. This is the ploy of deliberately setting up an easy target. But explorationists are more likely to be optimistic, particularly when production history is limited. These reality checks should be followed:

1. If the company budget counts income when it is booked for accounting purposes, remember that booking of income usually lags behind production by one month for operated properties or two months for outside-operated properties.
2. Initial rate estimates should be estimates of sustainable monthly production rates, which are often much lower then short-term production rates. The estimate should be further reduced for down time due to weather, maintenance, and market demand for gas.
3. Individual well rates often demonstrate appreciable decline over a yearly period. Quarterly or monthly estimates should reflect this.

Expense Budgets for Exploration

The estimation of exploration expense can be broken down into four main steps.

1. Start with the current status on each project—for example, projects in progress at the start or end of the year will require an adjustment for carry in/carry out expenditures.
2. Other than drilling costs, such expenses as general and administrative (G & A), geological and geophysical (G & G), and land acquisition can be projected from previous expenditures or estimated on a unit basis (dollars per acre, seismic cost per mile, etc.). Items such as salaries, office expenses, and allocated company overhead are included in the categories of G & A or G & G. Depending on company policy, salary costs may need to be adjusted to include overhead burden items such as benefits or expense accounts. Finally, if the proposed projects in the budget will require more personnel, be sure to provide for new employees or consultants.
3. Exploratory drilling expenditures are usually the largest component of the exploration budget and thus deserve the most scrutiny. The exploration manager should look very carefully at the proposed Authority for Expenditure (AFE) for each exploratory well. Be satisfied with the answers to these questions:
    a. Does the cost estimate include the amount of evaluation (logging, coring, and testing) appropriate to the particular prospect?
    b. Is the cost estimate comparable to previous experience (company and/or industry) in the trend or area?
    c. Do contingency expenses realistically represent possible problems? Remember that an estimate that is too low may come back to haunt you, but habitual padding will keep you from drilling more wells.
4. Many companies follow a budgeting rule that exploratory wells are budgeted at their dry hole cost while development wells are budgeted at completed well costs. If this is done, a contingency budget item should be included to cover the cost of the estimated number of completion attempts for exploratory wells.

Additional contingency items may be needed for follow-up of any discoveries if the follow-up begins during the budget year. These items could include additional land acquisition, seismic programs, delineation or development wells, and production facilities. It is unduly pessimistic to assume that all exploratory wells will be dry holes. Finally, in some exploratory plays such as fractured reservoirs, the exploratory well will require completion, stimulation, and extensive testing to evaluate its commercial potential. In such situations, all of these costs should be included in the exploration budget.

EVALUATING PROSPECTS FOR INCLUSION IN THE DRILLING BUDGET

From its prospect inventory, whether it be a filing cabinet or a sophisticated database, the company's explorationists prepare a list of prospects that will be ready for drilling in the budget year. The list may also include generalized prospects to represent those expected to be acquired from outside sources during the year. All prospects must be evaluated economically and ranked as to their fitness for inclusion in the company's drilling program.

The drilling decision is of critical importance, both for the relative amount of money at risk and as to the potential results. Yet we must deal with risk and uncertainty in virtually all of the data that goes into this decision. The major part of this discussion of budgeting covers how to use probabilistic concepts in making this decision. The decision requires the preparation of discounted cash flow economic analysis for the prospect. The risk cost side of the analysis should include all future exploration costs through plugging and abandonment. The return side should include all development costs, including completion of the discovery well, and all returns from development.

Almost all of the input variables to these calculations are probabilistic, and since the input is probabilistic, the output will be also. It is both difficult and misleading to represent economic indicators for a prospect, such as present worth, rate of return, or payout, as single-valued numbers. What needs to be portrayed is a range of values, with probabilities attached, that convey to the decision makers an accurate representation of the exploration department's "gut feeling" concerning the prospect along with their degree of uncertainty about that gut feeling. This is best accomplished by a Monte Carlo simulator, but can also be done with expected value methods. However, with expected value methods, some of the information that should be communicated will be lost.

General Principles

There are two points that apply in general to discounted cash flow calculations as used for budget decisions:

1. Perform the discounted cash flow analysis on a BFIT (before federal income tax) basis or an AFIT (after federal income tax) basis depending upon your company's policy. Many companies prefer to look at prospect analysis on a BFIT basis, utilizing the KISS (keep it simple, stupid) principle and assuming that the effects of federal income tax will average out over the entire

Table 1. Input Distributions for Volumetric Reserve Calculations

Variable	Type Distribution	Parameters
Productive acreage	Normal	Mean = 800 acres Standard deviation = 100 acres
Average net pay	Triangular	Minimum = 10 ft Most likely = 35 ft Maximum = 50 ft
Recovery factor	Uniform	Minimum = 250 bbls/acre-ft Maximum = 400 bbls/acre-ft

exploration program. But even assuming that a company has or will have enough taxable income from previously developed production or successful future exploration, the effects on AFIT profit of individual dry holes can vary greatly depending on such factors as acreage costs, drilling costs, and timing.

2. Include only money forward expenditures. Money that has already been spent on the prospect, such as acreage or seismic, should be treated as a sunk cost, except for its effect on federal income tax calculations if you are doing an AFIT analysis. This is consistent with the concept of cash flow, which recognizes expenditures *when they occur*. Previous expenditures have already been spent and are irrelevant to the decision to spend further money. Similarly, a decision to complete a well should consider only the costs of completion and ignore the costs already sunk in drilling.

Sources of Risk and Uncertainty

When preparing a drilling budget, that is, when deciding which prospects should be drilled during the coming year, the key consideration to keep in mind is that we are dealing with *specific prospects*. Risk adjustment or *risk weighting* is the process of adjusting the benefits (cash flow) from a discovery to account for a highly variable outcome, with the most likely result being a dry hole. We must also decide the amount we can afford to pay for the opportunity.

The challenge to the explorationist is to develop values for decision criteria while still getting across to the decision makers a full sense of the risks involved. Virtually all of the geological, engineering, and economic factors involved in reserve estimation and cash flow analysis are indeterminate. In a rigorous analysis, their values should be described by probability distributions rather than presented as single-valued estimates. The most important probabilistic variables are the prospect's reserves and its chance of success.

Lognormal Distribution for Reserve Estimates

It turns out that the range of reserve estimates for a single prospect are usually lognormally distributed. This arises from the estimation of reserves by a volumetric calculation involving multiplication:

Reserves = Productive acreage
x Average net pay thickness x Recovery factor

This multiplication can also be expressed as the sum of logarithms:

Log(Reserves) = Log(Productive acreage)
+ Log(Average net pay thickness)
+ Log(Recovery factor)

The central limit theorem, a basic law of probability, states that the sum of random variables, derived from any types of distributions, are distributed normally, that is, according to the well-known Gaussian bell-shaped distribution. So a random variable formed by the addition of random variables will have a normal distribution. A *lognormal distribution* is the distribution for a variable whose logarithm is normally distributed. It follows that multiplication involving random variables will result in a lognormal distribution.

Table 1 lists the input variables to a volumetric oil reserve calculation by Monte Carlo simulation. The simulation is performed 200 times. Each time, random samples are selected from each of the input distributions and the reserves calculated. The resulting 200 reserve values should be lognormally distributed. Figure 1, a cumulative frequency plot of the 200 values on lognormal probability graph paper, demonstrates this. On this graph paper, the y axis has a logarithmic scale and the x axis is scaled proportionally to the cumulative normal distribution. Samples from a lognormal distribution will plot as a straight line. The points on Figure 1 are the cumulative frequency distribution of the 200 reserve values, plotted at increments of 5%, from 5% to 95%. The straight-line fit is apparent; a closer fit can be obtained with more passes.

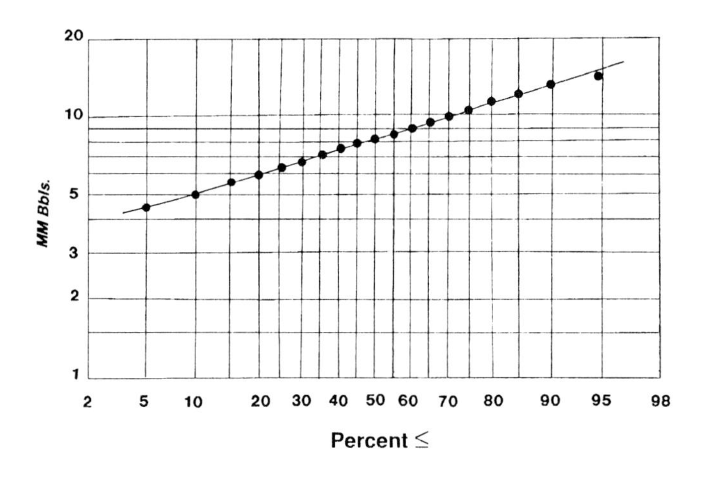

Figure 1. Monte Carlo reserve calculation.

Estimating the Chance of Success

A simple approach to estimating the chance of success for a prospect is to use the established success ratio for the applicable trend. If all prospects in a trend were alike, such that they were truly repeated trials of the same experiment, then in the long run, after drilling many wells, the actual results would approach this ratio. In the special case of budgeting for prospects that haven't been seen yet (that is, prospects expected to be obtained from outside sources during the budget year), this is the approach to use since nothing specific is known about the prospects at this time.

However, in the real world, all prospects, even those in the same trend, are neither quantitatively nor qualitatively alike. The *qualitative* aspects of the prospect are what influence the estimate of the prospect's chance of success. A judgment is needed on the quality and completeness of the data for the prospect, such as seismic, well control, net pay isopachs, shows, migration, and timing. Then those judgments must be converted into an estimated chance of success.

Every prospect has key geological factors that must all be present to result in success. For example, the key factors might be trap, reservoir, and hydrocarbon charge. It is much easier and more accurate to estimate the probability of each of these individual factors being present than it is to estimate the overall probability of success for a prospect. The probability desired is the probability for a yes answer to a simple yes–no question. It is *not* the probability for any specific quantity. For example, with regard to trap, the question to be answered is, "What is the probability that a trap will be present?", not "What is the probability for at least 500 acres of closure?" The answer to the second question is best entered into the analysis of potential discovery size, as done in Table 1 and Figure 1.

The answers to these questions, expressed as probabilities for each factor's occurrence, quantitatively convey the explorationist's gut feel for the quality of the prospect's data. To convert these individual factor probabilities to the prospect's probability of success, P_S, we assume that the factors are independent. Then the overall prospect probability of success is simply the product of the probabilities for each variable. For instance, if the explorationist estimates that the probability for the prospect's trap being present is 50%, the reservoir is 80%, and hydrocarbons is 75%, then the prospect's chance of success is $P_S = 0.50 \times 0.80 \times 0.75 = 0.30$, or 30%.

RISK ANALYSIS METHODS

Four factors enter into the application of the laws of probability to any decision under uncertainty:

1. The size of the reward if successful
2. The size of the loss if unsuccessful
3. The probability of success
4. The risk capital available for speculation

We have already discussed the estimation of the first three items. Probabilistic methods, as usually applied to exploration in expected value and Monte Carlo simulation, deal explicitly *only with these items*. In fact, these methods imply that unlimited risk capital is available. Practically speaking, this means that these methods should be used only by organizations with large exploration budgets. Later we shall see how to modify the analysis for a company with limited funds.

Expected Value Methods

In the absence of Monte Carlo simulation, an approximation can be made using the concept of expected value to represent the entire continuous probability distribution of an outcome parameter. Any company's long-range goals can be generalized as the maximization of present worth over the long run. So, if we must present only one value, we correctly use the expected value (EV) of present worth. Given that the company has large risk capital relative to the risk amount, this will lead to the correct decision. However, there are two different expected values, both of which need to be presented to the decision makers.

Since we have only one try at each prospect, the decision makers need to know what our expectation is *if we are successful this one time*. This is the expected value, *given a discovery*, incorporating only our uncertainty regarding the size of the outcome. Suppose, for example, that our decision parameter is 10% present value. We have made three estimates of

Table 2. Unrisked Expected Value

Case	10% Present Worth ($ million)	x	Probability	=	Weighted 10% Present Worth ($ million)
Low	3.2		0.25		0.8
Most likely	16.0		0.65		10.4
High	30.0		0.10		3.0
				Sum = Expected value =	14.2

reserve size, representing low, most likely, and high estimates, and we estimate the corresponding probabilities to be 25%, 65%, and 10%, respectively. (Note that the probabilities must add up to 100%.) *Holding all other variables fixed,* we have calculated the 10% present value profits for the three cases to be $3.2 million, $16 million, and $30 million. The expected value *decision matrix* is shown in Table 2.

Note that the expected value of $14.2 million is not our most likely estimate. Instead, it is our expectation of the outcome giving weight to our uncertainty over discovery size, but nothing else. Also note that we have made two types of approximations in this expected value calculation:

1. We have approximated the lognormal reserve size distribution with three points.
2. All other probabilistic variables such as production rates or well costs that enter into the calculation of present worth have been held constant at a single best estimate value.

Suppose that the prospect risk investment we are considering for the budget is $2.0 million. We can tell the decision makers that, if they spend the $2.0 million and the well is successful, our best estimate of the resulting 10% present worth is $14.2 million, or alternatively, the present value profit to risk ratio is $14.2 million/$2.0 million = 7.1:1. This value is often called the *unrisked present worth* or unrisked present worth profit to risk ratio because we have not yet taken into account the risk of drilling a dry hole.

The *risk-weighted* or *risked value* for the prospect does just that. Suppose that we estimate the probability of success, P_s, to be equal to 30%. Then, using this estimate and the unrisked expected value from Table 2, the risk-weighted present worth is calculated by another expected value, shown in Table 3. This calculation and expected value answer are equivalent to using an expanded decision matrix (decision tree) involving two probability steps.

If 10% is our desired minimum return rate and the risk-adjusted present value is positive, then if we could drill a large number of similar prospects, we would on the average earn more than a 10% discounted cash flow rate of return (DCFROR). If it is zero, we would earn exactly a 10% DCFROR, and if it is negative, we would earn less than our desired return. So the decision rule is simple. *If the risk-adjusted present value is positive or zero, drill; if it is negative, don't drill.*

The expected value of a sum is the sum of the expected values of its components. If we use this expected value approach on all prospects and *if we can make the estimates correctly often enough,* in the long run we will achieve the company's desired return or better. So with this risk-weighted approach we have made an approximation concerning the two major sources of risk—chance of success and range of discovery sizes.

A similar but easier method of accounting for the chance of success is the *carrying power* or *profit to risk* concept. This requires calculation of a required minimum probability of success based upon how many failures a successful find can carry with it and still make the company's desired return. For this approach, we use the unrisked expected present worth return ratio, which was 7.1:1 in our previous example. In probability terms, we need to have one success out of 8.1 wells (that is, one success at $14.2 million minus the loss from 7.1 failures at $2 million each equals $0, which is a 10% present worth from a program of 8.1 similar prospects). So the required probability of success is at least 1/8.1 = 12%. In mathematical terms, if p is the unrisked present worth expected profit and X is the risk cost (also in present value terms if necessary), then the minimum acceptable probability of success, P_m, is

$$P_m = \frac{1}{1 + \frac{p}{X}}$$

With this approach, instead of asking the explorationist to estimate the exact probability of success, all we have to ask is whether the probability of success is more or less than the required minimum probability of 12%. This is usually an easier question to answer. Again, the decision rule is simple. *If the explorationist estimates that the probability of success is equal to or greater than the required minimum value, then drill the prospect. If he or she thinks it is less than the required minimum, then don't drill.* If we can answer the question correctly for each prospect over a large number of

Table 3. Risk-Weighted Expected Value

Case	10% Present Worth ($ million)	x	Probability	=	Risk-Weighted 10% Present Worth ($ million)
Dry	−2.0		0.7		−1.40
Successful	14.2		0.3		4.26
				Sum = Expected value =	2.86

prospects (expected value), we will achieve our financial objective.

It is usually best to use the carrying power method because the probability estimate is easier to make. If the minimum probability of success from the carrying power method is used in a decision matrix such as in Table 3 for the risk-weighted expected value method, the expected value will be zero. The two methods are equivalent.

Sensitivity Analysis about an Expected Value Estimate

The number of variables making up a reserve estimate and its cash flow analysis leads to *sensitivity analysis*, in which any number of "what if" questions can be asked. For example, what if oil prices fall by 50%? Or what if they double? These types of assumptions can be analyzed while holding everything else constant. With inexpensive, fast computers almost everywhere, sensitivity analysis can easily overload the decision makers with endless numbers of cases, without any guarantee that all the right questions have been asked or that some significant combination of variables has been considered. Also, *there are no accompanying probability statements*. In the usual budget presentation there should be some sensitivity analysis, particularly for those financial variables such as product price estimates to which the decision makers are most sensitive—just don't overdo it.

Monte Carlo Simulation for Drilling Decisions

Expected value methods can provide valuable insight as "quick and dirty" approaches to analyzing a prospect. Compared with Monte Carlo simulation, however, they suffer from four disadvantages:

1. For any decision parameter, such as present worth in our examples, we have reduced what is really a continuous probability distribution of unknown shape into a discrete distribution with a small number of estimated outcomes. We provide no information concerning any other possible outcomes or their likelihoods.
2. We must start by making estimates of reserve size for which we calculate economic yardsticks. It is much easier to make estimates for the probability distributions of the individual factors that make up a reserve estimate, as shown in Table 1.
3. Our estimates for the probability of occurrence for each of the reserve estimates are generally even more arbitrary and subject to error than the reserve estimates themselves. In our example, how did we arrive at (and have we any confidence in) the estimated probabilities of 25%, 65%, and 10% assigned to the three reserve cases?
4. Of all the probabilistic variables that affect our decision parameter of present worth, only two—probability of success and discovery size—have been considered. All other variables are held constant, as if we were certain about them.

Monte Carlo simulation is the most thorough approach to incorporating uncertainty into the economic analysis of drilling prospects. The reserve size calculation and discounted cash flow analysis are performed as usual. The difference is that any input variable (for example, pay thickness, recovery factor, production decline rate, operating costs, etc.) can be treated either as a fixed value or as a probabilistic variable entered as a probability distribution of any type.

Even a probability such as P_s, the probability of exploration success, can be treated as a probabilistic variable, that is, entered as a probability distribution. In fact, given the range of estimates for P_s for a given prospect that will result from questioning several knowledgeable explorationists, it is usually best to do so.

Table 4 illustrates the input for Monte Carlo simulation of a prospect. The table does not list all of the necessary input, but provides examples of fixed and variable input parameters. For most variables a triangular probability distribution is easy to use and is adequate. After sufficient iterations of the Monte Carlo program, the output is the probability distribution function for any desired result such as reserves, payout, present value, or rate of return. Such a program identifies the whole range of outcomes and allows probability statements to be made at any level of outcomes, including the expected value.

Table 4. Input for Monte Carlo Prospect Evaluation

Reserve Data: As in Table 1.

Economic Data:

Oil price ($/bbl):	Triangular distribution with minimum value of $17.50, most likely value of $19.00, and maximum value of $23.00.
Gas price ($/MCF):	Triangular distribution with minimum value of $1.20, most likely value of $1.45, and maximum value of $1.90.
Taxes:	Severance tax on oil, 4.6% of sales; on gas, 9.0% of sales; ad valorem, tax 1.0% of sales; federal income tax not considered.
Operating costs:	Uniform distribution $1500–$2500 per well per month plus constant $1.00/bbl.

Well Costs ($ thousands, triangular distributions):

Type Cost	Minimum	Most Likely	Maximum
Exploration drilling cost	600	700	850
Development drilling cost	450	600	750
Development dry hole cost	400	500	650
Completion cost	200	250	350

Production Facilities Costs:

Maximum Throughput (bbls/day)	Cost ($ thousands)
1000	150
2000	250
5000	400

Production Data:

GOR:	1200 cf/bbl constant
Initial production rate:	Triangular distribution with minimum value of 30 BOPD/well, most likely value of 100 BOPD/well, and maximum value of 200 BOPD/well
Type decline:	Exponential
Decline rate:	20%/year
Abandonment rate:	5 BOPD

Ownership Data:

Working interest:	100%
Net revenue interest:	75%

Timing Data:

Spud exploratory well:	7/91
First production:	10/91

Probability of Success Data:

Exploratory well:	Triangular distribution with minimum value of 5%, most likely value of 10%, and maximum value of 12%.
Development well:	Triangular distribution with minimum value of 65%, most likely value of 80%, and maximum value of 85%.

Figures 2a, b, and c are examples of such output curves calculated for the unrisked case (successful only) and the risked case (expected value considering probability of success and amount at risk). The outputs are derived from a Monte Carlo simulation of the prospect described by the inputs in Table 4. Here are some examples of probability statements that can be made from these curves:

1. Considering risk, a 56% probability exists that the DCFROR will be less than 15% (point 1 in Figures 2a, b, and c).
2. Considering risk, only a 13% probability exists that the DCFROR will be more than 25% (point 2 in Figure 2a).
3. If successful, the median DCFROR will be 22% (point 3 in Figure 2a).
4. If successful, a 16% probability exists that the prospect will have a present worth profit of more than $10 million" (point 4 in Figure 2c).

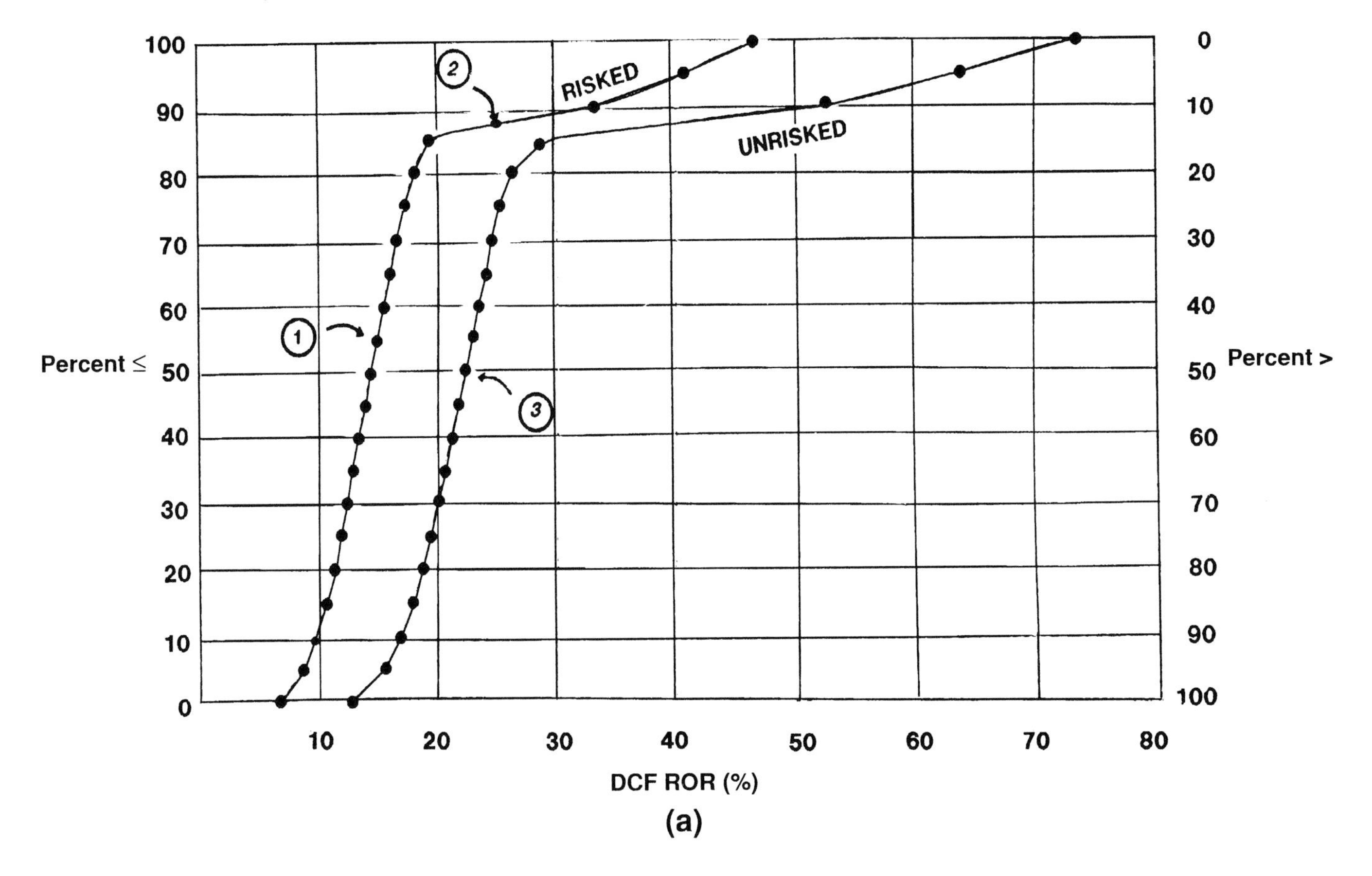
RISKED
UNRISKED
Percent ≤
Percent >
DCF ROR (%)
1
2
3
(a)

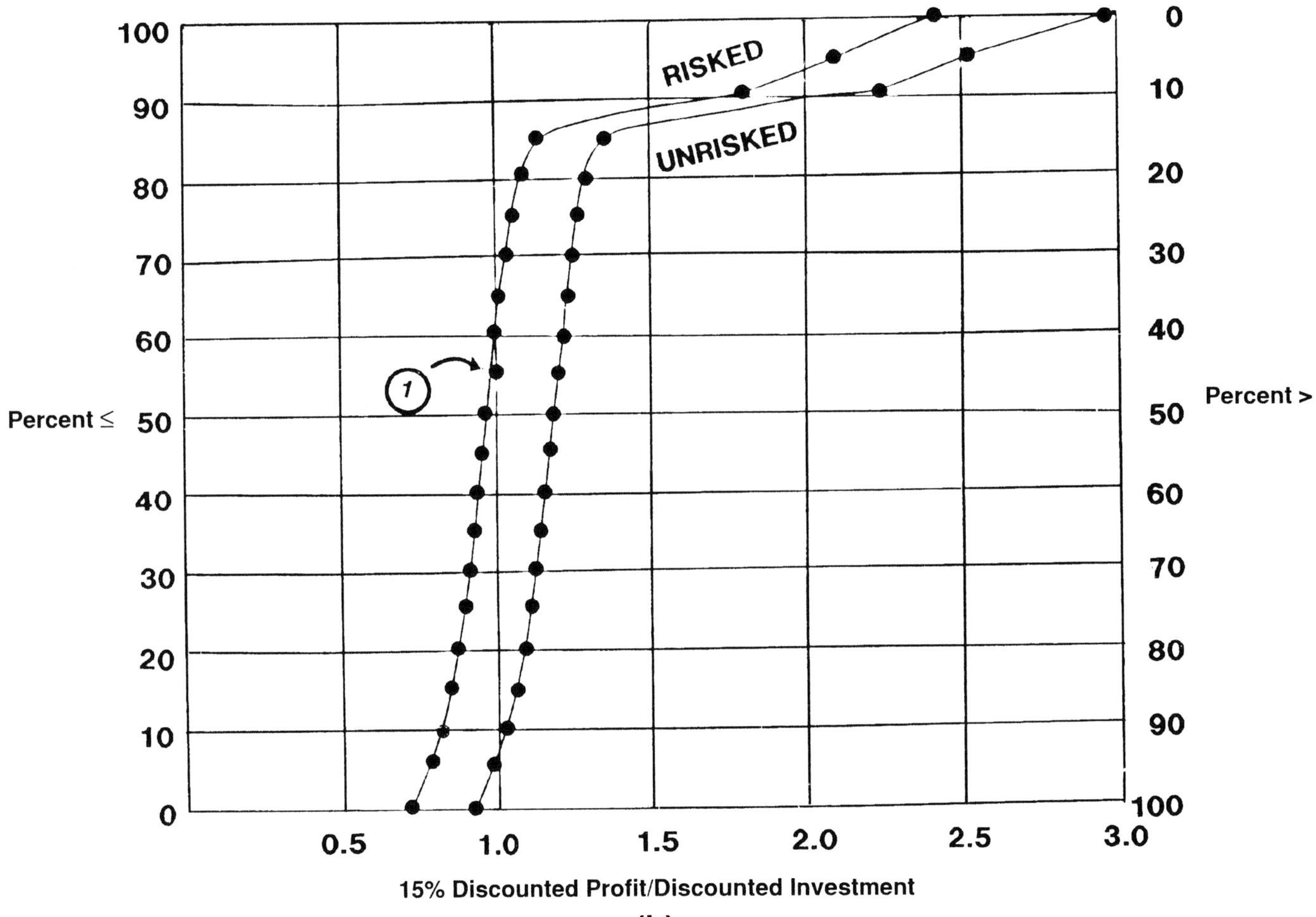
RISKED
UNRISKED
Percent ≤
Percent >
15% Discounted Profit/Discounted Investment
1
(b)

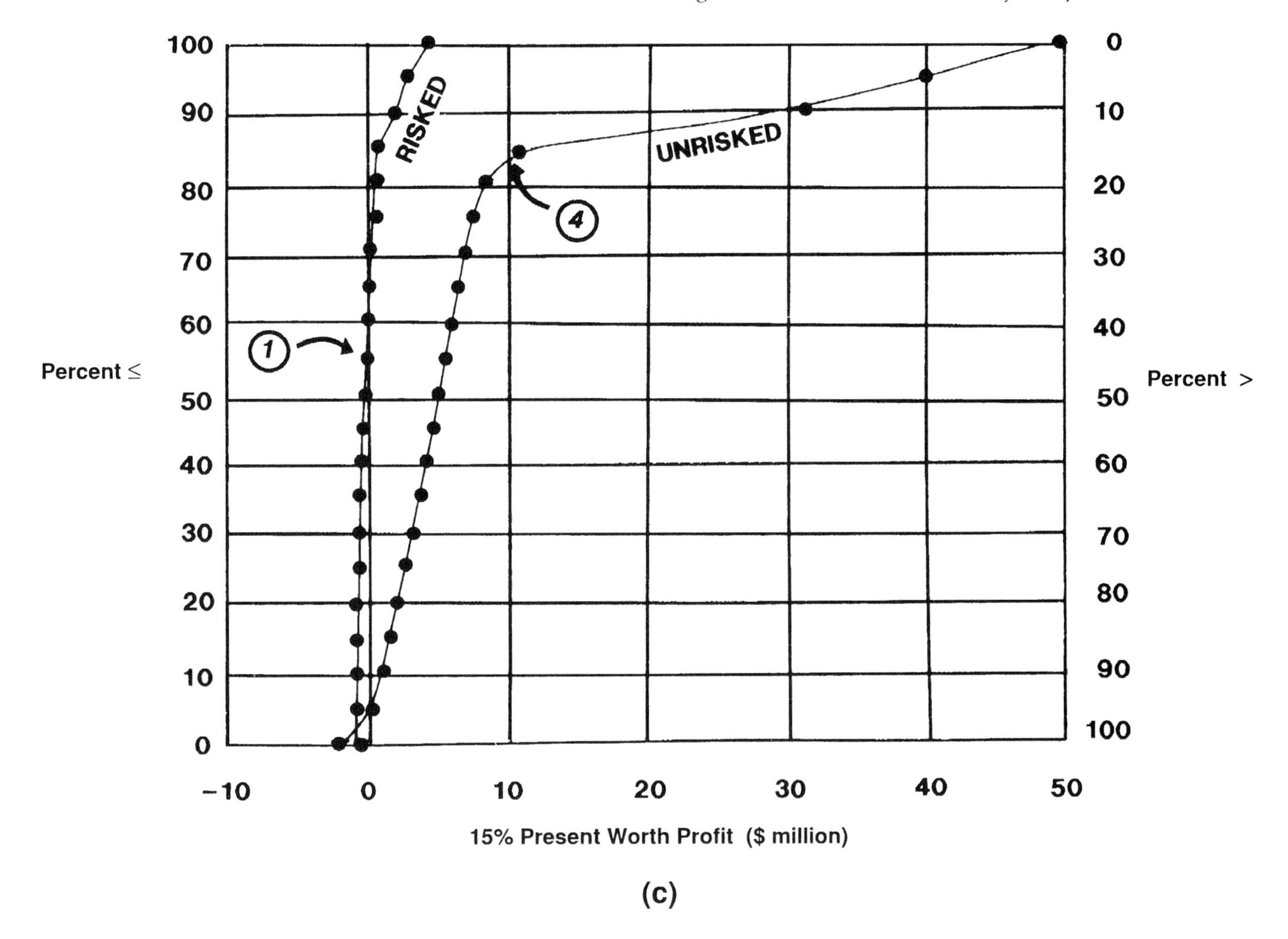

Figure 2. (a) Monte Carlo simulation output for discounted cash flow rate of return. (b) Monte Carlo simulation output for discounted profit to investment ratio. (c) Monte Carlo simulation output for present worth profit.

Note that because of the interrelationships between parameters, the same statement can be derived from different curves. For statement 1, a 15% discounted profit to discounted investment ratio of 1.0 (point 1 on Figure 2b) or a 15% present worth profit of $0 (point 1 on Figure 2c) are merely different ways of describing a 15% DCFROR (point 1 on Figure 2a). All three calculations yield the same cumulative probability of 56%.

The Monte Carlo approach can be elaborate or simple depending on how many variables are treated as probabilistic. As always, the accuracy of the results will be no better than the accuracy of our original estimates. But the advantage of such an analysis is that it allows us to express our knowledge, or lack of it, about each parameter separately and to combine our uncertainties in a consistent manner. The probability statements about the results allow us to communicate the uncertainties about each prospect to the decision makers in a consistent, quantitative fashion that fosters objective decision making.

Adjustments for Limited Risk Capital

The popular conception of expected value is that, if a gambler plays at a fair game and bets according to mathematical expectation, in the long run he will break even. But as far back as the 1600s, Bernoulli demonstrated that, unless the gambler has unlimited funds, he will inevitably lose. A slight chance exists that, at some time, the net losses (total losses from failures minus total gains from successes) will be larger than any bankroll the gambler brings to the game, however large. If the game is continued long enough, the probability that this will occur is certainty. So, to at least break even over the long run, a gambler must pay *less* than the expected value.

Another problem is the effect of limited funds in the short run—the gambler's ruin problem. With limited funds, the probability exists that the gambler's funds will be exhausted by a string of bad luck (dry holes). In addition, we know that different people (or organizations) have different attitudes toward risking a given amount of money for a given expectation. This has led to the use of utility or risk aversion techniques for decision making that are consistent with risk attitude.

Utility theory and its offshoots, such as *risk-adjusted value* (RAV) using a risk aversion function, are conceptually sound but difficult to apply in practice. There are two problems:

1. It is difficult to obtain a company's utility curve or risk aversion function, and even when this is done, it will change rapidly.
2. There is no guarantee that the utility curve or risk aversion function, as derived from past decisions or interviews with decision makers, is correct. The use of the technique merely ensures that decisions will be consistent with the risk attitude shown by the curve.

References are provided in the bibliography for the reader who wishes to pursue this topic further.

Long-Run Breakeven

The following equation states the long-run breakeven condition:

$$\left(1+\frac{R}{C}-\frac{X}{C}\right)^{P_s}\left(1-\frac{X}{C}\right)^{1-P_s}=1$$

where
R = potential value of reward if successful (unrisked present worth expected value),
C = total currently available risk capital,
X = risk amount (loss) if unsuccessful,
P_s = probability of success.

The equation states that, for every success, the available risk capital C is increased by a factor of

$$\left(1+\frac{R}{C}-\frac{X}{C}\right)$$

and for every failure, it is decreased by a factor of

$$\left(1-\frac{X}{C}\right)$$

Over the long run, the total increase will occur a number of times proportional to P_s, that is, the factor of increase is

$$\left(1+\frac{R}{C}-\frac{X}{C}\right)^{P_s}$$

and similarly for the decreases that will occur proportionally to $(1 - P_s)$. Since we have specified breakeven, that is, no change in the total risk capital, the product of these factors must be unity.

As in the expected value approach, the equation can be applied in two ways:

1. Use an estimate for the probability of success P_s and calculate X, the maximum amount that can be risked on the prospect given our currently available risk capital C. A solution for X, given that X is small compared to C, is

$$X=\frac{C(C+R)}{C+(1-P_s)R}\left[\left(1+\frac{R}{C}\right)^{P_s}-1\right]$$

The decision rule is simple. *If X is less than or equal to the estimated risk amount, drill the prospect; if it is more, don't drill.*

2. Use the estimated risk and reward amounts to calculate a minimum acceptable probability of success, P_m. The solution is

$$P_m=\left[1-\frac{\ln\left(1+\frac{R}{C}-\frac{X}{C}\right)}{\ln\left(1-\frac{X}{C}\right)}\right]^{-1}$$

As with the carrying power concept, *if the explorationist thinks the probability of success P_s is greater than or equal to this minimum acceptable value, then drill the prospect. If he or she thinks that the probability of success is less, don't drill.* Again, this is usually the easier method to use. It is easier to estimate whether the probability is more or less than a given value than it is to estimate a specific value of the probability.

Gambler's Ruin

For a single try at a prospect with probability of success P_s, the probability of failure is $(1 - P_s)$. For two independent tries, the probability of both failing is $(1 - P_s) \times (1 - P_s)$. Similarly, the probability α of a string of n failures in a row is $\alpha = (1 - P_s)^n$.

If an operator has C remaining risk capital and the risk cost of the prospect is X, then he can afford to take

$$n = C/X$$

tries before his risk capital would be depleted by a string of dry holes. So $\alpha = (1 - P_s)^{C/X}$.

Note that if C/X is not an integer, its value should be increased to $[C/X]$, the smallest integer that includes C/X. For example, 10.3 must be increased to 11 since we cannot drill a fractional number of holes.

As long as P_s is not certainty and C is limited, there will always be some probability, even if very small, that gambler's ruin can occur. So the operator must specify his attitude toward this risk by stating a value for α. This is the chance he is willing to accept that he could go broke by a string of dry holes. With a specific probability value for α, there are again two ways to use the equation:

1. Use the estimate for P_s, the currently available risk capital C, and the chosen value for α to calculate the risk amount X we can afford to pay. The solution for X/C is

$$\frac{X}{C} = \frac{\ln(1-P_s)}{\ln\alpha}$$

   Then adjust X/C to the smallest integer value that includes it. Multiply by C to obtain X. *If X is less than or equal to the risk amount for the prospect, drill; if it is greater, don't drill.*

2. Use the given risk amount X for the prospect, the currently available risk capital C, and the chosen value of α to calculate a minimum acceptable probability of success, P_m:

$$P_m = 1 - \alpha^{X/C}$$

   Again, X/C should be increased to the largest integer value containing it. This is the easier method, as before. *If the probability of success is considered greater than or equal to this minimum acceptable value P_m, then drill; if it is considered to be less, don't drill.*

Nomographic Solution

A proper evaluation of risk, given limited risk capital, considers both the long-run breakeven constraint and the gambler's ruin constraint. *Both should be met.* Figure 3 (from Arps and Arps, 1974) provides a nomographic solution for both constraints for $\alpha = 0.05$ in the gambler's ruin equation. (A similar graph can be constructed for a different value of α if desired.) The curves are the required values of X/R and C/R for various values of P_m.

The curved portions on the right of Figure 3 are for the long-run breakeven equation where it results in the larger minimum acceptable probability. The straight-line segments on the left are for the gambler's ruin equation where it results in the larger minimum acceptable probability.

For an operator with very large risk capital C, we can apply the carrying power concept discussed earlier. In that approach, the minimum acceptable probability of success, P_m, is

$$P_m = \frac{1}{1+\frac{p}{X}}$$

where
p = unrisked expected value of present worth profit,
X = risk cost.

The risk cost X is the *money-forward cost*. The reward R in the long-run breakeven equation is the present worth profit *plus* the risk cost, that is, $R = p + X$. It must be added back because the risk cost was subtracted when the present worth *profit* was calculated.

Substituting $X - R$ for p in the carrying power equation gives

$$P_m = \frac{1}{1+\frac{p}{X}} = \frac{1}{1+\frac{R-X}{X}} = \frac{X}{R}$$

This expresses the unlimited capital, minimum acceptable probability of success in terms of risk and reward, as in the long-term breakeven equation for limited capital.

Look at the right-hand side of Figure 3 where the long-run breakeven equation determines the minimum acceptable probability of success. For values of $C/R > 30$, the curves for P_m from the long-term breakeven equation asymptotically approach the value of X/R, which is P_m for unlimited capital. For practical purposes, if an operator's available risk capital C is 30 or more times greater than the reward R, then it can be considered infinite, and the simpler expected value approach is adequate.

Use of the nomograph is simple. For a prospect, plot the point (C/R, X/R). Then read the minimum acceptable probability of success, P_m, by interpolating between the curves. In the graphical application of the decision rule, *any prospect that plots above the minimum acceptable probability curve for the estimated probability of success, P_s, is unacceptable. A prospect that plots on or below the P_s curve should be drilled.* Let's consider some examples.

Example 1. A prospect has the following characteristics:

X = risk cost = \$560,000
R = reward = \$2 million
P_s = estimated probability of success = 30%
C = operator's available risk capital = \$20 million

This prospect plots at $C/R = 10$ and $X/R = 0.28$, which is point 1 on Figure 3. Since we have

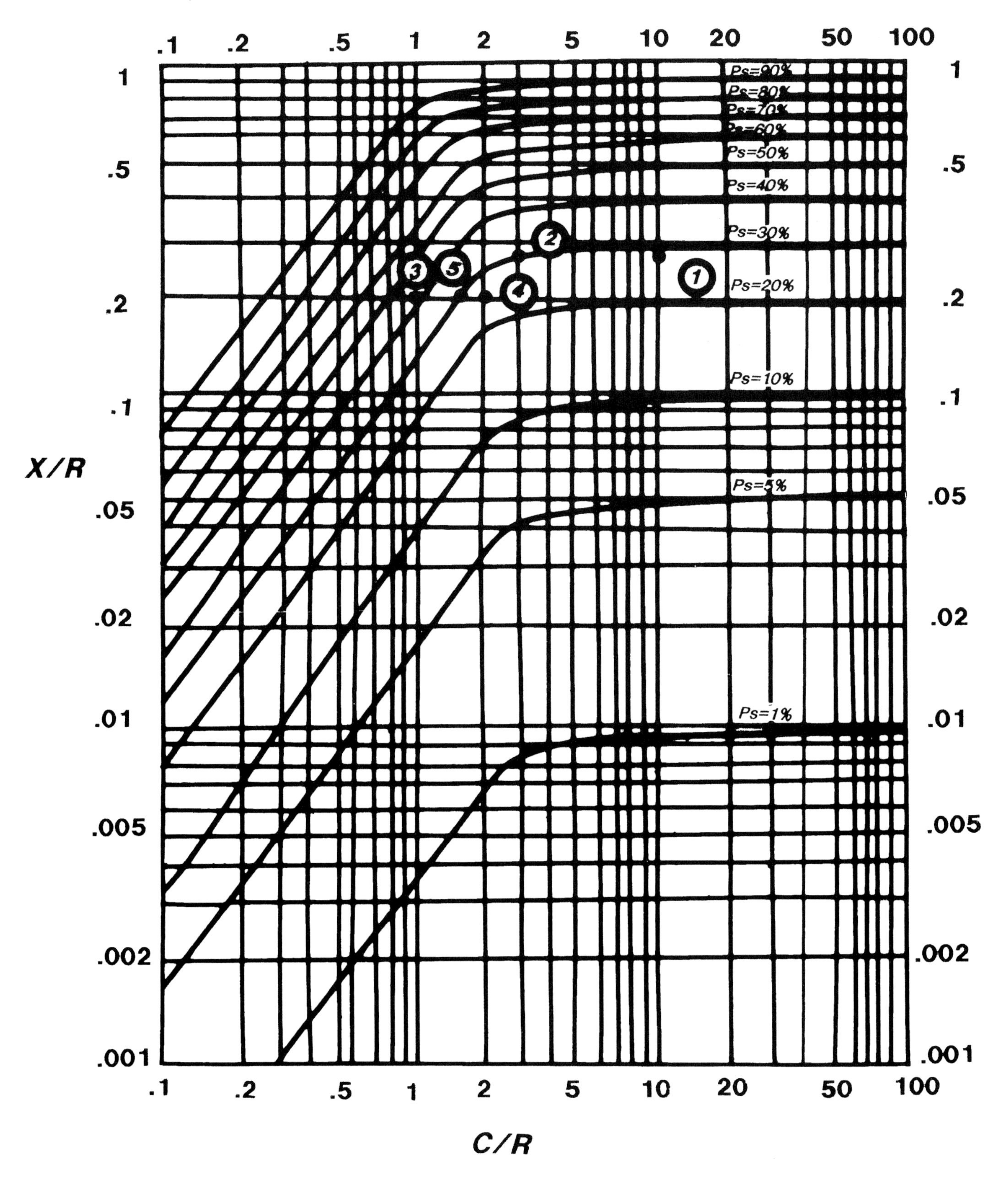

Figure 3. Minimum allowable probability of success, for $\alpha = 0.05$ in gambler's ruin equation.

estimated the probability of success at 30% and this point plots *below* the 30% line, the prospect should be drilled by this operator. The calculated value of P_m by the long-run breakeven equation is 0.29 and by the gambler's ruin equation is 0.08, confirming the nomograph-derived solution.

Example 2. Consider the same prospect as in Example 1. A different operator has only $6 million of available risk capital. The prospect plots at point 2 on Figure 3. The point lies at P_m = 32%, *above* the curve for P_s = 30%. This operator should not drill the prospect. Even though the carrying power minimum probability of success, $X/R = 0.28$, is less than the estimated $P_s = 0.30$, the capital is too limited to afford this risk.

Example 3. Another prospect has a risk cost of $200,000, a reward of $1 million, and an estimated probability of success of 30%. The operator's available risk capital is $1 million. This prospect plots at point 3 on Figure 3. Since the prospect plots at P_m = 45%, above the P_s value of 30%, this operator cannot prudently afford to drill this prospect. This time it is gambler's ruin that causes the problem; the long-run breakeven P_m is 27.5%. Although the unlimited capital carrying power, P_m, is $X/R = 0.20$, the operator's limited capital would only allow him to drill five prospects like this one. Keeping the probability of drilling five dry holes in a row to less than 5% requires a probability of success of at least 45%.

Example 4. Suppose the same operator in Example 3 takes only a 50% working interest in the same prospect. The risk to reward ratio, X/R, is now

$$\frac{\$100,000}{\$500,000} = 0.2$$

or unchanged. But the ratio of available capital to reward, C/R, is now

$$\frac{\$1,000,000}{\$500,000} = 2$$

This is point 4 on Figure 3. P_m is 25%; since P_s = 30%, the operator can prudently risk a 50% interest. The maximum working interest that can prudently be taken is found by moving point 3 to the right until it intersects the curve for the estimated probability of success, P_s = 30%. This is point 5 on Figure 3 at $C/R = 1.7$. Since C is still $1 million, the correct value of $R = C \div C/R$ = $1 million ÷ 1.7 = $588,000. The reward for 100% working interest is $1 million, so $588,000 represents a working interest of

$$\frac{\$588,000}{\$1,000,000} = 0.588$$

Therefore, the maximum working interest this operator should take in this prospect is 59%.

Improving Nonqualifying Prospects

Example 4 demonstrated the principle of "sharing the risk" by taking a partial interest in a prospect. This allows an operator with limited available risk capital to participate in a prospect *for which the expected value is positive*, but which is too big a risk. Use of the nomograph or of the previously discussed equations allow the operator to calculate just how much he can stand to risk in a prudent manner consistent with his goals.

Several methods are commonly in use in the industry by which an operator can improve prospects that fail to be prudent risks because of his limited capital:

1. Dilution of working interest, as previously discussed, which reduces the risk cost relative to limited capital, but keeps the same risk to reward ratio.
2. Obtain dry hole, bottom hole, or acreage contributions. For bottom hole contributions, this has the effect of reducing the risk cost while keeping the reward constant. Dry hole contributions also reduce the risk cost, but also decrease the reward slightly since no contribution will be received if the well is successful. As long as the additional acreage is productive if the well is successful, acreage contributions increase the reward while keeping the risk cost the same.
3. Obtain additional information (for example, seismic data) which, if favorable, allows an increase in the estimate of P_s, the probability of success. Even with no resulting increase in reward size, the prospect's required minimum probability of success can now be met.
4. Promote other investors. This allows the operator to reduce his risk cost proportionately more than his reward is reduced.
5. Obtain additional risk capital.

PROSPECT RANKING AND SELECTION

The discussion so far has been concerned with the analysis of individual prospects and their acceptance or rejection. Often a company has more acceptable prospects, including outside-generated opportunities, than it can afford to drill. Then it is necessary to rank the acceptable prospects to ensure that the best prospects are selected. The ranking process should consist of the following steps:

1. *Initial ranking.* The first step is to rank all the prospects on an unlimited budget basis. The ranking criterion should be the risk-adjusted expected value of present worth. The interest rate used to discount for present worth (*hurdle rate*) must be the same for all prospects, but its actual value is not important for the purpose of ranking prospects.
2. *Limited available risk capital adjustments.* The second step is the adjustment of each prospect's risk cost for the effect of limited total risk capital. At this stage of a prospect's development, the most likely correction is dilution of working interest, although the other methods can be used. Note however, that such risk-spreading tactics will usually involve further effort, such as the sale of the remaining working interest to other operators.

   To implement this step, first make an estimate of the total drilling budget. This becomes the initial value of the total available risk capital, C, in the limited budget calculations. Begin with the highest ranked prospect from the initial ranking, that is, with the prospect having the highest risk-adjusted expected value of present worth. Apply the limited capital equations or nomograph (Figure 3) to determine the maximum prudent working interest for the prospect. This correction is a step-wise process. After a prospect has been accepted, its adjusted risk cost must be subtracted from the total available risk capital to provide a new, lower value of C for the next prospect. Repeat this process until you run out of risk capital or prospects, whichever comes first.
3. *Final ranking.* If Monte Carlo simulation has been used to evaluate individual prospects, then the prospects from the second step can be ranked according to their ability to meet company goals. Suppose, for example, that the company's goals for exploration have been expressed as
   a. A primary goal of at least a 50% probability of achieving at least a 2:1 10% present value return ratio.
   b. A secondary goal of no more than a 10% probability of achieving less than a 1:1 10% present value return ratio.

   Then rank the prospects in order, according to the probability of achieving the primary goal first, eliminating all prospects that have less than a 50% probability. For those that survive this first cut, list the probability of the second goal, eliminating all prospects with more than a 10% probability. This is simply the portfolio analysis method of the "efficient frontier" (Quick and Buck, 1984).
4. If steps 2 and 3 result in a different ranking of prospects than step 1, it is necessary to repeat the entire process. Note that mere reduction of working interest will change expected values and Monte Carlo simulation results from the first step proportionately to the working interest reduction. The expected value of a 50% interest in a prospect, for example, is half that for a full interest.

   Any of the other methods used for the second step, such as obtaining bottom hole contributions, which change the risk to reward ratio or the probability of success, require recalculation of the Monte Carlo simulation or expected value of a prospect. Similarly, a revised ranking of prospects in the third step will lead to different values of remaining risk capital in the second step, requiring recalculation of the maximum prudent risk amount for the prospect. After a few iterations of the process, the results will not change enough to warrant continuing the iterations.

After the final iteration, consideration should be given to the factor that caused the end of the process. Did you run out of money or out of prospects? If you ran out of money while you still had good prospects left, then the remaining prospects should be postponed if possible or else you need to seek to increase the overall drilling budget.

In contrast, if you ran out of prospects first, then you can increase the allowance for prospects from outside sources or decrease the drilling budget. Prospects that fail to meet the company's decision criteria should be examined carefully as to *why* they fail, and then a decision should be made to postpone, drop, do further work, or farm-out.

FORMAT FOR BUDGET PRESENTATION

The format for budget presentation to the decision makers should be a spreadsheet tabulation of all recommended prospects, ranked according to the company's exploration goals. Estimated risk investment and probability of success, along with descriptive information such as name, location, TD, and working and revenue interests, should be included. Other parameters such as reserves, present value, and payout should be listed at their expected values.

The expected unrisked value *given a discovery* should be presented along with the risk-weighted expected value. The decision makers need to know what the expectation is if the project is successful. The spreadsheet can also present some information on the range of the key results and their probabilities, for example, the 10th and 90th percentile points. The specific make-up of the tabulation will vary from company to company, but these principles should be applied. Remember that the goal is to convey critical information to the decision makers without overloading them.

FINAL REVIEW OF THE EXPLORATION BUDGET

As a final step, a multiprospect Monte Carlo simulation of the proposed program should be calculated. The input to the program is the data from all the prospects in the budget. The output can be probability distributions for present value, rate of return, return ratios, or any other desired parameters. Even when such a Monte Carlo simulation is not done, the expected value of the total program for present worth can be obtained by summing the risk-weighted expected values for each prospect. Either of these approaches gives the exploration manager the ability to judge how well the proposed budget conforms to the principles of risk control, which include the following:

1. Diversification spreads risk money over many prospects and reduces the probability of loss. However, diversification also reduces the probability of achieving the maximum potential of the program.
2. Inclusion of prospects with high expected values improves the expected overall performance of the program. Unfortunately, such prospects are usually high risk (have a low probability of success) and are often high cost as well. It is the task of the exploration manager to balance these conflicting tendencies in the final drilling budget in such a way as to achieve the company's goals.

Semi-financial measures such as reserves replacement ratios or unit finding costs can be included. A distinction must be made here between the long-range results of the year's exploration program, that is, the ultimate reserves found, and what portion is expected to be credited to the company's proved reserves *this year*. Also, the drilling budget's average estimated probability of success, average well cost, average expected reserves, and so on should be calculated and reviewed.

Finally, the assembled budget needs to be given a few quick reality checks:

1. *Cost and timing*. Have we allocated enough or too much money for contingencies? Have we allocated enough time? Remember that one of Parkinson's laws is that everything takes longer than we expect.
2. *Long-term action plan*. How do the budget estimates for prospect costs, results, and success probabilities differ from those developed in our long-term action plan for exploration in the plays these prospects belong to? If they do differ significantly, why? And similarly, how do the average success ratio, average prospect reserves, and so on in this year's budget compare to the averages for the overall exploration plan and why?
3. *Past company performance*. Is this budget a radical departure from the company's level of effort and performance results in the recent past? If so, why? Or if not, why not?

BUDGET PERFORMANCE REVIEWS

A yearly budget, once approved, is not cast in stone. It is a working tool for management subject to periodic (usually quarterly) review, analysis, and adjustment for actual events. In many organizations, however, the annual plan, as reduced to particulars in the budget, is the standard against which performance and bonus compensation is measured. Therefore, it is necessary to retain the original budget for end-of-year performance measurement, while the adjusted budget is necessary for real-time scheduling of expected income and expenditures. So the budget review process described here often involves both original and adjusted budgets.

Variances

Budget review begins with variances or deviations that are over or under budget. For the exploration department, the most traumatic variance is a cost overrun usually involving exploratory well costs. Well costs are not only the most difficult to estimate but also the most difficult to control. An exploratory well by definition has more unknowns that can affect its cost than a development well. In addition, the control of daily operations is often in a drilling department not directly reporting to the exploratory management. A cost containment strategy that shows increasing appeal is the use of turnkey drilling contracts.

The key point for budget review of variances is to annotate the report with the *reasons* for the variances.

Timing

Budget variances often merely result from timing differences. Usually, the biggest problem is slippage—things don't get done in the time frame allotted in the budget. Again, the key is to annotate the budget review properly with the reasons for the slippages and to move the income or the expenditures to the updated time.

Results

A budget review must include measures of the results achieved compared to the original expectations. Here again the Monte Carlo simulation for multiple prospect programs is a useful tool. It can be used in three ways:

1. Results of the program to date can be expressed as a probability distribution using the data learned from the wells drilled. Dry holes obviously have a 100% probability of loss of the actual sum expended. Discoveries now have 100% probability of success and usually more is now known about the range of discovery reserves, development costs, and so on. Incorporating this new data as input provides a basis for comparison to the original budget projections.
2. Combining the results of the program to date with current projections for the remaining prospects provides the best estimate of the total results to be expected at the end of the budget period. Also, prospects are seldom drilled in the order of their merit ranking. It may be necessary to revise the maximum allowable risk expenditures for the remaining prospects, using the current actual value of available risk capital (remaining drilling budget dollars) as the initial value of C in a revised calculation of limited capital constraints.
3. If the projected results for the total program do not meet the company's goals, the multiprospect Monte Carlo simulator can be used to see what can be done to improve performance. By trial and error, beginning with the lowest ranked of the remaining prospects, it is possible to provide answers to such questions as
   a. Which prospects, if any, should be dropped from the budget?
   b. If possible, what kind of prospects would be needed to bring the program up to budget expectations?
   c. What would the impact be on the program of specific new opportunities that were not previously considered?

If a Monte Carlo simulation is not done, at least the expected values for results of the total program can be calculated. Again, this is done by summing the expected values for each prospect. Each of the steps above can be calculated for expected values only.

In such a budget review of measures of results (success rates, reserves, reserve replacement ratios, unit finding costs, etc.), it is necessary to sound two warnings. First, unless the company's annual budget includes a large number (50 or more) prospects to be drilled, binomial probability calculations easily show that the results from a small sample (this year's drilling program) will have a much greater uncertainty and hence variation of outcome than the total program (long-term action plan, total prospect inventory, etc.). This does not mean that such measures should be overlooked; it just means that they need to be evaluated in the context of a longer span, say, three to five years. The *trend* from year to year and the *cumulative results* of such measures are what should be used to measure the performance of an exploration program.

Second, if measurements depending upon reserve estimates (finding costs, reserves, replacement ratios, etc.) are used, it is important to be sure the appropriate reserve estimates are used. Reserves added to the company's books are a poor measure for an exploration program *in the short-term* for two reasons. (1) Most of the reserves will not be added, particularly to proved developed reserves, until the subsequent development wells are drilled. (2) Initial reserve estimates are highly uncertain and, in the case of larger discoveries, are often low. (Most large fields show continuing increases of estimates for their original recoverable reserves for the first 10 or more years after discovery. This is due to the justified conservatism on the part of the reservoir engineers "booking" the reserves. Overestimates will cause costly writedowns on the company's books, while increases can be a pleasant surprise or can be judiciously timed to offset reductions.) So, if a reserve estimate for a discovery must be made to assess budget performance, it should be directly comparable to the before-discovery reserve estimate used to justify the expenditure. In other words, it should be the *best current estimate of the total reserves* of the prospect, including future development.

Proper Use of Reserve-Based Performance Measures

Nevertheless, reserve-based performance measures, based upon the company's current booked reserves and expenditures, do have an important role in measuring the company's success over the short term, and even if they did not, they would often be required by short-term oriented senior management, investors, and/or financial analysts. The necessary caveats are as follows:

1. They should be applied to the company's total budget (including development and revisions), not just the exploratory budget, and should be viewed as a measure of total company success.
2. The long-term trend in such measures needs to be analyzed more closely than the current results.
3. The current results, in addition to the large variation from budget to be expected, must also be recognized to contain results that properly belong to both past and future years and, conversely, not to contain all the expenditures.

Measures of Technical Efficiency

A number of measures of technical efficiency can be derived from the drilling budget and/or prospect inventory database. Goals for these measures should be based upon the company's experience or industry experience. They can include, for example, the total number of prospects in inventory, the proportion of internally generated prospects out of the total inventory, the rate of adding new prospects, the number of prospects generated per explorationist, or the average size of prospects. In general, these types of performance measures are more of a management tool than an overall assessment of a program. Another important point about these types of performance indicators is that it is the year-to-year trend that is important. Are we generating more or fewer prospects? Is the average size increasing or decreasing?

CONCLUSIONS

A budget is a short-term action plan. It presents a schedule of income and expense items expected to result from actions taken in the budgeting period. The actions should be designed to advance the company's goals. The explorationist, and particularly the exploration manager, bears a large part of the responsibility to ensure that this occurs.

Exploratory drilling expenditures are usually the largest component of the exploratory budget, and they provide the funds for the final evaluation of all exploration projects. The critical decision is whether or not to include a prospect in the drilling budget. There are four factors involved in this decision: the risk cost, the reward size, the probability of success, and the total available risk capital.

Expected value methods provide a decision criterion that explicitly considers only the first three factors. They are appropriate only for companies with large exploration budgets. There are two methods for using the expected value approach—the classic risk-adjusted expected value and the easier to apply carrying power concept. Monte Carlo simulation provides an easy way to calculate expected value by considering all the probabilistic input variables. It also provides a range of possible outcomes and the probability of their occurrence.

Smaller companies must modify the expected value risk amounts by spreading the risk, that is, taking partial interests in a larger number of prospects. Utility theory provides decision criteria that will be consistent with a company's risk attitude, but quantifying the risk attitude is difficult. The practical approach is to consider explicitly two factors that affect the amount a company with limited total risk capital can prudently risk on a prospect. These two factors are the long-term maximum allowable risk investment and the short-term (gambler's ruin) risk of a string of dry holes. Equations to quantify these factors for a prospect can be incorporated into a budget analysis computer program or solved by a simple nomograph.

In preparing a drilling budget, prospects acceptable by expected value criteria are first ranked in order of expected value of present worth. Then, in a step-wise manner beginning with the top-ranked prospect, any necessary adjustments are made to each prospect's risk cost to ensure that the amount risked does not exceed the maximum that should be prudently exposed.

Budget performance reviews allow for mid-course corrections if necessary. They should answer the questions, "How are we doing?" and "What can be done to improve?" Here again Monte Carlo simulation, this time of the entire program rather than of a single prospect, provides a practical method to calculate the answers.

The budgeting process should be the culmination of the explorationist's efforts. The decisions as to which prospects to drill and how much to invest in them should not be left up to intuition, even if it is disguised as experience. Budgeting should be done with diligence and enthusiasm, but most importantly, it should be done on a sound technical basis.

Annotated Selected Bibliography

The comments after each entry relate the article to topics discussed in this chapter. The articles cited were selected from publications readily available to the average explorationist. Books were purposely omitted from the list.

Arps, J. J., and J. L. Arps, 1974, Prudent risk-taking: Journal of Petroleum Technology, v. 26, p. 711-716.
Key paper. Presents a soundly based, practical method for determining the minimum allowable probability of success for a prospect being considered by an operator with limited available risk capital. Long-term breakeven and short-term gambler's ruin both included. Simple to use nomograph. Improvement of nonqualifying prospects by contributions or dilution.

Busch, R. M., and S. S. Marsden, 1982, Bias in engineering estimates: Journal of Petroleum Technology, v. 34, p. 433-439.
Analysis of 640 production rate, reserve, and financial estimates related to installation of platforms in the Gulf of Mexico. All of these estimates were biased on the optimistic side compared to actual outcomes.

Campbell, J. M., 1986, Nontechnical distortions in the analysis and management of petroleum investments: Journal of Petroleum Technology, v. 38, p. 1336-1344.
Major problems are (1) inconsistent goals among various

groups in large organizations, (2) excessive optimism, (3) use of single-point estimates, and (4) ruler method of forecasting.

Capen, E. C., 1976, The difficulty of assessing uncertainty: Journal of Petroleum Technology, v. 28, p. 843-850.
Thought-provoking discussion of problems with "experts" estimating uncertainty.

Davidson, L. B., 1975, Investment evaluation under conditions of inflation: Journal of Petroleum Technology, v. 27, p. 1183-1189.
Explicit inclusion of inflation effects can change the rankings of a set of investments. Present value of future buying power must be considered.

Fang, J. H., and H. C. Chen, 1990, Uncertainties are better handled by fuzzy arithmetic: AAPG Bulletin, v. 74, p. 1228-1233.
Proposed alternative to Monte Carlo analysis using fuzzy set theory. Possibilities for input factors in place of probability distributions. Interesting idea, but doesn't explain how to generate "possibility" estimates.

Greenwalt, W. A., 1981, Determining venture participation: Journal of Petroleum Technology, v. 33, p. 2189-2195.
Working interest participation related to operator's available risk capital and the venture's profit to risk ratio. Considers gambler's ruin risk only; requires a method for determining fundamental aversion to risk, which is poorly explained and appears impractical.

Hayward, J. T., 1934, Probabilities and wildcats tested through mathematical manipulation: Oil and Gas Journal, Nov. 15.
"There is nothing new under the sun." Application of Whitworth's 1901 equation for the long-term breakeven price of a lottery ticket to the drilling of wildcats by operators with limited funds. Clear explanations and examples. A classic.

Kinchen, A. L., 1986, Projected outcomes of exploration programs based on current program status and the impact of prospects under consideration: Journal of Petroleum Technology, v. 38, p. 461-467.
Excellent conceptual discussion. Explains principles of risk control and effects of dilution. Suggested method of cumulants uses expected value calculations and gamma probability distribution, but is poorly explained. Suggested approaches for evaluating programs of multiple prospects are valid and can be carried out using Monte Carlo simulation.

McCray, A. W., 1969, Evaluation of exploratory drilling ventures by statistical decision methods: Journal of Petroleum Technology, v. 21, p. 1199-1209.
Triangular distribution for net profit outcomes. Monte Carlo simulation. Equations for resulting distributions for multiple tries at the same prospect assuming normal distribution for output. Gambler's ruin maximum participation.

Megill, R. E., and R. B. Wightman, 1984, The ubiquitous overbid: AAPG Bulletin, v. 68, p. 417-425.
Explanation for observed lognormal distributions of reserve estimates and sealed bids for offshore tracts.

Newendorp, P. D., 1976, A method for treating dependencies between variables in simulation risk-analysis models: Journal of Petroleum Technology, v. 28, p. 1145-1150.
Cross plots of variables to determine independence, complete dependence, or partial dependence. Calculation technique for partial dependence.

Newendorp, P. D., and P. J. Root, 1968, Risk analysis in drilling investment decisions: Journal of Petroleum Technology, v. 20, p. 579-585.
Discussion of problems with subjective probabilities, statistical inferences from past drilling, and mathematical models (determinate resource estimation models and binomial models). Explanation and example of Monte Carlo simulation technique.

Pirson, S. J., 1941, Probability theory applied to oil exploitation ventures: Petroleum Engineer, Feb.–May (four parts).
Fundamental probability rules. Application to estimate of probability of success assuming independent factors. Derivation of solution of Hayward's equation for maximum prudent risk exposure in a prospect. Application to evaluation of producing properties.

Quick, A. N., and N. A. Buck, 1984, Viability and portfolio analysis as tools for exploration strategy development: Journal of Petroleum Technology, v. 36, p. 619-627.
Viability defined as ability of proposed program to sustain itself. Gambler's ruin analysis using Monte Carlo simulation. Portfolio analysis of "efficient frontier" concept. Risk profiles to characterize risk nature of program (risk-averse, balanced, etc.).

Rose, P. C., 1987, Dealing with risk and uncertainty in exploration: How can we improve?: AAPG Bulletin, v. 71, p. 1-16.
Excellent discussion of risk versus uncertainty. Methods to improve consistency in risk decisions—gambler's ruin, risk aversion factors. Methods to reduce uncertainties (improve forecasts by eliminating biases)—cross plots, year-to-year trends, forecast reviews.

Schuyler, J. R., 1989, Modeling an exploration program: insensitivity to prospect ranking criteria: Oil & Gas Journal, Dec. 25, 1989.
Monte Carlo simulation of 100 prospect program demonstrates that, for optimizing net monetary gain, it makes little difference whether expected monetary value or profitability index is used as a ranking criterion. Neither does it matter significantly which discount factor is used to calculate present worth. The rankings only change for marginal prospects.

Schuyler, J. R., 1990, Decision rules using the EMV concept and attitude toward risk: An improved approach: Oil & Gas Journal, Jan. 15, 1990.
Risk considerations should be made at the end of the decision process. For risk adjustments, the expected utility approach is recommended. Exponential form of utility curve to convert EMV (Expected Monetary Value) into expected utility. Reducing the downside risk has a high cost in simultaneous reduction of expected net gain.

Smith, M. B., 1968, Estimate reserves by using computer simulation method: Oil & Gas Journal, March 11, 1968.
Monte Carlo simulation of reserve estimate (risked and unrisked) for a multiobjective prospect. Triangular distributions for area, thickness, and recovery factor. Probability of

occurrence estimates for trap, reservoir, and hydrocarbons. Tree diagram and explanation of computer model.

Smith, M. B., 1970, Probability models for petroleum investment decisions: Journal of Petroleum Technology, v. 22, p. 543-550.
Production, development, prospect, and trend models using Monte Carlo simulation. Detailed equations for triangular and uniform distributions and triangular approximation for normal distribution. Risked versus nonrisked curves. Good appendix of definitions of terms used in probability and statistical theory.

Wahlstrom, J. E., T. D. Mueller, and R. C. McFarlane, 1967, Evaluating uncertainty in engineering calculations: Journal of Petroleum Technology, v. 19, p. 1595-1603.
Earliest paper in generally available petroleum literature to discuss Monte Carlo simulation. Examples: (1) calculating water saturation and porosity from log data, (2) calculating recovery factor from PVT properties and relative permeability data, and (3) reserve calculation using volumetric equation. Uniform and triangular distributions for input variables and dependent variables.

Walls, M. R., 1990, Software uses risk attitude in prospect appraisal: Oil & Gas Journal, Jan. 22, 1990.
Utility theory approach. Cozzolino's method to calculate risk adjusted value (RAV) by exponential form of utility function. Calculation of risk aversion level for company (parameter r in utility function). Risk aversion level will change with changing corporate and industry environment. Examples of application to determining optimal working interest for a prospect and for prospect rankings. Derivation of RAV equation. Good graphic illustrations of results.

DONALD R. JONES, an independent consultant in petroleum evaluation in Houston, received a B.A. degree in Geology from UCLA in 1953 and an M.S. in Operations Research from the University of Houston in 1970. After serving 2 years in the U.S. Air Force as an intelligence officer, he was employed in 1955 by Marathon Oil Co. as an exploration geologist in Midland, Texas, and subsequently in Corpus Christi. In 1961 he became a log analyst at Marathon's Research Center in Littleton, Colorado. Mr. Jones was employed in 1966 by Offshore Operators Inc. (a predecessor of TransOcean Oil Inc.) as an evaluation engineer in Houston and later served as a reservoir engineer and Manager of Reservoir Engineering. In 1980 he became Vice-President of Exploration and Production of American Exploration Co. in Houston, and from 1987 to 1988 was Vice-President of Engineering for Oxoco Inc. He is a member of AAPG, SPE, SPWLA, and the Houston Geological Society.

Chapter 20

Funding Oil and Gas Ventures

Lee T. Billingsley
Sandia Oil & Gas Corporation
San Antonio, Texas, U.S.A.

In the ranks of independent geologists, external funding is what separates a dreamer with a good idea from a successful oil and gas producer who gets ideas evaluated on a regular basis.
—Lee T. Billingsley

INTRODUCTION

This chapter is addressed to geologists who must raise external funds for prospect acquisition, exploratory and development drilling, or to a limited extent, production acquisition. In the ranks of independent geologists, external funding is what separates a dreamer with a good idea from a successful oil and gas producer who gets ideas evaluated on a regular basis. The quantity of external funding usually dictates the level of drilling activity for independent geologists and small companies. Companies or individuals with high levels of internal cash flow set budgets to determine their drilling activity. However, most independents set goals for sales of prospects to determine their activity level. With the exception of the 1979–1981 time period, the quantity of external funds has been the limiting factor in activity level, not prospect ideas. Thus, raising external funding can be the most important step for a successful petroleum geologist.

The purpose of this chapter is to provide an independent geologist with an overview of raising external funding. The scope is purposely general. The real motivation behind this chapter is not to provide a cookbook for selling prospects. Instead, I hope to stimulate geologists to become students of salesmanship and thereby increase the effectiveness of their fund raising.

This chapter is organized into three parts. The first part describes funding types based on use of funds and various common deal structures associated with each type. The second part divides sources of funds by types of buyers and discusses attributes and drawbacks of selling to each type of buyer. The third part provides a general philosophy of salesmanship with some specific recommendations for petroleum ventures.

Before beginning, definitions and assumptions are necessary. The entity or person raising external funds will be referred to as the "Seller" and the funding source as the "Buyer." Also, two basic important assumptions are universally applicable to all ethical sales efforts. They can be adapted to petroleum ventures, as follows:

1. The Seller is raising funds for a project with sound technical merit that would provide the Buyer with an appropriate economic return, if successful.
2. The Seller has integrity.

TYPES OF FUNDING

Funding can be divided by its use into four types, which are (1) prospect acquisition, (2) exploratory drilling, (3) development drilling, and (4) production acquisition. Traditionally, petroleum geologists have raised funds of the first two, exploratory types. However, as more fields mature and larger companies divest properties, geologists will have an even greater opportunity in the near future to generate ideas, help evaluate, and raise funds for development drilling and production acquisition.

Prospect Acquisition

Prospect acquisition funding is the capital required to bridge the gap between a prospect concept and a prospect ready for drilling investment. Prospect acquisition funds can be used for seismic data, geochemical data, lease costs, or any other costs required to get the project to a drillable stage. Companies and independent geologists attempt to fund their own prospects in direct proportion to internal capital available. At some point, almost all explorationists need some external funding for prospect acquisition. Terms for prospect acquisition funding typically consist of cash plus an overriding royalty interest (ORRI) to the Seller. The amount of cash profit varies according to the size of the prospect and the amount of risk associated with getting the prospect to a drillable stage. The more risk taken by the Seller prior to selling the prospect, the greater the profit. For example, if the Seller has already purchased seismic data and secured the lease position, a larger profit can be demanded. Prospect fees may also be structured to be paid in stages, such as

1. initial concept,
2. successful seismic interpretation,
3. secure lease position, and
4. drilling initial well.

The amount of ORRI is also dictated by economics and local convention. On raw prospect concepts, the ORRI usually varies from 1 to 3%. With a more complete, ready-to-drill prospect, the ORRI may increase. Generally, the net revenue interest (NRI) available to the Buyer should be 75% or more. In areas of high economic potential and base royalties of 25%, Buyers may accept an NRI as low as 70%.

Many independent geologists sell their prospects to other entities, who actually acquire the drilling funds. This process is referred to as *wholesale prospect sales*. The Buyer then becomes a Seller and sells the prospect on a "retail" basis to raise necessary drilling funds. Wholesale prospect sales are time efficient for an independent geologist because (1) the prospect can be sold to one entity, (2) there is less liability in raising money and drilling the well, and (3) the geologist can retain an ORRI in the prospect without the burden of working interest cost. It can be a natural evolution for a successful wholesale prospect seller to convert to retail prospect sales. However, geologists who love the technical aspects of geology frequently remain as prospect generators and wholesale sellers. Raising drilling funds requires more business-related time and skills. Marketing, legal review, and accounting reduce the time available for technical work.

A major drawback to raising prospect acquisition funds is loss of control by the Seller. For example, if the Seller does not have a secure lease position in the prospect, an unethical Buyer may circumvent the Seller and acquire leases without compensation to the Seller. Assuming an ethical Buyer, the Seller must still rely on the Buyer to acquire the "proper" amount of seismic data and lease position. Differences may exist between the Seller's idea of how much is technically proper and the Buyer's idea of financially proper. Finally, the Buyer may need to raise drilling funds for the prospect. Even after buying the prospect, the Buyer may change plans or have difficulty raising drilling funds. In either case, the Seller loses control over when the test well is drilled, if it is drilled at all. A Seller should always have a provision in the sales agreement with the Buyer that allows control of the prospect to revert back to the Seller at a mutually agreeable time. Green (1989) provides an example of a confidentiality agreement, which requires the Buyer to hold the Seller's prospect data confidential for a specific time period.

Exploratory Drilling

After prospect acquisition, *exploration drilling funding* is the next step in the process. This can be defined as the money required to drill the initial well in a project. The Seller of an exploration drilling prospect attempts to recover all costs, including prospect acquisition, in proportion to the amount of interest sold. Exploration drilling funds are usually raised to cover the cost of drilling the initial well in the prospect to the production casing point, or through plugging, if a dry hole. Potential deal terms are numerous, but the Seller customarily receives a carried working interest to casing point in the initial well or even through completion in the initial well. Depending on the economics of the prospect, the Seller's carried interest usually varies from 12.5 to 25%. If the Seller receives a carried working interest, Buyers must pay a proportionately greater cost. For example, if the Seller receives a 25% carried working interest to casing point, Buyers must pay 100% of costs to casing point to receive 75% of working interest after casing point. In subsequent wells on the same prospect, the Seller must pay 25% of all costs and Buyers pay 75%.

The following table lists various carried working interests and corresponding factors that determine the before and after casing point interests of Buyers. The factors are calculated by the following formula:

$$\text{Factor} = \frac{100}{100 - \text{Carried interest (\%)}}$$

Carried Interest (%)	Factor
25	1.333
20	1.250
15	1.176
12.5	1.143

For example, if the Seller's terms are a carried 20% working interest to casing point and the Buyer wants to own 15% after casing point, then the Buyer must pay 15% x 1.25 (factor) = 18.75% before casing point. The term *a third for a quarter* refers to the Seller receiving a carried 25% interest, and therefore the Buyer must pay 33.33% (one-third) of costs before casing point for the right to pay 25% (one-quarter) of costs after casing point. The following example illustrates cost sharing in an exploration deal.

Example of Terms for Funding an Exploration Drilling Deal

Assumptions:

1. The Seller retains no interest in project except the carried working interest.
2. The Seller receives carried 25% working interest to casing point in the initial well and pays its share of completion costs.

Prospect acquisition cost	$120,000
Estimated initial well drilling cost	400,000
Total funds raised by Seller	$520,000
Completion costs	$200,000
For 25% working interest:	
Buyer's cost before casing point = $520,000 total cost x 33 1/3% =	$173,333
Buyer's completion cost = $200,000 x 25% =	50,000
Buyer's total cost =	$223,333
Seller's total cost = $200,000 x 25% =	$ 50,000

The value of the 25% carried working interest to the Seller can be calculated as $520,000 x 25% = $130,000. This represents the cost of the Seller's 25% working interest in drilling the initial well. However, the Seller has risked this compensation on the outcome of the well. The Seller must evaluate the actual cost of the compensation in terms of overhead expenses, marketing time and expenses, and finally, drilling results. Obviously, the carried working interest only has value in the cases of successful wells.

Prospect Acquisition and Exploratory Drilling Funding

In some cases, it may be more efficient for a geologist to combine the two types of funding. Certain Buyers may be willing to provide funds for prospect acquisition and exploratory drilling. For example, the author's company (Seller) had a year-long agreement with a large independent company (Buyer) in which the Seller agreed to show the Buyer all prospects generated within a certain area. If the prospect was accepted, the Buyer (1) paid the Seller a prospect fee, (2) reimbursed the Seller for prospect seismic costs, (3) acquired prospect leases, (4) granted the Seller a carried working interest to casing point in prospect's initial well, and (5) assigned an ORRI to the Seller's employees. This type of arrangement was possible for the Seller because they had completed a regional geological study of an exploration concept. It was advantageous to the Buyer because they were able to review several prospects at a set price on a first right of refusal basis. It was advantageous to the Seller because they received the carried interest from raising exploratory drilling funding without the financial burden of providing prospect acquisition funding.

Another combination of prospect acquisition and exploratory drilling funding is also possible. A prospect-generating entity can combine with a prospect acquisition funding entity, and together they agree to secure exploratory drilling funding. At first glance this described combination seems obvious and hardly worth mentioning. However, the author has found this type of combination funding to be one of the most successful and financially rewarding methods for an independent geologist to finance exploration. The advantage to geologists is that no prospect acquisition capital is needed, they still receive a carried interest (usually 10–12.5% working interest), and they have help raising drilling funds. The concept is easy to sell to a prospect acquisition funding entity because they receive help in drilling fund sales and get one-half of the prospect profit. The concept is even easier to sell to a Buyer of exploratory drilling deals, because the Buyer participates in the project at an early stage and pays less "promote" to the Seller. The obvious disadvantages to the Seller are the lack of control of the prospect and reliance upon the integrity of the Buyer to compensate the Seller.

One of the first questions arising when two entities share one prospect is how to divide profits and carried interest. A simple way is to divide the prospect into three equal parts: (1) prospect generation, (2) prospect acquisition capital, and (3) acquisition of exploratory drilling funding. Thus, each entity that provides one of these parts is entitled to one-third of both the profit and carried working interest. An obvious even split of a prospect is a prospect-generating geologist and a prospect acquisition partner who jointly raise drilling funds.

Development Drilling

Development drilling funds are the next logical step in the funding process. After a successful exploratory well is drilled, completion funds and development well funds are needed. In many deal structures, the Seller must pay its pro rata share of such expenses. The Seller may wish to "sell down" its interest to conserve capital or limit risk exposure in the subsequent expenses. Alternatively, the project could consist of infill drilling in an existing accumulation or secondary recovery. In either case, development drilling differs from exploration because it has a more defined risk to reward ratio and because multiple wells can be drilled and completed.

Deal structure for development funding does not follow a standard format like exploratory drilling. The Seller may view development funding as securing a loan and giving up some equity. The Buyer's only recourse on the loan is the subject property. The following deal structures are examples, but variations are endless:

1. The Buyer funds 100% of cost, receives 85% of revenue until payout, then 15% after payout. This is an example from a low risk, high return development project.
2. The Seller is carried for 10% interest on all costs until the Buyer has spent a prearranged amount. The economics on this project are less favorable, and the Seller receives less compensation.

Production Acquisition

Production acquisition funding means securing financing to purchase producing properties. In this case, the term *Seller* refers to the entity attempting to secure financing. The Seller is actually the bidder or buyer of the producing property. They are selling the concept of a particular property acquisition. The Buyer is the entity providing financing to close the acquisition.

Production acquisition funding has not been an area of involvement for many petroleum geologists. However, as domestic fields mature and major companies divest marginal properties, more geologists will become involved in evaluation and participation in such transactions. Geologists may view acquisitions as an opportunity to develop low risk prospects on producing acreage blocks.

Deal structure varies with economics, but generally the larger the funding amount the lower the percentage promotion to the Seller. One Seller of ideas for acquisition funding does not bid on properties with a value over $5 million, because in acquisitions over $5 million, the Seller is restricted to a small percentage of promotion. On these larger transactions, pension funds and insurance companies can bid aggressively for the properties, and thus squeeze the promotional interest that may be generated by a Seller of such funding. In acquisitions under $5 million, large, aggressive bidders are less active, and the same Seller receives a 5% upfront cash fee and a 15 to 20% working interest in the wells after the Buyer has received payout. In smaller deals, the Seller receives a 10% carried working interest, such that Buyer pays 100% of acquisition cost for 90% of working interest.

The advantage of acquisition funding is that many domestic U.S. properties are changing ownership. Consequently, large sums of money are changing hands, which presents opportunities for raising funding. However, acquisition deals are difficult to obtain. One active Seller of acquisition funding estimates that their bids on acquisitions are successful only 3 to 5% of the time. After a successful bid is accepted, they must secure the funding in a relatively short time, like 60 days. Thus, their funding must be virtually secured before the bid is made.

Trends in types of funding and the resulting deal structures can summarized as follows:

1. The more money received by the Seller upfront, the less equity available. As extremes, a geologist that receives a cash retainer each month must expect to receive a smaller equity position in his or her work. However, the geologist who risks lease, seismic and geological prospect costs, can expect to receive a larger share of equity.
2. Projects with well-defined risk and reward dictate well-defined profit margins for the Seller.
3. The Seller's actual cost in putting together and selling a project is not recoverable through sale of the project, but rather through the overall economic success of the project.

SOURCES OF FUNDS

The following is a summary of potential types of funding entities or Buyers. Buyers can be divided into the following categories: (1) petroleum industry companies, (2) non–petroleum industry companies, (3) individuals knowledgeable in petroleum industry, and (4) individuals not knowledgeable in petroleum industry. For each type of Buyer there is a description of variables associated with securing funds. The purpose of describing the variables associated with each type of Buyer is to provide a general guide to Sellers, so that Sellers develop a philosophy toward the type of Buyers they are seeking. There are tradeoffs associated with each type of Buyer.

Banks are not listed as a source of funds because bank financing depends on the individual circumstances of each loan and the borrower. It is beyond the scope of this discussion to describe the requirements, advantages, and disadvantages of bank financing.

The variables chosen for analysis of Buyers can be defined as follows:

Accessibility—The difficulty in establishing initial contact, scheduling an appointment and making a sales presentation.

Sales/Presentation Percentage—After a sales presentation is made, the chances of completing a sale with no regard to size of sale.

Funding Level—Compares Buyer's relative limits in size of potential sale.

Promotion Level—Relative amount of Seller's profit acceptable to each type of Buyer.

Potential Liability—Lawsuits that could be filed against the Seller by the Buyer. This variable assumes the Seller is honest, sales presentation is accurate, and contractual agreements have been properly prepared and executed. Potential misunderstandings may still occur, and this variable describes the relative chances of the Buyer suing the Seller.

Repeat Business—The degree of subjectivity that can be expected from the Buyer in evaluating the Seller's prospects. If the Seller's prospects have been successful, some Buyers are more prone to continue to buy prospects from the same Seller.

Companies in the Petroleum Industry

Accessibility

Companies are usually very visible through previous drilling and production operations. Companies that do not operate wells can be found through association with staff professionals and networking with other Sellers. Once contacted, they are straightforward in deciding if they will review a proposal.

Funding Levels

Usually the larger the project, the better. If the economics are justified, no project is too large. Smaller deals can be harder to sell to certain companies. Also, projects involving only one well or small reserves may not be acceptable.

Sales/Presentation Percentage

Companies are very knowledgeable buyers because they look at so many prospects generated either internally or externally. Consequently, they can afford to be very selective. They may only buy 1 to 10% of the prospects reviewed. Similarly, a Seller's sales to presentation percentage may be 1 to 10%.

Promotion Level

Restricted to "standard" industry deal terms. In exploration drilling, a 20–25% carried working interest to casing point is acceptable.

Potential Liability

Generally low chance of a lawsuit due to a misunderstanding because the company's personnel thoroughly review all agreements. Also, the Seller is not likely to be accused of misrepresenting the geology or potential because the company's own experts have had the opportunity to validate data and conclusions.

Repeat Business

Fair; because of their expertise, they will thoroughly and objectively evaluate each deal on its own merit. However, credibility from previous deals certainly counts and gives the Seller an edge on other Sellers that may be showing their deals to the company for the first time.

Non–Petroleum Industry Companies

These companies fund relatively low risk oil and gas projects, such as development drilling and production acquisition. Exploratory drilling requires a longer term joint venture agreement with the Seller. Typically, funding may originate from pension funds, insurance company investments, or university endowments. The funding entity hires experts to evaluate and manage their investments.

Accessibility

Relatively unknown to most petroleum geologists; mainly accessible through financial contacts. *Oil and Gas Finance Sourcebook* (1990) provides a listing of active companies. These companies require establishment of "business relationship" with the Seller. They need thorough knowledge, confidence, and trust in the Seller.

Funding Levels

Virtually unlimited on upside, including offshore development drilling or acquisition. Larger companies may prefer a $10 million minimum, but small companies have dropped minimum investments to about $1 million.

Sales/Presentation Percentage

Percentage of successful funding relative to number of attempts is higher than funding exploration drilling. Risk and reward may be better defined and agreed upon by both the Buyer and Seller. Funding potential is high relative to available deals. These companies have more funding than acceptable deals.

Promotion Level

Well defined by the Buyer's company policy. Usually, funding is provided through a loan to the Seller such that the Buyer's loan is secured only by the Seller's interest in the property. The Buyer receives a share of production such that the loan is paid off within 5 to 7 years, and the Buyer achieves an interest rate slightly over the prime rate. In addition, the Buyer receives equity in the form of an ORRI or carried working interest through the life of the property. This type of funding is called "mezzanine financing." The Buyer attempts to achieve an overall rate of return of 15–20% through combined loan payoff and equity.

Potential Liability

Generally low because the Buyer's experts thoroughly evaluate the Seller's project. Also, the Seller may receive an indemnification from the Buyer.

Repeat Business

Fair to good, because deal potential is well defined.

If the Seller performs according to expectations, high levels of funding are available.

Individuals Knowledgeable in the Petroleum Industry

This category includes retired or active employees in petroleum industry companies or simply individuals knowledgeable through years of investing.

Accessibility

Initial contact is difficult for two reasons. First, names of individuals investing directly in the petroleum industry are not well publicized. Second, these individuals may not invest in the industry full-time, nor do they have the time or inclination to develop a relationship with several Sellers. They prefer on-going relationships with mutual trust and loyalty. Experienced, successful Sellers develop groups of these investors over the years. The Seller may refer to "our group" or our "in-house" investors.

Funding Levels

Highly variable. As noted, a successful Seller usually has a group of individuals because individual Buyers may take no more than 25% of a deal and frequently much less. However, they may buy the same percentage of every deal from a given Seller over a period of time.

Sales/Presentation Percentage

Low in initial contact because previous relationships are so important to this group. However, after an initial sale is made and the Buyer is satisfied with the Seller's performance,the percentage may increase dramatically. Ideally, a Buyer develops trust and confidence in the Seller, such that the Buyer takes part of every deal presented by the Seller. A parallel from another high risk investment is that the Buyer is betting on the jockey, not the horse.

Promotion Level

Standard terms from similar industry partners. Loyalty is achieved by restraint on part of the Seller in setting terms of deals. The extra short-term profit possible from making a large profit on one deal is not worth the loss of long-term loyalty from this type of investor.

Potential Liability

Low, because they are knowledgeable. The Seller continues to submit agreements that are fundamentally similar to those previously agreed upon.

Repeat Business

High, because this group does not usually have the time or desire to review prospects from a large number of Sellers. They continue to buy from previously successful Sellers.

Individuals Not Knowledgeable in the Petroleum Industry

Accessibility

Initial contact is almost required to be based on previous personal relationships or recommendation. This group of Buyers is very skeptical. A low percentage of responses result, but if the number of contacts is sufficiently high, sales will occur. The most difficult step with this Buyer is gaining an opportunity for a presentation.

Funding Levels

Generally low per each entity. Since they are not knowledgeable, they try a small interest to gain experience. Funding levels should grow with success of operator.

Sales/Presentation Percentage

High percentage of sales once a presentation is made. Because the Buyer is not knowledgeable, they must evaluate the Seller, not the proposal. Successful projects can lead to numerous referrals.

Promotion Level

Higher level of promotion is possible with this type of Buyer. However, high ethics dictate that the Seller keep the terms the same for all Buyers in a given deal.

Potential Liability

High; because the Buyer is not knowledgeable, they do not understand the risk involved in oil and gas investments. If the project fails, such as a dry hole or marginal well, the Buyer may sue the Seller for misrepresentation.

Repeat Business

Highest, because the Buyer is evaluating the Seller, not the prospect. However, the Seller can only be as good as the result of the most recent prospect in the eyes of this type of Buyer. These Buyers are best suited for low risk prospects.

In summary, to raise funds effectively, the Seller must recognize certain tradeoffs associated with different sources of funds. The Seller must develop a philosophy for raising funds based on the above described tendencies. The philosophy must be consistent with the Seller's personnel, and the philosophy can be evolutionary. For example, a relatively new Seller without extensive initial contacts should probably concentrate on "industry-knowledgeable" investors because sales presentations are relatively easy to schedule. Often, the Seller is acquainted with the reviewing Buyer on a professional basis. As a Seller gains more contacts, knowledgeable individuals can provide a more cost-effective source of funds. They are better for repeat business and buy a higher percentage of projects presented.

MARKETING

Marketing is a significant part of the effort required from prospect acquisition through final sale. Marketing represents at least one-third of the effort in terms of the Seller subdividing the profit centers of a given deal. Marketing can represent a significant portion of the Seller's time, and if a project becomes too difficult or time consuming to sell, the Seller may lose money. Some Buyers and Sellers look upon the Seller's carried interest or promotional profit as "free." But because marketing time and expense and geological prospecting time are factored into the Seller's cost basis, Sellers usually have more at risk in a project than Buyers. The Seller may show a profit on a prospect that results in a dry hole. However, the Seller has paid for the carried working interest in the form of overhead associated with prospect generation and marketing. Thus, one of the earliest and most important decisions a Seller must make on a prospect concerns market reaction.

Marketing is a business term for selling. For some reason the word *sales* denotes a negative connotation. But petroleum geologists must be effective salespeople in order to sell their ideas to management or to partners outside their company. Great scientists who cannot persuade others or cannot "sell" their ideas are ineffective. Top business leaders are not necessarily the best technical people in their organization, but they are the best communicators, persuaders, and sellers. Petroleum geologists that cause the most wells to be drilled are also effective salespeople. In the United States, independents drill most exploratory wells and independents are dominantly financed by external funding. The most active independents are the best salespeople. Science is obviously important because a high level of activity can only be sustained through successful projects. The combination of technical expertise and salesmanship causes drilling activity and eventual success. Petroleum geologists must become effective salespeople.

Salesmanship can be learned from books, tapes, and seminars. Geologists expend considerable time, effort, and money to learn petroleum geology, but the ability to communicate their ideas is left to their own imagination. Some people are natural salespeople, while other personality types need a more concerted effort to become effective sellers. Many of the concepts on marketing in this section of the chapter are taken from a three-day seminar presented by Planned Marketing Associates (1990) in Hunt, Texas. However, these same principles are documented in various audio tapes and books. Table 1 provides a partial list of pertinent works.

Geologists must improve their sales skill to bridge the gap between a sound geological idea and an evaluated project. As an initial step, successful people set goals. The goals should be written and reviewed regularly. To be most effective, the goals should be written in first person and be positive, specific, challenging, and adjustable. Three types of goals include long-term, intermediate-term, and short-term. Long-term goals are our ambitions in life. They might be an answer to the question, "What would you do if you had a magic wand?" Long-term goals can afford to be somewhat ambitious and perhaps unattainable. They are visions of our future. Some examples might be

- I will become president of X company by age Y.
- I will own my own company with production of X barrels of oil per day by the year Y.

In contrast, intermediate-term goals are concrete, attainable within a few months or years, and consistent with long-term goals. Examples include

- I will sell X prospects this year.
- I will add X new names to my list of potential Buyers this year.
- I will sell X dollars worth of drilling prospects this year.

Short-term goals are set for completion within a few days or weeks. They keep us excited and on schedule to complete intermediate-term goals. Short-term goals relating to sales should be related to an activity level, not performance. Presumably, if activity level is high enough, sales will occur. Some examples are

- I will call 10 new clients today.
- I will show this prospect to 15 potential Buyers this month.

Another step in self-preparation is to develop a positive mental attitude. One's subconscious mind is well stocked with negative thoughts. Here are some

Table 1. Partial List of Suggested Publications on Salesmanship and Successful Living

Book	Author
Think and Grow Rich	Napoleon Hill
How to Win Friends and Influence People	Dale Carnegie
One Minute Sales Person	Larry Wilson
Secret of Closing Sales	Zig Ziglar
Secrets of Super Selling	John J. McCarthy
What to Say When You Talk to Yourself	Shad Helmstetter
The Magic of Thinking Big	David J. Schwartz
See You at the Top	Zig Ziglar
Changing the Game	Larry Wilson
In Search of Excellence	Tom Peters

examples:

- This deal is impossible to sell.
- No one is buying any deals.
- I am tired of showing this prospect.
- I don't feel like getting out of bed today.

The conscious mind can control the subconscious by feeding it positive thoughts, even if they are little white lies, such as

- I feel great today.
- I do what I ought to do, when I ought to do it, whether I want to or not. No debate.
- Some one wants to buy this deal, I need to find them.
- This prospect is perfect for this Buyer.

After a barrage of positive statements, the subconscious mind will succumb and turn positive. A positive mental attitude is the attribute that makes successful salespeople appear happy and optimistic.

Another aid to a positive mental attitude is to interact with successful people. Seek their advice. Successful people are flattered and usually willing to help someone who seeks their counsel. Eat lunch with optimistic peers. On the contrary, negative attitudes are plentiful, especially in difficult economic times. Too many geologists agonize and commiserate over the current economic plight of the oil and gas industry. They only reinforce each others negative attitudes.

The first external step in the marketing process is contacting potential Buyers. Sources for initial contacts include (1) networking with other professionals; (2) professional associations, such as AAPG, SEPM, SPE, SEG, SIPES, AIPG, IPAA, and local organizations; (3) publications that report drilling activity; (4) county records; (5) the recently published *Oil and Gas Finance Sourcebook* (1990); and (6) commercial listing services, such as Petroleum Listing Service, Inc. (P.O. Box 834, Midland, Texas 79702).

Networking is the *exchange* of information with other professionals. To be effective and receive good, potential contacts, the Seller must also provide information. To be effective at networking, professionals must be people-oriented. Effective networking can provide the Seller with up-to-date information on Buyers that are looking for a particular type of deal.

A Seller can look beyond simple networking by considering naturally existing economic relationships. For example, if a Seller needs names of qualified Buyers, why not ask people who will also benefit from the sale? Obvious beneficiaries include the Seller's accountant, lawyer, or other services that rely on the Seller's economic health and activity. More directly, drilling contractors, well service companies, and other Buyers in the project also benefit from the Seller completing funding of the given project. They are usually cooperative in providing names of potential Buyers. Furthermore, they may also provide a referral for the Seller through a letter or phone call.

Professional associations provide the best source of industry contacts. Sellers miss an excellent opportunity if they do not become actively involved in both national and local professional associations. Activities such as serving on committees, serving in office, and publishing articles create credibility and visibility with potential Buyers. The best advertisement for geologists or their companies is a published article. Initial contact becomes simple if the potential Buyer has read a publication by the Seller or perhaps served on a committee with the Seller. Also, the follow-up analysis by the Buyer is more likely to be favorable if the Seller has name recognition. The Buyer's attitude becomes, "This Seller has given his or her time unselfishly to aid an association from which I have benefitted. I will give this Seller every consideration in evaluating this proposal." If two project evaluations by the Buyer appear similar and the Buyer must choose between the two, the Seller's name recognition may make the difference.

County records and published drilling reports provide public documentation of Buyers. Drilling reports provide the names of well operators and the type, depth, and location of prospects they are drilling. The operator of a well is only part of the story on potential Buyers. A search through the county records reveals assignments of interest received by other Buyers in a particular transaction. Thus, Sellers can develop a list of potential Buyers for a particular project based on assignments that Buyers have received on similar deals in the past.

One publication provides a listing of hundreds of Buyers that invest in various types of petroleum ventures. The listing is published annually, and it is called *Oil and Gas Finance Sourcebook* (1990). The Buyers are first divided according to types of funding, such as drilling risk capital, reserve purchasers, and downstream risk capital. In addition, nonindustry Buyers are categorized, such as insurance companies, venture capital firms, pension funds, and others. Each Buyer listing contains a brief description of company background, contact person, prior years of activity, and direct quotes from the Buyer concerning the type and terms of deals in which they are interested.

Scheduling appointments can be one of the most frustrating aspects of marketing. Fortunately, the petroleum industry does operate somewhat like a brotherhood (or sisterhood). Petroleum industry professionals generally want to look at proposals from fellow professionals. Referrals are more essential with nonindustry Buyers. The Seller's use of referrals can range from simply using the name of a reference source to a letter of introduction from the reference source.

One of the simplest techniques to aid in scheduling

appointments is the use of a facsimile machine (FAX). Playing telephone tag and repetition of the same telephone description of the prospect are frustrating, time consuming, and inefficient for the Seller. Instead, the Seller can send a one- or two-page description of the project to potential Buyers via a FAX machine. The Seller can then request that the Buyer respond with a phone call to set up an appointment.

During actual sales presentations, the Seller should try to fulfill the following objectives:

1. Ask questions of the Buyer to determine potential problems or objectives.
2. Try to answer all questions or objections.
3. Ask the Buyer to buy.
4. Ask for referrals from the Buyer.
5. Try to make presentations to the decision maker.

Closing a sale is one of the most highly publicized aspects of salesmanship. Some professional salespeople do a great job of scheduling appointments and make great presentations, but they become weak when it becomes time to close the sale. They are afraid of rejection, of "no." They take "no" personally. They try to sell themselves to the potential Buyer (as well they should), but when the Buyer rejects the proposal, the salesperson feels personally rejected. Perhaps the best sales training advice on closing comes from Walter Hailey (pers. comm., 1990). He teaches that closing is simple if the earlier part of the presentation is effective. After all of the potential Buyer's questions have been answered, simply ask the Buyer to buy. Elaborate closing techniques can be especially ineffective on professionals.

Perseverance is certainly the ultimate key to successful funding of oil and gas ventures. To maintain the right frame of mind, the the Seller must realize that the word "no" from the Buyer is not necessarily personal. It might mean "not this time, perhaps next time." Research on sales calls indicates 15 presentations are necessary to efficiently eliminate a Buyer from serious future consideration (Planned Marketing Associates, 1990). The research is not specific to oil and gas funding, but the same principles should apply. After a Seller has identified the top 20% of the potential Buyers that represent 80% of the buying activity, then the Seller's percentages should be as follows:

Percentage that Buy	Number of Presentations
20	4
50	8
70	11
90	15

However, 50% of salespeople give up on the Buyer after the first presentation and 80% quit after the fourth. They quit just as the pecentages are turning in their favor. From the previous table, we can see that it takes eight presentations before 50% of the most qualified Buyers will have bought. If a Buyer has not bought after 15 presentations, the law of diminishing returns indicates that the Seller should forget that Buyer.

CONCLUSIONS

External funding is as important to the success of an independent geologist as good technical work. Funding can be divided according to use, such as prospect acquisition, exploratory and development drilling, and production acquisition. Each type of funding has its own acceptable deal structure. In general, profit available to the Seller on the sale of a deal is less than actual costs. The Seller must make a profit through a successful project.

The different sources of funds can be divided according to knowledgeable versus unknowledgable and industry versus nonindustry. Each source has its own attributes and drawbacks. For example, knowledgeable investors are easier to locate for initial contact, but they buy a lower percentage of prospects.

Geologists should study the principles of good salesmanship. These principles apply not only to raising funding for petroleum ventures, but also to all types of relationships with other people. Geologists must bridge the gap between technical expertise and effective communication.

Successful people have several common traits. Geologists should study these traits, evaluate themselves, and try to improve in deficient areas. Different publications provide different emphasis on desirable traits, but after studying Andrew Carnegie and other successful people for 20 years, Napoleon Hill (1983) listed 13 steps to success. These steps include desire, faith, knowledge, imagination, organized planning, decision, persistence, and the power of a group of minds working together. The real key to success lies within the mind of the successful person.

References Cited

Green, A. T., 1989, Confidentiality agreement: Society of Independent Professional Earth Scientists Newsletter, Aug., p 4-5.

Hill, N., 1983, Think and grow rich: New York, Fawcett Crest Books, 254 p.

Planned Marketing Associates, 1990, Power of persuasion seminar, with audio tapes: P.O. Box 345, Hunt, Texas 78024.

Oil and Gas Finance Sourcebook, 1990, Denver, CO, Hart Publications, Inc., 444 p.

LEE T. BILLINGSLEY, President of Sandia Oil and Gas Corporation in San Antonio, Texas, received a B.S. degree in Geology from Texas A&M University in 1975, an M.S. in Geology from Colorado School of Mines in 1977, and a Ph.D. in Geology from Texas A&M in 1983. He worked for both large and small exploration companies before becoming an independent in 1983. Dr. Billingsley formed Sandia, a privately held exploration company, in 1985. The company's focus is generation of oil and gas prospects, with current emphasis on the Wilcox Formation in the onshore Gulf Coast of South Texas. Dr. Billingsley's professional activities include past President of the South Texas Geological Society, member of SIPES, former AAPG delegate, and current member of the AAPG Publications Committee.

Chapter 21

Putting a Deal Together

Albert H. Wadsworth, Jr.
Wadsworth Oil Company
Houston, Texas, U.S.A.

The most difficult thing beginning independents must face is how to hatch a chicken when they have no egg.
—Albert H. Wadsworth, Jr.

INTRODUCTION

This chapter is offered as a guide for the beginning independent explorationist who wishes to assemble a drilling proposal or a production purchase. It is not intended for a land department's use, as their methodology would reflect their own company policy.

The ways to assemble a drilling deal from concept to spudding a well can be, and usually are, as varied as the many types of prospects. But there is one overriding criterion common to all oil deals that must be honored. The design of the deal must seek the greatest economic benefit commensurate with the risk for those taking the risk. If this criterion is not met, the deal is probably not marketable. This is especially true in the early 1990s when tax laws discourage risk taking.

An independent explorationist should know what kind of geological ideas and what producing trends are most marketable. Although this chapter will not deal with selling the deal, it should be understood that the product must be made marketable or time and money will be wasted.

What does one look for in assembling a marketable deal?

1. One needs to have a geological concept that is understandable, reasonable, in a favorable area, and with supporting data attractively organized. It must have a professional appearance.
2. The ratio of anticipated return to risk must be large enough to justify the risk. No standard formula exists because deals vary with cost, geology, net revenue interest, accessibility, and depth.
3. Acreage covering the project must be distributed equitably among risk takers and those with less or no risk.

Different geological provinces offer a wide variety of plays. One must remember that the marketability of a type of play depends upon timing. What is "popular" at one time may not be later. The Austin Chalk play in South Texas is a case in point.

IDEA DEVELOPMENT

Before the independent explorationist can make a sale, a geological concept must be transformed into a marketable idea set down in black and white. Every available resource to prove the point must be utilized. A few places may remain where subsurface work alone can suffice to get a well drilled. More than likely, the client will require the subsurface to be supported by seismic data. Additional data such as soil gas geochemistry, gravity, magnetics, and radiometrics are all useful and can strengthen one's concepts. Ideas might be better illustrated by cross sections, isopach maps, and computer-generated isometric drawings. The latter are very useful in explaining stratigraphic trap occurrences.

One must remember that one's submittal is in competition with perhaps a dozen others on any given day. Those selected are the ones that convince the client that they have the best chance for success. Contrary to the belief of many geologists, the winners are not always those with the soundest geological concept.

PRESENTATION

The presentation of one's ideas can be as important as the ideas themselves. Remember that the decision makers are often nonscientists, and carefully conceived scientific research may be lost on them. Nonetheless, they must gain from you the feeling that you are competent and have carefully and scientifically arrived at your conclusions. Your presentation must instill faith.

To do this, you must analyze what your audience will understand and what will be lost. This can be compared to the task of a teacher in a one-room country school, who must gauge the lessons according to the side of the room being addressed. This can be done without compromising fact or talking down to a client. Keying a presentation to one's audience is not conciliatory, as the tenets are not changed. They are simply stated in a manner more readily understood by that particular audience.

Explorationists have their own sytle and manner of report preparation. Such individuality is desirable within bounds. Clarity and concise syntax are desirable qualities in a written text, but not if ideas are lost for the sake of brevity. Forced brevity can muddle your thesis just as effectively as verbosity. Spell out the facts and state your opinions in a logical manner.

In presenting a deal to a potential buyer, many independent explorationists place too much emphasis on how much the project will earn. Not enough attention is paid to the hazards or to the risk factors involved. Understandably, few of us wish to emphasize the negatives—we prefer accentuating the positive. But this can be risky today. In recent years, the courts have upheld numerous claims filed by investors who lost money in a dry hole. Never mind that it was an honest attempt on a reasonable and legitimate play. By placing too much emphasis on the positive elements, explorationists can leave themselves vulnerable to unscrupulous legal criticism. The writer tries to follow the long established philosophy that if one takes care of the downside, the upside will take care of itself.

ACREAGE CONTROL

Early in our careers, we learn that whoever controls the acreage on a prospect controls that prospect. Therefore, acreage control becomes the basic determinant as to how a deal can be structured.

There are three conditions of acreage control: (1) the explorationist controls the acreage, (2) others control the acreage, or (3) the acreage is unleased. Each presents its own set of conditions that determine how a deal can be assembled.

In-House Control

The most ideal acreage situation is for the explorationist to own commercial leases covering the project. This allows the deal to be structured in a more favorable manner. Several options are available:

1. Farm-out. Leases can be farmed out at no expense or risk to the explorationist. A wide choice of participations can be arranged, such as overriding royalty, carried working interest, back-in for a larger share after payout, net profits, checkerboarded acreage, and so on.
2. Sale of the acreage for cash profit plus development and an override.
3. Syndicate with industry partners and drill.
4. Syndicate with nonindustry partners and drill.
5. Incorporate into a public offering.
6. Drill it yourself straight up.

Owning the acreage is by far the better way for putting a deal together, but it requires funds for lease bonuses that few independents have.

Others Control the Acreage

In former years, some companies bought leases intending to farm them out. With the increase in lease bonus costs as well as higher landowner royalty, this practice is less common now. Most companies buy leases today intending to drill. This leaves the independent with fewer and more costly choices:

1. Farm-in. Independent explorationists can farm-in acreage from the leasehold owner in return for development obligations. This means they must have or know where to get drilling money because the farm-in agreement has a time limit before operations must begin. If the well is successful, they will earn some previously stipulated interest in the acreage. A dry hole will usually earn nothing.
2. In some cases, acreage can be acquired in a checkerboard arrangement by the drilling of a successful well. This has the advantage of allowing the drilling parties to control their operations on their earned share of the lease block.
3. Partial farm-in. The lease owner farms out an agreed upon area around the drill site, but retains title to the remainder. This reduces the potential recovery to the risk takers.
4. Back-in. It is common today to grant the drilling party an interest in the acreage block with the grantor (leasehold owner) "backing in" for a larger interest after certain dollar amounts have been recovered from production by the drilling

party. Again, this reduces the risk takers' recovery.

5. Lease repurchase. Sometimes a company might have a reason to sell their lease block for cash, an override, or both. This is not common, but it has happened due to a change in management attitude, a dry hole, or some other internal reason.

The Land Is Unleased

Today many oil companies are putting most of their exploration resources into offshore or foreign plays. The result is that more acreage is available onshore for the independent who lacks the huge capital backing to go foreign or into offshore provinces. With the oil depression of the 1980s continuing into the 1990s, competition for acreage control is greatly reduced. Thus, the independent who develops a capital source has a rare opportunity.

1. The most desirable procedure is to acquire funds and purchase commercial leases from the landowners. This can often be done by seeking funds for lease purchase from a trustworthy partner by offering a participation in the deal.
2. Some companies use the exploration skills of idea-generating explorationists and compensate them only when and if they buy the idea. Many explorationists use this approach, but it can be risky. Few companies will actually "run around" the idea generator intending to steal from him or her, but it has happened. More likely, the idea might be seen and stolen by someone not connected with the company. In either case, the original idea generator has lost. To avoid this hazard, one should control the acreage or seek some kind of documentation that would protect the idea from usurption. The author strongly recommends using a confidentiality agreement with each submittal.
3. If the generating geologist cannot secure lease bonus funding and feels too vulnerable to use procedure 2 above, he or might consider "laying his cards on the table" with the landowner. At that stage, he probably has nothing to lose, and he may be surprised by the cooperation that the landowner will offer. The writer has used this technique, and it has worked more often than it has failed.

Within the framework just described, there are any number of combinations that can be used in assembling a deal.

PRACTICAL PHILOSOPHY

Few if any schools teach students how to earn a living as a geologist or geophysicist. It has always been assumed that once that title is earned, someone out there would take care of a skilled scientist. But for whatever reason, numerous qualified professionals have been thrown out of work with no knowledge or experience in how to practice geology. Printing stationery and hanging out a shingle does not constitute a viable business. One must be aggressive and call up creativity as well as science. Deal making is a product of those three talents: creativity, aggressiveness, and scientific knowledge. Other authors in this Handbook offer advice on ethics and integrity. These factors are so critical to the making of a deal that they deserve emphasis here. Most explorationists who enter the field are surprised at the high levels of integrity they find among independents. In the writer's opinion, one's integrity is more valuable to an independent career than one's scientific knowledge. Never compromise your integrity.

The most difficult thing beginning independents must face is how to hatch a chicken when they have no egg. A few might have a family fortune or a stable of voluntary investors anxious to pick their brains, but these are exceptions. In any business, the start-up money is at the highest risk and is the most difficult to obtain. As a rule, the independent works without pay while locating a drillable prospect. Acreage control and drilling funds then become his chicken to be obtained. The strength of the idea and his or her ability to sell it become the egg that will hatch the chicken. With the exception of a short period in the late 1970s, funds for this stage have always been limited, and competition for them is keen. This is where presentation becomes a factor. The presentation of the idea includes not only report preparation but the personality, demeanor, and attitude of the presenter. With some clients, the dress, manners, and tastes of the independent bear upon one's acceptability as a business venture partner. In all cases, the integrity of the individual overshadows all other considerations. In short, these and other subtleties all bear upon the delicate business of hatching a chicken without an egg.

An untested hydrocarbon prospect is a new concept that must be financed before it will produce. No new concept in business or science ever achieved acceptance without someone "selling" that idea. Thus, independent explorationists must include the ability to sell their ideas as part of the structure of putting a deal together (see Chapter 20). That ability will probably outrank scientific knowledge in determining the degree of career success in terms of earnings.

DEAL STRUCTURING

The beginner would be wise to offer deals through a confidentiality agreement describing the area, fees, and amount of overriding royalty. The agreement should run for a stated period of time and make the submittal subject to prior sale unless an exclusive right is offered. Unless wealthy, the beginner should avoid working interests. All agreements should be checked by an attorney familiar with security regulations as well as oil and gas law and taxation.

I have discussed the numerous variables that affect how an oil deal can be structured. Unfortunately, there are so many variables that it is impossible to set out standards of form that will satisfy all conditions. An attempt by the author to do this resulted in an impractical redundancy of anastomosing possibilities that was rightly discarded.

CONCLUSIONS

Whoever controls the acreage over a given petroleum prospect controls how, when, and if that prospect will be developed. I have offered guidelines covering how an independent explorationist might gain that control long enough to develop the prospect. Sometimes it simply is not possible, and one must sit back and wait. All of us have had to do this, and all of us can recite tales about the one that got away. That is the price we pay for independence.

This chapter began with comments about the marketability of a deal. In conclusion, it must be noted that the independent's success will be enhanced by his or her ability to judge what is saleable, the timing of its presentation, and how to sell the concept. Suffice it to say that the explorationist must somehow survive until he or she gains enough experience to learn "what the traffic will bear."

ALBERT H. WADSWORTH, JR., an independent geologist and oil producer in Houston, received B.S. and M.A. degrees in Geology from the University of Texas in 1941. After service with the U.S. Geological Survey during World War II, he returned to Texas and entered the oil business, moving to Houston in 1956. He has been a member of AAPG since 1941 and is a Certified Petroleum Geologist. He is a Founder of the Society of Independent Professional Earth Scientists and served as Vice-President in 1988–1989. He is currently President of the Association of Petroleum Geochemical Explorationists.

Chapter 22

Enhancement Techniques for Prospect Presentations

Robert F. Ehinger
Consultant
Tulsa, Oklahoma, U.S.A.

A good writer should make every word count.
—A. B. Guthrie, Jr.

INTRODUCTION

Approximately 170 prospects were critiqued for investment considerations and constitute the data base for this study. The prospects were initially evaluated on three criteria: economics, documentation of the trap, and reserve and reservoir parameters. During the study, it became apparent that many prospect generators could improve their prospect appeal by including certain essential data. These include a written geological synopsis of the prospect, an "analogy" field or well, documentation of production data, better conceived maps and cross sections, a current acreage map, and a valid authorization for expenditure (AFE). Additional items that would enhance the saleability of the prospect include regional maps that show the relationship between the specific prospect and its regional setting, block diagrams to illustrate pertinent geological features, applicable published articles, copies of field and regulatory rules, a topographic map of the prospect area, and a vita of the prospect generator.

The prospect generator should be prepared to answer questions on operations, type of drilling contract and AFE costs, casing point election, any unusual completion problems, and the current status of offset acreage. Questions that are specific to oil prospects include expected initial production rates, amount of casinghead gas, oil gravity, oil to water ratios, and the type and location of water disposal facilities. Typical questions on gas prospects include gas BTU, current price in the area, location of pipelines and their line pressures, type of gas contract expected, and the estimated lag time from completion to actual sales.

Data were compiled on the reasons for prospect rejections to establish some general trends. Prospects were rejected because of high geological risk (36.4%), deal structuring (16.9%), didn't meet required economics (16.2%), mismatch to company objectives (8.1%), and other various reasons (22.3%). One surprising observation was that in only 5.4% of the cases were prospects rejected because of differences in the geological interpretation.

Prospect generators can significantly enhance the appeal of their prospects by the inclusion of better documentation of the items just described. Carefully selected cross sections can provide much insight into the geological controls and reservoir parameters of the prospect. However, such cross sections seem to be the one ingredient that is often neglected.

This chapter evolved as the result of a consulting position in which the author's primary duties were to evaluate outside generated prospects for their economic and geological merit. During the course of this position, it became apparent that a significant number of geologists were encountering difficulty in selling their prospects because of a lack of documentation, or because the essential geological controls were not defined to the investing company. By its nature, a study of this type is highly subjective and may contain some regional bias. This chapter is an attempt to share a number of observations that may be of benefit to other prospect generators.

CRITERIA FOR EVALUATING PROSPECTS

Most companies that invest in outside generated prospects initially "screen" those prospects according to three common criteria. These are listed in order of priority:

1. Economic evaluation
2. Documentation of the trap
3. Reservoir and reserve parameters

With regard to the first criterion, most companies screen prospects based on whether the prospect as presented will meet their in-house economic guidelines. One effective way to focus their attention on your prospect is to make a summary statement at the onset regarding the prospect objectives, depth, expected reserves, AFE costs, and some economic criteria of the prospect such as a simple return on investment (ROI). The second criterion is best shown by the cross sections and maps, especially by the changes in the mapped parameters (structure, sand thickness, etc.) from the center of the trend to the "edge wells." The third screening criterion relates to the quality of the reservoir and its expected reserves. These items can be documented by the inclusion of porosity cross sections, production data (P/Z plots, decline curves, etc.), and key logs as part of the prospect package.

PROSPECT ESSENTIALS

The following items should be included in a quality, saleable prospect package. This list of items contains the *minimum* items that were included in a number of successful prospect packages. The widespread and diverse types of hydrocarbon traps may necessitate some regional modifications to the following list.

1. A well-written geological synopsis of the prospect.
2. An "analogy" well, field, or trend used as a model for the prospect.
3. A production zone map showing productive units, cumulative production, current producing rates, and expected ultimate recoveries.
4. Geological and geophysical maps and seismic lines.
5. Adequate cross sections—both parallel and perpendicular to the expected trap.
6. Prospect terms ("a deal sheet").
7. Documented production data, P/Z plots, monthly oil, and casing head.
8. A current land and lease plat.
9. An AFE or valid drilling and completion cost estimate.
10. A disclosure and disclaimer statement.

Because this list is important in getting your prospect sold, a brief discussion of each item is in order.

1. The geological synopsis should include a succinct summary of the prospect condensed into one or two pages. It should cover the major selling points of the prospect and address any obvious questions on key wells such as botched completion attempts. No matter how skillful you are in presenting your prospect to an outside company, your story is usually retold to senior management or internal investors, so it is essential that the person presenting your prospect story have all the facts at hand.

2. The analogy well, field, or trend is a graphic example of the exploration target that is the primary objective of your prospect. For a simple prospect that is predicated on the projection of a productive trend along strike, a simple cross section showing previously discovered wells will suffice. For more complex combination traps, it may be necessary to extend the search to several townships until you have found a similar field with the same combination of trapping elements as the one proposed for your prospect.

3. The production zone map should have symbol codes or color codes for the wells showing the productive zone(s). It should also contain current data on cumulative production, rates of production, and some estimates on ultimate production. Depending on the type of prospect, the map could contain the date of initial production, the initial potentials, and pertinent pressure data.

4. The geological and geophysical maps, along with the cross sections, are the most important elements of the prospect and should graphically show the trap to best advantage. Except for regional or locator maps, the major maps in the prospect should all be on the same scale, individually labeled, and include the name or initials of the geologist and a date. Only one or two mapped parameters should be shown on each individual map. If it essential to the prospect to show three or more trends, then the use of map overlays should be strongly considered. Seismic lines should be clearly marked as to the line designation along with a compass direction. Because the prospect evaluator may not be able to easily identify the stratigraphic units being shown on the seismic data, their stratigraphic nomenclature should be printed on the lines as well.

5. The cross sections should show the correlations and the changes in the mapped parameters out to the limits of the trap. Here, the geologist has to make some important decisions on the types of logs to be used, the scale, and the selection of a specific datum above or

below the zone of interest. The closer the cross sections depict the true geological setting, the more effective they become in selling the prospect. If key wells are not included on the cross sections, they should be included in the prospect packet.

6. The deal sheet constitutes a legally binding agreement, so it is essential that both parties clearly understand the terms and their obligations. Nothing sours a deal more quickly than when the investing group realizes that the prospect terms are different from their understanding at the time the prospect was presented.

7. The prospect prospectus should contain copies of the production data for the convenience of the prospect evaluator. This facilitates verification of the reserves that were stated as the objective of your prospect. Documentation of reserve data is essential when showing out-of-state prospects because the investing company may lack easy access to those records. If the investing company cannot readily locate the necessary production data needed for evaluation, they will usually reject the prospect and concentrate on others that do contain the necessary documentation.

8. Land and lease plats should be an integral part of the prospect package because they provide answers for a number of questions on land tenure. Other supporting land ownership data should also be made available if the company has a serious interest in acquiring your prospect. These data save the investing company a considerable amount of time, and they may see an "entry" into the land acquisition that is not known to the prospect generator.

9. A current AFE or valid drilling and completion cost estimate should accompany your prospect presentation. It is very important that the figures are accurate because most companies use your figures in their initial prospect evaluations. If your well costs are too high, you may, in effect, condemn your own prospect before it is drilled!

10. The disclosure and disclaimer statement should be an essential part of the prospect package when presenting a prospect to an unfamiliar group or company. Once a company agrees to buy your prospect, you have very little control over it after that point. The investing group may sell off parts of the prospect to internal investors or to other companies that may not be as knowledgeable about the inherent risks in the oil business. This statement is for your future protection should an unseen investor decide to bring legal action.

Although one might expect that the vast majority of these items would be included in every prospect package, examination of 170 prospects indicates otherwise. With regard to geological cross sections, approximately one-third of the submitted prospects had no cross sections at all or only had one cross section that showed only a few of the productive wells. A similar observation was made on the geological synopsis—approximately one-third of the prospects contained no geological summary.

PROSPECT ENHANCEMENT ITEMS

Again, using the same sample base (170 prospects), some prospect generators used a variety of techniques to enhance the appeal of their prospect and to educate the prospect evaluators. These included the following:

1. Maps showing the relationship of the specific prospect to the regional mapping.
2. Map overlays, block diagrams, and other graphics to provide a better three-dimensional perspective of the geological controls.
3. Pertinent published articles.
4. Copies of scout tickets, important offset logs, mud logs, and completion reports from previous exploration attempts in the prospect area.
5. Copies of field rules and spacing orders to resolve any questions that might arise.
6. A topographic map of the area where access or site preparation might be an issue.
7. A brief vita of the prospect generator (useful when showing a prospect to an unfamiliar company).

Again, some comments are in order regarding the items in this list.

1. Because a number of geologists work a specific formation over a broad area, the relationship of the regional geology to the specifics of the individual prospect can enhance the credibility of the prospect and put the geological risk factors into perspective. As a prospect generator myself, I realize the dilemma of "How much geology do I give away to sell a single prospect?" One solution to this dilemma is to show the regional maps as part of the prospect presentation, but put them in the same category as seismic records—no copies are allowed and they cannot be left with the investing group.

2. It is important to remember that some investing companies are not as comfortable with geological concepts when shown on two-dimensional maps, so any type of lucid illustrations can help to sell your prospect idea to that group.

3. Pertinent published articles that relate directly to your prospect can be an effective "selling" addition to the prospect package. Such articles can describe similar fields, document the depositional environment of the reservoir rock, or describe the geological history of the area that relates to trap development.

4. By providing the supporting documentation, it

becomes easier for the investing company to analyze your prospect and to convince themselves of the proposed interpretation.

5. Inclusion of copies of the field rules and spacing orders related to the prospect area can easily resolve many questions. The inclusion of such data will also impress the prospect reviewer with the thoroughness of your work.

6. A photocopy of the topographic map that covers the prospect area is a useful addition to the prospect packet. The map can allay investor fears about access, site preparation, and cultural features. A useful corollary would be the inclusion of a recent aerial photograph.

7. When a company buys an outside prospect, they expect the correlations and mapping parameters to be correct. They are relying on the ability of the prospect generator. If they are unfamiliar with your background, a brief vita included as a one-paragraph summary can be a useful addition. Certification as a Petroleum Geologist by AAPG or a similar organization can also be an asset, as it enables the outside company to familiarize themselves with your background by use of the biographical sketch provided in the directory.

It is important to remember that companies that rely primarily on outside generated prospects usually look at many prospects, so clarity and completeness of data will go a long way in putting your prospect ahead of the crowd. For companies that generate prospects and also invest in outside prospects, the outside prospects have to be more appealing than prospects they have in-house. In the case of a tie, the in-house prospect always wins. In today's environment, most companies will not remap or rework your prospect as they will have others of equal quality that contain all the necessary documentation. When showing your prospect to an investing company, you only get one chance to make a favorable impression, so adequate preparation is essential.

GENERAL PROSPECT QUESTIONS

As part of the prospect presentation, the investing company usually asks a number of questions. By being better prepared to answer these questions, you have enhanced your chances of selling them your prospect. Typical questions might include

1. Who will be the operator?
2. What is the percentage of the prospect being sold? Are there other participants?
3. What are the AFE costs? Do you have a drilling contractor? What is the type of contract? Day rate? Turnkey? Footage? What is the casing program? Is an intermediate string required? What is the size and type of production casing required?
4. How is the completion decision made? On the basis of logs only? Or logs plus DST?
5. Will the completion funds be escrowed?
6. Are there any unusual completion problems?
7. Are there any unusual lease terms? Are there any depth restrictions?
8. What is the current status of the offset acreage?
9. Are there any environmental restrictions or potential hazards?

Again, let's discuss some of these issues and questions.

1. In the present environment, the question of operatorship can make or break your prospect. It will greatly enhance the appeal of your prospect if you can sell a major portion to a company that has the reputation of being a prudent operator. In fact, that is probably the first group of companies to contact when you begin to show your prospect.

2. Investing companies are always more comfortable if the group selling the prospect will have some of their own funds at risk. The participation by other nonoperator companies can help the saleability of your prospect if those companies have a good reputation for being prudent investors. Conversely, the participation by nonprudent investors or ones with weak financial stability can hurt your chances for participation by other investing groups.

3. The questions on the AFE costs and the method of drilling the well can greatly influence the investment decision by outside groups. These companies will have more confidence in the project when reputable contractors have been selected and the AFE is acceptable for the area and depth. Attempts to hide an extra promotional fee in the AFE estimates will make the whole prospect suspect to outside investing groups.

4. The casing point questions are ones that the prospect generator should consider in advance regarding the relevance of additional testing. Here, a study of previous completion practices by prudent operators in the area can provide acceptable answers.

5. The escrowing of completion funds is a recent requirement arising from some companies inability to pay their share of the completion costs. A provision for the escrowing of such funds will actually help in selling a fractional interest in the prospect to several different groups.

6. If the prospect generator is aware of previous successes and failures in the area with respect to completion techniques, they should be able to answer any questions on the subject and impress the investing companies with the thoroughness of the prospect study.

7. In regards to lease terms, it is useful to have copies of the leases with you to be prepared to answer any unexpected questions.

Questions on Oil Prospects

Because oil and gas prospects differ considerably, some typical questions are usually asked that are specific to oil prospects.

1. What are the expected initial production rates?
2. Initially, will it be a flowing or pumping well?
3. Will it produce any casinghead gas? What are the expected rates?
4. What is the gravity of the oil? Any paraffin problems?
5. Is the well expected to make any water? If so, how much per barrel of oil? What are the type and location of water disposal facilities? What are the disposal costs per barrel of water?
6. What are the spacing rules? What are the well allowables? Are there any offset locations?

Most of these questions are familiar to the prospect generator and need no comment. However, the questions on water disposal can make or break your prospect economics if no disposal wells exist nearby and the water has to be trucked off location.

Questions on Gas Prospects

The following are some typical questions that are often asked on gas prospects:

1. What are the expected initial production rates?
2. What is the BTU content of the gas? Are there any associated liquids?
3. What is the current price per MMBTU in the area? What is the rate of "takes"?
4. Are there pipelines in the area? What are the line pressures?
5. What is the type of gas contract expected?
6. Can you estimate the lag time from completion to actual sales?
7. Is treatment or compression required?
8. Has the gas been dedicated previously?
9. What are the spacing rules? What are the allowables? Are there any offset opportunities?

Again, most of these questions are straightforward and need no additional amplification. However, with regard to the BTU content of the gas, upward BTU adjustment can significantly improve the overall economics in the currently depressed gas market. One observation here is that many prospect generators fail to document the associated liquids. Even a few thousand barrels of oil can significantly help the ROI, especially when dealing with wells in the 5000–7000 ft depth range.

Table 1. Reasons for Prospect Rejections

Reason for Rejection	Percentage
High geological risk	36.4
Deal structuring	16.9
Won't make required economics	16.2
Mismatch to investor objectives	8.1
Concern about offset depletion	5.4
Lack of geological control	5.4
Disagree with geological interpretation	5.4
Drilling and/or completion problems	2.7
Did not appeal to management	1.4
Lacked geophysical confirmation	1.3
Imminent lease expirations	0.7
Total	99.9%

REASONS FOR PROSPECT REJECTIONS

Up to this point, this chapter has emphasized positive items that will help the prospect generator do a better job of selling a prospect. However, to get a balanced view, one also has to examine the reasons why prospects are rejected by an investing company. In critiquing approximately 170 prospects, the author kept a one-line summary of each prospect and noted the one major reason for rejection for each. Of the 170 prospects, approximately 20 prospects were recommended to management for investment. Table 1 summarizes the reasons for rejection for approximately 150 prospects. This tabulation is not intended to represent a rigorous statistical analysis but rather to show some general trends as a learning device.

Because of the importance of the list in Table 1, some comments are in order. As the data indicate, the largest single reason for prospect rejection was the investing company's perception of the geological risk. The decision to reject certain prospects may have been reversed with a better prospect presentation, a demonstration of the relationship of the prospect to the regional geology, and better conceived cross sections and geological maps.

The next two categories in Table 1 ("deal structuring" and "won't make required economics") are interrelated, but I tried to distinguish between the two. Usually, on prospects rejected for deal structuring, the terms were not competitive with the current market or contained some hidden promotional costs in the AFE or in the acreage fees. Prospects rejected because of economics did not meet the investing company's objectives of a simple 4:1 ROI using $13–16 per bbl of oil and $1.20–1.40 per MCF of gas based on a most likely case for reserves and at the terms as stated on the deal sheet. No price escalations were used in the calculations, but to offset this, no monthly operating costs were included except in a few special cases. Many of the prospects fell into the 3:1 ROI category, so with higher product prices, they would have passed the ROI screening test. Again, it should be emphasized that this study was subjective and represented a single investing group at one point in time during a difficult period in the industry.

The next item in the table—mismatch to investor objectives—relates to the depth of the wells, geographic areas, the amount of front-in investment required, and length of payout. Here, the prospect generator should determine in advance the investing company's current objectives. If there is too much of a mismatch to the specific prospect, the prospect generator should concentrate efforts toward companies for which the match is better.

When working in mature provinces, the concern over previous depletion is an obvious problem that has to be addressed when generating and selling prospects. If the prospect generator has prepared an adequate explanation for this question before showing the prospect, it will help in getting the prospect sold.

One surprising item is that only 5.4% of the prospects were rejected because of differences in the geological interpretation. Based on the data that were presented during the prospect presentations, I would have mapped the trends in the same way in the majority of cases.

CONCLUSIONS

The following items seemed to have the most significance in the successful selling of a prospect to an outside company:

1. A well-written prospect synopsis.
2. Presenting an overview of the regional geology and its relationship to the specific prospect.
3. Inclusion of an analogous well or field as part of the prospect packet.
4. Cross sections that define the limits of the trap or of the mapped parameters.
5. Documentation of any hydrocarbon shows, DSTs, or production tests.
6. Hard copy documentation of prior production data.
7. The selection of a prudent operator.
8. Vita of the prospect generator.

If your prospect is rejected by a company, a follow-up phone call may provide some insight for necessary changes that will help the prospect generator sell the same prospect to another company.

Acknowledgments *I would like to express my appreciation to Farmers Energy Corporation for the opportunity to collect the necessary database and for their input during the conception of this paper.*

ROBERT F. EHINGER, a consulting geologist in Tulsa, Oklahoma, received a B.S. degree in Chemistry and Geology from the University of Alabama, an M.S. in Geology from the same institution, and a Ph.D. in Geology from the University of Montana. He worked for 10 years in lead-zinc exploration in the southeastern United States, was a Senior Geologist with Cominco American Inc., and a Staff Geologist for Cotton Petroleum. Dr. Ehinger's work experience has included prospect generation and evaluation in the Mid-Continent area, exploration in deformed rocks, and as a technical witness before oil and gas regulatory commissions. He is an AAPG Certified Petroleum Geologist, past President of the Tulsa Geological Society, and Vice-President and Director of the Oklahoma Well Log Library..

PART V

LEGAL, POLITICAL, ETHICAL, AND ENVIRONMENTAL ASPECTS OF THE BUSINESS

There are six chapters in this part. They cover oil and gas law in the United States, international petroleum contracts, analysis of international contracts, international exploration by independents, ethics in the exploration business, and environmental realities of petroleum exploration. Chapter 23 on U.S. oil and gas law explains some legal concepts, lease terms, and clauses, while Chapter 24 on international contracts gives background for doing business overseas. The next chapter outlines evaluation of international contracts, which is far more difficult than evaluating domestic contracts. Chapter 26 on international exploration by independents is written in a matter-of-fact style and is a practical follow-up to the previous chapters. The last two chapters give closure to this Handbook. Chapter 27 on ethics presents examples of situations that can happen and some ideas about courses of action to take. Chapter 28 on environmental realities recommends a set of principles for explorationists as a challenge and an obligation. A synopsis of each chapter follows.

Chapter 23, "Basic Oil and Gas Law in the United States," by Samuel B. Katz, is intended to give the working professional a feeling for the various legal problems and situations that may be encountered during the exploration for and production of oil and gas. It is not intended to take the place of legal counsel. Some of the major topics covered include (1) the oil and gas mineral estate, (2) trespass and adverse possession of the mineral estate, (3) divided ownership problems, and (4) the oil and gas lease.

Chapter 24, "Types of International Petroleum Contracts: Their History and Development," by Samuel B. Katz, portrays the various types of international contracts, including how and why they developed, and gives the explorationist an idea of why one form of contract may be preferred over another. It is not an exhaustive review, and interested readers should review the references cited for more details. Much of this chapter focuses on the formation and development of national oil companies, and their role in the evolution of concessions and contracts.

Chapter 25, "Analysis of International Oil and Gas Contracts," by Sidney Moran, aims at familiarizing petroleum geologists with general methods of analyzing contracts to assess whether the field qualities indicated by technical studies would produce acceptable economic results. Major topics covered include (1) the main types of international contracts, (2) a comparison of contract qualities and analyzing profitability of petroleum field models, (3) assessing tradeoffs among various terms of a contract, (4) U.S. taxation of international petroleum extraction income, and (5) special considerations that apply to the international arena. Figure 1 shows a particularly useful way to display an array of profitability data.

Chapter 26, "International Exploration by Independents," by Robert G. Bertagne, covers the nuts and bolts of living and doing business in a foreign environment and includes a detailed representative outline of a foreign project. The author thinks the foreign game is accessible to smaller independents and can be very rewarding if they can meet the following requirements: (1) a successful domestic track record, (2) constant profit over the last 3 to 5 years, (3) ability to commit $1 to $2 million per year for at least 3 years, (4) new play concepts in area of interest, (5) some staff with overseas operational experience, and (6) an open attitude toward a new technology, environment, and culture.

Chapter 27, "Ethics in the Business of Petroleum Exploration," by Robert W. Spoelhof, emphasizes the importance of high standards of conduct to individuals and companies. It considers behavioral practices in the exploration business, which are not addressed in formal law but are generally accepted standards of ethical behavior. The author states that petroleum exploration professionals, who have expertise in a particular body of knowledge, have an ethical responsibility to that body of knowledge. He lists 18 guidelines for ethical behavior. Codes of Ethics or Conduct for AAPG sister organizations are appended.

Chapter 28, "The Environmental Realities of Petroleum Exploration," by G. Rogge Marsh, reviews those historical events that caused the growth of environmental awareness. He discusses the impact of petroleum exploration on the environment and the impact of environmental concerns on exploration. The author concludes that environmental issues will continue to be a fact of life and recommends a set of principles as an action plan for a reasonable future. Explorationists should (1) tread as lightly as possible on the land, (2) clean up their own messes, (3) subject all environmental claims to close scrutiny and a test of reasonableness, (4) educate their regulators, (5) educate their legislators, and (6) educate the public.

Chapter 23

Basic Oil and Gas Law in the United States[1]

Samuel B. Katz
Attorney at Law
San Antonio, Texas, U.S.A.

You can't get the grease without a lease.
—Anonymous

THE LEGAL CONCEPT OF OIL, GAS, AND MINERALS

Since time immemorial, humans have recognized that land is valuable not only for agriculture and farming, but in addition, for the extraction of substances found lying in or under the surface of the land itself. To convey these substances, and by *convey* I mean, passing the legal ownership or "rights" to these substances from one individual to another, the generic term *mineral* has been applied. After the invention of the internal combustion engine and also because of their lubricating propensities, oil and gas began to be extracted in ever increasing amounts. Cases began to show up in the courts concerning the question of whether "oil and gas" were to be included in the definition of the term *mineral* when oil and gas were not specifically mentioned, in the deeds or leases, by the signatory parties (Hemingway, 1983).

The term *minerals* had traditionally been used in legal instruments as referring to something other than the soil or the land itself. When minerals were conveyed in a document, this created a *mineral estate* which may have been conveyed or reserved. The remaining rights, or the actual ownership of the surface of the land itself, has been referred to as *surface rights.* Individuals may, in some jurisdictions, have the legal right to the ownership of both the "surface estate" and the "mineral estate." In many instances, these rights are "owned" by different individuals, and the mineral estate may be considered to have been "severed" from the surface estate (Hemingway, 1983).

In determining whether the parties to a lease or deed intended to include oil and gas as part of the term *mineral,* if oil and gas are not expressly or specifically mentioned in the conveyance, the courts have taken three approaches:

1. That petroleum products were not included.
2. That the parties to the deed used the term *minerals* as is commonly understood, in the vicinity, at the time of the conveyance or reservation. These courts allowed the free use of parol evidence (meaning evidence not contained within the document itself) to determine such understanding.
3. That the term *minerals* included oil, gas, and petroleum products, with the court restricting the search for an contrary intent, to the conveyance or reservation of these items to the document itself unless the document was found to be ambiguous (Hemingway, 1983).

The early concept granting the rights of the mineral owner to the oil or gas underground was likened to the law dealing with wild animals. The rationale was that oil and gas was of a "fugitive" nature. If the mineral owner could "capture" the oil and gas, as one would capture a wild animal, the mineral owner could "keep" it. At that time, knowledge of the occurrence and

[1]This chapter is by no means to be construed as an exhaustive review of the totality of oil and gas law in the United States. Rather, it is an attempt by the author to give the working professional a "feeling" for the various legal problems and situations that may be encountered during the exploration for and production of oil and gas. This chapter is not intended to take the place of legal counsel. Furthermore, the author and AAPG accept no responsibility for the misuse of this chapter. Lastly, as the law is ever changing, the author and AAPG take no responsibility for any misrepresentations that may occur.

nature of oil and gas in reservoirs was an inexact science. Therefore, courts held that the mineral owner had title only to those products that were actually reduced to his or her possession (Weaver, 1989). As the science of geology and reservoir mechanics matured, these early analogies have been rejected. There are two clearly diverse concepts regarding the right to the oil and gas in the ground which have developed from these early cases.

The first theory is called the *ownership in place* or *absolute ownership concept*. Jurisdictions, such as Alabama, Arkansas, Colorado, Kansas, Maryland, Michigan, Mississippi, Montana, Ohio, Pennsylvania, Tennessee, Texas, and West Virginia follow the ownership in place or absolute ownership concept (Hemingway, 1983). These states view this ownership as being essentially similar to the ownership of hard minerals. In such jurisdictions, a corporeal (meaning tangible) estate in real property may be created in the oil and gas separate from the rest of the land. In these states, therefore, an owner can sever the mineral estate from the surface estate and convey the mineral estate to an outside interest.

Those states that follow the *nonownership* theory of oil and gas generally recognize that there is no ownership of oil and gas while they are in the ground. What does exist is only the right to search for and reduce this oil and gas to possession. These states do, however, recognize that a landowner has a sufficient interest in the oil and gas to allow him or her to have standing (meaning a right to sue in court) to prevent injurious or wasteful operations to the reservoir. This concept, known as *correlative rights,* is discussed in the following section. In the jurisdictions that follow the nonownership theory of oil and gas, a grant, or reservation, of the oil and gas subjects the land to the "burden" of an outstanding right to enter and remove the oil and gas. There is no creation, severance, or removal of a corporeal interest, that is, mineral estate from the rest of the land. This right to search for and remove oil and gas from the land is termed a *profit a prendre*. This proprietary right has also been called a *license, servitude, real right,* and *chattel real* (Hemingway, 1983).

RULE OF CAPTURE AND CORRELATIVE RIGHTS

The concept of ownership of oil and gas, particularly in those jurisdictions following the ownership in place theory, is contrary to the *rule of capture*. The rule of capture is a rule of *nonliability* for the drainage of oil and gas from other tracts (Hemingway, 1983). Simply put, it means that the owner of a tract of land that overlies an oil and gas reservoir has the right to capture and keep all of the oil and gas brought to the surface. This is true regardless of the fact that much of this oil and gas may come from adjacent underlying tracts. The consequences of the rule of capture are that it (1) encourages prolific drilling and rapid production, (2) encourages physical and economic waste, and (3) encourages state regulation to curtail prolific drilling, rapid production, and physical and economic waste (Weaver, 1989).

A limitation to the rule of capture is the concept of *correlative rights*. This simply means that while the owner of a tract of land has the legal privilege to remove as much of the oil and gas as he or she can produce, this owner also has the legal duty to the owners of adjacent and nearby tracts to not exercise this privilege in a manner that will "injure" the common source of supply. Surrounding landowners have the right to sue and be protected from the negligent operations of their neighbors (Texas Supreme Court, 1948). In the case of *Eliff v. Texon Drilling Co.*, the operator used a drilling mud of insufficient weight. This negligence caused the well to "blowout." The blowout caused gas to be drained from the adjacent property. This gas was therefore "wasted." The Texas Supreme Court held that the adjoining landowners could recover damages for the waste caused by the operator's negligence. The measure of damages (the amount the court determined the landowner could recover) was determined by calculating the value of the lost gas that had been in place under the surrounding tracts. Therefore, while the adjacent landowners owned the gas in place under their tracts (since Texas follows the absolute ownership theory of oil and gas), it can be captured by another and used. However, as this gas was wasted (burned) by the operator's negligence, the court found that the rule of capture did not insulate the operator from liability for his negligence.

Today, most "relative rights" situations and problems are taken up by state regulation. For example, in Texas, a landowner may bring suit challenging a particular order of the Railroad Commission because he or she believes the order fails to provide him or her with a "fair share" of the oil and gas in a reservoir. In other words, the order fails to protect their relative rights (Weaver, 1989).

OIL AND GAS MINERAL ESTATE

Depending on the particular jurisdiction, an owner of a mineral estate will either have a possessory corporeal estate in the minerals (absolute ownership theory) or the right to use and occupy the surface of the land for the exploration and production of the oil and gas (the nonownership theory) (Hemingway, 1983). In either jurisdiction, however, the owner of the mineral rights may drill their own well in the hopes of finding producible oil and gas. Due to the high costs of exploration and development, however, mineral owners ususally convey their exploration and development rights to another by use of an instrument with a defeasable term. *Defeasable* means that the instrument (contract) is subject to revocation upon the occurrence or nonoccurrence of certain conditions; these conditions are commonly expressed in the instrument itself. This instrument is called the oil and gas lease. Under the usual oil and gas lease, the costs of exploration, development, and production, with minor exceptions, are borne entirely by the lessee. The economic benefits which will be enjoyed or reserved by the lessor depends upon the agreement of the parties.

The mineral estate retains several attributes, which will be important to the land owner or lessor, in the oil and gas lease. The general attributes that belong to the owner of the mineral estate are as follows:

1. the power to lease,
2. the right to a bonus,
3. the right to delay rental,
4. the right to royalty, and
5. the right to shut in royalties and other rights (Hemingway, 1983).

These rights are discussed in the following sections.

Power To Lease

The power to lease, also sometimes called the *executive right,* is the right to transfer the development rights of the mineral estate to another individual or entity (Hemingway, 1983). For example, X, the owner of Fair Acres, transfers to Z by way of an oil and gas lease the exclusive rights to develop the mineral estate. This executive right can also extend as far as to allow the mineral estate owner to retain the right to lease out the mineral estate even after he or she conveys the mineral estate to another. For example, X, the owner of Fair Acres, sells Fair Acres to A and retains an undivided interest in the mineral estate. Furthermore, X reserves for herself the power to lease that portion of the mineral estate she conveyed to A.

Right to a Bonus

Bonus is the sum paid to the owner of the mineral estate for the execution of an oil and gas lease. This sum is commonly paid in cash or may be made on a deferred basis out of future production (Hemingway, 1983). The bonus approximately represents the market value for the sale of the minerals to the lessee for a limited term for development purposes. The amount paid usually depends upon the nature of the development activity in the vicinity.

Right To Delay Rentals

Delay rentals are generally defined as payments made by the lessee to the lessor, on a year-to-year basis, during the primary term of an oil and gas lease (Hemingway, 1983). These delay rentals allow the lessee to postpone the commencement of development operations on the leased land. This delay rental usually shows up as a clause in the oil and gas lease and will be discussed further in the section discussing lease clauses.

Right to a Royalty

A *royalty interest* is broadly defined as the right to receive a fractional share of production of the petroleum product free of cost or expenses incident to exploration, development, or production (Hemingway, 1983). Prior to the execution of an oil and gas lease, the right to a royalty from production is one of the attributes held by the owner of the mineral estate. However, this right to a royalty may be conveyed to another individual prior to an oil and gas lease arrangement.

There are three types of royalty interests, all of which share in the production from the mineral estate. These three types of royalty interests are as follows:

1. *Landowner's Royalty:* The landowner's royalty is a fractional share of the gross production payable to the lessor in the royalty clause of an oil and gas lease.
2. *Overriding Royalty:* An overriding royalty is that royalty interest that is carved out of the lessee's interest under an oil and gas lease.
3. *Nonparticipating Royalty:* A nonparticipating royalty is a royalty interest carved out of the mineral interest, which entitles the holder to a stated share of production. By *nonparticipating,* it is meant that the holder does not have the right to lease or participate in bonus and rentals (Weaver, 1989).

Right to Shut-In Royalties and Other Rights

Shut-in royalties and minimum royalties are contractual rights to receive certain payments instead of actual production (Hemingway, 1983). *Minimum royalties* are a minimum dollar amount expressly stated in the oil and gas lease. The stated amount is paid as royalty even though royalty payments from actual production may be less. *Shut-in royalties* are a stated dollar amount expressed in the oil and gas lease, which will be paid as royalty when a gas well, capable of producing gas in paying quantities, is shut-in due to lack of a pipeline connection or market for the gas (Hemingway, 1983).

The other possible type of contractual right is the

right to a production payment. A *production payment* is the right to receive a *sum-certain*, which is payable solely out of production. This production payment is similar to a regular royalty in that it is not burdened by the cost of production. The difference between the other types of royalty and the production payment is that royalty payments will continue as long as a well (or wells) is producing. A production payment ceases to pay after the sum agreed upon in the contract has been paid out to the lessor (Weaver, 1989).

CREATION OF INTERESTS IN THE OIL AND GAS ESTATE

It is impossible within the scope of this chapter to fully detail the various types of arrangements and variations that may be encountered in lease arrangements. Problems crop up when ambiguities exist within the lease itself and it cannot be determined whether a mineral or royalty interest has been created.

Generally, a conveyance within a lease of minerals or mineral rights designates a fully participating mineral interest (Hemingway, 1983). Further complicating the interpretation of terms contained within a lease is the fact that the various states follow different theories regarding the ownership of oil and gas under the ground. For example, a lease may state the following:

> An undivided 1/8th interest in and to all of the oil, gas, and other minerals in, under, that may be produced from the described land together with the right of ingress and egress for the purpose of entering, developing, and producing same.

In those jurisdictions following the ownership in place theory of oil and gas, courts will look for the presence of language in a lease granting or reserving minerals "in or under the ground" (or other similar language) as indicating the intent of the parties to convey the minerals in place (Hemingway, 1983). In those jurisdictions following the nonownership theory of oil and gas, such language purporting to convey ownership of minerals in or under the land is not possible. Therefore, the courts in those jurisdictions give little weight to this and other similar language in an oil and gas lease when determining what type of an interest is held (Hemingway, 1983). Regardless of the theory followed by the particular jurisdiction, however, the fact that the descriptive terms in the lease speak of "minerals," "mineral estate," or "mineral rights" is generally construed by the courts as indicating an intent to convey a fully participating mineral interest (Hemingway, 1983).

In most jurisdictions, the grant or reservation of a royalty interest is interpreted as creating a non–cost bearing interest that shares only in a fractional portion of the gross production. Furthermore, this interest does not participate in a bonus, delay rentals, or the power to lease. Royalty interests are carved out of a mineral interest and generally mean that a fraction of the gross production is being reserved (Hemingway, 1983). Furthermore, in the majority of jurisdictions, the word *royalty* has a definite and unambiguous meaning. Therefore, when the term *royalty* is used in the context of a lease, a royalty interest is deemed to have been conveyed rather than a mineral interest. For example, in the case of *Schlitter v. Smith* (Texas Supreme Court, 1937), the court held that a lease that conveyed a one-half interest in royalty rights did not include the right to receive a bonus or delay rentals and therefore conveyed a royalty, rather than a mineral, interest (Hemingway, 1983). The reason that these problems and ambiguities in conveyances are frequently litigated is shown by the following situation.

> If O conveys to A a 1/8th mineral interest, then not only does A have the right to lease or convey a portion of this mineral interest, but he also receives the right to receive bonus and delay rentals. Furthermore, A will receive a full 1/8th of all of the oil and gas produced from the lands.
>
> If, however, A receives a 1/8th royalty interest and O has retained for himself a 1/8th mineral interest, then A receives 1/8th of 1/8th, or thus only 1/64th, of the total production (Hemingway, 1983).

You must be aware of how the particular jurisdiction you are in interprets the term *royalty*. For example, in Kansas, *royalty* is apparently restricted to that which is payable under an existing lease. In a Kansas case, *Bellport v. Harrison*, the court held that the term *royalty* was definite and unambiguous and was related only to production under a specific lease. As the lease had expired, the right to receive royalty payments granted under that lease expired as well. Therefore, it is apparently not possible to create a perpetual nonparticipating royalty interest in Kansas (Hemingway, 1983).

TRESPASS AND ADVERSE POSSESSION OF THE MINERAL ESTATE

Trespass has been defined as an unlawful interference with one's person, property, or rights. At common law, trespass was a form of action bought to recover damages for any injury to one's person, property, or relationship with another (Black, 1979). The individual who wrongfully possesses another's mineral estate may be liable to the rightful owner (of the mineral estate) as a trespasser. The types of trespass that can occur include the following:

1. subsurface trespass,

2. trespass caused by injected substances,
3. conversion caused by negligence of surface operations,
4. geophysical trespass, and
5. slander of title (Hemingway, 1983).

Subsurface Trespass

For a subsurface trespass to occur, the trespasser must actually enter and possess the mineral estate. An example of such a trespass occurs when the lessor and lessee enter into a lease arrangement that has as a term certain "for three years and as long thereafter as oil and gas is produced." At the end of the three years, there is no oil or gas production; however, the lessee drills a well. This drilled well is a trespass to the mineral estate.

There are two different situations that can arise as a direct consequence of the drilling of this well. Either this well will be a dry hole or it will be a producing well. If the subsurface trespass results in no production, this may result in what is known as a *Kishi-type trespass*. In a Kishi-type trespass, the trespasser might be held liable for damages to the speculative value of the rightful owner's mineral interest.

In *Humble Oil & Refining Co. v. Kishi* (Texas Commision of Appeals, 1927), Humble Oil entered and began the drilling of a well under the express claim of right that the lease was still in effect as to all its interests. There was some question as to when the lease actually expired. One of the co-mineral interest owners contended that the lease had expired before Humble entered and began the drilling of the well.(Humble entered with the permission of the other co-owner). The well was a dry hole, and the speculative market value of the tract went from $1000.00 per acre to $0 per acre. Kishi, the co-tenant who Humble did not receive permission to drill from, sued and eventually collected for this difference in the market value of the acreage caused by the drilling of the dry hole.

Some other courts have held the same as Kishi and others have not. It appears, however, that when someone enters and wrongfully exercises a right they do not have, the landowner should be compensated for the exercise of that right regardless of whether or not oil was discovered (Hemingway, 1983).

The situation, however, is different when the subsurface trespass results in production or drainage. The extent of the damages owed by the trespasser to the rightful owners of the mineral estate depends on whether the trespasser acted in good faith. *Good faith* can be defined as honesty of intention and freedom from knowledge of circumstances that ought to put the holder upon inquiry (Black, 1979). When the trespasser has acted with the good faith belief that they had the right to enter and drill a well, and production is obtained, the courts will generally allow the trespasser to offset his or her reasonable costs for drilling, completing, and producing the oil and gas (Hemingway, 1983). This right is limited to the extent that the operations have enhanced the property of the true owner. Where there has been insufficient production, the good faith trespasser will not recover his or her costs (Hemingway, 1983). Likewise, a *bad faith* trespasser is liable for the gross value of the production from the well without any offset for cost. The burden of proof is on the trespasser to prove that he or she acted in good faith. Note that in Texas, a person who enters onto a disputed tract of land and drills during the pendency of a lawsuit is, as a matter of law, acting in bad faith (Weaver, 1989). Furthermore, a bad faith trespasser may have punitive damages assessed against him or her.

Trespass Caused by Injected Substances

There has been an increasing amount of litigation concerning the injection of substances into reservoirs as secondary recovery techniques become more commonplace in the oil and gas industry. The trespass occurs when the injected substances intrude, or migrate, into adjacent mineral estates. The question, therefore, is whether the owners of the adjacent mineral estates have a cause of action against the trespasser. When such an injection is done voluntarily without authorization from a state agency, and this trespass results in damage to adjacent wells, the injector, in at least one jurisdiction (Oklahoma), will be held liable for the damages to the surrounding wells (Hemingway, 1983). However, another Oklahoma case held that there was no actionable tort when (1) the amount of the injected material was negligible when compared to the size of the reservoir, (2) the material was injected into an adjacent salt water formation, and lastly, (3) no "communication" existed between the injected material and the adjacent producing horizons (Hemingway, 1983).

When secondary or tertiary injection programs are done under the approval of a state agency, it is generally held that this subsurface trespass is not actionable (Hemingway, 1983). This is done as a matter of public policy. The overwhelming public policy is to prevent waste of oil and gas products, and therefore this prevention of waste supersedes the right of the adjoining property owner to enjoin (meaning stop) the recovery program (Hemingway, 1983).

Conversion Caused by Negligent Subsurface Operations

A producer may be found guilty of conversion where, although a well was bottomed on his land, drainage of products from adjacent lands occurs as a result of negligent operations (Hemingway, 1983). This situation was discussed previously in this chapter in the section on correlative rights (see the *Eliff* case). It would be interesting to see how this type of trespass would be applied in a jurisdiction that followed a nonownership theory of oil and gas. Perhaps the argument could be made, citing the *Eliff* case, that the rule of capture only applies when the oil or gas is being captured through legitimate operations. Furthermore, it would appear that one could argue based on production history that his or her well was producing X amount of oil or gas and that this amount either was decreased or terminated due to the operator's negligence while drilling the well on adjacent property.

Geophysical Trespass

When a separation of mineral rights has occurred, that is, the mineral estate has been severed from the surface estate, it is generally held that for purposes of mineral development and exploration, the mineral rights are dominant (Hemingway, 1983). By *dominant,* we mean that the owner of the mineral estate has the right to use as much of the surface estate as is reasonably necessary for mineral exploration and development. (This right is subject to various limitations in different jurisdictions.) Generally, the owner of the mineral estate will recompense the surface owner for damages done while conducting geophysical operations.

Because of this severance of the estates, when there is an oil and gas lease on the property, the right to conduct the geophysical search from the surface belongs to the lessee if the lease grants an exclusive right to explore (Weaver, 1989). If the surface owner allows a geophysical crew from another operator to conduct geophysical surveys of the mineral estate from the surface, it appears that the mineral estate owner may be able to recover for the loss of the exploration rights to the mineral estate. For example, if T, the trespasser, shoots a seismic survey showing that there are no possible hydrocarbon-bearing structures underlying the surface of the land, then A, the owner of the mineral estate, could possibly recover for the diminution in value if it can be shown that the value of his land was $X per acre before the exploration survey was released and $0 to minimal dollars per acre afterward.

Slander of Title

Slander of title is defined as a false and malicious statement, oral or written, made in disparagement of a person's title to real or personal property or of some right of his causing him special damage (Black, 1979). Malice is an essential element to bringing any slander of title suit. An example of slander of title is when an operator, knowing that its lease has expired, refuses to release its lease on the mineral estate. As a result of this refusal, the owner of the mineral estate loses the possibility of leasing his or her estate to another operator and obtaining a per-acre bonus. To sustain his or her claim, the mineral estate owner must prove that there was publication of a false claim, which was done with malice, and that this resulted in the loss of a specific sale (that is, the conveyance of the mineral estate to another operator). If these conditions are met, then in all likelihood, the mineral estate owner will win a slander of title suit against the holdover operator (Weaver, 1989). One defense that the operator can assert in a slander of title suit is that it acted with good faith in believing that it had "good and clear title" to the property.

Adverse Possession of the Mineral Estate

Adverse possession is a method by which someone may acquire title to real property by "possessing" the property for the requisite statutory period of time and under certain conditions (Black, 1979). The statutory period of time to claim a property and obtain title by adverse possession differs from state to state. Commonly in oil and gas situations, the way to adversely possess the mineral estate depends on whether or not the mineral estate has been severed from the surface estate. For example, in the situation where the mineral estate has not been severed from the surface estate, one who enters upon the land claims it as his own, holds it for the requisite number of years, and follows the conditions set by his state will not only gain possession of the surface estate but will also adversely possess the mineral estate (Weaver, 1989). In those states that follow the nonownership theory of oil and gas, it would appear that this would be the only way to adversely possess the mineral estate since, if the adverse possessor gains control of the surface, he then gains the exclusive control to drill upon the land and convert any oil and gas to his possession when brought to the surface.

In those states that allow a severance of the mineral estate from the surface estate, adverse possession of the surface does not mean that the adverse possessor has gained ownership of the mineral estate. To adversely possess the mineral estate in such a situation, the adverse possessor must drill a well and produce hydrocarbons for the statutory period of limitations (Weaver, 1989).

DIVIDED OWNERSHIP PROBLEMS

It is entirely possible to create ownership in a mineral estate by more than one individual. As the potential lessee of the mineral estate, the oil and gas professional must be aware of the type and nature of the ownership or of the lessor in the mineral estate.

Take the following example:

> O is the owner in *fee simple* (an estate limited absolutely to a man or woman and their heirs and assigns forever without limitation or condition [Black, 1979]) of the mineral estate in Black Acre, a section of land containing 640 acres. O desires to convey to A 1/4 of the mineral estate. O's attorney then prepares a deed that grants and conveys an undivided 1/4 interest in the minerals for the purposes of exploration, development, and production.

This conveyance of a *undivided* fractional interest in land will create a concurrent ownership of the mineral estate with respect to A and O (Hemingway, 1983). This concurrent ownership, without any other words of limitations or of a specific intent, usually creates a *tenancy in common* (Hemingway, 1983). What this means is that each tenant is separately vested with a nonexclusive possessory right in all of the mineral estate and that this interest will pass to their heirs, devisees, or successors either through voluntary or involuntary conveyance (Hemingway, 1983). O and A each own a separate fractional share in all of the minerals in the estate. Without a formal partition, of the estate, neither tenant has the exclusive right to possess any particular or separate portion of the mineral estate (Hemingway, 1983). Any co-tenant is entitled to lease or produce the minerals without the consent of the others. However, the producing co-tenant must account to his or her nonconsenting co-tenants for their share of profits from the oil and gas produced (Weaver, 1989).

Using the example just cited, let's say that O leases her 3/4 interest to an oil company and retains a 1/8 royalty limited to the extent of her legal interest in the property (the lease has a proportionate reduction clause which we will discuss later). If and when a producing well is drilled, O will receive a cost-free royalty of 3/4 of 1/8 of production. A is entitled to the profits on 1/4 of the total production. The oil company is entitled to the remainder (Weaver, 1989). Note that A does not receive 1/4 of 1/8 royalty because A never entered into a royalty agreement with the oil company. Note further that while in most jurisdictions a nonconsenting co-tenant cannot be charged with any of the costs of drilling a dry hole, or conducting other operations that prove unsuccessful in the production of oil and gas, they also cannot receive any profit from the well until the costs chargeable to that interest have been recovered from production (Weaver, 1989). However, it is possible for the nonconsenting (unleased) co-tenant to ratify the lease and receive a royalty share from the date of the ratification (Weaver, 1989).

It is not always desirable to create a tenancy in common. In those jurisdictions where it is feasible, parties may create a joint tenancy or a tenancy by the entirety (Hemingway, 1983). In a *joint tenancy*, the surviving joint tenant has the right to the property interest held by the other, now deceased, joint tenant. In our example, if when O conveyed to A an undivided 1/4 interest in the mineral estate, and their jurisdiction allows the creation of a joint tenancy and it was expressly stated so in the lease, then upon O's death, A will not only keep his 1/4 interest in the mineral estate but will also acquire O's 3/4 interest in the mineral estate. A *tenant by the entirety* is a joint tenancy in which the joint tenants are husband and wife (Hemingway, 1983).

It is beyond the scope of this chapter to go into all the various problems that the oil and gas professional may encounter when leasing a mineral estate. Note that there can be problems in successive ownership when land becomes divided between a life tenant and remaindermen. These problems arise as to who may lease the mineral estate and how the proceeds of the lease are to be divided between them. Furthermore, problems may occur when there is more than one owner of a mineral estate and each individual owner creates separate lease arrangements with different oil companies.

OIL AND GAS LEASE

An oil and gas lease, by its nature, creates a fee simple determinable. A *fee simple determinable* is created when a conveyance, which contains words effective to create a fee simple (see previous definition), in addition contains a provision or provisions for the automatic expiration of the estate upon the occurrence of a stated event (Black, 1979). One of the most litigated areas in oil and gas law concerns the language used in determining when and exactly how the event, which acts to "cut-off" the mineral estate to the lessee, is to be construed.

In determining whether the instrument used in the conveyance is a deed or a lease, courts look to the presence of covenants (promises) that call for exploration and development and for the termination of the conveyance (Hemingway, 1983). The courts will also try to determine the intent of the parties from the face of the instrument by looking to the facts and circumstances that existed at the time of the execution of the instrument (Hemingway, 1983). Factors that courts have considered include the following:

1. the amount of monetary consideration for the instrument,
2. the form of the granting clause,
3. a provision for the reservation of payments out of production,
4. a provision stating an express duty to develop,
5. a provision that gives or states the term of the grant,
6. the presence of easements for the development in

use of equipment,
7. the presence of a covenant requiring an obligation to drill a well,
8. a provision regarding the right to remove fixtures, and
9. a provision to provide for termination of the lease (Hemingway, 1983).

The following discussion of the essential lease clauses, delay rental clauses, defense clauses, and other clauses is primarily taken from *The Law of Oil and Gas Leases* (2nd ed.) by Earl A. Brown and Earl A. Brown, Jr. (1990).

Essential Lease Clause

These clauses should be contained within any oil and gas lease. In fact, it would not be considered a lease without them.

Consideration and Granting Clause

Included in the Chapter Appendix is a typical modern oil and gas lease form (Exhibit A). An oil and gas lease is essentially a contract. Therefore, the oil and gas lease must be supported by sufficient consideration; otherwise, there is no mutuality of obligation, and the contract or lease is unenforceable. *Consideration* is defined as the cause, motive, price, or impelling influence that induces a contracting party to enter into a contract (Black, 1978). Furthermore, it is some right, interest, profit, or benefit that accrues to one party or some forbearance, detriment, loss, or responsibility that is given, suffered, or undertaken by the other (Black, 1979).

The consideration does not have to be a sum of money. In many instances, while a sum of money is listed as consideration in an oil and gas lease, the principle or basic consideration of the lease is the agreement by the lessee to develop the premises for oil and gas and pay royalties to the lessor (Brown and Brown, 1990). The granting clause, which usually immediately follows the recitation of the consideration, defines what rights are given by the lessor to the lessee (Weaver, 1989). The granting clause will typically state to what uses the lessee may put the property, such as

1. exploration and prospecting,
2. drilling for oil and gas,
3. laying pipelines,
4. building roads, and
5. erection of storage tanks and telephone lines (Brown and Brown, 1990).

The granting clause will also carry the legal description of the property under lease. The oil and gas lease must describe the land with such certainty that it can be located on the ground (Brown and Brown, 1990). The "Mother Hubbard" clause is another provision that follows the property description in an oil and gas lease. It is generally only used in Texas, but occasionally occurs in other states (Brown and Brown, 1990). This clause is intended to include everything that may not be included under the property description. For example, if an adjoining strip of land has been acquired through adverse possession and prescription, the Mother Hubbard clause is intended to include these other lands under the oil and gas lease (Brown and Brown, 1990).

Habendum or Term Clause

The *habendum clause* sets out the conditions for the primary and secondary terms of a lease. The most commonly used habendum clause, according to Brown and Brown (1990) is as follows:

> It is agreed that this lease shall remain in force for a term of ___ years from this date, and as long thereafter as oil or gas, or either of them, is produced from said lands by the lessee.

The *primary term* is a term whose duration is expressed and fixed by the lease. The *secondary term* is a period of time of indefinite length. The duration of the secondary term is dependent upon an interpretation of the word *produced*. To ensure that drilling and exploration for hydrocarbons occur during the primary term, the drill or pay clause (or delay rental clause) was developed, which is discussed later.

The secondary term of an oil and gas lease is created by the language "as long thereafter as oil or gas, or either or them, is produced." A great deal of litigation concerns the meaning of *produce*. A literal interpretation of this term would indicate that the production of any amount of oil or gas would satisfy this requirement. The problem with this interpretation is that it would enable a lessee to continue the lease for extremely long periods of time in the hopes that some day they will drill another well which will increase the amount of production (Brown and Brown, 1990).

The courts of many states now hold that the word *produce*, as used in the habendum clause, is to be construed as meaning *production in paying quantities* (Brown and Brown, 1990). This is the rule in Oklahoma, Texas, Montana, West Virginia, and Louisiana (Brown and Brown, 1990). The question that then comes up is how to determine what constitutes "production in paying quantities." The Texas Supreme Court listed several factors to consider in making that determination. In the case of *Clifton v. Koontz*, the court stated that some of the factors to be taken into consideration are as follows:

1. the depletion of the reservoir,

2. the price for which the lessee is able to sell the oil and gas,
3. the relative profitableness of other wells in the area,
4. the operating and marketing cost of the lease,
5. the operator's net profit,
6. the lease provisions,
7. a reasonable period of time under the circumstances, and
8. whether or not the lessee is holding a lease merely for speculative purposes.(Texas Supreme Court, 1959).

The court went on to state that in determining when there is production in paying quantities, there must also be a determination of whether a reasonable basis exists for the expectation of profitable returns from the well (Brown and Brown, 1990). Because the well in *Clifton v. Koontz* was a marginal well, the court went on to state,

> In a case of a marginal well, the standard by which paying quantities is determined is whether or not, under all of the relevant circumstances, a reasonably prudent operator would, for the purpose of making a profit, and not merely for speculation, continue to operate a well in the manner in which the well in question was operated (Texas Supreme Court, 1959).

Lastly, when reviewing items that are to be considered in determining whether the lease was producing in paying quantities, the court held the following:

1. Where production is reduced by order of regulatory bodies, proper allowance should be made in determining the "productive capacity of the lease."
2. Depreciation of the "original investment cost" is *not* to be taken as a loss or expense.
3. Overriding royalties are *not* to be deducted, but the "entire income attributable to the contractual working interest . . . is to be considered."
4. Temporary cessation of production, which the courts have held to be "justifiable" is *not* to be considered.
5. The good faith of the operator, in continuing to develop the property for the purpose of making a profit and not merely for speculation, is to be the "standard" (Brown and Brown, 1990).

The good faith of the operator has also been used as a standard in many cases, however, this rule is not universal (Brown and Brown, 1990). In one Illinois case, *Gillespie v. Ohio Oil Company*, Gillespie completed a well just a few days prior to the expiration of the primary term. Oil production began shortly before the end of the primary term and continued for several months thereafter. In holding that the lease had not expired, the court stated that,

> Oil was produced continuously after the drilling of the well. It is true that the quantity produced was so small as to make the venture unprofitable, but the strict letter of the lease was complied with and it has not expired by its own terms (Illinois Supreme Court, 1913).

The habendum clause in that lease stated "as long thereafter as oil or gas is produced thereon" (Brown and Brown, 1990).

It appears to be the law in many states that when interpreting the phrase "as long thereafter as oil or gas is produced," there must be actual production on the land at the date of the expiration of the primary term (Brown and Brown, 1990). It does not appear that the discovery of oil or gas will hold the lease past the primary term. This seems to be the law in the states of Texas, Illinois, Indiana, Kansas, Louisiana, Michigan, Montana, New Mexico, Ohio, Pennsylvania, and West Virginia. Apparently, however, Oklahoma follows a different rule. The Oklahoma rule appears to be that if a well is commenced during the last rental period after the rental has been paid, this well may be drilled to completion even if it is not completed until after the primary term, and if oil or gas is discovered, then the lease will continue as long as oil or gas is produced in paying quantities (Brown and Brown, 1990).

In some instances the courts may extend the primary term of a lease through the use of its equitable powers. Many of these cases seem to occur when an operator has discovered gas wells but, either because of the lack of a market for the gas or an extended development phase, production is not started until well after the primary term has expired. The courts appear to look at the conduct of the lessee and lessors in determining under those circumstances whether it would be fair to terminate the lease (Brown and Brown, 1990).

Royalty Clause

The *royalty clause* provides the amount and conditions for the payment of royalty on the production of oil or gas from the lease. Generally, the lessee will agree to deliver the oil, if oil production is occurring, into the pipeline to which the wells are to be connected. The purchaser of the oil will generally cut what is known as a *division order* and will pay the lessor in cash for his portion of the oil (Brown and Brown, 1990). Generally, the lessee is under a duty to market or sell the oil delivered on behalf of the lessor (Brown and Brown, 1990). Today, gas is treated like oil in the royalty clause. In the past, however, gas was considered a waste product and generally a flat rate clause was included for the production of gas (Hemingway, 1983). The oil and gas lease in Exhibit A (in the Appendix) includes specific types of gas that are to be included in the royalty. These various types are casing head gas, condensate, and other gaseous substances. Problems have arisen in the past when these specific types of gas are not expressly stated in the royalty clause of the oil and gas lease. Lessees have attempted to pay either a flat rate per year on this production or have contended that these various types of gases were not included under the lease and therefore the lessor was only entitled to the value of the products after the reasonable cost of production was deducted (Hemingway, 1983).

Delay Rental Clause

A *delay rental clause* is a provision that allows the lessee to pay an annual rental fee instead of drilling a well (Hemingway, 1983). Implicit within the lease itself is the implied duty or covenant to drill an exploratory well. The delay rental clause allows a lessee to pay a minimum rental rate rather than having to drill a well. There are two types of delay rental clauses—the unless clause and the or clause (Weaver, 1989). The oil and gas lease in Exhibit A is an unless clause.

The *or clause* states in general that the lessee should drill a well within a certain time or pay a stipulated sum as rental. This rental would postpone the lessee's obligation to drill a well for the stated period of time (Brown and Brown, 1990). The nature of the or clause makes it a contractual obligation of the lessee to either drill a well or pay the delay rental. As this is a contractual obligation, the lessor did not have the power to terminate the lease if the delay rental was not paid. In most jurisdictions, the lessor had to bring an action for the recovery of the rental price as his remedy (Brown and Brown, 1990).

The *unless clause* works as a limitation on the lease itself. The limitation is generally that the lease will terminate unless the lessee pays the delay rentals to the lessor (Weaver, 1989). Once again, this sets up a determinable situation. If a well is not drilled and the lessee does not pay the delay rental, the lease automatically terminates and reverts back to the lessor. Today, the most commonly used version of the delay rental clause is the unless clause. Problems arise, however, in determining when actual payment is due on the delay rental. Failure to pay the delay rental is a termination of the lease itself. Commonly, the rental must be paid on or before the anniversary date of the lease (during the primary term) to keep the lease alive.

Problems occur when the lessee waits to the last minute to mail the delay rental payment. The check may arrive after the anniversary date if the mail has been delayed. Sometimes, even if the check arrives in time, the check does not clear the bank until after the anniversary date or the check itself is dishonored by the bank it is drawn upon. Frequently, as can be seen in Exhibit A, provisions are written into the lease expressly stating the procedures to follow if checks are delayed and/or if, because of problems with the bank, the checks are dishonored and not paid. Jurisdictions vary in their treatment of the failure to receive delay rentals in a timely fashion (Brown and Brown, 1990). It is beyond the scope of this chapter to discuss the ways various jurisdictions address this situation.

Lastly, another problem arises as to the meaning of the term *commencement* of drilling (Weaver, 1989). Usually, this does not mean that the well actually has to be spud before the end of the first year of the primary term or any subsequent year of the primary term. Commencement usually means that there is some physical operation occurring on the surface of the land, be it the building of the access road, the caliche pad, or setting up the rig itself. These activities must be directly related to the drilling of the well, must be conducted in good faith, and must be diligently pursued to the completion of a well (Weaver, 1989).

Defensive Clause

Defensive clauses are added to oil and gas leases to modify the effect of the termination of the lease at the end of the primary term. Sometimes these defensive clauses may be stacked upon each other and will extend the lease past the primary term even if there is not production in paying quantities (Weaver, 1989).

Dry Hole and Lease Extension Clause

In general, a *dry hole clause* provides that after the drilling of a dry hole during the primary term, the lessee can still maintain the lease by resuming payment of the rental fees. Many modern oil and gas lease forms, such as the one in Exhibit A, contain a *lease extension clause* that will extend the primary term past its expiration date. This clause provides that if oil and gas are not being produced from the land at the end of the primary term, the lease will nevertheless remain in effect if the following conditions are met:

1. the lessee is engaged in operations for the drilling or reworking of any well on the lease,
2. no cessation of operations of more than 60 consecutive days occurs, and
3. the operations result in production (Brown and Brown, 1990).

Operations Clause

An *operations clause*, sometimes called a drilling operations clause, continuous drilling clause, or continuous operations clause, will prevent a lease from expiring at the end of the primary term as long as drilling is in progress (Weaver, 1989). Once again, operations that are deemed sufficient to commence drilling for the purposes of the delay rental clause are usually sufficient under the operations clause to maintain the lease (Weaver, 1989).

The dry hole, lease extension, and operations clauses may in many instances be stacked together. The effect of the stacking will allow the drilling of a well extending past the primary term. If that well is a dry hole but potential hydrocarbon production is indicated, then additional wells may be drilled provided that 60 days have not passed between the drilling of the dry hole and the drilling of these additional wells and that these wells result in the production of oil and gas in paying quantities.

Force Majeure Clause

The *force majeure clause* lists "acts of God" and other catastrophes that are beyond the reasonable control of the lessee. Therefore, this clause will excuse the lessee from an apparent breach of the lease (Weaver, 1989).

Shut-In Royalty Clause

The *shut-in royalty clause* addresses the situation when a well, capable of production in paying quantities, is shut-in due to the lack of a market for the gas. In this instance, the lessee makes payments of a royalty to the lessor in lieu of production and thereby extends the lease past the primary term (Weaver, 1989). Failure to make this royalty payment, in lieu of production, will generally mean the forfeiture of the lease.

Other Clauses

Several other clauses are included in the oil and gas lease to aid in the administration or maintenance of the lease itself.

Warranty Clause

The *warranty clause* allows the lessee to regain damages from the lessor due to the failure of title. Failure of title means that the lessor did not have the ability to convey to the lessee rights to the mineral estate.

Proportionate Reduction Clause

The *proportionate reduction clause* gives the lessee the right to reduce the royalty and interest paid to the lessor if it is determined that the lessor did not own the entire interest in the oil and gas or the entire fee simple (Weaver, 1989). For example, if the lessor only owns a 3/4th interest in the mineral estate, the clause will allow the lessee to reduce the royalty payment by 1/4th and not cause an automatic breach of the lease due to an underpayment of the royalty.

Notice Before Forfeiture Clause

The *notice before forfeiture clause* is the means through which the lessee ensures that the lessor must give notice to the lessee of any alleged breaches by the lessee. This therefore gives the lessee an opportunity to correct these alleged breaches before the lessor institutes a suit to terminate the lease (Weaver, 1989).

Pooling Clause

The *pooling clause* is a defensive clause that allows the lessee to combine acreage from adjacent tracts to create a unit and place the well upon these combined tracts (Weaver, 1989). As a general rule, the production from any well within the pooled unit operates as production from the leased tract, even though the producing well is not located on it. Royalties are generally apportioned on the basis of surface acreage between the various lessors (Weaver, 1989). A *Pugh* or *freestone rider clause* is a clause that may be inserted in a lease by the lessor so that the pooled acreage does not act as production from the rest of the acreage under the lease.

For example, without a Pugh clause, a lessee might pool 40 acres out of 500 acres under the lease into an adjoining tract. If a well is producing within this leased acreage, regardless of whether or not it is on the land of the lessor, this well will hold the entire 500 acres under lease, as long as the well is producing in paying quantities and will carry the lease past the primary term. In effect, the Pugh clause acts as a severance of the pooled acreage from the leased acreage. In this example, a Pugh clause would have severed the pooled acreage from the remainder of the 500 leased acres. To continue the lease on the remaining acres, the lessee would have to either drill a well or pay the delay rentals to prevent the lease from expiring. Once the primary term is over, the remainder of the unpooled acreage will revert back to the lessor and production will continue from the pooled unit.

IMPLIED COVENANT WITHIN OIL AND GAS LEASES

The general definition of an *implied covenant* is a promise or obligation that can be inferred or implied in law from the words used and that is based upon the presumed intention of the parties (Brown and Brown, 1990). In oil and gas leases, implied covenants are usually favored when construing oil and gas leases since they are in the interest of promoting development (Brown and Brown, 1990). It is generally considered that five covenants are implied in an oil and gas lease:

1. the implied covenant to drill an exploratory well,
2. the implied covenant to conduct additional development after paying production is obtained,
3. the implied covenant for diligent and proper operation of wells,
4. the implied covenant to market or utilize products, and
5. the implied covenant to protect the leased premises against drainage by wells drilled on

adjoining lands (Brown and Brown, 1990).

The implied covenant to drill an exploratory well occurs in those situations when a lease is granted that does not specify a time within which the work of exploration shall be commenced and also when there is no express provision for delaying the commencement of a well (Brown, 1990). The general rule in Texas is that where a lease specifically provides for a delay in drilling, such as payment of rentals, then there is no implied covenant to develop (Brown and Brown, 1990). In Oklahoma, it appears that the rule is stricter. An Oklahoma court held that there was an implied covenant to commence development within a reasonable time (Brown and Brown, 1990). In the situation before the court, the original terms of the oil and gas lease called for a primary term of 15 years. The lease did not contain a provision for the payment of delay rentals. After 13 years had passed without any drilling, the lessee brought suit to cancel the lease. The court held that there was an implied covenant to commence development within a reasonable time and that the failure of the lessor to drill a well within the first 13 years of the lease was a breach of this covenant and therefore the court canceled the lease (Brown and Brown, 1990). Other states that appear to follow this general rule are Kansas, Ohio, West Virginia, Illinois, Arkansas, and California (Brown and Brown, 1990). After reviewing the decisions from various states, Brown and Brown (1990) conclude that the rules regarding the implied covenant to drill an exploratory well appears to be as follows:

1. When the lease contract is signed as to when development shall be conducted, the law will imply a covenant to drill within a reasonable time. In Texas, it must be shown further that a reasonably prudent operator would drill a well.
2. When the lease contract contains an express provision for the payment of delay rentals, an implied covenant is applicable. The reason for this rule is that the parties are free to contract as they wish, and when development obligations are covered by an express provision, there is nothing left for implication (Brown and Brown, 1990).

The implied covenant to conduct additional development after production is obtained does not occur until oil or gas is being produced in paying quantities from the lease. Furthermore, this implied covenant does not arise unless there are no express provisions within the lease concerning such further development (Brown and Brown, 1990). The basic concept for this implied covenant is that generally an operator will fully develop the resources under his control within a "reasonable period of time" (Weaver, 1989). Note that in Texas, there is no covenant to conduct exploration other than that necessary to reasonably develop the premises. Generally, it is held that these operations need not be carried past the point of profitableness to the lessee (Brown and Brown, 1990). A further test appears to be that of the "prudent operator." In other words, whatever in the circumstances would be reasonably expected of operators of ordinary prudence, having regard to the interest of both lessor and lessee, is what is required (Brown and Brown, 1990).

The implied covenant for the diligent and proper operation of wells appears to be similar in concept to the implied covenant contained in service contracts that, when a party undertakes work, this work shall be performed in a proper and workmanlike manner (Brown and Brown, 1990). By way of this implied covenant, the lessee promises not to be negligent in the drilling and operation of wells on the leased tract, as negligence generally means the failure to exercise ordinary care (Brown and Brown, 1990).

The implied covenant to market generally means that the lessee is required to market production from the leased premises within a reasonable period of time and at the best available price (Weaver, 1989). After reviewing decisions by several state courts, Brown and Brown (1990) conclude the following, as regards the implied covenant to market:

1. There is an implied covenant that requires the lessee to market production from the lease, including the lessor's royalty, unless the lessor elects to market his or her own share of the production.
2. In marketing the lessor's royalty, oil or gas, the lessee acts as the agent of the lessor in accordance with trade usages in the industry.
3. In marketing the oil and gas, the lessee is required to show reasonable diligence and the time in which marketing is to be done is a reasonable time under the facts and circumstances of the case.
4. The price at which such production shall be sold is the reasonable market value of the product (Brown and Brown, 1990).

Finally, we have the implied covenant to protect the leased premises against drainage from adjacent tracts. In many oil and gas leases, this implied covenant is expressly stated within the lease and is designed to protect the lessor from having his or her oil and gas drained by the adjacent land owners and their lessees. After reviewing decisions by several states concerning this implied covenant, Brown and Brown (1990) conclude

> There is inherent in an oil and gas lease an implied covenant of the lessee to protect the leased premises against drainage from wells on adjoining

premises by drilling offset wells. This covenant, however, is subject to the following qualifications:

1. Where the lease contains express provisions defining the lessee's duty with regards to drilling offset or protection wells, no implied covenant exists.
2. Where the implied covenant is applicable, it must be shown that the well or wells demanded by the lessor would be profitable to the lessee, that is, that they would pay the lessee a profit over and above the drilling and operating costs.
3. Where the lease provides for the payment of delay rentals in lieu of drilling, and the lessor accepts a rental with knowledge of that drainage of his or her land is occurring by offsetting wells, he thereby waives the implied covenant, at least for the period for which the rental is paid (Brown and Brown, 1990).

STATE REGULATION

Many states, as a matter of public policy, regulate the production of oil and gas within their jurisdiction to prevent the waste of the state's natural resources, namely, oil and gas. Many of these regulations concern spacing regulations for wells, exceptions to these spacings, and the conducting of secondary recovery operations. Furthermore, many states regulate and control production. In Texas, for example, the Railroad Commission sets a state-wide, field-wide and well by well production allowable cap (Weaver, 1989). It is beyond the scope of this chapter to examine in detail the kinds of regulations that each state provides, and the reader should contact the proper agency or authority in his or her state regarding such regulations.

References Cited

Black, H. C., 1979, Black's law dictionary: Definitions of the terms and phrases of American and English jurisprudence, ancient and modern: St. Paul, MN, West Publishing Co., 1511 p.

Brown, E. A., and E. A. Brown, Jr., 1990, The law of oil and gas leases, 2nd ed.: New York, Matthew Bender, 1064 p.

Hemingway, R. W., 1983, The law of oil and gas, 2nd ed., West Publishing Co., 543 p.

Illinois Supreme Court, 1913, Northeastern Reporter 2nd.: St. Paul, MN, West Publishing Co., v. 102, p. 1043.

Texas Commission of Appeals, 1927, Southwestern Reporter 2nd: St. Paul, MN, West Publishing Co., v. 299, p. 687.

Texas Supreme Court, 1937, Southwestern Reporter 2nd.: St. Paul, MN, West Publishing Co., v. 101, p. 543.

Texas Supreme Court, 1948, Southwestern Reporter 2nd.: St. Paul, MN, West Publishing Co., v. 210, p. 558.

Texas Supreme Court, 1959, Southwestern Reporter 2nd: St. Paul, MN, West Publishing Co., v. 325, p. 684.

Weaver, J., 1989, Oil and gas, Bar/Bri Texas Bar Law Review: Orlando, FL, Harcourt Brace Jovanovich Legal and Professional Publications, 40 p.

Chapter Appendix

EXHIBIT A: FORM (TEXAS) WITH SHUT-IN AND POOLING CLAUSE

(Modified from *The Law of Oil and Gas Leases*, 2nd ed., Brown and Brown, 1990.)

OIL AND GAS LEASE

THIS AGREEMENT made and entered into the _____ day of __________, 19___, by and between ____________________, Lessor and __________________________, Lessee.

WITNESSETH:

GRANTING CLAUSE

1. Lessor, in consideration of the sum of _______________ Dollars ($__________), in hand paid, receipt of which is hereby acknowledged, and the royalties herein provided, does hereby grant, lease, and let unto Lessee for the purpose of exploring, prospecting, drilling, and mining for and producing oil and gas and all other hydrocarbons, laying pipe lines, building roads, tanks, power stations, telephone lines, and other structures thereon to produce, save, take care of, treat, transport, and own said products, and housing its employees, and without additional consideration, does hereby authorize Lessee to enter upon the lands covered hereby to accomplish said purposes, the following described land in ________ County, Texas, to wit:

PROPERTY DESCRIPTION

____

____

____

____

MOTHER HUBBARD CLAUSE

This Lease also covers and includes any and all lands owned or claimed by the Lessor adjacent or contiguous to the land described hereinabove, whether the same be in said survey or surveys or in adjacent surveys, although not included within the boundaries of the land described above. For the purpose of calculating rental payments hereinafter provided for the lands covered hereby are estimated to comprise __________ acres, whether it actually comprises more or less.

HABENDUM CLAUSE

2. Subject to the other provisions contained this Lease shall be for a term of __________ years from this date (called "primary term") and as long thereafter as oil and gas or other hydrocarbons are being produced from said land or land with which said land is pooled hereunder.

ROYALTY CLAUSE

3. The royalties to be paid by Lessee are as follows: On oil, one-eighth of that produced and saved from said land, the same to be delivered at the wells or to the credit of Lessor into the pipe line to which the wells may be connected. Lessee shall have the option to purchase any royalty oil in its possession, paying the market price therefor prevailing for the field where produced on the date or purchase. On gas, including casinghead gas, condensate or other gaseous substances, produced from said land and sold or used off the premises or for the extraction of gasoline or other products therefrom, the market value at the well of one-eighth of the gas so sold or used, provided that on gas sold at the wells the royalty shall be one-eighth of the amount realized from such sale. While there is a gas well on this Lease, or on acreage pooled therewith, but gas is not being sold or used Lessee shall pay or tender annually at the end of each yearly period during which such gas is not sold or used, as royalty, an amount equal to the delay rental provided for in paragraph 5 hereof, and while said royalty is so paid or tendered this Lease shall be held as a producing Lease under paragraph 2 hereof. Lessee shall have free use of oil, gas and water from said land, except water from Lessor's wells, for all operations hereunder, and the royalty on oil and gas shall be computed after deducting any so used.

4. Lessee, at its option, is hereby given the right and power to voluntarily pool or combine the acreage covered by this Lease, or any portion thereof, as to the

oil and gas, or either of them, with other land, lease or leases in the immediate vicinity thereof to the extent hereinafter stipulated, when in Lessee's judgment it is necessary or advisable to do so in order to properly develop and operate said leased premises in compliance with the Spacing Rules of the Railroad Commission of Texas, or other lawful authorities, or when to do so would, in the judgment of Lessee, promote the conservation of oil and gas from said premises. Units pooled for oil hereunder shall not substantially exceed 80 acres each in area, and units pooled for gas hereunder shall not substantially exceed 640 acres each in area plus a tolerance of ten percent thereof in the case of either an oil unit or a gas unit, provided that should governmental authority having jurisdiction prescribe or permit the creation of units larger than those specified, units thereafter created may conform substantially in size with those prescribed by governmental regulations. Lessee under the provisions hereof may pool or combine acreage covered by this Lease, or any portion thereof as above provided for as to oil in any one or more strata and as to gas in any one or more strata. The units formed by pooling as to any stratum or strata need not conform in size or area with the unit or units into which the Lease is pooled or combined as to any other stratum or strata, and oil units need not conform as to area with gas units. The pooling in one or more instances shall not exhaust the rights of Lessee hereunder to pool this Lease, or portions thereof, into other units. Lessee shall file for record in the county records of the county in which the lands are located an instrument identifying and describing the pooled acreage. Lessee may at its election exercise its pooling operation after commencing operations, for or completing an oil or gas well on the leased premises, and the pooled unit may include, but is not required to include, land or leases upon which a well capable of producing oil or gas in paying quantities has theretofore been completed, or upon which operations for drilling of a well for oil or gas have theretofore been commenced. Operations for drilling on or production of oil or gas from any part of the pooled unit composed in whole or in part of the land covered by this Lease whether or not the well or wells are actually located on the premises covered by this Lease, and the entire acreage constituting such unit or units, as to oil and gas or either of them as herein provided, shall be treated for all purposes except the payment of royalties on production from the pooled unit as if the same were included in this Lease. For the purpose of computing the royalties to which owners of royalties and payments out of production and each of them shall be entitled upon production of oil and gas, or either of them from the pooled unit, there shall be allocated to the land covered by this Lease and included in said unit a pro rata portion of the oil and gas, or either of them, produced from the pooled unit after deducting that used for operations on the pooled unit. Such allocation shall be on an acreage basis, that is to say, there shall be allocated to the acreage covered by this Lease and included in the pooled unit that pro rata portion of the oil and gas, or either of them, produced from the pooled unit which the number of surface acres covered by this Lease and included in the pooled unit bears to the total number of surface acres included in the pooled unit. Royalties hereunder shall be computed on the portion of such production, whether it be oil or gas or either of them, so allocated to the land covered by this Lease and included in the unit just as though such production were from such land. The production from an oil well will be considered as production from the Lease or oil pooled unit from which it is producing and not as production from a gas pooled unit; and production from a gas well will be considered as production from the Lease or gas pooled unit from which it is producing and not from the oil pooled unit.

POOLING CLAUSE

5. If operations for drilling are not commenced on said land, or on acreage pooled therewith as above provided for, on or before one year from the date hereof, the Lease shall terminate as to both parties, unless on or before such anniversary date Lessee shall pay or tender to Lessor, or to the credit of Lessor in the __________ Bank at __________, Texas, (which Bank and its successors shall be

DELAY RENTAL CLAUSE

Lessor's agent and shall continue as the depository for all rentals payable hereunder regardless of changes in ownership of said land or the rentals) the sum of __________ Dollars ($__________), herein called rentals, which shall cover the privilege of deferring commencement of drilling operations for a period of twelve (12) months. In the like manner and upon like payment or tenders annually the commencement of drilling operations may be further deferred for successive periods of twelve (12) months each during the primary term hereof. The payment or tender of rental under this paragraph and of royalty under paragraph 3 on any gas well from which gas is not being sold or used may be made by check or draft of Lessee mailed or delivered to Lessor, or to said Bank on or before the date of payment. If such Bank, or any successor Bank, should fail, liquidate or be succeeded by another Bank, or for any reason fail or refuse to accept rental, Lessee shall not be held in default for failure to make such payment or tender of rental until thirty (30) days after Lessor shall deliver to Lessee a proper recordable instrument, naming another Bank as Agent to receive such payments or tenders. Cash payment for this Lease is consideration for this Lease according to its terms and shall not be allocated as a mere rental for a period. Lessee may at any time or times execute and deliver to Lessor, or to the depository above named, or place of record a release covering any portion or portions of the above described premises and thereby surrender this Lease as to such portion or portions and be relieved of all obligations as to the acreage surrendered, and thereafter the rentals payable hereunder shall be reduced in the proportion that the acreage covered hereby is reduced by said release or releases.

DRY HOLE AND LEASE EXTENSION CLAUSE

6. If prior to discovery of oil, gas or other hydrocarbons on this land, or on acreage pooled therewith, Lessee should drill a dry hold or holes thereon, or if after the discovery of oil, gas or other hydrocarbons, the production thereof should cease from any cause, this Lease shall not terminate if Lessee commences additional drilling or re-working operations within sixty (60) days thereafter, or if it be within the primary term, commences or resumes the payment or tender of rentals or commences operations for drilling or reworking on or before the rental paying date next ensuing after the expiration of sixty (60) days from the date of completion of the dry hole, or cessation of production. If at any time subsequent to sixty (60) days prior to the beginning of the last year of the primary term, and prior to the discovery of oil, gas or other hydrocarbons on said land, or on acreage pooled therewith, Lessee should drill a dry hole thereon, no rental payment or operations are necessary in order to keep the Lease in force during the remainder of the primary term.

OPERATIONS CLAUSE

If at the expiration of the primary term, oil, gas, or other hydrocarbons are not being produced on said land, or on acreage pooled therewith, but Lessee is then engaged in drilling or reworking operations thereon, or shall have completed a dry hole thereon within sixty (60) days prior to the end of the primary term, the Lease shall remain in force so long as operations are prosecuted with no cessation of more than sixty (60) consecutive days, and if they result in the production of oil, gas or other hydrocarbons, so long thereafter as oil, gas or other hydrocarbons are produced from said land, or acreage pooled therewith. In the event a well or wells producing oil or gas in paying quantities shall be brought in on adjacent land and draining the leased premises, or acreage pooled therewith, Lessee agrees to drill such offset wells as a reasonably prudent operator would drill under the same or similar circumstances.

EQUIPMENT REMOVAL CLAUSE

7. Lessee shall have the right at any time during or after the expiration of this Lease to remove all property and fixtures places on the premises by Lessee, including the right to draw and remove all casing. When required by the Lessor, Lessee shall bury all pipe lines below ordinary plow depth, and no well shall be drilled within two hundred (200) feet of any residence or barn located on said land as of the date of this Lease without Lessor's consent.

8. The rights of each party hereunder may be assigned in whole or in part, and the provisions hereof shall extend to their heirs, successors and assigns, but no

ASSIGNMENT AND CHANGE OF OWNERSHIP CLAUSE

change or division in the ownership of the land, rentals or royalties, however accomplished, shall operate to enlarge the obligations or diminish the rights of Lessee; and no change or division in such ownership shall be binding on Lessee until thirty (30) days after Lessee shall have been furnished with a certified copy of recorded instrument or instruments evidencing such change of ownership. In the event of assignment hereof in whole or in part, liability for breach of any obligation issued hereunder shall rest exclusively upon the owner of this Lease, or portion thereof, who commits such breach. In the event of the death of any person entitled to rentals hereunder, Lessee may pay or tender such rentals to the credit of the deceased, or the estate of the deceased, until such time as Lessee has been furnished with the proper evidence of the appointment and qualification of an executor or an administrator of the estate, or if there be none, then until Lessee is furnished satisfactory evidence as to the heirs or devisees of the deceased, and that all debts of the estate have been paid. If at any time two or more persons become entitled to participate in the rental payable hereunder, Lessee may pay or tender such rental jointly to such persons, or to their joint credit in the depository named herein; or, at the Lessee's election, the portion or part of said rental to which each participant is entitled may be paid or tendered to him separately or to his separate credit in said depository; and payment or tender to any participant of his portion of the rentals hereunder shall maintain this Lease as to such participant. In the event of an assignment of this Lease as to a segregated portion of said land, the rentals payable hereunder shall be apportioned as between the several leasehold owners ratably according to the surface area of each, and default in rental payment by one shall not affect the rights of other leasehold owners hereunder. If six or more parties become entitled to royalty payments hereunder, Lessee may withhold payment thereof unless and until furnished with a recordable instrument executed by all such parties designating an agent to receive payment for all.

NOTICE BEFORE FORFEITURE CLAUSE

9. The breach by Lessee of any obligations arising hereunder shall not work a forfeiture or termination of this Lease nor cause a termination or reversion of the estate created hereby nor be grounds for cancellation hereof in whole or in part unless Lessee shall notify Lessor in writing of the facts relied upon in claiming a breach hereof, and Lessee, if in default, shall have sixty (60) days after receipt of such notice in which to commence the compliance with the obligations imposed by virtue of this instrument, and if Lessee shall fail to do so then Lessor shall have grounds for action in a court of law or such remedy to which he may feel entitled. After the discovery of oil, gas or other hydrocarbons in paying quantities on the lands covered by this Lease, or pooled therewith, Lessee shall reasonably develop the acreage retained hereunder, but in discharging this obligation Lessee shall not be required to drill more than one well per eighty (80) acres of area retained hereunder and capable of producing oil in paying quantities, and one well per six hundred forty (640) acres of the area retained hereunder and capable of producing gas or other hydrocarbons in paying quantities, plus a tolerance of ten per-cent in the case of either an oil well or a gas well.

WARRANTY CLAUSE

10. Lessor hereby warrants and agrees to defend the title to said lands and agrees also that Lessee at its option may discharge any tax, mortgage or other liens upon said land either in whole or in part, and in the event Lessee does so, it shall be subrogated to such lien with the right to enforce same and apply rentals and royalties accruing hereunder towards satisfying same. Without impairment of Lessee's rights under the warranty in event of failure of title, it is agreed that if Lessor owns an interest in the oil, gas or other hydrocarbons in or under said land, less than the entire fee simple estate, then the royalties and rentals to be paid Lessor shall be reduced proportionately. Failure of Lessee to reduce such rental paid hereunder or over-payment of such rental hereunder shall not impair the right of Lessee to reduce royalties payable hereunder.

FORCE MAJEURE CLAUSE

11. Should Lessee be prevented from complying with any express or implied covenant of this Lease, from conducting drilling, or re-working operations thereon or from producing oil or gas or other hydrocarbons therefrom by reason of scarcity

of, or inability to obtain or to use equipment or material, or by operation of force majeure, or because of any federal or state law or any order, rule or regulation of a governmental authority, then while so prevented, Lessee's obligations to comply with such covenant shall be suspended, and Lessee shall not be liable in damages for failure to comply therewith; and this Lease shall be extended while and so long as Lessee is prevented by any such cause from conducting drilling or re-working operations on , or from producing oil or gas or other hydrocarbons from the leased premises; and the time while Lessee is so prevented shall not be counted against the Lessee, anything in this Lease to the contrary notwithstanding.

IN WITNESS WHEREOF this instrument is executed on the date first above set out.

______________________ ______________________

______________________ ______________________

______________________ ______________________

SAMUEL B. KATZ is an attorney with emphasis on oil and gas law and litigation in San Antonio, Texas. He received a B.S. degree in Geology from Wayne State University and an M.S. degree in Geology from the State University of New York at Stony Brook. He began his career as an exploration geologist for Marathon Oil Company in 1980, with various assignments ranging from exploration in West Texas to development geology in Tunisia, Egypt, and Norway. Mr. Katz received a Doctor of Jurisprudence Degree from the University of Houston in 1989.

Chapter 24

Types of International Petroleum Contracts: Their History and Development[1]

Samuel B. Katz
Attorney at Law
San Antonio, Texas, U.S.A.

The equation—oil equals power—had already been proven on the battlefields of World War I, and from that conflict emerged a new era in relations between oil companies and nation-states.
—Daniel Yergin, in *The Prize*

INTRODUCTION

The emphasis in petroleum exploration has steadily shifted away from the domestic "oil patch" and toward the international arena. Today an estimated 577.3 billion bbl, or approximately 83% of the world's proven oil reserves, lie outside the western hemisphere. The distribution of these proven oil reserves can be broken down into three major socioeconomic categories—(1) developed market economies, (2) centrally planned economies, and (3) developing countries—indicating that fully 76.7% of the world's proven oil reserves occur in developing countries (Halbouty, 1982). Estimates of the total ultimate recoverable hydrocarbon reserves of the world are 2275 billion bbl of oil and 10,900 TCF of gas. Discounting what has already been discovered, this still leaves an estimated 1166 billion bbl of oil and 4256 TCF of gas as potential reserves (Halbouty, 1982). The majority of the underexplored and unexplored sedimentary basins of the world lie outside the United States. Therefore, to continue to replenish their oil supplies, oil companies must concentrate their efforts on a worldwide basis.

As these multinational oil companies come into contact with the various host countries around the world, their course of action, and the course of the host country's reaction to them, are unfortunately colored by events that proceeded the initial contact. Fears of exploitation, pollution, loss of national pride, and traditions stem from the treatment of host countries at the hands of multinational oil companies in the past. Likewise, the multinational oil company harbors a fear of the expropriation of oil and privatization of their investment which stems from similar incidents occurring in the past.

These concerns have led to the increased involvement of the developing host country in their energy negotiations with multinational oil companies. This involvement has led to the formation of the state or national oil company. The first national oil company (in a developing country) was formed as early as 1922. In 1938, Mexico expropriated and nationalized the foreign oil companies operating within its borders (Grayson, 1980). However, as a reflection of the

[1]This chapter is intended as a review of the history and development of international petroleum contracts for the explorationist. Some of this may be familiar to the explorationist who regularly works in the area of international exploration. What I have attempted to do is to portray the various types of contracts, how they developed, why they developed, and to give a feeling for the reasons why one form may be preferred over another. It is by no means an exhaustive review and is in some ways overly broad. However, it is impossible within the space of one short chapter to go into great detail concerning the various types of contracts and the history of worldwide petroleum exploration. The interested reader is urged to review the references cited for more detailed overviews. Throughout this chapter, I have referred to the modern predecessors, or parent company, of the companies that entered into the various contracts or concessions mentioned. For example, I refer to Exxon even though the contract may have been entered into by Humble or a subsidiary corporation. Lastly, there will be contract types that are not mentioned or new forms of contracts by the time this chapter appears in print.

increased interest in international operations, it is only in the last 30 years that the trend to develop national oil companies has hit full stride. By 1977, there were approximately 80 national oil companies operating in developing countries.

The primary motive for the establishment of a national oil company is the desire by the host country to control their own economic growth by ensuring that their supply of oil is developed and produced in a prudent manner. The means by which a national oil company ensures that this occurs varies from one country to another. Some of the mitigating factors that determine how the host country will act are the socioeconomic level of the host country, the amount and geological setting of the petroleum reserves, and the size of the host country domestic market.

The purpose of this chapter is to examine the delicate relationships between the multinational oil company, the host country, and the national oil company and will primarily focus on the formation of the national oil company and how they developed, in part, as the result of the frictions between the multinational oil company and host country, what role it has played in the evolution of concessions and contracts, and how it aids today in the joint exploration and development of the host country as a partner of the multinational oil company.

OIL EXPLORATION AND THE EVOLUTION OF THE OIL CONCESSION

The search for oil internationally and the involvement of host countries can be broken down into three periods (Turner, 1980). These periods relate to major shifts in the power balance between nations. Two of the shifts were the results of World Wars I and II; the third shift occurred during the 1970s as a result of the changing relationships between the traditional western powers of Western Europe, the United States, and Canada on the one hand and the Third World on the other. The first two shifts resulted in a major expansion of the seven Anglo–American–Dutch major oil companies known as the "Seven Sisters" into territories formerly held by the losers of the two world wars. The Seven Sister companies were Exxon (formerly Standard Oil of New Jersey), Mobil (formerly Standard Oil of New York), Chevron (formerly Standard Oil of California), Royal Dutch Shell (Dutch–British Joint Venture), Gulf (now part of Chevron), Texaco, and the British Petroleum Co. (BP). Not coincidentally, the parent governments of the Seven Sisters were on the winning sides of these two wars (Turner, 1980).

The last shift resulted in a retreat of the Seven Sisters and other multinational oil companies from the older production areas and a redistribution of the wealth generated by the oil industry. Along with this shift has been the increasing importance in the petroleum industry of the national oil companies.

Pre–World War I

Prior to World War I, the oil industry was generally concerned with exploration in the United States, primarily Pennsylvania, the midwest, and Texas; Russia, primarily the Caucasus mountain area; and Indonesia (then the Dutch East Indies) (Turner, 1980). The parent governments of the large oil companies were not generally concerned with the strategic nature of petroleum. Automobiles were not widely used and the world's navies were principally coal fired. The British Royal Navy first experimented with fuel oil as a propulsion system in 1899 (Turner, 1980). By 1904, the First Sea Lord of the British Royal Navy, Admiral Fisher, was convinced of its superiority as a fuel and urged that the British government take a more active role in oil exploration (Turner, 1980). With the threat of war looming, the British government acquired a 53% interest in the Anglo-Persian Oil Company (APOC). The U.S. government was not particularly active in the support of its oil industry abroad and, in fact, had ruled against the Standard Oil Company in 1911 and had fragmented it into pieces.

Post–World War I

The aftermath of the war placed the remnants of the Ottoman Empire within the sphere of British influence. In 1920, ratification of the San Remo treaty by France and England gave the British a mandate over Mesopotamia and assigned the German interests in the Turkish Petroleum Company to France (Turner, 1980). This mandate greatly aided the British oil companies operating within the area. Similarly, the political influence the United States held in Latin America greatly benefited the U.S. oil companies wishing to explore in those regions.

One aspect of World War I was the realization that modern warfare requires large and secure supplies of oil. The interests of the U.S. Navy, and its concern that British dominance over world reserves would prevent it from obtaining the oil it needed in the event of war, spurred American governmental interest in the international oil market. This increased national concern raised oil issues to the diplomatic level. American oil companies complained of exclusionary tactics by the British and Dutch. It became necessary for their parent governments to reach an accord before the oil companies would enter into joint exploration efforts or into agreements prohibiting exclusions.

An example of the influence of diplomatic activity was the flurry of negotiations regarding the entry of

American oil companies into Iraq. The Senate began an investigation which concluded that American interests were being systematically excluded from foreign oil sources (Blair, 1976). Negotiations between the foreign and American companies began in 1922 and culminated 6 years later with some American interest granted in the Iraq Petroleum Company (IPC). American companies were allocated 23.75% of the shares in IPC (Mikesell, 1984). In 1928, the infamous Red Line Agreement decreed that each of the participants in IPC—the Anglo-Iranian Oil Company, Cia Française des Petroles, and the Near East Development Corporation (Mobil and Exxon)—would refrain from entering into separate concession agreements within the area that had once formed the bulk of the Ottoman Empire.

The direct consequence of this agreement was to preclude or greatly limit the expansion of American interests within Iraq and Iran, areas where England had dominant positions. However, some American companies, barred from exploration within the area delineated by the Red Line Agreement, were able to expand into Saudi Arabia and eventually formed the Arabian–American Oil Company (ARAMCO).

The initial Saudi Arabian concession was obtained in 1933 by SOCAL, which created the California Arabian Standard Oil Company (CASOC). A half interest in CASOC was later obtained by Texaco in 1936, and the name was changed to ARAMCO in 1944 (Mikesell, 1971). Additional interest in ARAMCO was also acquired by Exxon and Mobil.

The original Arabian concession ran for 60 years, later amended to 66 years, and covered 318,000 sq. miles. There was also an agreement to grant CASOC the preferential right to meet any offer placed on an adjoining 177,400 sq. miles. As a result of this liberal arrangement, ARAMCO retained a direct and indirect concession for approximately 75% of the total land area of Saudi Arabia. The concession terms were similar to the traditional concessions of the time.

In general, the ARAMCO concession and other early concession contracts, while differing from area to area, had similar features. The following list of common elements is compiled from a list by Mikesell (1971):

1. a definition of the concession area within which the company was given the right to carry an exploration and oil development;
2. a minimum amount of drilling that must be done over a period of time before oil is found in commercial quantities;
3. the duration of the concession, usually 60 to 75 years;
4. the financial obligations of the company, which usually required lump-sum payments, an annual rental fee, and royalties for each barrel of crude oil produced;
5. a provision to supply the domestic oil requirements of the host country either at cost or at discounted rates below world market prices;
6. installation rights and the right of eminent domain; and
7. other express contracted rights, such as freedom of taxation and freedom from production controls.

Oil prices prior to 1950 were set by the multinational oil companies. Royalties to the host countries were determined by the price of crude sold by the multinational oil companies to their downstream operations. Prices set were generally not the same across the board. For example, in 1947 ARAMCO contracted to sell crude to France for $0.95 a barrel, to Uruguay for $1.00 a barrel, and to the U.S. Navy for $1.23 a barrel (Sampson, 1975). In addition, the oil companies had fixed the price of oil to an indexed "Gulf plus" system which had been agreed to in the 1920s. Under this system, the price of delivery of crude to Europe was equivalent to the price of crude in the Gulf of Mexico plus freight at the published rates to Europe less the cost of freight of crude from either the eastern Mediterranean or the Persian Gulf (Sampson, 1975). This practice meant that the relatively inexpensive crude from the Middle East was sold in Europe at the higher American cost.

Under the traditional concession pricing agreement, the producing countries had no role other than to wait for the periodic royalty payments made by the multinational oil company. Venezuela began the break with this system in the late 1930s. Venezuela demanded that oil contracts be revised to allow for higher royalties and taxes in return for a 40-year renewal (Sampson, 1975). As a coercive measure, they threatened nationalization of the multinational oil companies if they would not comply. The oil companies, under the prodding of the U.S. State Department, gave in to these demands. In 1945, Venezuela had a new and radical party in power and a new oil minister, Perez Alfonso (Sampson, 1975). Under his direction, Venezuela demanded a 50/50 share in all oil profits and passed a law in 1948 that established the Venezuelan government as a partner with the multinational oil companies in its oil industry.

Other countries dealt with their problems with the multinational oil companies in different ways. As previously mentioned, Mexico nationalized its foreign-owned oil concessions in 1938. This nationalization occurred as a result of the apparent arrogance of the multinational oil companies. The view of many of the concession owners in Mexico was that they owned the mineral rights under their concession grants, which adhered to the 1884 mineral laws of Mexico. Following the Mexican Revolution, the Mexican Constitution of 1917 was ratified. Article 27 of this constitution described the right of the nation to the ownership of its petroleum deposits as "inalienable and imprescriptible"

(Sampson, 1975). In 1927, the Mexican Supreme Court ruled that those companies, which had held concessions before 1917, had perpetual rights on their concessions (Sampson, 1975). In 1934, Mexico elected a new president, Cardena, who was sympathetic to the labor forces of Mexico. The petroleum workers formed a union in 1935, and in 1936 this union petitioned for a collective bargaining contract (Sampson, 1975). The conflict between the oil workers' union and the oil trusts led to the expropriation of the property of 17 American and European corporations on March 18, 1938, and the formation of Pretroleos Mexicanos (PEMEX) on June 7, 1938 (Sampson, 1975).

Argentina approached the problem of dealing with the multinational oil companies in a different way. Its national oil company, Yacimientos Petroliferos Fiscales Argentinos (YPF), was founded in 1922 (Ghadar, 1983). Under its charter, YPF was granted the exclusive right to all future oil exploration and production. Exxon, Shell, and others were already exploring for and producing oil in Argentina under concessions dating back to 1910 (Ghadar, 1983). Argentina tried other ways to enlarge its share of its domestic oil industry. It constructed a refinery and cracking plant and contracted with Auger y Compania (a local company) to market Argentinian oil domestically.

By 1928, YPF had obtained a 14.6% interest in the domestic retail gasoline market. Along with YPF, Exxon controlled 45.9%, Shell 27.6%, and others 11.9% (Ghadar, 1983). These percentages remained fixed for several years due to the Achnacarry Agreement of 1928 (Ghadar, 1983). Among the seven principles of this agreement, it was agreed that each member accept and maintain the market status quo. Argentina continued to increase YPF's market share by the ingenious method of granting YPF additional reserves and sole exploration and production rights. The multinational oil companies began to lose interest in Argentina since they were unable to explore for new production. By 1937, YPF had more than doubled its market share to 32.47%. Exxon's share declined to 21.17%, Shell's share increased to 29.83%, and the others increased to 16.53% (Ghadar, 1983). In this sense, Argentina obtained control of its oil industry, not by expropriation, but by passive legislation.

Post–World War II

After World War II, the nature of the international oil business began to change as the host countries and their fledgling national oil companies began to exert more control over their natural resources. The effect of the emergence of the host country into the control phase of the oil business reduced the relatively overwhelming bargaining position that the multinational oil companies had enjoyed in its dealings with the host countries when negotiating concession agreements. In addition, the entrance of the independent oil company into the international arena also affected the way the multinational oil companies and host countries would interact. These independent oil companies are companies that were not part of the Seven Sisters and are smaller than a typical multinational oil company. Some examples of these companies are Marathon Oil, Conoco, Phillips Petroleum, and Amerada-Hess. Generally, to compete with the Seven Sisters companies, the independent oil companies would negotiate agreements that were much more advantageous to the host country.

An example of this liberalization of contract terms was the concession waivers given by the independent oil companies in Libya. In 1955, Libya had drafted its Petroleum Law to provide a legal framework for the development of the Libyan oil industry (Waddams, 1980). The Petroleum Law divided Libya into four zones that would be further subdivided into concessions. The third and fourth zones were in the most remote areas, and concessions in these zones carried more lenient rental terms and working obligations (Waddams, 1980).

Libya intended for its 1955 Petroleum Law to induce oil companies to pursue exploration activities there. However, they were concerned that their natural resources would become dominated by the major multinational oil companies. To ensure that this did not happen, the Libyan government decided to grant what was considered its most favorable concessions to the independents. Furthermore, major oil companies could not hold contiguous areas in the most favored areas (Waddams, 1980).

The Petroleum Law set the fees and rents initially at low levels to encourage applications. Initial fees, in British pounds sterling, were set at £500.00, payable upon the awarding of the concession (Waddams, 1980). The rents per 100 km^2 ranged from £5 to £10 for the first 8 years, £10 to £20 for the following 7 years or until petroleum was discovered in commercial quantities, and £2500 for the remainder of the concession (Waddam, 1980). The amount of rent incurred was dependent upon which zone the concession was located in.

The Libyan government soon realized that the terms it had granted were extremely generous. Beginning with the 73rd concession, the oil companies began to offer various incentives to the Libyan government. These incentives took the following forms: (1) waiver of depletion allowances, (2) increased royalties, (3) bonus payments, or (4) offers of government participation in the event of commercial development (Waddams, 1980).

The final and major blow to the supremacy of the multinational oil company in the international arena was the 1973 oil embargo by the Middle Eastern OPEC countries. OPEC, having been founded in the early

1960s, remained relatively ineffectual as a market force while oil was in plentiful supply. The petroleum shortages of the 1970s brought the OPEC countries into prominence as the marketplace determinant of the price of crude oil. In addition, no longer did these countries have to rely on the multinational oil company to market their crude. They began to negotiate their own contracts for the sale of crude with importing countries. By the end of the 1970s, the majority of the Middle Eastern OPEC countries nationalized their petroleum industries.

These changes in the nature of the petroleum industry forced changes in the ways the multinational oil company contracted with the host countries. The traditional concession contract gave way in most cases to service agreements that included *sharing contracts, pure service contracts,* and *risk service contracts.* The concession agreement still exists today, but not in the traditional sense. The major difference between concession agreements and the various types of contracts used today occurs in the way production is divided between the multinational oil company and the host country. The following discussion covers the major contract types, their structures, and examples from various contracts.

CONCESSION AGREEMENTS

The traditional concession agreement is no longer in use. However, the concession agreement in some form is still used in some areas of the world. For example, the concession agreement is found in Tunisia, Pakistan, and the Sudan.

Under the terms of the *modern concession agreement,* the host country generally has no direct involvement in the management of petroleum operations and petroleum. It receives its pay in the form of royalties, income taxes, bonuses, and other minor taxes (Mikesell, 1984). The main features of the modern concession agreement are as follows (Blinn, 1986):

1. The multinational oil company, at its own risk and expense, generally has the exclusive right to explore for and exploit petroleum reserves in the concession area.
2. The multinational oil company owns the production from within its concession area and is free to dispose of it subject to its contractual obligation to supply the host country's domestic market.
3. During the exploration and exploitation phase, the multinational oil company is subject to pay surface rentals to the host country.
4. The multinational oil company, at the election of the host country, pays the royalty either in cash, production, or a combination of the two.
5. The multinational oil company pays taxes to the host country on profits it derives from the production.
6. The multinational oil company owns the equipment and installations used in its operations.

A great disadvantage to the host country of the concession agreement is that it greatly limits the involvement by the host country. Under the concession agreement, the host country generally has no say in the development of its natural resources. Furthermore, its citizens generally do not get a chance to acquire the technical skills the host country sorely lacks. These skills are necessary for a host country to obtain if they want to pursue an independent energy course. Therefore, under the concession agreement, the host country remains dependent upon the multinational oil company to manage and develop its petroleum industry. This one particular disadvantage can be overcome by the host country requesting that the multinational oil company employ and train its nationals in the necessary skills.

Joint Venture Agreement

A variation on the modern concession agreement is the *joint venture agreement.* The joint venture agreement, unlike the modern concession agreement, provides for the host country to share with the multinational oil company the risk and expense of the exploitation phase. Generally, the multinational oil company will carry the project solely through the exploration phase.

The joint venture agreement evolved as a means to address some of the deficiencies inherent in the modern concession agreement. The primary aspect was to have a managerial say in the day-to-day operations of the producing fields. As oil production increased, royalty payments to the host country increased and the host country's economy became more dependent on the generated revenues. For example, between 1966 and 1969, Saudi Arabia received 90% of its government revenues from oil (Stevens, 1976). As the proportion of oil revenues increased, the host country would want to have more control over this vital portion of their economy. In addition, the host country's desire for control over managerial decisions would also stem from a desire for the host country to see that the oil sector play an important role in the development of its domestic economy (Stevens, 1976). This could take the form of the training of nationals by the multinational oil company to take jobs in the oil sector or a requirement by the host country that the multinational oil company purchase supplies from local firms. Finally, there is the political question of control. Host countries have felt threatened by the domination of this vital sector of their economy by foreign companies.

An example of a host country utilizing the concession and joint venture agreement is Colombia.

Commercial production in Colombia began as early as the 1920s. The first commercial oil was produced by the Tropical Oil Company, a subsidiary of Exxon (Mikesell, 1984). This concession eventually expired, and the assets were taken over by the Colombian national oil company (Ecopetrol). Prior to 1969, the multinational oil companies operated under a concession agreement. Royalties ranged from 11.5% to 14.5%, and income taxes were collected from the multinational oil company on their net profits (Mikesell, 1984). Concession agreements dictated exploration phases ranging from 3 to 4 years and development phases from 30 to 40 years with a provision allowing for a 10-year extension (Mikesell, 1984). The Colombian government passed Law 20 in 1969, which stated:

> The government will be entitled to declare any petroleum area of the country as a national reserve and assign it to Ecopetrol, either for direct exploitation or for development with either public or private capital (Mikesell, 1984).

This law also authorized Ecopetrol to enter into contracts of association with the multinational oil companies in which Ecopetrol would receive royalties and a share in the production. These association contracts are, in reality, joint venture agreements. The contracts provided that all development and operating expenses be shared 50/50 by Ecopetrol and the contracting company (Mikesell, 1984). Ecopetrol would reimburse the company for 50% of the exploration cost of the discovery well. However, the contract requires that production is to be distributed on a 50/50 basis after payment of a royalty to the host country, which ranges from 16% to 30%. The multinational oil company receives 40% of the output, but the royalty paid is calculated on the full output (Mikesell, 1984).

PRODUCTION SHARING CONTRACTS

The concept of the production sharing contract evolved in Indonesia. In 1960, the Indonesian government passed Oil and Mining Law No. 44, which basically stated that all oil and natural gas production was to be conducted by the state. The state-owned oil company was authorized to engage in petroleum exploration and production on behalf of the state (Mikesell, 1984). The concession agreements then in existence were, after negotiation, converted to "contracts of work" with the multinational oil company operators acting as contractors for the state.

The first *production sharing contract* was entered into in 1966 between Permina (later Pertamina) and the Independent Indonesia American Petroleum Company (Iiapco). This contract contained the basic principles that have been incorporated into every production sharing contract entered into by Indonesia ever since. These principles have been stated by Mikesell (1984) as follows:

1. Pertamina has responsibility for the management of petroleum operations, and the contractor is responsible to Pertamina for the execution of such operations in accordance with provisions of the contract.
2. The contractor provides all financing and technology required for the operations and bears the risk of production costs.
3. During the term of the contract, total production, after allowance for operating costs, is divided between Pertamina and the contractor in accordance with provisions of the contract.
4. The contractor must submit annual work programs and budgets for scrutiny and approval by Pertamina.

As can be expected the provisions of the production sharing contract have changed over the years. The provisions most susceptible to change have been the production sharing ratios (which determine what percentage of production the state and the multinational oil company will receive), the method of determining the production costs, and the method of handling and paying the Indonesian income tax (Mikesell, 1984).

The production sharing contract concept has spread widely since its inception in Indonesia and has become the most widely used type of petroleum contract in non-OPEC countries (Mikesell, 1984). The production sharing contracts have several basic features that remain the structural backbone of the contract. These features, as discussed in Blinn (1986), are as follows:

1. The multinational oil company is appointed by the host country as the contractor for a certain area.
2. The multinational oil company operates at its sole risk and expense under the control of the host country.
3. Any production belongs to the host country.
4. The multinational oil company is entitled to a recovery of its costs out of the production from the contractual area.
5. After cost recovery, the balance of production is shared on a pre-determined percentage split between the host country and the multinational oil company.
6. The income of the multinational oil company is liable to taxation.
7. Equipment and installations are the property of the host country, either at the outset of production or progressively in accordance with agreed upon amortization schedules.

The production sharing contract offers several advantages for the host country. First, it allows the developing host country with little monetary resources the ability to have its natural resources developed at no cost to itself, unlike the joint venture contract. Second, the host country gets to retain any equipment and installations in the country. For a third world developing host country, this keeps in the country important infrastructure with which it can build an independent industry. For the more developed host country, the production sharing contract can and has been modified to allow for a joint venture arrangement (Blinn, 1986). In these arrangements, the national oil company shares in all phases of the venture, as outlined previously. The incentive for this type of arrangement is that in proved low risk areas, the host country can increase its share of the revenues. In addition, the host country can benefit by having its national personnel employed in its national oil company work directly with personnel of the multinational oil company and gain invaluable experience (Mikesell, 1984).

SERVICE CONTRACTS

The term *service contract* encompasses those various contracts in which the host country contracts with a service company or a multinational oil company for the performance of a task. There are two basic types of service contracts—the *pure service contract* and the *risk service contract*. There are fundamental differences between these two types and each represents an end-member in a spectrum of agreements in which the contractor either performs a service for a set fee or entirely assumes the risk of the project.

Service contracts originated in Argentina. The Argentineans have contracted with both types of service contracts, the production sharing contract and the risk service contract. Unfortunately, Argentina petroleum policy practically changes with each change in administration. Argentina had sought to increase its domestic oil production so as to cut imports and conserve foreign exchange (Mikesell, 1987). In 1958 and 1959, the President of Argentina, Frondizi, signed a series of contracts known as the Frondizi contracts with private companies (Mikesell, 1987). The contracts called for the companies to commence a program of oil exploration, drilling, and development, which was beyond the scope of the national oil company, Yacimientos Petroliferos Fiscales (YPF).

The problem stemmed back to the 1920s. YPF was first organized in 1922. Soon afterward, the government granted YPF certain oil lands, known as the state reserves, and the necessary capital to enter the petroleum industry. However, by 1934, 60% of the total Argentinian production was still being produced by outside companies (Mikesell, 1984). In 1935, the state reserves were increased in size and private concessions were limited. All new exploration and development was to be solely handled by YPF. By 1957, the private companies were producing only half of the amount they were in 1935, but YPF had only managed to increase its production from 26% of the total in 1935 to 33% of the total in 1957. As a direct consequence of the production fall off, oil imports to Argentina doubled between 1947 and 1955 (Mikesell, 1984).

In 1953, President Peron passed Law 14,222, which established conditions for allowing foreign capital to aid in the development of energy resources (Mikesell, 1984). Tentative negotiations took place, but nothing had been done by the time Peron was removed by a military coup in 1955 (Mikesell, 1971). In 1958, Frondizi, who had advocated oil self-sufficiency as part of his campaign platform, took office as President. He had advocated that the work to bring Argentina to petroleum independence should be performed by YPF. That position was later abandoned. Between 1958 and 1961, the Argentinian government contracted with numerous multinational oil companies (subsidiaries of Exxon, Shell, Marathon, Tenneco, Standard Oil of Indiana, Cities Service, Kerr-McGee, and Union Oil of California) to perform three basic services; drilling, development, and exploration/development. Under the drilling contracts, the multinational oil company was responsible for the drilling of a specified number of wells, in a specified time, for a given area. Payment was calculated on the amount of footage drilled and the per-hour completion costs. Once completed, the wells were to be turned over to YPF.

Under the development contracts, the companies drilled development wells in known oil-producing areas for YPF. The companies were paid by YPF under a prearranged fee schedule. The fee was to be determined by the amount of crude produced and delivered to YPF. Finally, under the exploration and development contracts, the work was to be performed for YPF in rank wildcat to semi-proven areas. The risks of exploration were borne by the contracting companies, and payment was conditioned on whether oil was found in commercial quantities. For example, in the case of Shell, payment was to be in crude, while most other contracts were for payment in money (Mikesell, 1984).

Two types of service contracts originated in Argentina and have spread to other areas. Other countries that use service contracts are Brazil and Peru, among others. The two different types of service contracts are discussed here.

Pure Service Contract

A pure service contract is a simple arrangement whereby a host country contracts for a service from a multinational oil company. This oil company acts as a contractor in the performance of the service to the host country. One example of a pure service contract is the ARCO contract with Saudi Arabia (Barrows, 1983). ARCO runs and manages the production operations for the government and receives in return a net fee of U.S. $15 per bbl (plus inflation) after taxes (Blinn, 1986). ARCO also conducts exploration for Saudi Arabia, and the risk is paid back with the additional crude found, plus an additional U.S. $6 per bbl on the production.

There are two types of pure service contracts—the service contract *with* crude purchase and the service contract *without* crude purchase. The former type is an offshoot of the takeover of concessions by a host country. The multinational oil company is in a position of producing the crude oil or gas for the host country while reserving the right to purchase part of the petroleum produced. In some cases, for providing services to the host country, the oil company may elect to be paid with a percentage of the crude produced while providing the service. An example of this arrangement occurs in Qatar (Blinn, 1986). The multinational oil company performs their services, which consist of providing trained personnel and technical advice to the national oil company through a local wholly owned company. This company is remunerated by receiving a set fee per barrel produced. The second type of pure service contract, the service contract without crude purchase, occurs when a national oil company needs technical expertise and assistance, but does not wish to grant a share of the petroleum produced.

Risk Service Contract

The risk service contract places all the risk and investment on the contractor. The contractor provides the capital for all exploration and production (Blinn, 1986). If no commercial discovery is made, the risk service contract ceases to exist. If a commercial discovery is made, the contractor completes the discovery, develops the field, and places the production on stream. Depending on later negotiations, the day-to-day production operations may be performed by the national oil company or by the contractor (in which case it is really a pure service contract). The capital invested is reimbursed with interest and a fee. In some cases, a set established fee may also be contracted for the petroleum and the contractor may purchase at a predetermined price.

PROBLEMS FACING THE DEVELOPING HOST COUNTRIES

What is the rationale for a host country to invite foreign companies in to explore and develop its natural resources? The contracts previously discussed are, after all, merely a means to achieve that end. The type of contract the host country will enter into with a multinational oil company differs from country to country and from company to company. In fact, the type of contract entered into may even vary between the same host country and multinational oil company merely because circumstances have changed with time.

Some of the reasons a host country contracts with a multinational oil company have been summed up by Alleyne (1982):

1. Exploration is risky. The chances of finding crude are slim.
2. An enormous amount of financial resources must be available to fund even a modest exploration effort.
3. To increase the odds of finding crude, it is generally necessary both to expand the area undergoing exploration and to increase the intensity of effort.
4. Most exploration efforts, except for the occasional lucky ones, will take a long time to produce positive results. Therefore, with each passing year, the total amount of capital investment increases. This places an enormous burden on a cash-poor host country.
5. Even when success has been achieved, it is usually only after a long period of time. Meanwhile, exploration activity continues and generally involves testing less promising areas, thereby requiring more capital investment by a host country.

Technology

In addition to the factors just listed, the host country has an interest to see whether it can negotiate for the transfer of technology to aid in its development of its natural resources. Many host countries are lacking in both the technical expertise and skilled personnel needed to run modern petroleum enterprises.

The term *transfer of technology* is a misnomer. Technology is an intangible entity that represents an ever-changing flux of knowledge, tools, and equipment. Alleyne (1982) states that the best way to accommodate a developing host country is for them to contract for:

1. the right to acquire and use certain aspects of the relevant technology;
2. the right to develop, within their own spheres of operation, the atmosphere conductive to a familiarization with the basic scientific and technical ideas and operations relevant to the industry; and
3. the right to have their own personnel trained in the petroleum industry.

The last two points are the most critical. They would allow the host country to derive the most benefits from the rights to use the technology. The host country can contract with a multinational oil company for both patents and services through any of the contracts previously discussed. However, the benefits derived cease when the contract expires and the multinational oil company leaves.

Financial Problems

As stated previously, enormous amounts of capital are needed to fund even the most modest exploration effort. Considering that the average per capita income in many of these countries is only between $300 and $400, a significant problem is evident. To finance and develop their resources, the host country must contract with the multinational oil company. The multinational oil company, for its own sake, will only generally contract if the following conditions, listed by Alleyne (1982), are met:

1. The multinational oil company believes that there is a chance of success. The multinational oil company will rely on the knowledge and expertise of its own specialists—geologists, geophysicists, engineers, and risk analysts.
2. The multinational oil company receives a concession, or a license with a concession possibility, if the prospect is justified by the evaluation of its specialists. The host country may view the concession idea as unfavorable in light of the more favorable production sharing contract, but may treat this in terms of a justified loss of a small amount of its territory in the hope that if petroleum is discovered, it can negotiate for more favorable terms in the adjacent areas and/or attract other companies.
3. Depending on the particular host country, it may try to contract for a joint venture contract, pure service contract, or service contract between its national oil company and the multinational oil company.

The developing host country must generally obtain the necessary funds for financing through its capital budget, from either external or internal loans, or from grants from other countries or institutions. This places a further strain on already strained economies. The richer host countries will be the ones that are able to negotiate for and obtain the more favorable contracts and contractual terms, while with the poorer host countries, the multinational oil company is in the better position for contractual negotiations.

References Cited

Alleyne, D. H. N., 1982, Transfer of Petroleum Exploration Technologies: Views of the Developing Countries, *in* Petroleum exploration strategies in developing countries: London, Graham & Trotman Limited, in cooperation with the United Nations, p. 127-146.

Blair, J. M., 1976, The control of oil: New York, Pantheon Books, 441 p.

Blinn, K. W., 1986, International petroleum exploration and exploitation agreements: Legal, economic, and policy aspects: New York, Barrows, 346 p.

Ghadar, F., 1983, The petroleum industry in oil-importing developing countries: New York, Free Press, 217 p.

Grayson, G. W., 1980, The politics of Mexican oil: United Nations Symposium, Onstate Petroleum Enterprises in Developing Countries, 211 p.

Halbouty, M. T., 1982, Petroleum exploration strategies in developing countries: in Proceedings of the United Nations meeting in Dehay, March 16–20, 1981, United Nations Natural Resources and Energy Division, Department of Technical Cooperation for Development, 372 p.

Mikesell, R. F., 1971, Foreign investment in the petroleum and mineral industries: Case studies of investor–host country relations: Washington, D.C, Johns Hopkins Press for Resources for the Future, Inc., 459 p.

Mikesell, R. F., 1984, Petroleum company operations and agreements in the developing countries: Washington, D.C., Johns Hopkins Press for Resources for the Future, Inc., 147 p.

Sampson, A., 1975, The Seven Sisters: The great oil companies and the world they created: New York, Viking Press, 334 p.

Stevens, P. J., 1976, Joint ventures in Middle East oil, 1957–1975: Beirut, Middle East Economic Consultants, 205 p.

Turner, L., 1980, Oil companies in the international system, 2nd ed.: Winchester, MA, Allen & Unwin, 254 p.

Waddams, F. C., 1980, The Libyan oil industry: Baltimore, Johns Hopkins University Press, 338 p.

SAMUEL B. KATZ is an attorney with emphasis on oil and gas law and litigation in San Antonio, Texas. He received a B.S. degree in Geology from Wayne State University and an M.S. degree in Geology from the State University of New York at Stony Brook. He began his career as an exploration geologist for Marathon Oil Company in 1980, with various assignments ranging from exploration in West Texas to development geology in Tunisia, Egypt, and Norway. Mr. Katz received a Doctor of Jurisprudence Degree from the University of Houston in 1989.

Chapter 25

Analysis of International Oil and Gas Contracts

Sidney Moran
Consultant
Houston, Texas, U.S.A.

If terms look too good to be true, they probably are.
—Sidney Moran

INTRODUCTION

The purposes of this chapter are (1) to describe the main types of international petroleum contracts, (2) to give the reader methods for comparing contract qualities and analyzing the profitability of petroleum field models under basic contract terms, (3) to describe methods for assessing tradeoffs between various terms of a contract, (4) to describe briefly U.S. taxation of international petroleum extraction income, and (5) to discuss certain special considerations that apply to the international arena.

There is extensive literature on the description, analysis, and comparison of specific international contracts. This chapter is aimed at familiarizing the petroleum geologist with general methods of analyzing contracts in order to assess whether the field qualities indicated by technical studies would produce acceptable economic results.

BACKGROUND

Arrangements between host countries and multinational petroleum companies have evolved over many decades toward increasing host country involvement in petroleum operations (often through a government petroleum company). In addition, host countries have tended to take increasingly large shares of gross profits, a trend especially noticeable after the dramatic price rises of 1973 and 1979. However, following the oil price collapse of 1986, there have been improvements in many contracts to encourage companies to pursue petroleum operations in a lower total gross profit environment.

Among the advantages obtained by host countries in dealing with multinational companies are (1) avoiding the financial risks of petroleum exploration, (2) obtaining the use of a highly sophisticated technology (often transferred to host country nationals), (3) financing of all phases of exploration and production, and (4) obtaining marketing arrangements for produced hydrocarbons. These advantages are obtained while the host country still receives the lion's share of gross profits. The multinational companies in turn obtain exploration rights to large areas and, when successful, may produce larger volumes at lower unit costs than would be typical in a domestic venture.

BASIC CONTRACT TYPES

There are many styles of petroleum exploration and production contracts around the world. Almost all are variants of two basic types: *tax royalty participation* and *production sharing*. A third type—the service contract, in which the contractor receives a fee for exploration and/or production services—is rarely attractive to exploration companies. However, where a per-barrel fee on very large volumes is involved, this contract structure can provide acceptable results. We will deal with the two more widely used forms in this chapter.

Tax Royalty Participation Contract

The tax royalty participation contract was used almost exclusively until recent decades because it conforms to the style of most other business arrangements. Under this system, the contractor explores for, develops, and produces hydrocarbons. The production revenues are subject to royalty payments (sometimes in kind) and to taxes of the host government or its agencies. In many countries, the government may participate as a partner in the development and production phases, sometimes on a carried basis.

Variations include sliding scale royalties and special profits taxes. The contractor generally controls the disposition of his or her working interest share of hydrocarbons, but many contracts include assurances that domestic needs are met (sometimes on a reduced price basis). Examples of countries using this form of contract include the United States, Canada, Australia, Colombia, the United Kingdom, Norway, France, Italy, most other European countries, and New Zealand.

Production Sharing Contract

In the production sharing contract, the contractor explores for, develops, and produces hydrocarbons. An agreed upon share of gross hydrocarbon revenues, often referred to as *cost oil,* is made available to the contractor for recovery of exploration, development, and operating costs. Production capital, and sometimes exploration costs, may be recovered on an amortized basis. Cost oil is usually figured on a contract area basis rather than a field by field basis. Frequently, a royalty to the government is taken from the gross. The remaining hydrocarbon revenue, called *profit oil,* is then split according to an agreed upon formula. This formula may involve a volume-related sliding scale. The contractor's share of profit oil may be subject to an income tax, but it may also be received on a tax paid basis. The host government sometimes retains the right to participate in development. Examples of countries using this contract type include Indonesia (which pioneered it), Malaysia, Egypt, China, and Angola.

Most production sharing contracts have a clearly stated cost oil limit, or maximum share of petroleum sales that can be used to recover costs. If the contractor's actual costs exceed this limit in a given year, unrecovered costs can be carried forward, but if there is a cumulative unrecovered cost balance, no mechanism for recovering the shortfall exists. It is essential that a contractor negotiate an adequate cost oil limit.

The type of contract offered is not as important as the specific terms. The production sharing contract carries a cost oil risk, but it also has the advantage of making available unused cost oil (within the contract-wide cost oil limit) for the development of high cost fields that would otherwise have been marginal.

Table 1. U.S. Federal Offshore Gross Profit Split

Gross revenue	100	
Less costs		30
Gross profit	70	
Less royalty		16.7
Contractor profit before tax	53.3	
Less tax (34%)		18.1
Contractor profit after tax	35.2 (35.2/70 = 50%)	
Government–Contractor split	50%–50%	

The geologist plays a key role in assessing the potential volumes and possible field qualities. These are essential in analyzing and negotiating contract provisions. The geologist bears a major responsibility in the decision of whether the contract terms justify a major exploration and production effort.

GROSS COMPARISONS OF CONTRACT QUALITY

Gross contract qualities can be compared for radically different contract types by carrying a standard model through the contract terms to arrive at a bottom line profit split between the host government and the contractor. The calculated split varies depending on the ratio of total exploration, development, and production costs over the life of the contract to total gross field revenues. Selecting a fairly typical cost to revenue ratio, such as 30%, permits a reasonable comparison of contract quality. In applying this method, revenues are expressed as 100 units and other elements are expressed as fractions of revenues. Table 1 illustrates this procedure for a U.S. federal offshore lease having the following parameters: (1) royalty of 16.67%, (2) corporate tax rate of 34%, and (3) cost to revenue ratio of 30%. The bottom line split of gross profits between the government and the contractor is 50%–50%.

Table 2 illustrates the same procedure for examples of the two major types of international contracts. Though radically different in style, these contracts have the same bottom line profit splits. The profitabilities of fields would differ, however, depending on such factors as amortization rates and, even more important, whether the host country's participation is straight-up or carried. Arraying the contract qualities of the host countries of interest to the exploration contractor is a useful first step in international contract analysis. While a company's share of gross profits in a moderate cost field is about 50% in the United States, in most international contracts, the shares range generally from 10% to 40%, with 20–30% being common.

Table 2. International Profit Split Examples

	Terms of Country A Tax Royalty (%)		Terms of Country B Production Sharing (%)	
Royalty rate	20		10	
Tax rate	55.4		35	
Participation	30		None	
Profit oil share	—		40	
Cost oil limit	—		40	
Calculation of Government–Contractor Profit Split				
Gross revenues	100		100	
Less costs		30		30
Gross profit	70		70	
Less royalty		20		10
Remaining share	50		60	
Contractor profit oil share (40%)	—		24	
Less tax		27.7 (55.4%)		8.4 (35%)
Profit on 100%	22.3		15.6	
Less government participation		6.7 (30%)		—
Contractor profit	15.6		15.6	
Contractor share of gross profits: 15.6/70 = 22.3%				

Table 3. Production Models

Year	CAPEX[a] ($ million)	OPEX[b] ($ million)	Production (MMbbl/Year) Case A	Case B
1	40			
2	50			
3	60			
4	50	14	7.00	14.00
5		14	7.00	14.00
6		14	5.95	11.90
7		14	5.06	10.12
8		14	4.36	8.72
9		13	3.65	7.30
10		13	3.10	6.20
11		13	2.64	5.28
12		13	2.24	4.48
13		12	1.91	3.82
14		12	1.62	3.24
15		11	1.34	2.68
16		10	1.17	2.34
17		10	1.00	2.00
18		10	0.85	1.70
19		8	0.72	1.44
20		5	0.39	0.78
Totals	200	200	50.00	100.00

[a]Capital expenditures.
[b]Operating expenses.

MODEL FIELD PROFITABILITY ANALYSIS

Once an attractive technical opportunity has been identified, geologists, geophysicists, and engineers should jointly estimate potential volumes and prepare field models of possible development projects. International contract areas often encompass hundreds of thousands or even millions of acres and may contain multiple plays of various prospect sizes. The profitabilities of an array of field models should be tested against contract terms. Often, the terms are subject to negotiation or competitive bidding, and the impact of changes in contract variables must be assessed.

Contract Analysis Example

Let's assume that an exploration and development team has estimated that individual field volumes in an area of interest should be in the 50–100 MMbbl range. Development and operating costs in constant dollars are expected to be about $200 million each over the life of the field. Costs and volumes can vary over a wide range. We can analyze the effect of these variations by applying multipliers to the cost data, as illustrated here. The base case models proposed for analysis are given in Table 3.

The contracting company uses the following economic premises:

- Year 1 oil price: $21 per bbl, increasing 5% per year
- Development and operating cost inflation: 5% per year
- Contractor discount rate for total E & P effort: 11%

Assume that contract terms must be negotiated with the host government. In the tax royalty analysis, the variable is the effective tax rate, and in the production sharing analysis, the variable is the profit oil share. The ranges of terms are indicated in Table 4.

In a tax paid production sharing contract with no royalty, the contractor's share of profits is equal to the profit oil share. The tax rates used as variables in the tax royalty case were chosen to produce the same bottom line profit split as in the production sharing case when costs amount to 30% of revenues. (The profit split in the tax royalty contract will change slightly in response to moderate changes in this ratio.)

The field models were analyzed by a computer, applying the contractor's economic premises. Sensitivities to faster capital amortization and to the impact of exploration expenditures were examined. An extra low cost case was also created. Tables 5, 5a, 6, and 6a summarize the results. Referring to Tables 5 and 6, note that the 50-MMbl field model produces only a 17–18% internal rate of return, even without considering exploration expenditures. If cost estimates are reasonable, then better terms are required for profitable development of fields of this size.

Table 4. Terms of Contract Analysis Example

Terms	Tax Royalty	Production Sharing
Capital amortization	10% per year	10% per year
Royalty	20%	None
Tax rates	58%, 65%, 72%	Tax paid contract
Profit oil shares	—	30%, 25%, 20%
Cost oil limit	—	40%

For the models considered, production sharing results turned out to be worse than tax royalty results when costs were high. Note, for example, that for the 50-MMbl base case, the tax royalty discounted profit at a 58% tax rate is about $57 million (Table 5). The comparable production sharing result (30% profit oil share) is about $49 million (Table 6). The difference is due to a shortfall in cost recovery. When revenues generated within the contractual limit on cost oil are inadequate, there is no provision to make up the shortfall. For the 100-MMbbl case with the same terms, the percentage difference in results is much smaller and is in fact due to a better bottom-line profit split (the contractor share is 32% in the tax royalty and 30% in the production sharing contract) rather than any inherent advantage in the contract type.

The inclusion of $50 million in exploration expenditures for a 100-MMbbl field (spent uniformly over a 2-year period prior to development) reduced the total internal rate of return by about 10 percentage points. The impact on discounted profit, however, was not as dramatic, ranging from about $35 to $40 million. It is simpler to consider the impact of exploration expenditures on a total play basis rather than on a field-by-field basis. (Compare Tables 5 and 5a with Tables 6 and 6a.)

THE JOHNSON PLOT

A useful way to display an array of profitability data has been devised by R. K. Johnson, who was manager of Exploration and Production Economics at Pecten International until his retirement in 1989. His plot makes use of the fact that the cash flow of a field model can be disaggregated into three elements—(1) revenues, (2) capital, and (3) operating costs—with revenues being positive and capital and operating costs being negative.

Capital is spent early in the project life, and tax write-offs or cost recovery come later from a revenue stream. Operating costs are spent, on average, near the mid-life of the project and are written off or recovered immediately. It was found that if the discounted per-barrel field profit (in inflated dollars) is plotted against the sum of unit capital plus one half of unit operating

Table 5. Tax Royalty Contract: Field Model Profitability Summaries Using 20% Royalty with Various Tax Rates

Case	Undiscounted Profit ($ million)	Internal Rate of Return (%)	Profit, disc. @11% ($ million)
Base			
50 MMbbl/58% tax	283.6	18.0	56.9
50 MMbbl/65% tax	236.3	15.3	34.5
50 MMbbl/72% tax	189.1	12.6	12.1
100 MMbbl/58% tax	804.8	39.5	282.0
100 MMbbl/65% tax	670.7	34.2	222.0
100 MMbbl/72% tax	536.6	28.5	162.1
100% CAPEX, 150% OPEX			
50 MMbbl/58% tax	213.8	15.5	33.2
50 MMbbl/65% tax	178.2	13.0	14.7
50 MMbbl/72% tax	142.5	10.6	(3.8)
100 MMbbl/58% tax	732.6	38.1	257.9
100 MMbbl/65% tax	610.5	32.8	202.0
100 MMbbl/72% tax	488.4	27.3	146.0
150% CAPEX, 100% OPEX			
50 MMbbl/58% tax	237.0	10.8	(3.0)
50 MMbbl/65% tax	197.5	9.1	(21.9)
50 MMbbl/72% tax	158.0	7.4	(40.8)
100 MMbbl/58% tax	758.3	27.2	222.0
100 MMbbl/65% tax	631.9	23.4	165.6
100 MMbbl/72% tax	505.5	19.4	101.2
150% CAPEX, 150% OPEX			
50 MMbbl/58% tax	167.2	8.5	(27.0)
50 MMbbl/65% tax	139.3	7.0	(41.9)
50 MMbbl/72% tax	111.5	5.7	(57.0)
100 MMbbl/58% tax	686.0	26.0	197.9
100 MMbbl/65% tax	571.7	22.3	145.5
100 MMbbl/72% tax	457.3	18.4	93.1

Table 5a. Sensitivity Studies for Tax Royalty Contract: Field Model Profitability Summaries at 20% Royalty with Various Tax Rates

Case	Undiscounted Profit ($ million)	Internal Rate of Return (%)	Profit, disc. @11% ($ million)
Low cost 50% CAPEX, 50% OPEX			
100MMbl/58% tax	923.7	70.8	366.0
100MMbl/65% tax	769.7	62.3	298.5
100MMbl/72% tax	615.8	52.9	231.1
Base with 5 yr CAPEX amortization			
100MMbl/58% tax	804.8	42.7	296.9
100MMbl/65% tax	670.7	37.8	238.7
100MMbl/72% tax	536.6	32.5	180.6
Base with $50 million exploration program			
100MMbl/58% tax	783.8	27.4	243.7
100MMbl/65% tax	653.2	24.1	185.9
100MMbl/72% tax	522.6	20.5	128.0

Table 6. Production Sharing Contract: Field Model Profitability Summaries at 40% Cost Oil with Various Profit Oil Shares

Case	Undiscounted Profit ($ million)	Internal Rate of Return (%)	Profit, disc. @11% ($ million)
Base			
50MMbl/30% PO	256.1	17.2	48.8
50MMbl/25% PO	204.5	14.3	25.4
50MMbl/20% PO	156.0	11.5	2.5
100MMbl/30% PO	761.0	37.3	259.6
100MMbl/25% PO	634.2	32.2	203.4
100MMbl/20% PO	507.4	26.9	147.2
100% CAPEX, 150% OPEX			
50MMbl/3O% PO	154.8	14.0	16.1
50MMbl/25% PO	112.5	10.1	(4.7)
50MMbl/20% PO	71.8	5.0	(25.7)
100MMbl/30% PO	705.4	36.2	241.8
100MMbl/25% PO	586.8	31.3	188.4
100MMbl/20% PO	468.3	26.0	130.0
150% CAPEX, 100% OPEX			
50MMbl/30% PO	185.0	9.3	(17.9)
50MMbl/25% PO	137.0	7.1	(39.5)
50MMbl/20% PO	90.6	4.8	(60.9)
100MMbl/30% PO	727.8	26.0	205.7
100MMbl/25% PO	606.5	22.3	152.0
100MMbl/20% PO	458.2	18.6	98.3
150% CAPEX, 150% OPEX			
50MMbl/30% PO	71.4	4.0	(56.4)
50MMbl/25% PO	31.1	2.9	(75.9)
50MMbl/20% PO	(7.8)	—	(95.1)
100MMbl/30% PO	672.0	25.0	187.6
100MMbl/25% PO	559.1	21.5	136.7
100MMbl/20% PO	446.1	17.8	85.8

Table 6a. Sensitivity Studies for Production Sharing Contract: Field Model Profitability Summaries at 40% Cost Oil with Various Profit Oil Shares

Case	Undiscounted Profit ($ million)	Internal Rate of Return (%)	Profit, disc. @11% ($ million)
Low cost 50% CAPEX, 50% OPEX			
100MMbl/30% PO	845.9	66.2	330.7
100MMbl/25% PO	704.9	58.1	269.1
100MMbl/20% PO	564.0	49.2	207.5
Base with 5 yr CAPEX amortization			
100MMbl/30% PO	761.1	41.2	277.6
100MMbl/25% PO	634.2	36.4	222.6
100MMbl/20% PO	507.4	31.2	167.7
Base with $50 million exploration program			
100MMbl/30% PO	746.1	26.3	224.9
100MMbl/25% PO	621.7	23.1	170.2
100MMbl/20% PO	497.4	19.6	115.5

costs (expressed in uninflated dollars), the values would lie close to a straight line. The slope of the line represents the impact of cost changes. When field profit is tested against a variable term, the plot points can be represented by a family of subparallel lines whose spacing represents the impact of changes in the variable.

Furthermore, field model internal rates of return form a family of lines crossing the unit profit lines, radiating from the origin of the plot. Johnson plots of field model data from Tables 5 and 6 are given in Figures 1 and 2.

These plots are excellent tools for estimating the profitability of untested field models based upon unit costs only, and the total venture can be assessed by considering total expected volumes and average unit costs. If an exploration and production model analysis is desired, the unit exploration cost can be added to the unit capital cost component of the plot.

Discussion of Johnson Plot Results

Referring to Figures 1 and 2, note that the straight lines drawn through the data points fit the actual data closely. These lines were produced by a linear best-fit graphics program, but a hand-drawn fit would be satisfactory. An even better fit would be obtained by eliminating the high cost data points from consideration. These may reflect the impact of delays in tax write-offs due to inadequate operating income and, more significantly, cost oil shortfalls (see Figure 2). The high cost data points are important in indicating profitability problems associated with high cost field development, but they need not be allowed to influence the trendlines of the mainstream fields.

The spacing between the lines indicates the impact of the variables of tax rates or profit oil shares. For example, note in Figure 1 that at a $4 unit cost level, a 7% change in the tax rate produces a $0.55 per bbl change in the unit profit, or almost $0.08 per 1% change in the tax rate. Note in Figure 2 that a 5% change in the profit oil share also produces a $0.55 per bbl change in the unit profit, or about $0.11 per 1% change in profit oil share. The $0.55 per bbl total change is not coincidental because the changes in the variables of tax and profit oil had been selected to produce about the same effect on the bottom line.

The plot provides guidance about maximum tolerable unit costs for achieving satisfactory profitabilities for a range of contract terms. For example, in the production sharing contract, if costs in the $4 range are anticipated, and if the contractor believes that a unit discounted profit of $2 per bbl is required to justify the risks, then a profit oil share of about 30% must be negotiated. If this cannot be done, the contractor must realize that he must discover some lower cost fields if he is to succeed.

Similar considerations apply to the tax royalty contract. To analyze an international venture, it is crucial to understand the relationships among contract terms, economic projections, and the field qualities that can be anticipated. A company may decide to gamble on discovering a very low cost, high volume field in a tough contract area, but it should clearly understand that most of the expected range of results may not be economically attractive.

TRADEOFFS OF CONTRACT TERMS

Many countries do not have fixed petroleum contract terms. Instead, they negotiate various elements of the contract or call for competitive bids. You must understand the relative impact of various terms to be able to negotiate effectively the best possible contract. The more common variables within the two basic contract styles are listed on Table 7.

In negotiating a contract, you may need to know how much additional royalty you could offer to reduce the amortization rate from 10 to 5 years, or how a 1% royalty change compares in value to a 1% tax rate change. Tradeoffs should optimize the profit, discounted at the company rate, or the discounted percent profit (discounted profit divided by discounted capital investment) without incurring undue risk. The tradeoff estimates depend on economic assumptions, field models, and in some cases, the initial values for the terms to be changed. During contract negotiations, guidelines should be given to the front line negotiators.

Because of the variety of styles within the basic contract types, it would be difficult to cover all of the tradeoffs that might be considered, but these tradeoffs can be tested by sensitivity analysis of field models. Often, elements in a tradeoff can be quickly assessed from Johnson plots. Some examples follow.

Profit Oil Share Versus Amortization Rate

This tradeoff requires the estimation of (1) the worth of more rapid cost recovery versus (2) a reduction in profit share. Referring to Tables 6 and 6a, note that for the 100-MMbbl field for which an amortization sensitivity was analyzed, total discounted profit increased by about $18 million when amortization was reduced from 10 to 5 years. The unit capital cost in the model is $2 per bbl in uninflated dollars. Thus, the total improvement is seen to be $9 million per unit dollar of capital for 100 MMbbl, or $0.09 per dollar per bbl. The 100-MMbbl field would not necessarily be considered typical of the venture. (Furthermore, our example assumes that all capital costs are amortized, but many contracts provide for immediate recovery of intangible capital costs.) Let's assume that we expect the average barrel to have $2.50 in amortized capital. Thus, the benefit of the reduction in amortization period would be 2.5 x $0.09, or $0.225 per barrel.

We had previously noted from Figure 1 that each 1% change in profit oil affected the discounted value per barrel (at the anticipated $4 per bbl unit cost level) by about $0.11. Therefore, the reduction in the amortization period is worth about 2 percentage points of profit oil.

There are several points to consider in this tradeoff. First, tradeoff values are very sensitive to amortized capital, and if costs are much lower or higher than anticipated, the tradeoff values can change significantly. Second, the value of profit oil will be sensitive to future prices, which could differ greatly from projections. Third, there is normally excess cost oil available within the cost oil limit in the early years of production, and higher amortization rates can be accommodated. As production declines, unit costs rise and capital amortization may have to compete with operating costs for the available cost oil revenue. Finally, rapid amortization increases the contractor's cash flow in the early years. This reduces payout time and can minimize project cash exposure as well.

Royalty Versus Working Interest in a Tax Royalty Contract

Assume in our tax royalty model that a 58% tax rate has been agreed upon and that the contractor has offered a 20% royalty and a 15% straight up working interest. The host government insists either that the royalty must be raised from 20 to 25% or that the working interest must go from 15 to 25%. Which is preferable to the contractor?

Let's assume once again that the contractor expects the cost to be about $4 per bbl on the Johnson plot (Figure 1). To estimate the impact of a 1% royalty change, we note that the unit profit at the 58% tax level projected back to the ordinate is $4.50 per bbl. This would be the discounted profit per barrel, at zero cost, subject only to a 20% royalty and a 58% tax rate (that is, the value of an 80% royalty). From this relationship, we find that the discounted profit per barrel, based on 100% of the revenue subject to a 58% tax, is $4.50/(1 – 0.2 royalty), or $5.62 per bbl, which is the value of a 100% royalty. Thus, each 1% royalty is worth $0.056 per bbl, so that a 5% increase in royalty would cost $0.28 per bbl on a 100% working interest basis. Since at this point in the negotiations, the contractor has an 85% working interest, the cost to the contractor would be 0.85 x $0.28, or $0.24 per bbl.

Next, we note from Figure 1 that at a $4 unit cost, a 100% working interest share generates a profit of about $2.25 per bbl. Thus, a 10% increase in the government working interest share would decrease the contractor's profit by $0.22 per bbl. We can also see from Figure 1 that increasing the working interest of the government

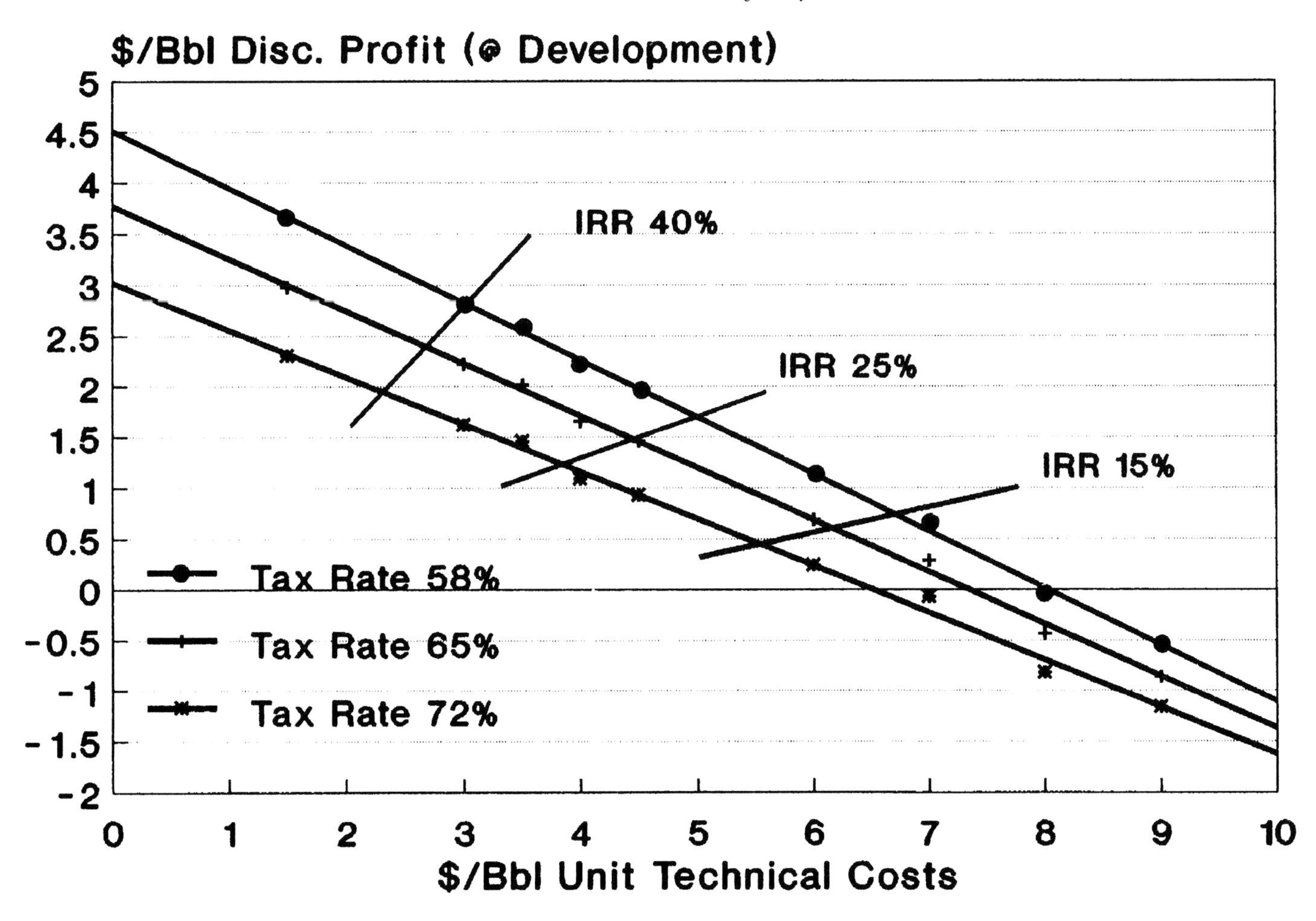

Figure 1. Graph of tax royalty contract profitability calculations. Unit discounted profit is shown versus unit technical costs, along with tax rates.

is a better choice down to a cost of about $3.70 per bbl. This is about the point where a 10% working interest is worth $0.24 per bbl.

Another consideration in this tradeoff is that, by selecting the higher working interest alternative, the profit earned by the contractor is generated by a smaller capital investment since the government is a straight-up partner in this example. Also, a high royalty burden increases the field size that can be profitably developed. For these reasons, the working interest option would be preferable for the contractor.

Tradeoff Between Royalty and Tax Rate in a Tax Royalty Contract

From the Johnson plot (Figure 1), we noted that at the $4 per bbl cost level, each 1% change in the tax rate resulted in an $0.08 change in discounted profit, and from our last example, we found that a 1% change in royalty affected profit by $0.056 per bbl. From this relationship, it is evident that a 1% change in the tax rates equals a 1.4% change in royalty (in unit profit impact on working interest).

Royalties, which come off the top, are one of the most damaging elements in a contract. If the financial tradeoffs involving a royalty are close to even, it is best to reject a royalty increase in favor of an element able to adjust itself to field quality, such as a tax rate.

Many other tradeoffs arise during contract negotiations and it is hoped that these examples will prove useful to those involved in analyzing them.

U.S. TAXATION OF INTERNATIONAL EXPLORATION AND PRODUCTION VENTURES

U.S. tax rules covering international ventures of U.S. based companies are complex, so experts in the field should be consulted. One of the most important rules is that results of all U.S. incorporated international subsidiaries (engaged in petroleum extraction) of a company are aggregated for U.S. tax computations. Properly styled international taxes can be credited (or,

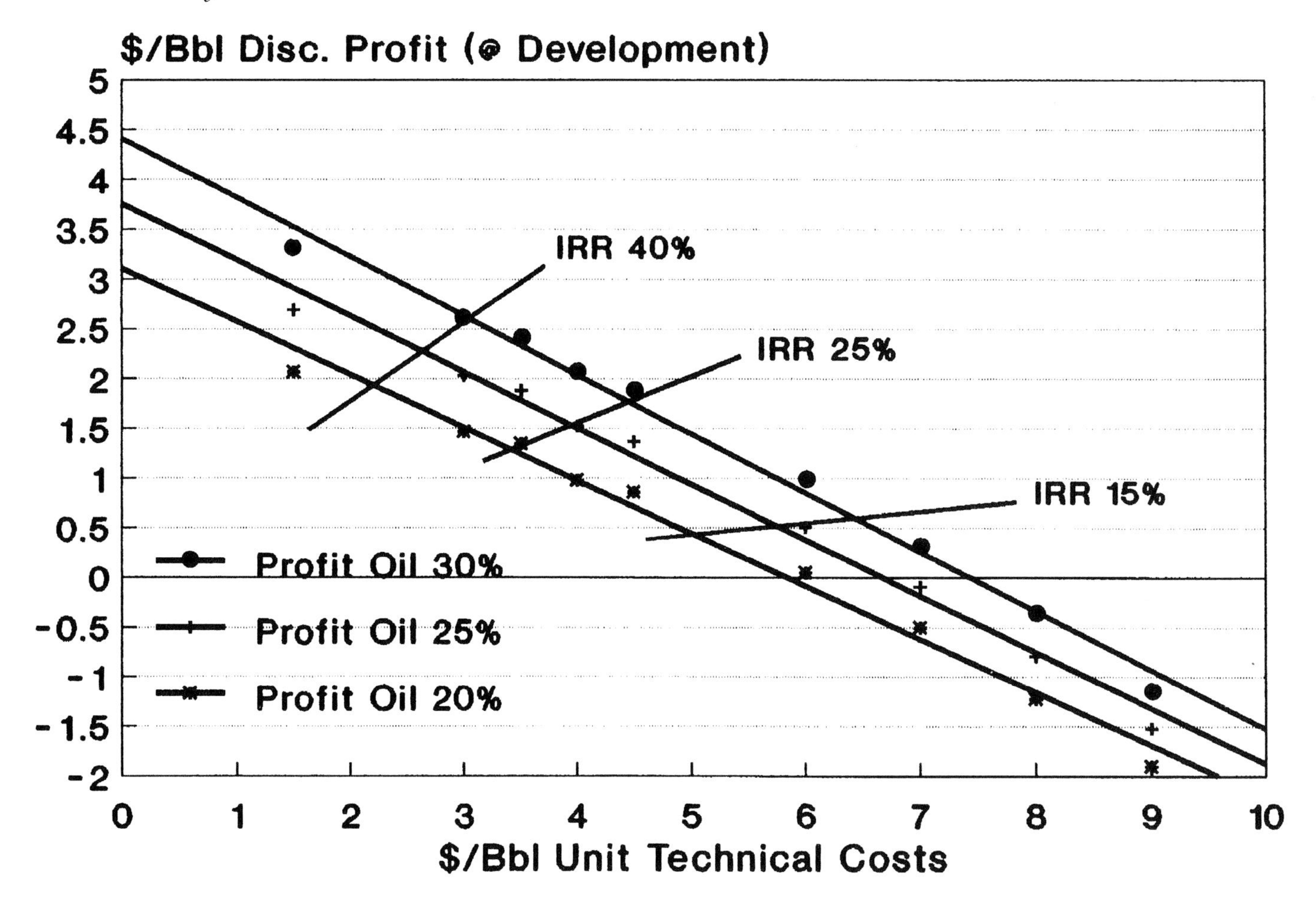

Figure 2. Graph of production sharing contract profitability calculations. Unit discounted profit is shown versus technical costs, along with profit oil shares.

alternatively, deducted) in calculating U.S. tax liabilities. If the aggregate of international taxes paid is greater than the U.S. taxes computed on the aggregate result, then no U.S. taxes would be due. (Further considerations involve the ratio of income from international sources to total company results.)

If a company elects to credit international taxes paid against computed U.S. tax liabilities on international operations, these international taxes cannot also be deducted in computing the U.S. tax liabilities. Furthermore, when crediting of international taxes is elected, U.S. taxes can never be negative (that is, a credit against domestic income tax cannot be generated). If, after deducting international taxes paid, aggregate results show a loss, this loss can be used to offset domestic income. Under these circumstances, an election to deduct would be advantageous.

Another rule requires that international losses be cumulated and kept on the books. If aggregate results later turn positive, there is a mechanism under which the U.S. tax system in effect recovers the prior deduction benefits by limiting the use of international tax credits.

Table 8 gives examples of aggregates for which the choice of crediting or deduction could be made in a given tax year. In Case 1, U.S. net income on the total before deducting international taxes is $75 million. If deduction of international taxes were elected, then taxable income would be $25 million and a U.S. tax (at the corporate rate of 34%) of $8.5 million would be due. If crediting of international taxes were chosen in Case 1, a U.S. tax of $25.5 million would be computed (34% of $75 million). Since international taxes of $50 million had been paid, this credit would completely offset the U.S. liability and no taxes would be due. Therefore, crediting is advantageous in this case.

In Case 2 in Table 8, U.S. net income before deduction of international taxes is zero and thus no U.S. taxes would be due. However, if international taxes were deducted, there would be a U.S. taxable loss of $50 million. This loss could be used to offset domestic income, resulting in a tax savings of $17 million, so that election to deduct would be advantageous.

International tax rules have changed many times over the years. At present, the bottom line is that the overall U.S. tax impact on long-term international operations should be neutral to slightly positive, provided that (1) international taxes are properly styled

Table 7. Common Contract Variables

Tax Royalty	Production Sharing
Tax rate	Tax rate
Royalty rate	Royalty rate
Government participation	Profit oil share
Amortization	Cost oil limit
Local supply obligation	Local supply obligation
Rate of return triggers for special taxes	Volume triggers for changes in profit oil share
Bonuses	Government participation
	Amortization
	Bonuses

so as to be creditable in the United States and (2) international income is not from countries affected by special U.S. legislation.

EXPLORATION COSTS IN INTERNATIONAL VENTURES

The preceding discussion of taxes implies that the costs of unsuccessful international exploration can ultimately cost more than domestic exploration. This is because when a company is successful overall, early deductions for unsuccessful exploration are later recaptured for U.S. tax purposes. Thus, such costs are not deductible either in a host country or (permanently) in the United States. Total international exploration costs, however, may not be very different from those in the United States since exploration costs in successful ventures are largely recoverable in the host countries.

Consider 100 units of exploration spent in the United States. At a tax rate of 34%, the after-tax cost will be 66 units. Assume that in an international program, ultimately half of the exploration costs are in unsuccessful ventures and half are in successful ventures. (This does not imply that half of the ventures undertaken are successful since the total costs of an exploration program in an ongoing successful venture are usually many times that of the cost of failure in an unsuccessful venture.) In addition, assume that half of the successful exploration is in tax royalty countries with average tax rates of 55% and that half is in production sharing countries with average profit oil shares of 30%. (This implies a cost of exploration of $0.30 per dollar since unused cost oil goes into the profit oil share.) The aggregate after-tax cost of international exploration would then be

(0.50 unsuccessful exploration)
+ (0.25) x (0.45 after tax in tax royalty countries)
+ (0.25) x (0.30 after cost recovery in production sharing countries)
= 69 units per 100 before tax units.

This is not very different from U.S. costs. U.S. deductions on domestic exploration would normally come sooner than tax and cost oil recoveries in international ventures, making the disparity in the discounted cost of exploration somewhat higher than the undiscounted cost. Overall, international after-tax costs should be roughly comparable to those of domestic exploration.

SPECIAL CONSIDERATIONS FOR INTERNATIONAL CONTRACTS

There are many additional considerations for international contracts beyond those normally considered for domestic exploration and production ventures. These are discussed in the following sections.

Political Risks

Some host countries face internal and/or external threats that can affect the long-term stability of petroleum ventures. Governments may unilaterally change contract terms, especially if petroleum prices rise suddenly. (The windfall profits tax had a similar effect domestically.) Quantitative allowance for political risk is a specialized field beyond the scope of this chapter, but this risk must be considered in assessing ventures.

Some countries that do not yet have production may have very old petroleum laws that are unrealistic by today's standards. If terms look too good to be true, they probably are. Thus, it is best to negotiate updated versions of a contract before beginning exploration rather than wait for a forced renegotiation after discoveries have been made. If outdated laws are not changed, then the contractor should test his field models against newer contracts in the area to be sure that the venture would be viable.

Repatriation of Funds

The contractor should insist that profits can be promptly repatriated to avoid the prospect of an otherwise successful venture's generating funds that can only be reinvested in the host country. Payment procedures, currency exchange rate determinations, and other fiscal matters should always be clearly set forth.

Dollar Accounting

In countries with high inflation rates, it is essential that costs for later recovery (either through cost oil or tax writeoffs) be recorded in dollars, lest their value

Table 8. Examples of Aggregate International Source Income (in $ million)

	Case 1			Case 2		
	Country A	Country B	Total	Country A	Country B	Total
Revenue	200	—	200	200	—	200
Expense	100	25	125	100	100	200
Net income	100	(25)	75	100	(100)	—
International tax	50	—	50	50	—	50

literally melt away. Countries with relatively stable currencies may not be willing to agree to this, and hedging techniques are available that can minimize currency exchange rate risks if desired.

Overheads

It is very important that a contract contain realistic provisions for recovery of overhead charges. In some cases, items such as engineering and design can be included in capital costs. However, many legitimate overhead costs in international operations should be recoverable. If they are not, a large unrecovered overhead burden can serve as a significant drag on the project cash flow.

Price Caps

Some countries impose petroleum price caps on the theory that prices above those cap levels would generate windfall profits. With these caps, excess values for petroleum go directly to the government. Such caps may not take into account that much of the rise in prices can be attributed to inflation, nor that the government (which generally gets the lion's share of profits in any case) would automatically benefit from such rises. Prices are a two-way street, and there are no provisions for relief if they drop below expected levels. If a price cap is unavoidable, it is very important that it be indexed for inflation.

Local Supply Obligations

Some contracts state that the contractor must supply a portion of its produced petroleum to satisfy domestic needs. If a concessionary price is also included, such provisions have the effect of a hidden royalty. A contractor should attempt to negotiate for market prices.

Sliding Scale Terms

Many production sharing contracts have sliding scale profit oil shares related to total contract producing rates on the theory that larger volumes usually mean higher profits. Thus, the increasing share to the government automatically adjusts the gross profit split as production results justify it. The contract must be worded so that only the incremental volume above the trigger point is subject to the change.

A sliding scale profit oil provision can increase the marginal field size since a small field discovered late can be subject to tougher provisions than earlier, better fields. However, this effect may be offset by the benefits of contract-wide cost recovery, which can provide for full, rapid capital recovery.

Some tax royalty contracts have special taxes, triggered when certain overall contract area rates of return are achieved. Such provisions need to be analyzed very carefully by the contractor, and it is essential that the rates of return be indexed for inflation. Even then, there is a danger that a small but prolific low cost discovery can achieve a high rate of return early in the contract life. This can adversely affect the profitability of all later projects by subjecting them to stiffer terms.

Furthermore, since the impact of this type of provision can be controlled in part by the phasing of investment and production, there could be conflicts with the government on these matters. Nevertheless, a contract with such provisions, if carefully drafted, can still be attractive.

As overall results improve, sliding scale and rates of return provisions will trigger automatic increases in the share of profits for the host government. Therefore, including these provisions should help to forestall unilaterally imposed contract changes when unexpectedly good results are obtained. In the absence of such provisions, there is a risk that host governments will unilaterally impose tougher terms.

Contract Life

The duration of the contract should fully cover typical exploration ventures. There is a tendency to build field models with up to 20 years of production, but many fields actually produce much longer than this. Also, new development may continue for many years. Profitability analysis places little discounted value on income received many years hence. Late stage production can nonetheless be very important to the contracting company and especially to those running the company when the contract expires. Every effort should be made to get contracts lasting to the end of economic production, or at least 25–30 years.

Farm-In Agreements

When taking a farm-in for international exploration, an overriding royalty to the originating party should be avoided. This is because, if terms are toughened by the host country, the override can become increasingly burdensome and can damage the venture substantially. Granting back-in participation rights on development, net profit shares, or cash payments are preferable choices.

Sensitivity Analysis

Significantly lower hydrocarbon prices than those projected for base cases can have devastating effects on profitability, particularly in production sharing contracts, if cost oil is inadequate for full cost recovery. Unused cost oil in most contracts goes into the profit oil share, so that the actual effective recovery of costs is 100% minus the profit oil share after tax. This usually amounts to 70–80%. However, if available cost oil is inadequate, then excess costs cannot be recovered. It is important that a cost oil limit that is adequate under the contractor's reasonable minimum price scenario be negotiated.

All contract provisions that may have discontinuous impact (such as cost oil limits, sliding scale trigger levels, and rate of return triggers) should be carefully examined by computer analysis of models.

Work Program

Many countries grant contracts through competitive bidding or by negotiation. They may place great value on the work program offered (seismic, drilling, and so on) and may trade better fiscal terms for greater exploratory effort. It is a good strategy to get the most mileage possible out of the work program by bidding or offering the full amount necessary to test the exploration concept adequately.

Contract Structure

Although a contractor may not be able to affect significantly the structure of a host country contract, maximum development of hydrocarbons benefits both the host country and the contractor. Therefore, a contract that encourages development of all significant discoveries is desirable for both parties. Contracts that give a large share of revenues off the top to the host country or that provide inadequate cost oil limits preclude the development of potentially profitable fields. One producing country in Latin America has a contract providing for a production split of 50–50 between the contractor and the government, leaving the contractor responsible for all costs. This can produce excellent profitability for low cost fields, but as costs rise, the contracting company's profitability deteriorates rapidly. A neighboring country offers the contractor a biddable share of gross revenues, along with recovery of capital costs with interest. This contract form can still provide the lion's share of profit to the host government in good fields, but it also encourages the development of lower quality fields.

Maximum Cash Sinks

In many international areas, a contractor may be seriously concerned about exposing any more cash than is necessary to undertake an exploration and production project. The maximum cash sink can often be controlled by project phasing. In countries having high political risk, a contractor may be willing to forego some profit to minimize cash at risk. Insurance programs are also available to reduce risk.

Posted Prices

Some countries use "official prices" instead of market prices in the settlement of royalty and tax provisions. This should be avoided if at all possible. A posted price 7% above market price in a tax royalty contract, having a 15% royalty rate, a 55% tax rate, and costs equal to 30% of revenues, can reduce the contractor's cash flow by about 20%! Provisions for determining market value should be set forth in a contract.

Documentation of Costs

It is essential that the contractor and the host government have a clear procedure for documenting costs for cost recovery or tax writeoff and that the contractor carefully maintain such documentation.

Official Language

For U.S. companies, it is advantageous to have an official English version of the contract, but this is often not possible. In this case, a very carefully translated copy is most important. When a contract contains complex sliding scale provisions, a numerical example illustrating the operation of such provisions should be included to avoid later misunderstanding.

Arbitration

There should be clear provisions for arbitration to settle disagreements between a contractor and the host government.

Special Provisions

Most contracts provide for the training and employment of nationals and for some host country involvement in planning and operations. Finally, it is important that a contract be perceived as fair and reasonable by both parties. Negotiation with knowledgeable government representatives who fully understand petroleum economics is preferred. If host country representatives do not understand petroleum economics, or if they are dissatisfied with the final results of negotiations, the contact is not likely to be honored over the long term.

ANALYSIS OF INTERNATIONAL VENTURES

There is no difference in principle between analysis of domestic and international ventures. The bottom line is that the venture must have a positive expected value. This means that it must meet the following condition:

$$P_S \times \text{(Discounted profit on E \& P)} > (1 - P_S) \times \text{(Discounted cost of failure)}$$

where P_S is the probability of success. Once volume, cost, and profit estimates have been made as carefully as possible, the required minimum probability of success can be compared with the actual assessment to see if there is a comfortable margin for error.

Although rules of thumb involving rates of return on successful ventures are sometimes used, they do not properly account for differing costs of failure and probability of success between ventures of comparable potential. A full-scale probabilistic approach is recommended for each venture.

Beyond individual venture analysis, a company must have an overall strategy consistent with its exploration philosophy and its capital strength. A series of high risk, high reward ventures can wipe out an exploration program through "gambler's ruin," and thus a mix of ventures should be undertaken.

CONCLUSIONS

Success in international exploration and production ventures requires not only recognizing attractive technical opportunities but also carefully analyzing the contracts under which the operations are to be undertaken. Specific contract provisions that are properly matched with the technical opportunity are the ingredients of a successful international venture. Extensive computer analysis of field models, singly and in aggregate, should be made (1) to determine acceptable terms, (2) to understand profitability tradeoffs between contract elements, (3) to understand the sensitivity of profitability to changes in contract elements or economic assumptions, and (4) to understand the impact of trigger points for special provisions related to volumes, rate of return, or other elements. In addition, it is important that there be a clear understanding between the contractor and the host government concerning accounting procedures, repatriation of funds, overheads, applications of all contract provisions, and other essential elements for a long-term, mutually beneficial venture.

Acknowledgments *The author wishes to acknowledge the generous assistance of the Pecten International subsidiary of Shell Oil Company for performing the many computer profitability analyses of field models that were required for this chapter. R. K. Johnson and R. Steinmetz reviewed the text and offered many helpful suggestions. Ione Moran contributed to the clarity through her editing assistance.*

SIDNEY MORAN is a consultant in Houston on international projects and evaluates and acquires domestic oil and gas properties and royalty interests for investors. He received a B.A. degree from the University of Chicago in 1951 and B.S. and M.A degrees in Geology from the University of Texas in 1953 and 1955. Mr. Moran joined Shell Oil Company in 1955 in Midland, Texas, and was engaged in Permian Basin exploration until 1969, when he was transferred to Shell's Southwestern Exploration and Production Economics department in Houston. In 1973 he was assigned to the International Ventures group (later changed to Pecten International) and was engaged in international venture evaluation until his retirement as Geological Advisor in 1988.

Chapter 26

International Exploration by Independents

Robert G. Bertagne
Consultant
Houston, Texas, U.S.A.

Toto, I have a feeling we're not in Kansas anymore.
—Dorothy, in *Wizard of Oz*

INTRODUCTION

Recent industry trends indicate that the smaller U.S. independents are looking at foreign exploration opportunities as one of the alternatives for growth in the new age of exploration. To get involved overseas, however, requires an adaptation to different cultural, financial, legal, operational, and political conditions. Generally, foreign exploration proceeds at a slower pace than domestic exploration because concessions are granted by the government or are explored in partnership with the national oil company.

The problems of communications and logistics caused by different cultures and by geographic distances must be carefully evaluated. A mid-term to long-term strategy tailored to the goals and the financial capabilities of the company should be prepared and followed by a careful planning of the operations.

This chapter addresses some aspects of foreign exploration that should be considered before an independent ventures into the foreign field. It also provides some guidelines for conducting successful overseas operations. When properly assessed, foreign exploration is well within the reach of smaller U.S. independents and presents no greater risk than domestic exploration; the rewards, however, can be much larger. Furthermore, the *Oil & Gas Journal* surveys of the 300 largest U.S. petroleum companies show that companies with a consistent foreign exploration policy have fared better financially during difficult times.

DOMESTIC VERSUS FOREIGN EXPLORATION

An analysis of foreign and domestic exploration includes a comparison of the following factors.

Finding Costs

It is accepted that the finding costs per barrel are substantially lower internationally. While this cost is about $4 domestic (or $6 to purchase reserves), it costs on the average between $1 and $3 to discover a barrel of oil overseas, with the average being about $2 (Figures 1 and 2).

Reserve Potential

Large reserve potential is offered in many international plays while becoming increasingly limited domestically. Multiple field discoveries ranging from 50 to 500 MMbbl recoverable reserves are typically sought in international exploration.

Acreage Position

Domestic acreage position is often a patchwork of leases leading to small positions in established plays. International exploration permits are traditionally large (250,000 to 1 million acres), allowing for multiple plays and fields within a given exploration license.

Competition

Intense competition prevails domestically for leases in productive basins. International competition is more limited.

Rentals and Bonuses

Contrary to widely reported and isolated instances,

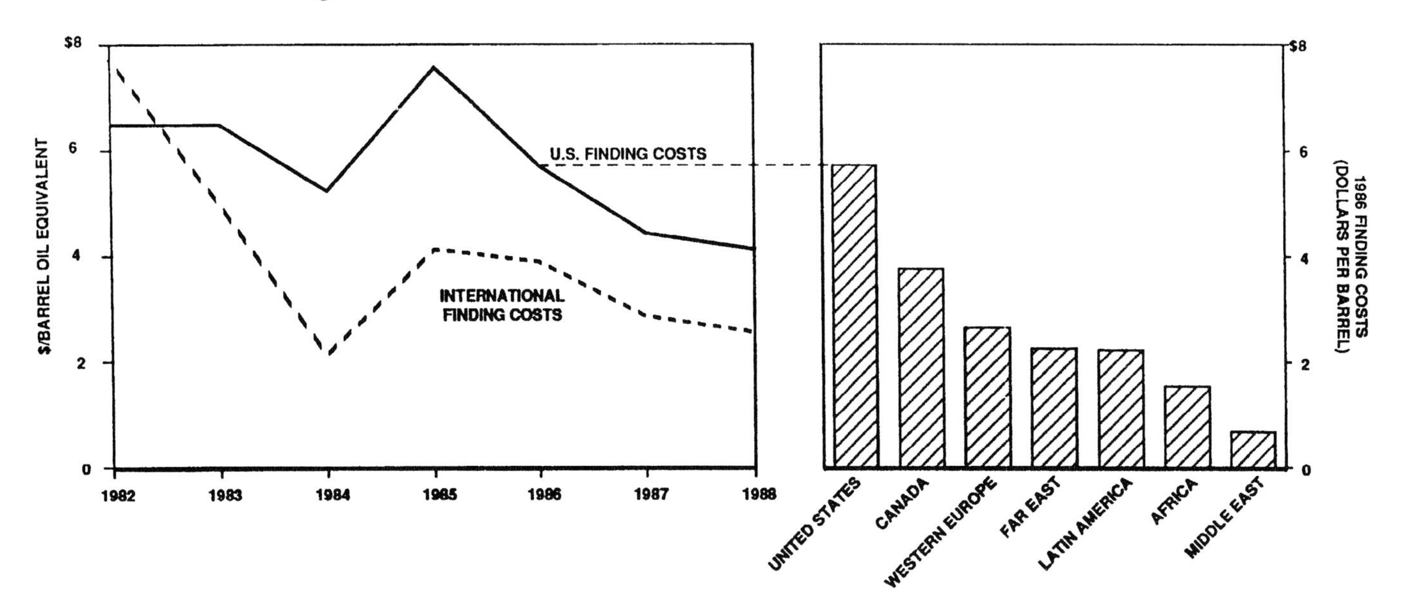

Figure 1. Hydrocarbon finding costs in the United States versus selected foreign countries from 1982 to 1988. (Source: E. Deffarges, Annual Reports, Booz, Allen & Hamilton.)

large signature bonuses are not the norm in international acquisitions.

Foreign Exchange Controls

Agreements with the host government provide for repatriation of funds, provided the work program expenditures are in hard currency.

Operational Requirements

Foreign exploration operations do not necessitate the creation of a full-blown oil company abroad. A two- or three-person nucleus team can normally start up and fully operate a foreign exploration office prior to commercial discovery.

FACING THE FOREIGN ENVIRONMENT

To assist independents in better understanding the business and operational aspects of international exploration, the various problems they may face overseas are discussed here.

Political Risk

This is probably the first and most frequently mentioned term when one talks about foreign involvement. One should not minimize this risk, as evidenced by the mid-1990 crisis in the Middle East. However, the definition of political risk, as originally used in the 1950s, is much different today. In the 1950s, a company was mostly concerned about "nationalization." In effect, we have already gone through that phase since, presently, the control of oil is with a few exceptions in the hands of the host country. The production sharing agreement, in which the company is a contractor to the government or the national oil company, removes the risk of nationalization since from the very onset of the contract, the oil belongs to the host country. Under a production sharing agreement, the company is solely responsible for the geological (exploration) risk, but the host country and the company both share the profits as related to the sale price of crude on the world markets.

It is less known that they also share the operational costs. If the host country considers the company to be an inefficient operator, it might lead to a dispute, but one that could be resolved by normal legal means. Assuming the differences were very large, one might even consider international arbitration (as spelled out in any international contract). However, it is unlikely under any circumstances that the dispute would lead to nationalization or to expelling the company from the host country.

With regards to an unpredictable international event, such as the Persian Gulf crisis, it can even affect the host countries themselves, as in the case of Kuwait. The only way to be protected against that eventuality is the time-proven approach that spreads the political as well as the geological risk: once production has been brought on stream and the cash flow starts coming in, the best protection is to start prospecting in a different part of the world.

We should also mention the existence of OPIC (Overseas Private Investment Corporation), which is a semi-official organization based in Washington, D.C., offering a full range of insurance, financing, and preinvestment assistance programs in selected host countries throughout the world.

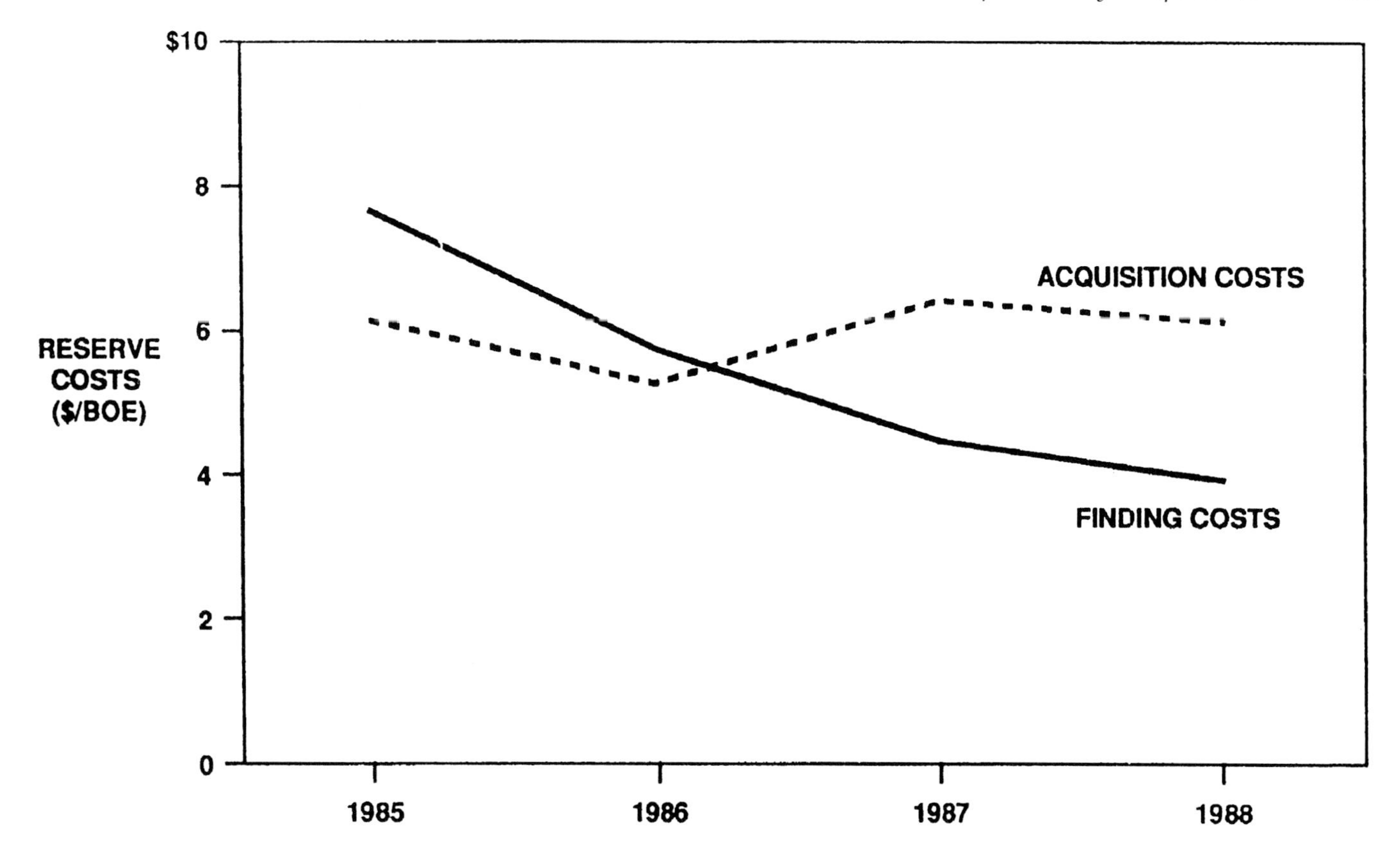

Figure 2. U.S. hydrocarbon acquisition versus finding costs from 1985 to 1988. (Sources: Strevig & Associates; E. Deffarges, Annual Reports, Booz, Allen & Hamilton.)

Geological Risk

In international exploration, the aspects of geological risk take a different dimension in comparison to domestic exploration. Structures are usually larger, but control is less dense, and the economic parameters to reach profitability vary greatly from those in the United States because of distance to markets, more expensive operations, and longer time needed to bring a discovery on stream. It is recommended that you use explorers with foreign experience to evaluate these opportunities realistically and see them in their proper context.

Prospect Generation

Because foreign concessions or exploration permits are much larger than comparable U.S. acreage, prospect generators usually work on delineating several prospects at a time, if they have access to the data. Some host governments sometimes refuse to make available large amounts of data on the concession until a contract has been signed. This explains the limited amount of generators for foreign prospects. In this case, the prospective explorer has no other choice than to approach an established independent or a major to obtain data on the country of interest. They will, however, charge the newcomer a high promotion fee.

Culture Shock

When we are exposed to a different business environment, legislation, business mentality, culture, language, and logic, as is often the case overseas, we all suffer from culture shock. This applies to companies as well as individuals. The initial reaction is insecurity brought about by a lack of understanding, which in turn leads to mistrust. This first reaction usually wears off as one becomes more familiar with the new environment and set of rules.

The Instant Rapport Belief

Because of the dynamic nature of the rapport between independents, the attitude prevails, when engaging in overseas operations, that a similar rapport can be established with the host government representatives. Because these representatives are administrators rather than risk takers (in the business sense), one usually finds that a considerable amount of time is needed to enroll their cooperation and trust in seeing things the way the explorer (and risk taker) does.

Priorities

When a company goes foreign, there is the inevitable feeling that the domestic business comes first. This is normal since it is the part of the company's operations that generates the cash flow to finance the overseas work. However, a niche must be carved for the overseas operations. One must face the fact that an overseas project requires more time than a similar domestic project, even in the early stages. It has been our experience that once an exploration permit or a concession is obtained overseas, it rapidly becomes a full-time effort by one to three peeople to attend to the foreign business.

Time Loss Factor

Because of more difficult communications brought about by time differences, distance, and different systems and mentalities, it takes more time to do business overseas. One has to "buck" the time loss factor constantly. In foreign operations, one needs to "ride shotgun" constantly to follow through with the job and complete it within the allocated time.

Overseas Operations

Overseas operations differ from domestic in that constant planning, coordination, and supervision for all aspects of operations—not solely the technical aspects—are needed by the concession holder. Since the host government considers him or her to be ultimately responsible for everything taking place on the concession, he or she is forced to intervene on behalf of the contractor in all administrative matters involving the local authorities, including the police, customs, and local taxes. The concession holder must also do public relations work in the field, such as liaising with the local prefect or chief of province (equivalent to our state governors). This is the reason why a seasoned veteran with direct on-site experience in various countries and different types of geographic and climatic environments is needed to conduct overseas operations successfully.

Following the Majors

Overseas, the majors have written the book on foreign exploration and operations. Through their early presence in the host countries, they also have been involved, willingly or not, in the evolution of legislation and basic regulations applying to foreign work. Consequently, the approach to foreign exploration is generally more methodical and slower than in the United States.

When an independent decides to work overseas, he or she must necessarily follow the path traced by the majors. This means that some learning and educating may be necessary to adapt to the new set of rules.

PERSONNEL REQUIREMENTS

U.S. based independent oil companies clearly have the financial and technical resources (either individually or collectively) to explore, develop, and produce large petroleum reserves. To accomplish this internationally requires a team of experts to apply the relevant geological, operational, economic, and political parameters to the prospect. Developing this in-house talent is often an uncertain and costly process and leads to missed opportunities. Alternatively, independents can associate with the majors and larger internationally established independents. However, these companies tend to group themselves in developing new plays and farm-out less attractive prospects, or to seek disproportionate entry terms. To expand effectively and successfully in the international arena requires the following personnel and capabilities.

Technical Expertise

A team of international explorationists is needed who are experienced with a majority of the worldwide sedimentary basins. They must be capable of applying the full range of new geological concepts and innovative geophysics to the development of new prospects either in close proximity to established production or within new, potentially productive trends.

New Play Concepts

The exploration team must be abreast of new exploration developments worldwide and must be capable of rapidly evaluating host governments' data banks (seismic, well logs, and previous reports and interpretations) and selecting drillable prospects.

Operational Expertise

A practical knowledge of overseas operational problems and logistics is necessary, coupled with the ability to assess exploration and development operations accurately. Also, the team personnel must have the ability to evaluate the profitability of a venture in accordance with the host government's contractual and fiscal terms, including tax and accounting regulations.

Effective Public Relations

The team must have the ability to establish and maintain effective communications with the host government and national oil company (ideally from managerial level to technical services staff). This is necessary to ensure maximum access to the data banks, reproduction of data, expeditious negotiations, and efficient operations. Foreign operations are usually conducted under the supervision of an agency of the government or in partnership with the national oil company. Good communication leads to increased operational efficiency through access to large amounts of exploration and production data available from the host country's data files. It also leads to close coordination with technicians of the national oil company having first hand experience in exploration and production problems pertinent to that country.

Risk Assessment

The team must be able to assess accurately the host

country's business and political risk from an oil industry investment perspective. Too often this expertise comes from people familiar only with geopolitical considerations.

Competitive Advantage

In all phases, the small independent must be able to proceed expeditiously once a play concept has been developed and to stay ahead of the competition, particularly the majors and established independents. These established companies often owe their success more to competent and motivated professionals than to their financial or technical resources.

Management

Next comes the problem of whom to choose if one does not have the required experience. Independents usually look for a retired manager trained by a major with 20 plus years of experience. That may not always work. Many of these managers, as their title implies, were working through other technical personnel and did not have long-term, direct field experience. *Only a person who has been on the "firing line" can understand and appreciate the many subtle and often hidden details that make the difference between success and failure in overseas operations.* Another concern is "staying power." Majors can devote a much larger cash flow to a foreign exploration projects, prior to making a profit, than independents can. Consequently, the strategies used by the majors for overseas operations must be modified and adapted to the needs of the smaller independent.

GUIDELINES FOR FOREIGN WORK

The following sections will acquaint companies and personnel, who have not been previously involved in foreign work, with certain aspects of overseas involvement.

Attitude of Foreign Governments Toward Independents

In the United States, the independent generally obtains leases from private landowners. In all other countries (with some exceptions in Canada and Colombia), the government owns the mineral rights and one must either deal with the Ministry of Mines and Hydrocarbons (or a similar organization) or form a joint venture with the national oil company. As a rule, these organizations are more inclined toward bureaucracy, and a certain protocol must be observed to approach them. Furthermore, although their staff may have worked with the majors or may have studied in the United States (or sometimes in eastern countries), they do not have a hands-on knowledge of how independents operate. Consequently, an awareness of their attitude toward independents, whether based on fact or fiction, will increase the chances of success when getting involved overseas. Generally, foreign governments have the following traits:

1. They prefer to deal with a major, but in certain specific cases will deal with independents that have a good domestic track record and display a mature attitude toward foreign work.
2. They know very little about independents. Their knowledge is often limited to the old cliches and makes them very cautious toward independents. The company must be prepared to do quite a bit of public relations work to demonstrate their financial and technical ability to perform overseas operations.
3. They are very conscious of their "sovereignty." Most developing nations still have a fresh memory of being dominated during the colonial era or have had a rough time defending their interests with the majors after their independence.
4. They are slow at reacting to a commercial interest because of the bureaucracy that has been left in place by the colonizing power.
5. They trust their advisors explicitly (whether locals or foreigners), which in most cases are totally dedicated to the interests of the country they serve.
6. They prepare themselves for negotiating from a position of strength, although they realize they may not be familiar with the latest technology and progressive financing schemes.
7. They want to ascertain very early the seriousness of the candidate both from a technical and financial standpoint. In particular, they want to avoid "promoters," although they understand the spreading of risk through joint ventures.
8. They want to establish that the candidate can adapt mentality to a different culture and different operational conditions. What can you do for my country and can we work together as a team?
9. They jealously guard their data base, which in many cases has had to be painfully recovered from the previous operators.

Farming into an Overseas Venture

If a smaller company farms into a concession held by a major, it greatly relieves the host government from the task and *responsibility* of establishing the financial and technical capacity of that company. In turn, since the major who is farming out has done all the basic homework for establishing the venture in the country, has conducted the early exploration work, and is responsible to the host country for the fulfillment of the contract terms, it will charge the smaller company a sizeable fee.

By farming into an established foreign venture, the company farming in may pay a high cost because of previous inefficient operations by the company farming out or unknowingly inherit bad relations with the host government due to poor relations between the previous management team and government members. This rule applies even if the previous operator was a major or an internationally established independent.

Farming into a venture overseas with a major or an established independent does not guarantee that the deal will go through. The host government will ultimately decide whether the terms of the farm-in are acceptable to them, although in most cases they will approve.

Maintaining a Rapport with the World Bank

The World Bank is a multimillion dollar organization involved in practically all aspects of industry, agriculture, and trade in numerous developing countries. It maintains a large staff, both in Washington and in the host countries, who can provide assistance in evaluating the technical and financial aspects of an overseas venture, as well as business intelligence allowing reduction of front-end costs in starting an operation. The World Bank staff is usually very cooperative, and it is recommended that a rapport be developed and maintained with them while operating in a foreign environment. Specifically, the World Bank does the following:

1. It maintains an active interest in the host countries since it participates with them in many projects other than those related to petroleum.
2. It is a very good source of information for both contractual terms and technical data. However, it is careful to maintain confidentiality whenever required and to observe the sovereignty status of the country.
3. It can assist in pointing out good investment opportunities and can lend its support to the conduct of such projects.
4. Through its own evaluation of the financial and technical capacity of the company, it can provide valuable assistance in being acceptable to the host government.
5. Through the International Finance Corporation (IFC), its affiliate dealing with the private sector, it can lend assistance with the financing of selected projects or join as an active partner.
6. As a general rule, it is a good organization to have on your side, particularly if your company's financial strengths, expertise, and efficiency in new technology can be demonstrated to their technical staff.

Basic Rules Applicable to Overseas Operations

1. Have patience; it takes time to do things overseas.
2. Keep a low profile. When applying for a foreign exploration permit or concession, observe total secrecy with outsiders until the permit has been granted.
3. Read, observe, and apply the directives of the U.S. Foreign Corrupt Practices Act and, if applicable, the trade embargo restrictions applying to selected countries.
4. Do your homework with a rigorous, methodical, and thorough approach. Run through your assessment of all the exploration, operational, and contractual parameters. Make comparisons with neighboring countries in the same geological environment and see where the best opportunities and economics are.
5. As a corollary of point 4, when zeroing in on a new geological province, always survey the maximum number of prospects in various host countries to carry the subtle message that you are "looking around" and therefore are not tied to any particular country. This improves your chances at the negotiation table and gives you the opportunity to negotiate for a permit or a concession from a position of strength.
6. Listen to the opinion of others in the host country and be willing to learn. Use the front door/back door approach. It works most of the time.
7. Show flexibility. Contrary to the attitude found in the United States, things are not black and white overseas, but shades of gray.
8. Demonstrate respect and understanding for others. In some countries where the culture and traditions are different from ours, civil servants are particularly sensitive to the assertiveness of Westerners; they consider it a form of arrogance and a lack of culture.
9. **Communicate.** We stress the need for coordination, planning, and teamwork to attain our goals. Overseas these needs increase at least 10-fold; in some cases they are more like 100-fold.
10. Do not limit yourself to one type of prospect. Just because the exploration team has had extensive success by playing one particular prospect type (for example, HCIs).does not mean that when looking at overseas acreage, other attractive prospects such as reefs should be down played at the expense of the first ones.
11. Do not be overly optimistic. Just because a structure produces 2000 bbl of oil per day does not necessarily imply that the field is commercial. In fact, overseas, this figure often constitutes the threshold between commercial development or abandonment. But be realistic. When prospecting a new trend, do not shy away from drilling a larger, deeper prospect with greater potential reserves. This is why you are going foreign.
12. Reach a decision. The most frequently heard phrase from foreign managers is "I want to be responsible for a discovery that will have an impact on the company's reserves position." But are they also willing to commit themselves to the company's future by putting their name on the

line and being an innovator rather than a follower?

13. Use seasoned consultants. They may be considered by some as highly paid mercenaries with obsolete expertise and limited loyalty to their client. In actual fact, they are dedicated, hard-working professionals with a wealth of experience and industry-wide contacts who believe in getting the job done and who concentrate on getting results.
14. Think long range. It is not possible to spend 3 days in an overseas geological province and come back with a concession in hand. The company must methodically record an inventory of available acreage in targeted provinces and/or host countries and patiently wait until quality acreage becomes available (sometimes as long as 5 to 10 years).
15. Most importantly, have a strategy or a business plan. It is necessary to go through an "initiation period," when the game is learned and mistakes are made. But one should attempt to avail oneself of the early irreparable mistake from which one never recovers.

Desirable Profile for an Independent Considering Overseas Operations

1. A successful independent understands and adapts to the ways of the host country. He works closely with the government agency responsible for hydrocarbons rather than confronting them or ignoring them. When negotiating with the representative of the host government, he must think of common interests rather than divergent objectives.
2. Although experience acquired in the United States is most valuable, the independent admits that there may be as good a way or a better way to do things by listening to and working with others in the host country.
3. Because of his smaller size, the independent knows that he can be more efficient than the majors. At the same time, he realizes that he has to learn the basics of organization for overseas operations from the majors.
4. A direct application of point 3 is that the independent is willing to learn more about coordination, planning, and communicating. These skills are not as critically needed in a small domestic organization, but they are absolutely necessary to work overseas.
5. To conduct a successful exploration program overseas, the successful independent does his own homework in exploration and closely supervises his own exploration and operations rather than depending on other operators.
6. As a consequence of point 5, the independent avoids "patchy" operations that usually cost more or sometimes result in disastrous losses.
7. The successful independent tones down his ego. Most independents are individualists and have reason to believe they are good businesspeople and great oil finders. This attitude may be appreciated at home, but it can be considered arrogant in the host country. It can often lead to failure in attempting to obtain exploration permits or concessions.

REQUIREMENTS FOR EXPLORATION PERSONNEL TRAVELING OVERSEAS

The list that follows is not intended to be all inclusive. It highlights items that are applicable to all overseas travelers in general, but some are more specific to the oil industry.

Personal Documents

- You will need a valid passport and, where applicable, an entry visa.
- Take extra passport photos with you.
- Keep a list of your traveler's checks numbers separate from the checks themselves in the event of loss, but also for declaration to the authorities when entering certain countries having restrictions for foreign exchange.
- Also keep a list of credit card numbers, serial numbers of valuables (watches, radios, binoculars, and portable computers), not only in case of loss but also for customs declaration.
- Leave with family and at the office a list of people's addresses and telephone or telex numbers through which you can be contacted on short notice. Also leave a copy of your travel itinerary and your hotel's telephone and telex numbers.

Medical Documents

- Most countries no longer require vaccination against smallpox.
- In selected areas, it is advisable to have inoculations for tetanus and to have a gamma globulin shot.
- For certain areas in the Middle East and Asia, cholera and yellow fever shots are advised.
- On your medical card (Yellow Card), be sure to include your blood type and, if applicable, your glasses prescription, special drug prescriptions that you may need, and any necessary information about your particular health condition.
- If any medical prescription you take with you contains barbiturates, make sure to have a doctor's certificate to avoid problems with the

police upon entering the territory.
- If going to a malaria-infested area, medication should start before entering the area and should be continued from 3 weeks to 1 month after leaving the area.
- An overseas medical travel kit can be purchased from most clinics for taking care of minor emergencies.
- Carry a set of spare prescription glasses if applicable.

Work Documents and Equipment

Before leaving overseas, you should do the following:

- Obtain copies of standard travel guides and maps.
- Attempt to read selected books on history, culture, and new political developments in the destination country.
- Obtain from the host country embassy in Washington the names and addresses of ministries, government officials, and the national oil company. Also obtain a travel visa if needed and a list of applicable restrictions if any.
- Establish telex contact with officials to be visited and await their reply. Do not arrive unannounced or unauthorized.
- Obtain copies of articles on the geology and oil production from, for example, the Library of Congress or the U.S. Geological Survey.
- Research technical publications for articles of interest (for example, *Oil & Gas Journal*, *AAPG Bulletin*, or *GSA Bulletin*).
- If applicable, order geological maps, topography sheets, air navigation charts, and satellite photography for area of interest.

Lawyers and negotiators should take the following materials and equipment overseas:

- Company literature and financial statements or annual reports
- Company letterhead stationary
- Corporate seal
- Power of Attorney
- HP 12C calculator or equivalent with one set of spare batteries
- Lap-top computer for certain countries

Engineers and explorationists should include these materials and equipment:

- Tracing paper
- Graph paper (various types as required)
- Colored pencils and markers
- A protractor and a ruler for converting English to metric units
- HP 11C calculator or equivalent with one set of spare batteries

If you are planning a field trip, be sure to include the following:

- Compass
- Signaling mirror for emergencies (World War II type with aiming hole)
- Binoculars
- Camera
- Flashlight with adequate supply of batteries
- Field survival kit (can be obtained locally)
- Adequate clothing and a hat
- Salt tablets if in desert environment
- Sunglasses and sun screen lotion
- Mosquito repellent for jungle environment

General Information for Traveling Overseas

- Arrange for access to international telex facilities. TWX is not used overseas. In certain countries, overseas telephone service is poor, and telefax facilities are rare or nonexistent.
- Teach your U.S. secretarial staff international telephone and telex discipline, particularly with regards to different languages, procedures, and time zones.
- Although not required, a world time chart is helpful.
- In some countries a shortwave transistor is the only way to keep in touch with the outside world.
- Do not pack personal, legal, or work documents in your suitcase unless absolutely necessary. Always carry them in your briefcase. If they represent a large volume, use a pilot's map case.
- Use luggage tags that cover your name and the company's name.
- Maintain a low profile in your dress, your luggage, and your behavior. Tell yourself you are just another anonymous traveling businessperson.
- Carry an extra set of luggage keys.
- Carry enough clothes with you to appear businesslike at a meeting in case your suitcase arrives late or is lost.
- In some countries, be prepared for a thorough search of your person and your belongings upon arrival and departure.
- Be ready to account for all cash, traveler's checks, credit cards, and valuables upon entering and leaving certain countries. Your business files may also be inspected.
- If carrying out technical documents provided by the national oil company or the Director of Energy, a letter of authorization may be needed to take them out of the country.
- In Muslim countries, do not bring alcohol or girlie magazines with you.
- In certain countries, some books and magazines may be censored.

- If applicable, carry 110/220 volt adaptor plugs and transformer equipment.
- Carry a small camera, but not so small (minox) so as to look like a "spy."
- In some countries, one should not take pictures of humans, particularly children, because the camera is thought to be an instrument of bad luck. In many Muslim countries, one should not take pictures of the unveiled face of an adult woman. Yet in other countries, a Polaroid is definitely an ice-breaker. Inquire ahead of time.
- Avoid VCRs unless you have been advised ahead of time that it is OK to use them. The same goes for tape recorders.
- In some countries, it is advisable to carry black and white film so that it can be developed, and the pictures checked by the proper authorities, before you leave the country.
- When arriving in some countries, a visit to the U.S. Embassy may be advisable. Inquire ahead of time.
- When traveling within a foreign country, always carry an I.D. or passport, a medical card, and an address that can be contacted in case of an emergency.
- Never exchange dollars or other hard foreign currencies at black market rates.
- Some countries do not readily accept credit cards.
- Make it a practice to use the hotel safe for valuables and confidential documents.
- Do not take expensive jewelery with you.
- Be aware of unannounced "wildcat" holidays.
- Take into account religious or local customs that may slow down the work schedule (for example, Ramadan, summer hours, etc.).

REPRESENTATIVE OUTLINE OF A FOREIGN PROJECT

Although the conditions described here cover various geographical, cultural, operational, and political environments, it is obvious that only a certain set of conditions applies to a particular country. Basically, in Europe and certain parts of Asia, the conditions of work are very similar to those in the continental United States, with only some variations. In parts of Latin America, Africa, and the Far East, the conditions may vary from average to very difficult. (Note: An asterisk (*) indicates that the business or legal language may not be English.)

Strategy and Planning

The following points must be considered in the planning of your foreign project strategy:

- Budget size as a function of the company cash flow.
- 100% company or joint venture.
- Operator or partner.
- Manpower requirements.
- Geographic location and climatic conditions.
- Onshore or offshore.
- Political risk.
- Economic status of the host country.
- Terms of the petroleum law and/or national oil company joint venture contract.
- Petroleum geology.
- Petroleum exploration operations.
- Cost and financing of a development program, including environmental restrictions in some countries.
- Logistics and economics for moving the oil or gas once discovered.
- Marketing of the hydrocarbons discovered.
- Do not accept at face value a venture agreement already concluded by another company or operator. Conduct your own evaluation of the geology, logistics, economics, and contractual aspects of the project.
- In the host country, be careful of middle men, whether they be "influential" businessmen or "relatives" of high government officials.
- Obtain a copy of the FCPA (Foreign Corrupt Practices Act) and associated documents and trade embargo restrictions applying to selected countries.

Initial Approach (2 to 6 weeks)

- Installation of international telex facilities in the home office.
- *A visit with the Energy Ministry (or equivalent); familiarize yourself with the petroleum law.
- *A visit with the national oil company; familiarize yourself with the contract terms.
- A visit with other oil operators.
- *Gathering and studying of data bank, including geological and geophysical reports, well reports and logs, and operational reports.
- Field trip to determine operational factors.
- Analysis of results and a recommendation to management.

Concession Aquisition (6 months to 1 year)

- *Negotiation of a contract with the national oil company or with the Director of Energy (or equivalent).
- *Alternately, application for an exploration permit or concession.
- Hiring a local lawyer and a CPA.
- Several trips to the host country to follow up on the progress of the application.
- A study by the Council of Mines.
- *A new round of negotiations if active competition is present.
- Signature by the Energy Minister.

- Approval by Parliament or Head of State.
- Concession or permit award.

Opening Local Office (1 to 3 months)

- *Staffing with expatriate general manager and administrative manager/controller.
- Selection of office quarters. Relocation of staff and families.
- *Hiring of local support staff.
- *Registering company in host country. Obtaining residence and work visas for expatriates.
- *Obtaining local telephone and telex facilities.
- *Selection of bank and transit agent. Survey of local contractors for availability and repairs of field equipment.
- *Establishing close contact with representatives of various government organizations whose approval is needed to conduct operations, including Director of Energy, Director of Transportation, labor department, customs, police, and ports and airport authorities.
- *Obtaining a full set of exploration, operations, and production data for concession area from national oil company for transmittal to home office (geological, geophysical, well, drilling, and operational reports, as well as seismic records and tapes, well logs, well samples, and selected well cores).
- *Preparation for seismic operations (preliminary phase). Applying for authorization to carry seismic and gravity/magnetic surveys. Applying for import permits for field equipment and for communications licenses for radio equipment. Preliminary coordination with customs.

Preliminary Study Phase (home office, 3 to 6 months)

- *Study of reports by pertinent departments.
- Reinterpretation of magnetic, gravity, and seismic data and reprocessing of selected seismic lines. Choose equipment and design of new acquisition parameters. Study of field or marine logistics and call for seismic bids. Selection of seismic contractor.
- Reinterpretation of geological surface and subsurface data. Restudy of well samples and cores. Geochemical analysis. Regional bio- and litho-stratigraphic and paleogeographic studies. Study of regional structure and tectonics and integration with satellite imagery, surface data, and subsurface data. Construction of selected stratigraphic and structural cross sections and geoseismic sections. Isolation and classification of prospective areas. Layout of additional seismic control, magnetic, gravity, and surface geology programs.
- Prospect studies. Integration of the preceding data on a local scale. Preparation of pertinent additional geological, geophysical, structural, and stratigraphic maps, aspects, and sections. Preparation of prospect montages and/or brochures. Determination of prospect economics. Layout of additional detailed seismic program.
- Preliminary study of drilling, engineering, and logistics. Well cost analysis. Investigation of the availability of drilling equipment and sources of supply equipment and drilling supplies.
- Search for additional partners, if applicable.

Seismic Operations (lead time 3 to 6 months)

Marine Operations (1 week to 1 or more months as a function of program length)

- Arrival of geophysicist.
- *Preparation for arrival of seismic vessel and navigation equipment. Obtaining navigation, shooting, and communications permits. Obtaining import permits for navigation equipment and customs clearance for seismic vessel.
- *Arrival of navigation equipment and personnel (contracted). Arranging for transportation, supply, and logistics of personnel and equipment. If applicable, arrange for police and security clearance for personnel and issuance of pertinent documents. Installation and test of equipment.
- *Arrival of seismic vessel (contracted), customs clearance. Moving to prospect area and conducting program.
- *Customs and police formalities for departure of seismic vessel. Offloading seismic tapes and air shipment to processing office after obtaining pertinent permits.
- Processing of new data and integration with existing geophysical and geological data.
- Recommendation to drill or (if applicable) to abandon acreage.

Land Operations (6 months to 1 year)

- Arrival of geophysicist.
- *Preparation for arrival of seismic equipment and personnel (contracted). Obtaining shooting and communications permits. Obtaining permits for import of equipment and work permits for personnel.
- *If explosives are to be used, placing order for explosives and arranging for import and police permits, storage of explosives, and surveillance by police or army unit (lead time 3 to 6 months).
- *If applicable, start of bulldozer or jungle trail building crew (contracted).
- *Arrival of seismic equipment and crew. Relocation of selected personnel. Crew move to prospect area. Setting up camp (if applicable). Start of experimental program followed by field operations.
- Activating crew supply operations by land, boat, or aircraft (whichever is applicable).
- Local processing of data or shipment to

processing office after obtaining applicable permits.
- Interpretation of data and integration with existing geophysical and geological data.
- Recommendation to drill or (if applicable) to abandon acreage.
- *If applicable, customs and police formalities for departure of seismic crew.

Preparation for Drilling Operations (lead time 6 months)
- *Study of previous reports on drilling and engineering operations.
- Planning of drilling program based on geological objectives. Design of mud and logging program. Choice of ancillary equipment. Preliminary survey, land or sea environment, climate, and logistics.
- *Trip to host country by drilling team personnel to survey availability of drilling rigs adequate for the project, local supplies of machinery and equipment, and repair facilities. Study of transportation needs and availability of local equipment for land, sea, or air transport and bases of operation.
- Alternatively, contact foreign or U.S. drilling contractors with previous experience in the host country or in an operational environment similar to that of the project.
- *Survey of local medical facilities in case of emergency and/or medical evacuation procedures.
- *Visit with national oil company drilling and engineering departments and/or other operators.
- *Visit with Director of Energy, customs department, local police authorities, and regional dignitaries.
- Field trip to study on-site conditions, availability of local equipment, and personnel if applicable.
- Final choice of equipment. Survey of available contractors and suppliers. Request for bids. Choice of drilling contractor and, if applicable, supply boats, air transportation (helicopters and airplane), and land transportation.
- Order mud products, drilling bits, pipe, and casing. If applicable, order blowout preventors and other ancillary drilling equipment.

Move-In Phase (1 month)
- Arrival of company drilling supervision team: drilling superintendent, tool pusher, materials man, engineers, and well site geologists.
- Activating marine base or land warehouse facilities. Installing communications network. Hiring of local staff for base facilities.
- Arrival of consumable drilling goods and clearance through customs. Storage at marine base or land warehouse.

Marine Operations
- Arrival of rig positioning team and helicopter(s). Arrival of jackup, semi, or drilling vessel and clearance of all equipment through customs.
- *Hiring of local labor for drill crew and maintenance crew.
- Activating local catering (contracted) and all ancillary services.

Land Operations
- If applicable, preparation of beach or river offloading facilities and arrival of landing crafts (LCTs).
- Construction of road to drill site. Preparation of drill site.
- Construction of air strip and camp facilities. All preceding work contracted (lead time 3 to 6 months).
- *If using foreign contractor, arrival of drilling rig and clearance through customs.
- If using contractor established in host country, final inspection of equipment and acceptance of same.
- Move of drilling equipment to field location.
- *Hiring of local labor for drill crew and maintenance crew.
- Activating catering and ancillary services.

Drilling Operations

Once the previous tasks have been performed and the backup organization is in place, the project proceeds along the same lines as in the United States. Rotation of personnel is usually back to the United States or to a foreign point of origin, and depending on distance and climatic conditions, it may be on a different schedule from that used in the United States.

Field Logistics and Scheduling of Operations
- If the field operations take place in a desert environment, working throughout the year is usually possible, except during the peak period of hot weather. In this case, an early morning shift for seismic crews is possible. Drilling operations are usually uninterrupted.
- If working in a tropical environment, the work schedule is usually a function of the weather window, and several months of work may be lost during the rainy season unless preparations are made ahead of time. Seismic crews usually shut down during the rainy season. Drilling operations can continue, provided an all-weather road or landing strip has been built and selected supplies can be brought in by aircraft.
- In certain areas of dense jungle or rough terrain, helicopter support for seismic and geological operations may be needed, adding to the cost of the project.
- In swampy areas, a portable seismic crew using

several hundred local laborers may be necessary and entails very complex logistics.
- In nearshore areas and bays, the combination of a mothership and a light craft can be used for seismic surveys.
- For drilling operations in tropical countries, a combination of LCTs (Landing Craft Tanks), all-weather roads, landing strips, supply by aircraft (DC-3 or C-130), and helicopter support may be necessary, adding to the cost of the operation.

Closing Down Foreign Operations (1 to 3 months)
- *Obtain clearances and various authorizations for personnel to leave the country, in particular, from police authorities, tax authorities, customs, banks, and utilities companies.
- *Obtain the necessary permits from customs to reexport the equipment that was imported temporarily. Account for all consumables and materials that have been damaged or destroyed, or else pay the applicable customs duties.
- *In many countries, give advance notice to local personnel and pay them a severance bonus if applicable.
- After closing down the operation, two or three trips of one or two weeks each by a company representative are necessary to tie up loose ends.

When You Make a Discovery

You will face a new set of conditions and challenges if you make a discovery. However, the experience gained during the early exploration phase will be of definite use in solving later problems. Besides, at this stage, you will have all the help and guidance you will need from the numerous service and management consulting organizations operating in the foreign field .

CONCLUSIONS

The foreign game is accessible to smaller independents and can be very rewarding. However, certain requisites are necessry, and the rules of the game traced by the majors must be followed. Generally, a small company wanting to get invoved overseas must do the following:

1. Demonstrate a succesful domestic track record as an oil finder and producer to be acceptable to the host government.
2. Show a constant profit over the last 3 to 5 years to qualify for obtaining a concession (or exploration permit) that will usually require a work program consisting of seismic plus one or several wells.
3. Have a stable cash flow and be able to commit the sum of $1 to $2 million each year for a minimum period of 3 years.
4. Come up with new play concepts in their area of interest and be in a positioon to demonstrate the validity of these concepts to the exploration staff of the Director of Energy (or the national oil company).
5. Have on the team a minimum number of staff experienced with overseas operations and having first-hand experience on site in various foreign countries.
6. Be free from technical dogma and have an open attitude toward a new environment and culture.

Acknowledgments *The publication of this chapter was made possible through numerous brainstorming sessions and discussions with Stan Dur, Triton Energy, Dallas, TX, who also contributed to the editing and streamlining of the text, as well as Spence Collins, Consultant, Houston, TX, and Marlan Downey, ARCO International, Dallas, TX. Credit is also due to the authors of various articles that have appeared in the* Oil & Gas Journal: *Henry Dejong, Consultant, Litteton, CO; C. J. Gudim, Consultant, Evergreen, CO; Etienne Deffarges, Booz, Allen & Hamilton, San Francisco, CA; and John Mc Caslin, Exploration Editor for the* Oil & Gas Journal *(now retired), who provided the first opportunity for publication of the material that preceded this more complete version. I also want to thank the members of the AAPG group who participated in the publication of this Handbook, whose comments and reviews were much appreciated. I am particularly indebted to Richard Steinmetz, Editor, for his valuable guidance, and to Ted Beaumont and Norman Foster for their vision and continued encouragement. Finally, special thanks to Don Rusk, a great petroleum explorer and esteemed friend, who made the publication of this chapter possible through his early recommendation to the Editor.*

ROBERT G. BERTAGNE is a consultant in Houston, Texas, for various U.S. petroleum companies operating in the international field. He has also consulted for the U.S. government, the World Bank, various foreign governments, and national oil companies. Mr. Bertagne has a baccalaureate in Math and Physics from Aix-Marseille University in France and a degree in Geology from the University of Tulsa. He worked 19 years for Conoco and its affiliates, 8 years for independent companies, and the last 9 years as a full-time consultant. Since the mid-1950s, he has participated in most major overseas exploration plays, both as an explorer and in management positions, and he resided overseas for 25 years.

Chapter 27

Ethics in the Business of Petroleum Exploration

Robert W. Spoelhof

Pennzoil Company
Houston, Texas, U.S.A.

If you always tell the truth, you never have to remember what you said.
—Anonymous

INTRODUCTION AND DEFINITIONS

This chapter is intended to emphasize the importance of high standards of conduct to individuals and companies in the business of petroleum exploration. The questions to be addressed include

- What are ethics and how do they relate to professionalism?
- What are the bases for ethical behavior?
- What are some typical ethical issues in the exploration business?

Ethics comprise a set of standards that control and measure relationships between individuals and companies in social and business interactions. These standards apply to questions of whether folks have done the moral right by the other person, whatever the nature of their interaction might be. Standards of behavior involving ethical issues take one of several forms. Many such rules are unwritten, but are generally and intrinsically recognized by society. They include the group of truths cited by the U.S. Constitution: "We hold these [ethical] truths to be self evident . . . ." Such standards include the fundamental rights that each member of society grants to others.

The means of granting rights to others are included in the mores of society. These are the unwritten, yet valid, rules that dictate, for example, how bargaining is conducted between buyer and seller or how properly to greet people of different social stature. Mores constitute the protocol and etiquette of personal and business interaction.

Still other standards of behavior are formally written as codes of ethics and are formally agreed to by all who wish to join the group to which the code applies. A contractual relationship is nearly established between the individual and the subset of society. Codes of ethics specify the attitudes and actions that the group will allow its members in their conduct of the group activities.

The most explicitly rigid ethical standards are codified as law and are officially imposed on everyone regardless of their desires. Codified law is generally incorporated into professional codes of ethics. Thus, a breach of the law during the practice of a profession will be regarded as impinging on professional codes of ethics. This is certainly true for members of AAPG.

This chapter is specifically intended to consider behavioral practices in our business that are not addressed in formal laws, but that are nevertheless generally accepted standards whether or not we have personally agreed to honor them.

Ethics govern interpersonal relationships (Brown, 1985). Rules of ethics spring from the acknowledged rights and duties of both parties in a business or social exchange. Both rights *and* duties are important. One does not behave ethically if the other person is prevented either from doing his duty or from retaining his rights. Nor is it ethical for one to fail in his own duty or to yield his rights without exchange. Drucker (1981) requires that "each side be obligated to provide what the other side needs to achieve its goals and to fulfill itself." That is, both sides need to retain their rights and obligations.

In the exploration business, a company would thus behave unethically if, for example, management required a certain level of performance from its employees, but at the same time prevented them from doing their job, from fulfilling their obligations. The company would also be unethical if it denied its employees a safe work environment, which is one of their rights. At the same time, employees would be unethical if, through office gossip, they prevented the company from providing adequate management, which is one of its obligations.

BASES OF ETHICAL BEHAVIOR

Motivators of ethical behavior (see Kohlberg, 1971) might include the following:

- adherence to high moral principles,
- expectation of enhanced profit,
- enhancement of productivity,
- fear of sanction,
- demands of society, and
- requirements of professional affiliations.

Principles

The overriding reason to be honorable in our personal and business dealings is that it is simply the right thing to do. Ethics must arise, fundamentally, out of personal conviction based on an individual philosophy, a moral position, and high ideals.

There are, to be sure, as many personal convictions as there are people. The important thing must be for each of us to decide on and even to formalize a personal view of things. A person's perception of the world forms the basis from which even minute interpersonal business decisions logically arise. Logic is the key word. It suggests an interlocking matrix of actions, thoughts, and ideas, all of which are mutually interdependent and integrated with one another. One's integrity is the measure of the extent to which his or her actions are based on and integrated with a solid foundation of personal moral rightness.

For example, consider the occasions on which it is noticed that someone has integrity. Usually, honor and integrity are most noticeable when a person does something or decides something that appears to put him or her at personal risk. We assume that the underlying cause of the action or decision must be high principles and a firm moral commitment, something that surmounts the apparent short-term negative results of the action and must relate to unseen aspects of the person's life.

Make no mistake, everyone does have an overall view of the way things work. It can be imagined that if you fundamentally believe that your employees will not produce unless you hound them continually, you probably also believe that people in general are selfish, lazy, and unprincipled. Indeed, you may even believe that the physical world is incapable of self-regulation. Your fundamental philosophy is a reflection of the second law of thermodynamics, that natural systems tend toward entropy and disintegration, unless (your?) outside "energy" is infused.

Alternatively, if you basically believe that the world is characterized by order and integration, you likely accept uniformitarianism, that physical and chemical processes work the same way now as they always have, as well as Walther's law, that vertical changes in structure and stratigraphy are mimicked laterally. You will also likely believe that people perform in consistent ways and that your incessant personal interference is not necessary to order the actions of others.

The discussion here, however, is not about blind, inexorable application of a single standard in all situations. Our paradigm for human behavior can be no more rigid than our models for geological processes. After all, we explorers are in the business of looking for anomalies in what seem to be predictable patterns in our data sets.

There will, of course, be employees who do not perform up to expectation, even if we believe that most of them are creative, industrious, and trustworthy. But the way we deal with the wayward individuals will be based on our basic principles and on our perception of people generally. We can assume that the laggard truly wants to cheat the company or we can assume that he really does want to do a good job, but simply needs guidance and counseling to know how to do it. Your first steps in correcting the situation will be based on fundamental assumptions. Decide on your fundamental assumptions. The next step is to make each move in keeping with your basic precepts. That will be integrity.

Profits

The business of petroleum exploration must be operated on the basis of high moral standards and ethical behavior because our business deals primarily in ideas, innovations, and insights. We do not deal, in essence, with hard currency or tangible products. The heart of our business, the added value that we provide and that produces our profits, is creative insight. Our investors do not buy anything that can be examined directly; they buy into exploration *ideas* (coupled, of course, to the lease and the drill bit).

If an exploration idea should prove to be defective, a dry hole cannot be returned for a refund, nor can the product be repaired. The only recourse for the buyer is to buy or be given a whole new idea. For investors (or the company pay master) to continue to buy our ideas, they must be convinced that they are getting the best possible ideas for their money. In most instances, and especially for frontier projects, their confidence rests solely on our individual proficiency and honesty, that is, on our professionalism, in evaluating all available data, in projecting and interpolating the data, and in portraying the risks involved. In all cases, they are investing in our personal integrity as well as our capability. They are buying our ideas on the basis of our professional ethics.

If it is true that our essential product depends on our ethics in its production, then we have nothing to offer if our ethical standards should fail. Presumably, all else being equal, the stronger our ethics, the better our products will be and the more our personal or corporate profits will increase.

The likelihood of improved profits from increased sales resulting from ethical standards in our prospect and deal preparation has been noted. This idea can be expanded to include the improved availability of capital funds if our ethical standards are beyond question. We are all aware that the acquisition of a loan is easier when our banker knows us; when he or she knows that we will repay on time, that our business practices are sound, and that our products will sell. In short, our banker will more readily fund our projects when there is no question as to the ethics of our business.

Integration of ethical standards with everyday business practices will not only promote the sale of our products and simplify funding, both of which enhance profit, but it can also be a safeguard against the day when the quality of our prospects may be questioned. Johnson (1989) credits the strict adherence to a company code of ethics as having a large part in the Johnson & Johnson Company comeback from the tainted Tylenol® incidents in 1982 and 1986. Similarly, an explorationist may expect to continue in business after a particularly disastrous dry hole if it is known that he or she has a commitment to fair play and quality prospects.

Productivity

High ethical standards can also affect personal and corporate profits by enhancing productivity. Wantland (1989a) rightly believes that a well-defined personal standard of behavior can enhance productivity by removing distractions caused by the inability to resolve ethical dilemmas (see also Chapter 17, this volume). Megill (1985) believes that ". . . integrity affects your productivity. The undivided person has a unity of purpose. His energies are not dissipated in internal conflicts about the right ethical decision, the right path to follow, or the proper course of action." Surely also the time spent in litigation resulting from unethical behavior could be better spent in exploring and producing.

The construction of a strong foundation based on ethical values and overall personal or corporate purposes can also provide a basis for surviving major disruptions in productivity of the industry (Wantland, 1989b). In the face of rapidly changing oil prices, exploration budgets, and staff levels, the strength necessary to survive and prosper cannot depend solely on your company or the business climate. It must be an integral part of the explorationist's daily behavior. Individuals with strong ethics will already know what their responsibilities and responses should be in their relationships with family, company, society, and themselves. It will be clearly understood why a particular course of action was chosen in the face of calamity. A clear understanding of why an individual has chosen his or her present circumstance will ease feelings of dissatisfaction, uselessness, and lack of direction. The retention of drive and controlled direction will positively affect other people's perception of the explorationist, thereby enhancing his or her survivability and continued advancement.

Likewise, ethical behavior toward employees in the exploration business will enhance productivity. It will help us to retain the best people even if we are unable to pay top wages. When employees are given responsibilities, a company manager behaves ethically when he or she also provides all that is necessary to meet the responsibilities since ethics involve provisions for both rights and responsibilities. When this happens, employees feel an increased sense of purpose and usefulness to the company. If all of the actions of the company relative to the employee are equally ethical, the stability of the staff will be increased.

Sanctions

That unethical behavior leads to opposite results seems obvious. Unethical behavior will likely result in eventual lost sales, dismissal, serious law suits, and other horrors. Unethical practices can produce added work loads if we are "requested" to retrace our footsteps to acquire the rest of the data, to switch to a better area, or to recontour. We may be "reassigned" or dismissed altogether. Our best contacts may desert us for fear of being deceived themselves. Our professional society may remove us from the rolls, regarding our behavior as reflecting badly on other members of the profession.

Society

In addition to our personal and corporate motives for ethical behavior, such behavior is almost universally demanded by our society. In petroleum exploration, we must now restore land surfaces, properly dispose of salt water, protect unsophisticated investors, and the like.

For most entities that now apply strong ethical pressures, the early motivation was preservation of the group in the face of onslaughts from the outside, even if the good of the whole meant sacrifices on the part of a few of its members and even if those few who were sacrificed did so unwillingly or unknowingly (Goldman,1980). For example, it can be imagined that a group of people accidentally isolated in rough terrain and in dangerous weather might decide not to deplete their shared security by leaving whatever protection they had, even if one of their members was missing and might be saved if everyone went out to search.

In more secure communities, society becomes less and less tolerant of sacrificing the few for the sake of the many. Society becomes civilized. The result has been that our current society will not tolerate health risks to a few folks who happen to live near a faulty nuclear reactor even though the reactor can provide cheap, abundant energy to the greater body. We are not willing to risk the despoiling of a few beaches in the quest for adequate petroleum supplies (see Chapter 28). Nor are we willing to allow the labor force to be exploited, even though doing so could contribute a less expensive commodity to everyone.

In recent years, we have come to realize that all aspects of our social, natural, and economic environment are closely interrelated. We have become less and less tolerant of any individual, group, or company that soils some part of our environment, even if the action produces some offsetting gain. Careful consideration of the impact and possible restraint with loss of potential profit in our individual endeavors are necessary for ethical behavior.

Profession

The restraints imposed by the ethical standards of society can apply to everyone. Even tighter limitations may be imposed on people in responsible leadership or service positions where their actions impact a large segment of society. An example is petroleum explorers, who have unusual capabilities that are necessary to the whole of society. We have chosen to be singled out as professionals to handle essential aspects of service. If we do our job poorly or at the expense of others, society in general, and specifically other members of our profession, will be adversely affected. Thus, petroleum explorers are bound by both general ethical rules and the standards of our profession.

According to the dictionary, a *profession* is defined as "an occupation or vocation requiring training in the liberal arts or sciences and advanced study in a specialized field." Furthermore, a *professional* is "one who has an assured competence in a given field or occupation." The Taft–Hartley Act, which governs U.S. labor relations, similarly defines a professional on the basis that his or her work is nonroutine, that it requires consistent exercise of discretion and judgment, and that it involves advanced knowledge in a field of science or learning (Campbell, 1990).

Those who are professionals in the petroleum exploration business, who have some special expertise in a particular body of knowledge, also have an ethical responsibility to that body of knowledge. That is, we are not only representatives and servants of society at large, but also of our science or discipline. Generally, misconduct by "professors" of a discipline not only brings disapproval down upon themselves, but also down upon the body of knowledge that they profess.

Geologists view with displeasure the inappropriate or incompetent application of geological principles because (1) no geologists wish to have aspersions cast upon them because of the actions of another who also bears the title of geologist, and (2) geologists regard geology as a rational, predictable, orderly body of knowledge. Misuse of the science casts doubt upon the integrity of the discipline of geology itself, which, if substantiated, would question the presupposition that science can, in fact, yield insights on the basis of logic. The alternative, that geology is a collection of random observations and events unconnected by logic or reason, and without any keys to understanding, is intolerable. That alternative means that geology is not a science and cannot yield insight into the natural world or yield useful predictions. Thus, unethical behavior cannot be tolerated in the geological community.

Weimer (1984) believes that professionalism involves not only the acquisition of specialized learning and the exercise of discretion but also a sense of dedication and a devotion of time to becoming as technically able as possible. He also equates professionalism with doing a job right regardless of economic considerations and with upholding high standards that cannot be discredited. His use of such words as "right" clearly reflects that he considers professionalism to be a part of ethical behavior for explorationists.

That all geologists view geology as a profession, a rational, orderly body of knowledge, is embodied in Section 1 of the Code of Ethics of the American Association of Petroleum Geologists (AAPG, 1991) (see other codes provided in the Chapter Appendix):

SECTION 1. General Principles

(a) Geology is a profession, and the privilege of professional practice requires professional morality and professional responsibility.

(b) Honesty, integrity, loyalty, fairness, impartiality, candor, fidelity to trust, and inviolability of confidence are incumbent upon every member as professional obligations.

(c) Each member shall be guided by high standards of business ethics, personal honor, and professional conduct. The word *member* as used throughout this code includes all classes of membership.

Note that Section 1 describes (in Subsection c) the bases for behavior, namely, "high [personal] standards" that are expressed in the business place as "professional morality and . . . responsibility" (Subsection a) and that produce individual "honesty, integrity, loyalty, fairness . . ." (in Subsection b) and other desirable traits.

The next three sections of the Code deal with the geologist's relationships to other individuals. Note that each attribute is an expression of at least one of the desirable personal traits as listed in Section 1, Subsection b.

SECTION 2. Relation of Members to the Public

(a) Members shall not make false, misleading, or unwarranted statements, representations or claims in regard to professional matters, nor shall they engage in false or deceptive advertising.

(b) Members shall not permit the publication or use of their reports or maps for any unsound or illegitimate undertakings.

(c) Members shall not give professional opinions, make reports, or give legal testimony without being as thoroughly informed as reasonably required.

SECTION 3. Relation of Members to Employers and Clients

(a) Members shall disclose to prospective employers or clients the existence of any pertinent competitive or conflicting interests.

(b) Members shall not use or divulge any employer's or client's confidential information without their permission and shall avoid conflicts of interest that may arise from information gained during geological investigations.

SECTION 4. Relations of Members to One Another

(a) Members shall not falsely or maliciously attempt to injure the reputation or business of others.

(b) Members shall freely recognize the work done by others, avoid plagiarism, and avoid the acceptance of credit due others.

(c) Members shall endeavor to cooperate with others in the profession and shall encourage the ethical dissemination of geological knowledge.

Finally, AAPG demands rather specific responsibilities and relationships between the Association and each individual member.

SECTION 5. Duty to the Association

(a) Members of the Association shall aid in preventing the election to membership of those who are unqualified or do not meet the standards set forth in this Code of Ethics.

(b) By applying for or continuing membership in the Association, each member agrees to uphold the ethical standards set forth in this Code of Ethics.

(c) Members shall not use AAPG membership to imply endorsement, recommendation, or approval by the Association of specific projects or proposals.

SECTION 6. Discipline for Violation of Standards

Members violating any standard prescribed in this Article shall be subject to discipline as provided by the Bylaws.

Note once more that AAPG bases the entire Code on "high standards." These high standards spring from a personal philosophy that forms the foundation of *all* decisions, actions, and personal relationships, not just those involved in business. The word *all* is stressed because ethical behavior and integrity are close kin. Integrity or integration suggests a wholeness of an individual. This means that connecting strands between actions and motives are woven together into a cloth whose beauty and durability depends both on the quality of the strands and on the pattern and uniformity of the weave. One cannot create a life of beauty without a strong connection between actions and personal beliefs or without a pattern that relates actions to one another.

Less clear, however, are the issues of how to be ethical in petroleum exploration, of what is covered by ethical considerations, and the extent to which ethical principles should be applied. The next section deals with a few such considerations.

QUESTIONS AND ISSUES

Ethical behavior in preparing a prospect involves several aspects. The first and fundamental question revolves around attitude. We all know that we can choose to approach any task expecting either to produce a "good enough" result or to arrive at a conclusion that is the very finest our minds can conceive. When we are exploring, we can find a "prospect" almost anywhere in short order by stretching or generalizing the data and by making improbable assumptions. The same is true if we are making a sales call, writing a contract, or dealing with employee relations.

Alternatively, we can decide to restrict our prospecting only to what has a high probability of actuality given the setting and data available. Taking such an approach will likely yield fewer prospects, but ones that are more firmly grounded in the data and science. Consequently, they will have more potential for success and favorable economic return. Our clients are far more likely to return to us if they know that all of our products arise from a point of view that emphasizes quality at the possible expense of quantity.

Having decided to pursue only quality prospects, we need to extend our decision further to include the gathering of all pertinent data. Furthermore, the perception of what data are pertinent will be based on our experience and training. More of each will enhance the quality of our products. Training and experience are clearly linked to the concept of professionalism, and according to the opening phrase of the AAPG Code of Ethics, professionalism is in turn linked to ethics.

Experience is gained by doing. Practice in a particular geographical area will define what data have been useful in the past, and experience in other areas may suggest other data that might be applied to the current project with favorable results.

As an aside, if it is agreed that enhanced experience is necessary to increasingly improve prospects or other products, then during times when no one is buying prospects or when the company is not exploring as aggressively as we might wish, at the very least we ought to be gaining experience. That can be achieved without great expenditure of capital by such activities as field mapping and section measuring, completing producing field studies, regional mapping of subsurface or seismic data, and careful reading of professional journals and textbooks.

Training must be an ongoing process regardless of whether or not it is demanded by professional societies or state certification boards. For example, how would a prospector know that pressure data could contribute to the value of a prospect unless something is known about fluid flow and hydrodynamics? Or how would one know that total organic carbon data are important without knowing something about petroleum geochemistry?

Training and experience are necessary in knowing which data are needed for a given project and are also important in the synthesis of the data. Ability must be coupled with creativity and tempered with fairness and honesty at this point. Knowing what is a fair and honest interpretation of the data involves part intuition and part fact. The reality and credibility check on intuition and creativity most often includes the study of analogs from recent or ancient settings. An interpretation that involves major thrust faults is acceptable in an area where such faults are known, but ought not to be called upon, from an ethical or professional standpoint, where such a structural setting is not likely. Similarly, it would be unethical, albeit highly creative, to suggest the presence of a shale-encased distributary channel fill in an area of otherwise blanket eolean sand deposition.

The data will have been ethically and professionally interpreted if they at least allow and even suggest, if not strongly support, the interpretation that is made. If so, then we will have treated the data with integrity and our ethics will have enhanced the quality of the prospect.

What Is Whose?

In the exploration industry, it is clearly accepted that hardcopy maps and cross sections belong to the company and not to the employee who generated them. Thus, if an explorationist changes companies, no reports, maps, documents, devices, or cross sections may ethically be carried along to give to the new employer or to release to the public. The same is true of techniques that are developed while in the employ of another. For example, if a geologist finds a new relationship among several log responses that, when mapped, leads to a new play concept in the Powder River basin, commonly accepted ethical standards demand that the new technique remain the "intellectual property" of the employer. The discoverer of the technique may not use the method for the benefit of another company. To quote from the agreement signed by all applicants at Pennzoil:

> . . . I agree:
>
> 1. All inventions, discoveries or improvements made or conceived in whole or in part by me during my employment and relating in any way to products, apparatus, methods, manufacturing procedures, research projects and investigations of the Company or its affiliates, or resulting from or suggested by any work which I may do for the Company shall belong to the Company, and I hereby assign, set-over and convey to the Company my entire right, title and interest in and to the said inventions, discoveries and improvements.
> 2. I will promptly disclose all such inventions, discoveries and improvements to the Company and I will assist the Company in every way during and after my employment to obtain for the Company's benefit patents in any and all countries, and I specifically agree that I will execute any and all documents, patent applications, continuations of such applications or substitutes thereof, assignments of patents, and when requested, will give testimony, all without any expense to me and all without further compensation to me.
> 3. I will, during and after my employment, maintain in strict confidence all inventions, discoveries and improvements belonging to the Company until the same are disclosed either by the issuance of a patent or a printed publication or by public use in the United Sates of America. I further agree that during and after my employment I will not use for my own benefit or the benefit of others nor will I disclose directly or indirectly to persons outside the Company any confidential information or trade secrets owned by the Company or disclosed in confidence to it.

But what of general exploration skills? The standard here seems to be that if the skills are already available to the industry outside of the company, then it is acceptable to use those skills for the benefit of a new employer. Thus, although a geologist may have learned to recognize stratigraphic sequences, to identify pollen, or to calculate reserves while with a former company, these skills of the trade are generally recognized to be available to successive employers.

An exception to this seems to arise when an employee has recently been given specialized training by the company. Most explorationists agree that the trainee needs to let the company have the benefit of his new expertise for some undefined period of time to allow the company to recapture some of its investment. A rough gauge of the "required" minimum tenure after company-provided training or other expenditure on an individual might be 1 year for each month of training received. So, for example, if the company sends a young manager off to Harvard for 6 months of business education, the company ought to expect at least 6 years of service in exchange. Similarly, 3 months of service in exchange for a one-week training session also seems reasonable (1/4 year for 1/4 month). To be sure, this is not a firm rule, rather one that seems to operate in practice.

When Is a Deal a Done Deal?

Folklore has it that in the "good old days" of the petroleum business, deals were struck over a cup of coffee or across a fence line and were consumated with a handshake. Supposedly, nobody ever backed out of any part of a deal made in this way unless there was mutual agreement between the principals.

Today, explorers bemoan the fact that lengthy and convoluted contracts are required that cover every eventuality and that cannot be prepared or even understood except by a specialist. It seems that the deal is not "done" until all have signed off on a formal contract. We seem to have gotten the idea that there is some progressive binding among the parties to an arrangement as we move from verbal negotiation to a letter agreement onto a formal contract and, perhaps, finally to a judicial mandate. Yet nothing is inherently different between a verbal agreement and a letter of intent or between such a letter and a formal contract. Each document in the chain looks more impressive, to be sure, but only because someone has spent the time necessary to provide for more contingencies. A contract does have more detail, but we all know that no contract covers every eventuality or else disagreement among the parties would never occur. The same is true of a verbal commitment.

The main difference between a handshake deal and a carefully prepared contract is that a contract is tangible and can be evaluated by individuals who are not parties to the agreement. A verbal commitment can only be interpreted by those who have made the agreement. If a question should arise about the intent of one party in a verbal agreement, the other party can only ask the other about his or her meaning. In the asking, the questioner must assume that the respondent will accurately recall and decide his or her intentions at the time of the agreement, not as modified by subsequent events. The assumption of accuracy depends on acceptance of the fact of honesty and integrity on the part of the other party.

Any question about the integrity and ethics of parties to a contract demands that written documents be prepared. The expense in time and money of document preparation and subsequent interpretation or litigation is a direct consequence of reduced emphasis on ethics in the exploration business. Vast improvements in efficiency of exploring could be made if all parties to all agreements would place greater value on their word than on their wallet. That is, if all would stand behind their original intent in a deal even if that meant a loss in profits as a result of succeeding events or the revelation of more data.

Should We "Blackball" Unethical Explorers?

The increasing use of courts of law in the exploration business suggests that unethical folks lurk among us. Any opportunities we have to remove such hindrances to the effective practice of our science and business ought to be exercised simply as a matter of good organizational procedure. Sentimentality or even concern for an individual who calls himself a professional but is not, or who calls himself a businessperson but is a hindrance to the effective conduct of business, should not be confused with our responsibilities to profession and business. At the same time, our business responsibilities may not cloud our ethical responsibilities to that person. Nevertheless, all ethical means should be used to exorcise the lack of ethics in our sphere of influence.

The available means of repair include counseling and persuasion of the individual, the use of increased pressure by an expanded group, and procedurally correct removal of the offensive individual from organizations with which we are associated (see Chapter Appendix). In a company, the removal stage can include reassignments to a less sensitive area or complete separation from the company. In a professional society, the removal may include prohibitions against publication or may involve the mechanisms specified by most associations for removal of the individual from the society rolls. In the arena of independent and consulting explorationists, the steps of removal may include personal dissociation from the offensive individual or may involve carefully considered dissemination of data regarding the offender.

Clearly, we in the exploration business need to police our own ranks, but we need to do so in carefully considered, progressive, ethical steps. The remedy for a problem must include education, persuasion, pressure, and only in the extreme case, removal.

How Should Unethical Situations Be Handled?

Regardless of whether or not one believes that a peer or a supervisor has behaved unethically, the first step in dealing with the situation should be to try to change the unethical behavior. A personal visit appears to be in order. The visit should include an expression of concern, the reason for the concern, and the perception of the unethical behavior; then the party should be given a chance to state his or her opposing perceptions. The visit can be nonthreatening to either party, but at the end of the visit, both parties should clearly understand what the meeting was about.

In most cases, such a simple meeting may resolve the problem. If not, a committee of responsible individuals should be established to review the situation and to make recommendations to the allegedly unethical person. A suitable way to establish such a committee would be to contact an uninvolved person in the company. The Director of Human Resources, the chief scientist, or a supervisor in another department would be good choices to continue the process. Failure of the investigating committee to correct the situation will result in involvement of line management above the level of the parties involved. Their disposition of the problem will be final.

The reason for using less-involved individuals in the intermediate stage is to protect both parties against excessive "whistle blowing" and/or paranoia. Objective consideration of any problem is always worth the time spent, but it is especially important when a person's character, integrity, and ethics are being questioned. Sticks and stones can break bones, but words may utterly destroy.

Typical Scenarios

Admittedly, the preceding considerations are not very specific. The lack of precision is a necessary consequence of the complexities of ethical considerations, most of which are beyond the scope of one short chapter. Nevertheless, a few easily described scenarios are given here that may stimulate your thinking and perhaps suggest others more applicable to your individual situation. You will perhaps find that the "best" choice is not given as an option. The objective is for you to consider your own choices.

1. You have looked at a prospect and materials have been left with you at the end if the presentation. You are no longer interested. Do you
   - keep the material?
   - copy the maps and return the originals?
   - return all materials?
   - make no copies, but note the location of the prospect?

2. You and a prospective partner have verbally agreed to an area of mutual interest (AMI), but the contract is not yet drawn up. Do you
   - buy offsetting acreage just outside the AMI?
   - continue looking for a better partner?
   - pass along all of your company's information to your partner?

3. Your company and your partner have a firm AMI agreement and you, while working on other projects, have found a new exploration technique. Do you
   - pass the new technique on to your partner knowing it will enhance exploration of the AMI?
   - not tell them, but wait for the AMI to expire and then apply the technique in the area and make a big discovery for your own company?

4. You have several clients exploring in the same area. Your contacts with the companies give you quite detailed information on the activities and capabilities of each of them. Do you
   - gently give suggestions to your best contact?
   - not tell your best contact that their chosen course of action will lead to disaster?
   - combine all the information in some novel way to the benefit of your own company?

5. You have left one company to join another. Do you
   - carry along maps and models?
   - pass on only what is in your head?
   - never let the new company know about anything you learned at the old company?

6. You are in a position to be aware of proposed changes in the personnel policies of your company. Do you
   - modify your own posture to take advantage of the new policy when it becomes effective?
   - let some of your friends know so they can do as they wish?

7. You have occasion to believe that a fellow geologist or geophysicist has stolen prospects from someone in Midland. Do you
   - volunteer the information to company officials?
   - tell the chap in Midland?
   - confront the thief?
   - inform your professional society?
   - wait until you are asked?
   - never admit any knowledge?

8. Your company has "asked" you to under-report production from a field for some unspecified reason. Do you
   - assume they know what they are doing?
   - report the matter to very high level management?
   - ask for a transfer?

- quit your job?
- delay and hope the order goes away?
- report the matter to civil authorities?

9. Your company is doing less and less in your area of expertise. Do you use your extra time to
   - keep up with company rumors?
   - compile your private list of prospects?
   - go get a haircut?
   - take some sick days?

10. You have an opportunity to speak privately and off the record with a senior company manager. You wish to take the opportunity to advance your career. Do you
    - stress your own qualities?
    - compare your performance with that of someone else?
    - suggest a reorganization that will cut someone else out?

11. You need to reduce staff and you initiate an early retirement plan. The law requires that you cannot give guidance to anyone in their decision to reject the package, but you know that many who do will be terminated with much poorer benefits. The company will be justifiably sued if you give anyone guidance. Do you
    - give hints only to those you are sure would not sue the company?
    - give suggestions only to your friends?
    - disregard the consequences and tell everyone just where they stand?

12. You have submitted your expense statement. Did you
    - slightly inflate the cost of meals to cover the price of a book you bought at the airport?
    - buy better meals than were necessary?
    - arrange your flight schedule to achieve the minimum time away from your work?

13. An inexpensive continuing education seminar is being offered for two evenings. Will you
    - pass it up because you already work 8 hours a day?
    - get the notebook from someone who did go?
    - attend only one night to get the certificate?

14. Your company has sent you to a national convention. Will you
    - register but never attend?
    - spend your time talking to old acquaintances?
    - report new advances to others back home?
    - develop contacts that will provide new opportunities for your company?

15. You have an exclusive agreement with your employer to work for him alone, but a friend with a different company needs a quick interpretation in an area where your employer is not currently working. Do you
    - flatly refuse your friend, forcing her to hire help when she can least afford it?
    - give your interpretation but for a large fee that would cover any legal action your company might take against you if they found out?
    - give your interpretation for free?

16. You have a producing well on the property of a relatively unsophisticated royalty owner. You have just discovered that you have not been meeting all the requirements of the lease. Do you
    - make restitution for missed royalties?
    - continue on as before hoping the lessor will never catch the omission?
    - don't admit anything, but increase royalty payments slightly so it will all balance out in a couple of years?

17. You have an employee who is constantly late in reporting for work. Do you
    - fire him now?
    - provide a free bus pass?
    - reassign him to another supervisor?
    - get the company counselor to find out what the problem is?

Guidelines for Ethical Behavior

No list of rules can possibly cover all ethical questions that arise between people because ethical standards are attitudes and fundamental approaches to life. Nevertheless, I yield to the temptation to emphasize what seem to be important guidelines in petroleum exploration.

- Decide on your fundamental assumptions.
- Separate the facts from the opinions. No one will be misled if the data are identified and the conclusions are separated and identified as such.
- Develop a commitment to the highest quality work.
- Devote the time and energy necessary to produce outstanding work.
- Do not cheat on time. Give full time while in the office, and do not double bill your clients. Invoice for a half day if a half day was spent (Geer, pers. comm., 1990).
- Maintain confidentiality. Do not use knowledge gained from one source to the unfair benefit of another.

- Do not defame other companies or other explorationists.
- Maintain the ethics of exploration by following proper channels to remove unethical individuals.
- Do not plagiarize. Give credit to earlier studies.
- Do not use "inside" information for unfair personal gain.
- Avoid conflicts of interest. Do not take bribes.
- Make no unilateral "adjustments" to any part of a agreement.
- Be worthy of trust—call a situation just as you see it (such as in an employee evaluation or an estimation of risk).
- Be at the leading edge of technology.
- Do reject some conclusions—not everything is a good deal.
- Change company policy if you must, but do not circumvent it.
- Do not be a bigot or a chauvinist.
- Do what is right before you receive a court order.
- Recognize and credit employees and supervisors who do a good job.

CONCLUSIONS

Ethical questions are exclusively questions about relationships between and among people. One's ethical position is defined by attitudes and actions with regard to other people, be they suppliers, creditors, clients, superiors, employees, partners, or society in general as represented by the government. In all these situations, Shakespeare's admonition "To thine own self be true" is expanded by the Scriptures' lesson to "love your neighbor as yourself." People who have constructed a firm foundation of moral principles have the basis for integrity and for ethical behavior. If an individual has a sound footing of personal beliefs and has then established an orderly pattern for the relationship among events in his life, he will also have a basis for weaving new events into his life.

Whether the state of the industry is characterized by stock options or cutbacks, the integrated individual will have built a rationale for his or her behavior in any new situation. Very few decisions will be required in new situations—or at least the scope of the questions will be reduced and problem situations will be more strongly focused. Wantland (1989a) recently claimed that

> Good times or bad, the very process of setting goals based on personal values makes achievement not only possible but probable. The alternative is to drift with no sense of control over events in one's career or one's life.

References Cited

American Association of Petroleum Geologists, 1991, The American Association of Petroleum Geologists constitution and bylaws: AAPG Bulletin, v. 75, n. 12, p. 1918-1925.

Brown, D. J., 1985, Business ethics: Is small better?: Policy Studies Journal, v. 13, n. 4, p. 766-775.

Campbell, J. M., Sr., 1990, Improved professionalism: A critical need: Journal of Petroleum Technology, v. 42, n. 3, p. 338-341.

Drucker, P. F., 1981, What is "business ethics"?: The Public Interest, n. 63, p. 641-656.

Goldman, A. H., 1980, Business ethics: Profits, utilities, and moral rights: Philosophy and Public Affairs, v. 9, n. 3, p. 260-286.

Johnson, C. H., 1989, A matter of trust: Management Accounting, Dec., p.12-13.

Kohlberg, L., 1971, From is to ought: How to commit the naturalisitic fallacy and get away with it in the study of moral development, *in* Mischel, T., ed., Cognitive development and epistemology: New York, Academic Press, p. 151-235.

Megill, R. E., 1985, How to be a more productive employee: Tulsa, OK, Pennwell Books, 122 pp.

Wantland, K. F., 1989a, Personal goals generate results: AAPG Explorer, May 1989, p. 9.

Wantland, K. F., 1989b, Survivors plan to manage crisis: AAPG Explorer, Dec. 1989, p. 18.

Weimer, R. J., 1984, Response to conferring of the Sidney Powers memorial medal: AAPG Bulletin, v. 68, n. 8, p. 1058-1061.

Chapter Appendix

GRIEVANCE PROCEEDINGS OF THE AMERICAN ASSOCIATION OF PETROLEUM GEOLOGISTS

(From American Association of Petroleum Geologists, 1991)

SECTION 1. Investigation

Charges of misconduct in violation of Article IV of the Constitution [the code of ethics] shall first be submitted in writing to the Executive Director at Association headquarters, by a member in good standing, with a full statement of the evidence on which the charges are based. If, in the judgement of the President, they merit further consideration, the President shall refer them to the Chairman of the Advisory Council, who shall appoint an investigating committee of three (3) members of the Advisory Council, including at least one (1) former president of the Association, to examine into the charges. If, in the judgement of said investigating committee, the facts warrant, the committee shall prepare and file with the Advisory Council at Association headquarters formal charges against the accused member.

SECTION 2. Notice of Hearing

As soon as may be after the receipt of such formal charges the Advisory Council shall fix a date and place for hearing thereon, and shall give to the accused member notice thereof in writing, mailed to the member by registered mail at the member's last known post office address not less than thirty (30) days before said date, accompanied by a copy of the formal charges, and a copy of this Article.

SECTION 3. Hearing

On the day fixed for the hearing the attendance of the Chairman and at least two-thirds (2/3) of the members of the Advisory Council shall constitute a quorum and full representation of the Advisory Council for the conduct of the hearing provided in this section. The accused member may appear with legal counsel before the Advisory Council, hear any witnesses called in support of the charges and at the member's option, cross-examine the same, present witnesses in the member's behalf, and submit oral or written statements in the member's behalf. The Advisory Council may likewise present witnesses and have the right to cross-examination. At the member's option, the accused member may, by registered letter addressed to the Chairman of the Advisory Council at Association headquarters, postmarked not less than ten (10) days prior to the date of the hearing, waive personal appearance and request the Advisory Council to adjudge the matter on the basis of a written statement of the member's defense accompanying such a letter.

SECTION 4. Decision of Council

After the conclusion of the hearing of study of the written defense submitted in lieu thereof, the Advisory Council shall consider and vote to sustain or dismiss the charges. If, by a two-thirds (2/3) vote of those present, the Advisory Council shall declare sustained the charges against the accused member, it shall recommend to the Executive Committee alternatively that it:

(a) admonish the member; or
(b) suspend the member for a stated period of time; or
(c) allow the member to resign; or
(d) expel the member.

Failure of the accused member to appear, or to submit a waiver letter and a written defense, as in this section provided, shall not prevent the Advisory Council from rendering final advisory judgement and the Executive Committee from action on the basis of the evidence available to it on the hearing date.

SECTION 5. Executive Committee Action

The decision of the Advisory Council in all matters pertaining to the interpretation and execution of the provisions of Sections 1, 2, 3, and 4 of this Article shall be submitted to the Executive Committee for final action. A report of the Executive Committee action shall be published in the Bulletin.

SECTION 6. Resignation

Resignation by the accused member from the Association, at any stage in the foregoing prescribed proceedings, shall automatically terminate the proceedings. Following resignation, the accused person so resigning shall not be eligible for reinstatement to membership under any circumstances in the future.

SECTION 7. Expulsion

Persons expelled from the Association under these proceedings shall thenceforth be ineligible for reinstatement to membership under any circumstances in the future.

SECTION 8. Alternative Procedure

Any member convicted of a misdemeanor involving moral turpitude or of any felony may be suspended from membership in the Association upon a majority vote of the full Executive Committee of the Association. A member whose conviction is reversed on appeal or which is the subject of an executive pardon shall be reinstated to membership.

A member whose conviction is upheld on final appeal may be expelled from membership in the Association upon a majority vote of the full Executive Committee of the Association.

In the event that suspension or expulsion of a member so convicted is proposed, a date shall be set for a hearing thereon and for consideration by the Executive Committee of such proposed suspension or expulsion. The member shall be given notice in writing of the date and place for the hearing, mailed to the member by registered mail to the member's last-known post office address not less than thirty (30) days before said date, accompanied by a copy of any applicable order of an appellate court, and a copy of this Section. At the hearing the member may appear before the Executive Committee with legal counsel, may submit oral or written statements to the Executive Committee, and may present witnesses to testify on the member's behalf before the Executive Committee. The Executive Committee shall have the right to cross-examine the member and any witnesses presented by the member on the member's behalf. At the member's option, the member may, by registered letter addressed to the President of the Association at Association headquarters, postmarked not less than ten (10) days prior to the date of the hearing, request the Executive Committee to consider the matter on the basis of a written statement by the member accompanying such a letter without the personal appearance of the member before the Executive Committee. The Executive Committee, if such oral or written statements or testimony of witness are presented, shall consider said statements and testimony prior to voting on the suspension or expulsion of the member.A member expelled from the Association under the procedure stated above shall be ineligible for reinstatement to membership unless reinstated by a unanimous vote of the full Executive Committee of the Association.

CODE OF ETHICS OF THE AMERICAN ASSOCIATION OF PETROLEUM LANDMEN

(From American Association of Petroleum Landmen, 1990)

The Code of Ethics shall be the basis of conduct, business principles and ideals for the members of the American Association of Petroleum Landmen; and it shall be understood that conduct of any member of the Association inconsistent with the provisions set forth in this Article shall be considered unethical and said individual's membership status shall be subject to review for possible disciplinary action as prescribed in Article XVI of these Bylaws.

In the area of human endeavor involving trading under competitive conditions, ethical standards for fair and honest dealing can be made increasingly meaningful by an association organized and dedicated not only to the definition, maintenance and enforcement of such standards, but to the improvement and education of its members. Such is the objective of the American Association of Petroleum Landmen and such is its public trust.

Section 1. It shall be the duty of the landman at all times to promote and, in a fair and honest manner, represent the industry to the public at large with the view of establishing and maintaining good will between the industry and the public, and among industry parties.The landman, in his dealings with landowners, industry parties and others outside the industry, shall conduct himself in a manner consistent with fairness and honesty, such as to maintain the respect of the public.

Section 2. Competition among those engaged in the mineral and energy industries shall be kept at a high level with careful adherence to established rules of honesty and courtesy.

A landman shall not betray his partner's, employer's or client's trust by directly turning confidential information to personal gain.

The landman shall exercise the utmost good faith and loyalty to his employer (or client) and shall not act adversely or engage in any enterprise in conflict with the interest of his employer (or client). Further, he shall act in good faith in his dealings with industry associates.

The landman shall represent to others in his area of expertise and shall not represent himself to be skilled in professional areas in which he is not professionally qualified.

STANDARDS OF PRACTICE OF THE AMERICAN ASSOCIATION OF PETROLEUM LANDMEN

(From American Association of Petroleum Landmen, 1990)

The Bylaws of the American Association of Petroleum Landmen (AAPL) provide that a Code of Ethics has been established "to inspire and maintain a high standard of professional conduct" for the members of the Association. The Code of Ethics is the basis of conduct, business principles and ideals for AAPL members. This standard of professional conduct and these guiding principles and ideals mandated by the Code of Ethics within the AAPL Bylaws are summarized as follows:

A. Fair and honest dealing with landowners, industry associates and the general public so as to preserve the integrity of the profession (Article IV, Section 1);

B. Adherence to a high standard of conduct in fulfilling his fiduciary duties to a principal (Article IV, Section 2);

C. Avoiding business activity which may conflict with the interest of his employer or client or result in the unauthorized disclosure or misuse of confidential information (Article IV, Section 2);

D. Performance of professional services in a competent manner (Article IV, Section 2);

E. Adherence to any provisions of the Bylaws, Code of Ethics, or any rule, regulation, or order adopted pursuant thereto (Article V, Section 5);

F. Avoiding the aiding or abetting of any unauthorized use of the title "Certified Professional Landman" (Article V, Section 5); and

G. Avoiding any act or conduct which causes disrespect for or lack of confidence in the member to act professionally as a landman (Article V, Section 5).

The masculine gender used herein shall refer to both men and women landmen.

Preamble

Under all is the land. Upon its wise utilization and widely allocated ownership depend the survival and growth of free institutions and of our civilization, The Code of Ethics shall be the basis of conduct, business principles and ideals for the members of the American Association of Petroleum Landmen.

In the area of human endeavor involving trading under

competitive conditions, ethical standards for fair and honest dealing can be made increasingly meaningful by an association organized and dedicated not only to the definition, maintenance and enforcement of such standards, but to the improvement and education of its members. Such is the objective of the American Association of Petroleum Landmen and such is its public trust.

Such standards impose obligations beyond those of ordinary trading. They impose grave social responsibility and a duty to which the landman should dedicate himself. A landman, therefore, is zealous to maintain and improve the standards of his calling and shares with his fellow landmen a common responsibility for its integrity and honor. The term "Landman" has come to connote competency, fairness, integrity and moral conduct in business relations. No inducement of profit and no instruction from clients can ever justify departure from these ideals.

In order to inform the members of specific conduct, business principles and ideals mandated by the Code of Ethics, the Association has adopted the following Standards of Practice, and every member shall conduct his business in accordance therewith:

1. In justice to those who place their interests in his care, a landman shall be informed regarding laws, proposed legislation, governmental regulations, public policies, and current market conditions in his area of represented expertise, in order to be in a position to advise his employer or client properly (D, E).[1]

2. It is the duty of the landman to protect the members of the public with whom he deals against fraud, misrepresentation, and unethical practices. He shall eliminate any practices which could be damaging to the public or bring discredit to the petroleum or mining industries.

3. In accepting employment, the landman pledges himself to protect and promote the interests of his employer or client. This obligation of absolute fidelity to the employer's or client's interest is primary, but it does not relieve the landman of his obligation to treat fairly all parties to any transaction, or act in an ethical manner (A, B).

4. The landman shall not accept compensation from more than one principal for providing the same service, nor accept compensation from more than one party to a transaction, without the full knowledge of all principals or parties to the transaction (B, C).

5. The landman shall not deny equal professional services to any person for reasons of race, creed, sex or country of national origin. The landman shall not be a party to any plan or agreement to discriminate against a person or persons on the basis of race, creed, sex or country of national origin (A, F).

6. A landman shall provide a level of competent service in keeping with the standards of practice in those fields in which a landman customarily engages. The landman shall not represent himself to be skilled in nor shall he engage in professional areas in which he is not qualified, such as the practice of law, geology, engineering or other disciplines (D).

7. The landman shall not undertake to provide professional services concerning a property or a transaction where he has a present or contemplated interest, unless such interest is specifically disclosed to all affected parties (C).

8. The landman shall not acquire for himself or others an interest in property which he is called upon to purchase for his principal, employer or client without the consent of said principal, employer or client. He shall disclose his interest in the area which might be in conflict with his principal, employer or client. In leasing any property or negotiating for the sale of any block of leases, including lands owned by himself or in which he has any interest, a landman shall reveal the facts of his ownership or interest to the potential buyer (C).

9. If a landman is charged with unethical practice or is asked to present evidence in any disciplinary proceeding or investigation, or has direct knowledge of apparent unethical misconduct by another member, he shall place all pertinent facts before the proper authority of the American Association of Petroleum Landmen (E).

10. The landman shall not accept any commission, rebate, interest, overriding royalty or other profit on transactions made for an employer or client without the employer's or client's knowledge and consent (B).

11. The landman shall assure that monies coming into his possession in trust for other persons, such as escrows, advances for expenses, fee advances, and other like items, are properly accounted for and administered in a manner approved by his employer or client (B).

12. The landman shall avoid business activity which may conflict with the interest of his employer or client or result in the unauthorized disclosure or misuse of confidential information (C).

13. The landman shall at all times present an accurate representation in his advertising and disclosures to the public (A).

14. The landman shall not aid or abet the unauthorized use of the title "Certified Professional Landman."

15. The landman shall not participate in conduct which causes him to be convicted, adjudged or otherwise recorded as guilty by any court of competent jurisdiction of any felony, any offense involving fraud as an essential element, or any other serious crime.

SOCIETY OF PETROLEUM ENGINEERS GUIDE FOR PROFESSIONAL CONDUCT

(From Society of Petroleum Engineers, 1990)

Preamble Engineers recognize that the practice of engineering has a direct and vital influence on the quality of life for all people. Therefore, engineers should exhibit high standards of competency, honesty, and impartiality; be fair and equitable; and accept a personal responsibility for adherence to applicable laws, the protection of the public health, and maintenance of safety in their professional actions and behavior. These principles govern professional conduct in serving the interests of the public, clients, employers, colleagues, and the profession.

The Fundamental Principle The engineer as a professional is dedicated to improving competence, service, fairness, and the exercise of well-founded judgement in the practice of engineering for the public, employers, and clients with fundamental concern for the public health and safety in the pursuit of this practice.

Canons of Professional Conduct

1. Engineers offer services in the areas of their competence and experience, affording full disclosure of their qualifications.

2. Engineers consider the consequences of their work and societal issues pertinent to it and seek to extend public

[1] References are to the foregoing summary of the standards of professional conduct and guiding principals and ideals mandated by the Code of Ethics and AAPL Bylaws.

understanding of those relationships.

3. Engineers are honest, truthful, and fair in presenting information and in making public statements reflecting on professional matters and their professional role.

4. Engineers engage in professional relationships without bias because of race, religion, sex, age, national origin, or handicap.

5. Engineers act in professional matters for each employer or client as faithful agents or trustees disclosing nothing of a proprietary nature concerning the business affairs or technical processes of any present or former client or employer without specific consent.

6. Engineers disclose to affected parties known or potential conflicts of interest or other circumstances that might influence—or appear to influence—judgement or impair the fairness or quality of their performance.

7. Engineers are responsible for enhancing their professional competence throughout their careers and for encouraging similar actions by their colleagues.

8. Engineers accept responsibility for their actions, seek and acknowledge criticism of their work, offer honest criticism of the work of others, properly credit the contributions of others, and do not accept credit for work not theirs.

9. Engineers, perceiving a consequence of their professional duties to adversely affect the present or future public health and safety, shall formally advise their employers or clients and, if warranted, consider further disclosure.

10. Engineers act in accordance with all applicable laws and the canons of ethics as applicable to the practice of engineering as stated in the laws and regulations governing the practice of engineering in their country, territory, or state and lend support to others who strive to do likewise.

CODE OF ETHICS OF THE SOCIETY OF EXPLORATION GEOPHYSICISTS

(From Society of Exploration Geophysicists, 1990)

The Constitution of the SEG, Article IV, Section 1, states that "Membership of any class shall be contingent upon conformance with the established principles of professional ethics.

As an elaboration of these established principles of professional ethics, the following Code of Ethics is enunciated. It shall be your duty as a geophysicist, in order to maintain the dignity of your chosen profession:

1. To carry on your professional work in a spirit of fidelity to clients and employers, fairness to employees and contractors, and devotion to high ideals of personal honor.

2. To treat as confidential your knowledge of the business affairs, geophysical or geological information, or technical processes of clients or employers when their interests require secrecy.

3. To inform a client or employer of any business connections, interests, or affiliations which might influence your judgement or impair the disinterested quality of your services.

4. To accept financial or other compensation for a particular service from one source only, except with the full knowledge and consent of all interested parties.

5. To refrain from associating yourself with, or knowingly to allow the use of your name by, an enterprise of questionable character.

6. To advertise only in a manner consistent with the dignity of the profession, to refrain from using any improper or questionable methods of soliciting professional work, and to decline to pay or to accept compensation for work secured by such improper or questionable methods.

7. To refrain from using unfair means to win professional advancement and to avoid injuring unfairly, or maliciously, directly or indirectly, another geophysicist's professional reputation, business, or chances of employment.

8. To cooperate in building up the geophysical profession by the interchange of general information and experience with your fellow geophysicists and with students and also by contributions to the work of technical societies, schools of applied science, and the technical press.

9. To interest yourself in the public welfare, and to be ready to apply your special knowledge, skills, and training in the public behalf for the use and the benefit of mankind.

References Cited in Appendix

American Association of Petroleum Geologists, 1991, The American Association of Petroleum Geologists constitution and bylaws: AAPG Bulletin, v. 75, n. 12, p. 1918-1925.

American Association of Petroleum Landmen, 1990, Bylaws, American Association of Petroleum Landmen: Landmen's Directory and Guidebook 1990–91, p. 24A-28A.

Society of Exploration Geophysicists, 1990, Code of ethics: The Leading Edge, v. 9, p. 33.

Society of Petroleum Engineers, 1990, Society of Petroleum Engineers guide for professional conduct: Journal of Petroleum Technology, v. 42, n. 5, p. 625.

ROBERT W. SPOELHOF, Exploration Advisor for Pennzoil in Houston, received an A.B. degree from Calvin College, and M.S. and Ph.D. degrees from the Colorado School of Mines. In 1974 he joined Shell Development Company as a research geologist and in 1976 was transferred to Shell Oil Company as an exploration geologist with assignments in the Mid-Continent and Rocky Mountains. In 1981 Dr. Spoelhof joined Pennzoil, where he has been Manager of the Frontier Evaluation Group and Denver District Exploration Manager and is currently assigned to a new ventures group in domestic exploration.

Chapter 28

The Environmental Realities of Petroleum Exploration

G. Rogge Marsh
Exxon Exploration Company
Houston, Texas, U.S.A.

Treat the earth well . . . it was not given to you by your parents . . . it was lent to you by your children.
—Kenyan Proverb

INTRODUCTION

The American explorationist has succeeded in finding enough endemic petroleum to sustain the greatest, most dynamic economy the world has ever seen. The oil we found fueled two world wars; the technology we devised has spread across the globe; and our efforts continue to provide the world with the energy to move about from place to place and run the machinery to get the world's work done.

Yet in the final decade of this century, we are assured by the Cassandras of the environmental movement that humans will end by destroying the planet. We must halt economic growth, energy use growth, population growth, and all other growth (except possibly government growth). Environmental pollution is a sign of major incompatibility between free enterprise and the earth that supports it. The world is running out of resources. And the industry most at fault for the world's ills is the oil industry! Our society appears to have become afraid of itself, a view typified by biologist Paul Ehrlich's (1975) statement, "Giving society cheap, abundant energy at this point would be the equivalent of giving an idiot child a machine gun."

Clearly we are dealing with a public who has little apparent gratitude for our efforts. The environmental movement—arguably the most significant social feature of the second half of this century—has cast our industry as one of the major contributors to what many apparently believe to be the impending demise of the planet.

This is obviously an exaggerated indictment. We recognize our industry as being at least as responsible as any other and much better than most. Nevertheless, as of now, the indictment stands. Mobil's Herb Schmertz (1986) observed that if you consciously set out to choose an industry whose public image had all of the ingredients of the perfect villain, ours would be the one you would pick. Our champions are few, especially in Congress and in the media, where champions count the most. Therefore, it is up to us to see that our own name is cleared and that we get a fair public hearing. To do this, we need to answer four questions:

1. How did we end up behind this environmental eight ball?
2. What is exploration's impact on the environment?
3. How do these environmental issues affect our operations?
4. How can we approach the future in a positive manner so that we can get more reasonable treatment, understanding, and respect for our efforts?

ENVIRONMENT AS AN ISSUE: THE ROAD TO TODAY

People have always been concerned about the earth around them; it is their sole support. However, humans are unique among animals; rather than adapt to the environment, we adapt the environment to ourselves. As we have studied our surroundings and gained experience with the world, we have created complex systems for living and have invented machines for transforming the world around us, allowing us not only to live longer but also to live in immeasurably greater comfort. With all of these machines and systems and transformations, we have sometimes overdone it and had to correct matters.

History overflows with examples of humankind's environmental mistakes against nature. Plato (Arnold, 1987) spoke of the needless stripping of trees from the hills of Greece: "The rich, soft soil has all run away leaving the land nothing but skin and bones." The Roman philosopher Seneca (Whelan, 1985) wrote of getting away from "the heavy air of Rome and from the stink of the smoky chimneys thereof, which being stirred poured forth whatever pestilential vapors and soot they had enclosed in them." Ancient and modern hunters have exterminated wild animal species such as the passenger pigeon. City life in the nineteenth century suffered from frequent bouts with malevolent air problems arising from coal burning. By the turn of the century, Gifford Pinchot was spreading a conservation gospel to forestall what he believed was the impending loss of America's forests and timber reserves (Whelan, 1985). His approach was "the use of foresight and restraint in the exploitation of the physical resources of wealth as necessary for the perpetuity of civilization, and the welfare of the present and future generations."

Perhaps the first appearance of widespread general public concern for the environment can be traced to the almost universal and understandable fear of the radioactive debris from nuclear testing in the early 1950s. The worry over its effects on the environment, and especially human health, transcended all political and cultural boundaries.

However, the birth of what can be called the "modern" environmental movement, at least in the United States, took place with the appearance of Rachel Carson's (1962) book *Silent Spring*. The book is written as a fable of how human use of pesticides could interfere with the food chain, killing untold numbers of birds, wildlife, livestock, and insects. The implication is that, without drastic action, the fable could become reality. Because Carson was an exceptional writer, her book received high praise from nonscientists. However, the scientific reviews tended to emphasize the inaccuracies, distortions of fact, and overstatements. Norman Borlaug (1972), winner of the Nobel Peace Prize for pioneering the green revolution in Third World countries, said,

> The gravest defect of *Silent Spring* was that it presented a very incomplete, inaccurate, and oversimplified picture of the needs of the interrelated, worldwide, complex problems of health, food, fiber, wildlife, recreation, and human population.

Nevertheless, this enormously popular book is given credit for EPA's decision to ban DDT in 1972, as well as passage by Congress of the Toxic Substances Control Act in 1976.

The watershed environmental calamity for our industry began on January 28, 1969, when the fifth well drilled from Platform A in the Dos Quadras field, five miles from the beach at Santa Barbara, California, became the most infamous blowout in recent history. Over the next 10 days, between 10,000 and 77,000 bbl of crude oil escaped into the Santa Barbara Channel. Some of this oil landed on Santa Barbara's beaches, receiving avid and unparalleled press coverage. By the eighth day, the spill covered 600 square miles, and more than 100 miles of channel beaches were fouled. Kallman and Wheeler (1984, p. 71) characterized the significance of the event:

> The 1969 Santa Barbara oil well blowout did more than any other single incident in the nation's history to polarize public attitudes concerning protection of the environment. . . . From accepting the exploitation of the sea, the tenor of the times changed to the preservation of the sea.

Other mileposts marking environmentalism's progress could be included here, such as the publication of *The Limits to Growth* (Meadows et al., 1972), the Three Mile Island incident, James Watt's tenure as Secretary of the Interior, and various tanker accidents. Many problems, perceived as well as actual, have conspired to put the oil industry where it is today—on the defensive. Overall, though, we and the rest of the world's industrial sector are victims of a reaction to the growing recognition that humankind can affect the quality of the world's supply of air and water. This is a disturbing enough thought to earth scientists such as ourselves. Imagine how much more frightening it must be to the vast majority of people who are not in the resource business, have not studied the earth, have little scientific background, and get most of their information from electronic and print media more interested in bad news than good.

The growth of environmental activism has brought about laws and regulations governing almost all aspects of our business. It is unquestionably in our best interest to know where and to what extent our activities impact the environment or, perhaps even more important, environmental perceptions.

EXPLORATION'S IMPACT ON THE ENVIRONMENT

Exploration operations onshore bear little resemblance to the same activity when carried out on the water. Each milieu is unique. The problems of each and their solutions are largely unrelated to the other and thus are treated separately.

Onshore

Freshwater Protection

Perhaps the oldest and most meaningful onshore environmental problem explorationists have faced is protection of fresh groundwater resources. Aquifers close to the earth's surface are subject to contamination by drilling fluids. Therefore, we have long used surface and other shallow casing strings to isolate any freshwater aquifers from a drilling well. Reserve pit wastes can also impact fresh groundwater resources and must be disposed of according to methods approved by the various state regulatory agencies and commissions. Good housekeeping is a must. If, for instance, the wrong waste is inadvertently tossed into a pit, the entire contents could require special (and expensive) offsite disposal.

The subject of drilling waste disposal is extensive, requiring entire books, and thus is well beyond the scope of this brief survey of exploration's environmental issues. One word of caution, however: the well operator retains ultimate responsibility for whatever drilling wastes are created. Thus, for materials requiring offsite disposal, the operator must ensure, for his or her own protection, that the contractor or whoever disposes of the waste does it in a proper manner at a permitted and approved commercial disposal site. Some operators will reserve (and exercise) the right in the drilling contract to track the disposal methods and facilities used to get rid of these wastes.

Surface Disturbance

Exploration operations, regardless of how careful they are of the environment, inevitably result in temporary and localized disturbances of the ground on which they are located or across which they pass. Paradoxically, any disruption takes on greater significance as exploration's operations move to lands less inhabited and less used by humans. The more "natural" the area of operations, the more concern may be voiced over any disturbance. *Disturbance* can be very broadly defined to include noise, ground and vegetation impact, visual impact, human presence, atmospheric emissions, and almost anything else you can name that causes a change from the way things were before you arrived.

Of all exploration activities, perhaps the one with the least associated disturbance is surface geological investigation: the "rock-knocker knocking rocks." However, because its impact is perceived to be negligible, that does not mean that the response to it will always be positive. On public or local jurisdiction lands where permits are required for any commercial activity no matter how benign, the presence of oil company geologists may not be welcome. There have been instances (A. K. Scott, pers. comm., 1990)—at least one of which resulted in geologists actually going to jail—where seemingly trivial violations of the most strict interpretation of the terms of a permit for surface geological investigation have been prosecuted. Keep in mind that, while an oil company geologist may think of himself or herself as a research scientist, in some parts of the world this same research scientist may be perceived as a harbinger of industry, ugliness, and environmental degradation.

Seismic surveying, whether by thumper, vibroseis, or portable explosives, also subjects the earth to some disruption simply because vehicles and people require access. In areas that are environmentally "sensitive" because of their scenic beauty, wildlife habitat, or ecological uniqueness, seismic permits may require heli-portable drill rigs and perhaps even air shots (explosives mounted on stakes 3 ft above the ground) to keep the surface disturbance at an absolute minimum. Wetland areas, such as the southern Louisiana marshlands, can be especially sensitive and critical. Perhaps the only onshore environment where a geophysical crew can pass with virtually no lasting evidence of its passage is the flat tundra lands of Alaska's Arctic coastal plain. Exploration here takes place in the winter months when the ground is frozen solid and protected by ice and snow. Even here though, while flying over the land in the summer, the careful observer can detect faint traces of a previous winter's passage of the large-tired rollogon vehicles.

Exploratory drilling can be highly disruptive, depending on the amount of construction required for a road to the well site and the clearance necessary for the drill site, personnel camp, and/or possible helicopter pad. In remote areas, the road to the rig can account for the greatest surface disturbance, requiring more acreage than the drill pad itself. Table 1 illustrates these impacts. As the overall relief of the region increases, the area of disturbance also increases. In addition to localized loss of vegetation, the road and the drill pad will cause direct mortality to nonmobile wildlife populations through excavation, burial, overturning, clearing, and grading. In wetlands, canals may be the only practical means of access to well sites. These canals and associated spoil banks can, in some cases, impede marsh drainage, and if located close to the sea, they may promote saltwater intrusion into the freshwater marsh.

Wherever the activity, whether high desert, Arctic tundra, or marshlands, we, our colleagues, and our contractors must remain aware that, although the landscape may not be hospitable to human activities, it provides habitat for many nonhuman species whose well-being we, as stewards of the earth, must safeguard to the extent practicable.

Table 1. Estimated Acreage Requirements for Oil and Gas Activities

Activity	Acreage
Seismic exploration	1.6 acres per mile
Drilling/production location	
0–10% slope	2.77 acres per location
11–24% slope	3.66 acres per location
25–39% slope	4.46 acres per location
40%+ slope	5.38 acres per location
Access road[a]	
0–10% slope	2.28 acres per mile
11–24% slope	3.12 acres per mile
25–39% slope	5.00 acres per mile
40%+ slope	8.68 acres per mile
Power line[b]	2.4 acres per mile
Pipeline[b]	2.4 acres per mile
Compressor station, gas plant, water disposal plant	5 acres per plant

[a]For access roads, the disturbance necessary to make a 16-ft-wide driving surface.
[b]Assuming a 20-ft swath of disturbance within a 50-ft right-of-way.
Source: Bureau of Land Management, 1980.

Wildlife Habitat

Human presence, and especially any scientific or commercial activity, inevitably produces noise. It may be brief such as the noise from seismic air shots in a mountainous area; conversely, it may also be the relatively constant noise from drilling an exploration well over a period of weeks to months. Most wild animals are highly sensitive to noise and may act in unusual and atypical ways when exposed to it. Thus, noisy human activities, at the least, result in what is considered a "taking."

"Take," as defined in the Endangered Species Act, means to "harass, harm, pursue, hunt, shoot, wound, kill, trap, capture, or collect, or to attempt to engage in any such conduct." Most reasonable people will agree that a single geophysical crew or drilling rig is a local and very temporary source of disturbance. However, consider the case of multiple disturbances of this nature by competing seismic and exploration companies. An area may be reshot because of different seismic contractors or different companies wishing proprietary data; also, new well information frequently calls old data into question. In areas where discoveries lead to subsequent production, drilling can become a long-term activity that may have a noticeable impact on the wildlife habitat of a region.

Physical Presence

Public lands in the western United States and Alaska have largely remained the property of the federal government because they were thought to be relatively unfit for human use or habitation. In the nineteenth century, almost all of the land in the western states was owned by the federal government. To promote settlement and civilization's westward movement, the government made the land available to those with the fortitude to broach the wilderness. Whatever land was habitable, arable, or ranchable was settled and developed. The land that failed these tests remained in the hands of the federal government, generally because it was very rugged, remote, and/or desolate.

In today's more crowded world, society feels differently about these lands. There is a growing tendency to view them not as wasteland, as we did in the nineteenth century, but as highly desirable, pristine natural areas, some of which should remain uninhabited and undeveloped because of these now rare attributes of ruggedness and remoteness. As a result, the more scenic areas have become national parks, monuments, forests, wild and scenic rivers, and so on. The remote and desolate areas have become "wilderness areas," prized for exactly those characteristics. In the Wilderness Act, Congress defined wilderness as an area "where the earth and its community of life are untrammeled by man, where man himself is a visitor who does not remain."

Unfortunately, because these areas are so rugged, desolate, and remote, their petroleum potential has not been evaluated as thoroughly as have other more accessible areas of the United States. Thus, a conflict arises between those who want to keep the desolation desolate and those who believe the energy and mineral resource potential must be assessed. And wilderness has become, in one writer's words (Tucker, 1982), ". . . one of the highest products of civilization. It is a reserve set up to keep people out, rather than a 'state of nature' in which the inhabitants are truly free."

Air Quality

Perhaps the only reason for concern over air emissions is the likelihood that the subject will arise if environmental activists oppose drilling. Exploration drilling operations release relatively low levels of air pollutants. A National Petroleum Council survey (Ethridge, 1984) surveyed 24 land-drilled wells having an average depth of 5840 ft. The study indicated that these drilling operations emitted an average of 14 lb of nitrogen oxides per well per day. For comparison, that is about the equivalent emission rate of 54 average cars and trucks being driven 50 miles a day. As we shall see, exploration air emissions have been a much greater problem for certain offshore areas where opposition to drilling has been intense.

Offshore

In the United States, drilling for oil under the ocean dates back to 1897 when wells were drilled from piers built out over the Pacific at Summerland in California. By the mid-1930s, the technology for drilling offshore had reached a stage where it was no longer unusual to think about exploring this watery domain. Following passage of the Outer Continental Shelf (OCS) Lands Act in 1953, the first lease sale was held in the federal waters of the Gulf of Mexico. Since then, more than 7200 exploratory wells have been drilled in U.S. waters, three-fourths of them on the OCS.

Oil Spills

Most of the environmental concern over offshore exploration operations, and probably all of the paranoia, centers on fear of oil spills. Even so, in the 35-year history of drilling for oil and gas on the OCS, there have been few big spills. The last one, and the only one to reach land, occurred more than 20 years ago in the Santa Barbara Channel. Perhaps because of that single event, the American public associates oil spills with blowouts occurring while drilling. Yet drilling's safety record on oil spills has been excellent.

Since 1971 (which marks the advent of accurate reporting and records on blowouts and oil spills in U.S. OCS activities), 37 blowouts attributable to exploration well drilling have occurred (Mineral Management Service, 1989). However, these 37 blowouts have accounted for zero barrels of oil spilled into the marine environment. The record for development well drilling and other "nondrilling blowouts" is almost as exceptional. Since 1971, 72 blowouts have spilled a total of 900 bbl of oil. Over this same period, production from the OCS was over 6 billion bbl. For perspective, if you used this same ratio of 900 to 6,000,000,000 for a martini recipe, you would need only one drop of vermouth for 3-1/2 gallons of gin!

The President's Outer Continental Shelf Leasing and Development Task Force (1990) was charged with studying environmental concerns linked with OCS leasing off northern and southern California and southwestern Florida. It evaluated the oil spill risks in these areas from all potential sources (such as tankers and barges) as well as from leasing and development. Their findings, shown in Table 2, indicate that leasing and subsequent drilling increase the chances of an oilspill from 1 to 8%. The task force concluded that the risk of an oil spill of significant size from OCS-related exploration and development activities in these areas is "quite small" compared to the risk from other existing sources, notably tanker and barge traffic unrelated to OCS development.

Although offshore oil continues to be the prime target of the public's greatest concern for a pollution-free marine environment, the industry's record shows that the public's worries over oil spills have little basis in fact. This record results from more rigorous regulations and improved technology introduced into the offshore industry after the 1969 spill. Offshore oil operations are among the most highly regulated of all industrial activities. Companies working in the ocean must comply with 74 sets of federal regulations and secure up to 17 major permits. Exploration, development, and production plans must all be consistent with state coastal management programs. More than 30 federal laws concerned with environmental protection and navigational safety affect offshore oil and natural gas operations.

Table 2. Chances of an Oil Spill over a 30-Year Period (Greater Than 10,000 bbl)

OCS Lease Sale Area	Chance Without Leasing	Chance With Leasing	Added Risk of Oil Spills from Drilling
Sale 116, Part II			
Southwestern Florida	96%	97%	1%
Sale 95			
Southern California	93%	94%	1%
Sale 91			
Northern California	77%	85%	8%

Source: President's Outer Continental Shelf Leasing and Development Task Force, 1990.

Mud and Cuttings

Disposal of drilling mud and cuttings and their potential effects on marine life are also controversial environmental aspects of offshore operations. However, a growing body of scientific evidence indicates that these discharges pose insignificant threats to the environment. Regulations allow only oil-free, water-based mud to be discharged overboard in U.S. waters, and mud toxicity is tightly regulated by the Environmental Protection Agency (EPA). A National Academy of Science (1983) study on OCS drilling discharges found no evidence that these discharges posed a threat to ocean life. Some of their findings include

- Over 96% of drilling fluids tested were classified as "slightly toxic" or " practically nontoxic."
- Most of the drilling fluids discharged "have low acute and chronic toxicities to marine organisms" and are quickly diluted in the open ocean.
- Bioaccumulation of metals from drilling fluids is quite small, is restricted to barium and chromium, and has been demonstrated only in the laboratory.

Despite studies such as this, the EPA is currently developing revised guidelines for OCS discharges of mud and cuttings which may be more stringent than existing regulations.

Marine Life

The concern over oil industry marine activities and their impact on marine organisms is as old as the OCS program itself. Attention focuses on geophysical activities and, to a lesser degree, on exploratory drilling. Geophysical data gathering on water usually involves surface vessels in the 50- to 300-ton class. Virtually all energy sources used in the OCS today are air guns, which release compressed air to generate a shock wave. These devices have been designed to avoid the negative impact on sea life that dynamite had in the early days of marine geophysics. However, the effect air guns may have on fish directly, on their feeding habits, or on fish eggs and larvae is still the most critical environmental issue facing marine geophysics.

Offshore California studies have shown varying degrees of impact (Bowles, 1990). One study dealing with a claim that seismic activity affected fish and shellfish distribution showed no noticeable dispersal of fish, but indicated that fishermen seemed to catch less fish after a seismic vessel had passed the study location. Work done 2 years later indicated that startle responses and some subtle changes in fish behavior could result. Although the catch per unit effort declined when the airgun was operating and circling the hook and line fishing boat at a very close range, the fish returned to their original behavioral patterns within minutes after the airgun stopped.

OCS seismic activities are commonly timed to avoid conflicts with animal migration patterns. For example, seismic work was curtailed in 1981 in Alaska's Beaufort Sea because of concern that the use of compressed air guns might hamper the fall migration of the bowhead whale. The bowhead is the preferred target for Eskimo subsistence whaling. Native whalers feared that seismic activity might cause the whales to alter their migratory pattern. As a result, all marine seismic work was forbidden when the whalers were out to sea and when their prey were either expected to be in the area or actually sighted. A 2-year study (Richardson et al., 1985) by oil industry, whaling, and government personnel concluded that marine seismic operations do not significantly or adversely affect the behavior of whales. Nevertheless, the response of migratory animals such as the bowhead whale remains to this day a highly controversial issue.

Air Quality

Petroleum industry operations offshore California have come under intense scrutiny and criticism in recent years from environmental groups and others increasingly dedicated to maintaining a rig-free oceanscape. Southern California's severe air pollution problems have provided one avenue and slim justification for further opposition. Information developed by the Santa Barbara Air Pollution Control District has been used to infer that OCS development represents 39% of the emission inventory affecting Santa Barbara air quality—nearly as significant as the emissions from all cars and trucks (44% of the total). Industry testimony (Harper, 1989) placed the actual emissions from offshore oil and gas operations at less than 5% of the inventory. Industry's *net* effect is probably closer to zero because of offsets, distance from shore, and wind movements.

The air quality issue is, however, quite real to areas such as southern California even if it cannot realistically be blamed on petroleum operations. The Clean Air Act amendments passed in 1990 placed the authority for air in the OCS under the EPA, in all areas except the Gulf of Mexico. More restrictive regulations, especially for the California OCS, are a strong likelihood.

Physical Presence

Offshore drilling opponents claim that offshore structures visible from the beach will likely impact tourism. Once again, however, the facts indicate otherwise. The Santa Barbara Channel is an excellent example of a densely developed offshore area. If offshore oil will drive tourists off, surely this area provides a test case. However, the Santa Barbara Conference and Visitors Bureau claims that 81% of their tourists are repeat visitors. The bureau also estimates that tourism revenue in Santa Barbara County increased more than 200% between 1979 and 1984. Yet during this same 5-year period of increased tourism, the number of platforms installed in the channel rose from seven to twelve. Evidently tourists attracted to the county are unaffected by the presence of the offshore platforms.

Exploration operations are noticeable. We have yet to find means of hiding a 135-ft-tall jackknife rig in piñon pine or a semisubmersible 5 miles off a California beach. We leave roads in the wilds and canals in wetlands. We cannot conceal our presence, nor can we avoid leaving some evidence of our passage. Our activities are not without some cost to the environment. It appears, however, that exploration has probably suffered more from environmental concern than the environment has suffered from exploration.

ENVIRONMENTAL CONCERNS: THEIR IMPACT ON EXPLORATION

Congress has responded to the environmental activism of the past 20 years by passing numerous laws that have had direct effects on petroleum exploration. These laws and the regulations that implement them must be reckoned with at every turn, especially when exploring on federal lands. An explorer's ability to obtain a permit to drill on public lands depends on whether the project conforms to these many laws. The most important are

- National Environmental Policy Act (1969)
- Marine Mammal Protection Act (1972)
- Federal Water Pollution Control Act (1972)
- Endangered Species Act (1973)
- Resource Conservation and Recovery Act (1976)
- Outer Continental Shelf Lands Act Amendments (1978)

Although obtaining permits is critical to exploration on public lands, our most serious problem related to environmental concerns, and the one that seems to occur with ever-increasing frequency, is the complete withdrawal of lands from leasing.

Land Access

The exploration function has a paradoxical relationship to the broad question of environmental concern and the oil business. As we saw in the previous section, our activities—geophysics, leasing, and exploration drilling—either leave behind little imprint of their passage or are so brief as to cause negligible concern among reasonable observers. However, exploration activities are too often viewed as the first steps in an implacable progression of increasingly noticeable and destructive events leading to development of whatever resources may be found. Therefore exploration, although in itself substantially benign, too frequently suffers the one punishment from which it cannot recover: denial of access to the land.

Public land withdrawals can result from a variety of government actions. The laws that are responsible for most of these include

- National Wilderness Preservation Act (1964)
- Wild and Scenic Rivers Act (1968)
- Marine Protection and Sanctuaries Act (1972)
- Federal Land Policy and Management Act (1976)
- National Forest Management Act (1976)
- Alaska National Interest Lands Conservation Act (1980)

Federal lands can also be withdrawn by the Secretary of the Interior (by Public Land Order), the Congress (by using the annual appropriations process or by creating more wildernesses), or the President (by Executive Order). Exploration has suffered denial of access from all of these mechanisms, both onshore and in state and federal waters.

Onshore

The Federal government owns about one-third of the land in the 50 states, about 688 million out of a total of 2271 million acres. Yet, according to a Rocky Mountain Oil and Gas Association (RMOGA) estimate, more than 40% of these lands are off limits to drilling (Crow, 1990). Table 3 lists the various land categories that have resulted in a set-aside of 301.5 million acres of land, an area equal to the states of Montana, Wyoming, Colorado, and New Mexico. The Bureau of Land Management (BLM) has historically been, and is today, the primary onshore public land manager directly responsible for about 342 million acres of public land. The U.S. Forest Service is the next largest (190 million acres of national forests), followed by the Fish and Wildlife Service (90 million acres) and the National Park Service (80 million acres).

Table 3. Onshore Federal Lands Off Limits to Drilling

Category	Acres (Million)
National parks	80.0
Wilderness	
National forests	90.0
National forest study areas	4.1
BLM	0.4
BLM study areas	26.7
Instant study areas[a]	0.5
Fish and wildlife refuges	90.8
Areas of critical environmental concern	6.5
Wild and scenic rivers	0.5
Future planning areasa	1.8
Primitive areas	0.2
Total off limits	301.5
Total onshore federal lands	688.3
Percentage off limits	43.6%

[a]Administrative withdrawals

Source: Rocky Mountain Oil and Gas Association, 1990, unpublished data (quoted in Crow, 1990).

Areas in the most restrictive public land management categories, such as national parks and wilderness areas, are off limits to all mineral activity. Since an area has to be rather special and unique to qualify for national park status, opponents of development throughout the west have in recent years used the Wilderness Preservation System as the prime instrument for limiting development on public land. Table 3 shows just how successful the preservationists have been.

The Wilderness Act, passed in 1964, mandates that wilderness areas be administered "for the use and enjoyment of the American people in such a manner as will leave them unimpaired for future use and enjoyment as wilderness." At that time (in 1964), 9.1 million acres were set aside as the initial contribution to the wilderness system. The U.S. Forest Service was directed to evaluate its remaining primitive areas as possible additions. In 1971, it commenced a Roadless Area Review and Evaluation. The service's standards for wilderness determination were challenged in the courts. This finally led to a redetermination in 1978. As a result, today more than 90 million acres, primarily in the west and Alaska, are dedicated to wilderness.

Wilderness designation is not the only means by which public lands have been withheld from petroleum exploration. Over 30 million acres of U.S. Forest Service and BLM lands are currently designated as "study areas," that is, they are under consideration for inclusion as wilderness. Most of these lands have yet to be evaluated for energy potential. Other laws effectively foreclose millions of acres of other lands to exploration. These include lands affected by the Alaska Native Claims Settlement Act, the Endangered Species Act, and the Clean Air Act. Further closings result from federal land managing agencies, which from time to time restrict access or withdraw areas either through direct action or, most often, by failure to act.

The environmental concerns that can affect access to these lands are seldom driven by actual problems since, in most cases, no entry has taken place. At this stage, we must deal with people's perceptions of the harm that may accrue from "opening the door" to petroleum exploration. The concern is ostensibly rooted in the primary environmental value of the area to be leased or opened to exploration activity. For instance, perhaps the greatest concern over leasing the Arctic National Wildlife Refuge in northeastern Alaska is the possible harm to the caribou and the loss of a pristine and complete ecosystem, almost untouched by human presence. And the concern driving the call for a buffer zone around Yellowstone Park appears to be prevention of further contraction of grizzly bear habitat due to encroachment by human activity.

Offshore

The offshore areas of the United States are divided into zones. The coastal states' territorial waters extend from the beach outward 3 nautical miles (10 miles for Texas and Florida in the Gulf of Mexico). The Federal domain, the OCS, starts at this point and extends outward to 200 miles, designated as the Exclusive Economic Zone of the United States.

Access to state waters varies widely among the states that have established production. California, where offshore drilling originated late in the nineteenth century, has closed all of its offshore state waters to further leasing. Attitudes in the other coastal producing states of Alaska, Alabama, Louisiana, Mississippi, and Texas vary from reluctant to congenial.

The Minerals Management Service (MMS) of the Department of the Interior is charged with administering all the public's resources, living and nonliving, in the OCS. Exploration access to these lands has been through the OCS leasing program since it began in 1954. As of March 1990, 99 sales have resulted in leasing more than 34 million acres, 82% of which was in the Gulf of Mexico. The program has always had opposition from those concerned over the possible negative impacts that OCS petroleum development may have on the marine environment. Activists have litigated each of the three 5-year leasing programs as well as most lease sales outside the Gulf of Mexico, but with little success.

In 1981, however, congressional opponents of offshore leasing and development devised an effective strategy for subverting the leasing program. Working in the Appropriations subcommittees of the House and Senate, they added language to the 1982 Department of the Interior Appropriations bill which denied any funding for leasing OCS areas off central and northern California. Since that time, leasing moratoriums have been passed annually by Congress to forestall leasing in various areas of the OCS (Figure 1). The Congressional moratorium for 1990 includes more than 84 million acres.

Opposition to offshore exploration and development has risen both in its clamor and its effectiveness. On June 26, 1990, President Bush created the largest, longest moratorium thus far by placing more than 192 million acres of the OCS off limits to leasing until 2001. Considering that this was done despite decreasing domestic oil production, increasing domestic oil consumption, and increasing foreign crude and product imports, it speaks eloquently of the political power exerted by industry's opponents. The Presidential moratorium also raises very real questions concerning exploration's future offshore in the face of the wave of environmental concern currently sweeping the country.

Resources Withdrawn from Exploration

How significant are these withdrawals? What have they meant to the nation in terms of foregone resources—oil and gas we are not allowed to explore for? The U.S.G.S. and the MMS divide the United States into 13 petroleum regions for assessment purposes. Regions 1, 2, 3, and 4 (North Dakota, South Dakota, Colorado, and western New Mexico, and all states westward of them, including Alaska) hold almost 90% of all federal onshore lands. By multiplying the percentage of public lands in each region (Table 4) by the most recent U.S.G.S. assessment of undiscovered, economically recoverable oil and gas, we obtain an estimate of the explorable resources under public lands onshore. (This analysis employs the simplifying assumption that resources are evenly distributed between public and private lands throughout each region.) Multiplying these western public land resources by RMOGA's estimate of 43.6% of all federal lands off limits to exploration, these withdrawals mean approximately 5.8 billion barrels of oil equivalent (BBOE) for which we cannot explore.

Offshore, the Pacific OCS is almost entirely withdrawn from exploration until 2001. The MMS estimates that the undiscovered economically recoverable resources for this area are about 3 BBOE.

In summary, federal land withdrawals have set aside areas that could yield almost *9 BBOE*, about as much as 3 years worth of domestic oil production, at current producing rates. Considering the nation's need for energy, this set-aside does not seem reasonable. Is there anything we explorers can do to ensure a more reasonable future?

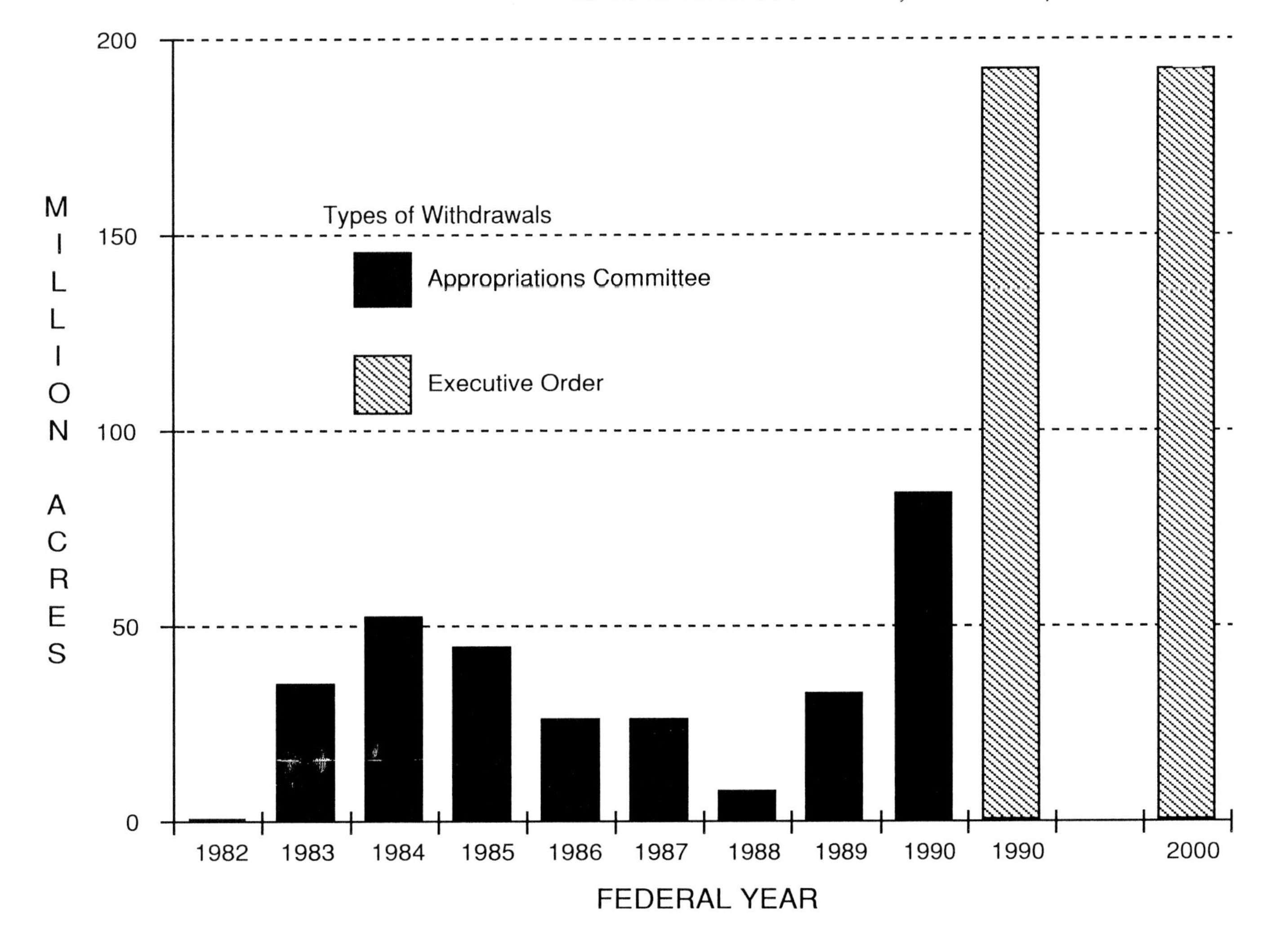

Figure 1. Leasing moratoriums for the Outer Continental Shelf. (Source: Mineral Management Service.)

ACTION PLAN FOR A REASONABLE FUTURE

Remember the Kenyan proverb:

> Treat the earth well . . . it was not given to you by your parents . . . it was lent to you by your children.

We are all environmentalists. Only outlooks differ. Some see today as the beginning of the end—the apocalypse. For others, today is the beginning of a promising future. Most explorers will likely number themselves in the latter category; optimism is perhaps our leading virtue. Although the future holds promise, it will assuredly be different, too. A concern for, and wise use of, the earth—its soil, water, air, and even its space and its scenery—must become axiomatic in all oil operations.

Environmental issues are and will continue to be a fact of life. Therefore, it is appropriate for us to consider a canon—a set of principles—with which geologists can meet and help shape a reasonable future.

I suggest the following:

1. We will tread as lightly as possible on the land. This has always been a good business practice, but nevertheless, the smaller our footprint, the greater the acceptance of our activities.
2. We will clean up our own messes. Again, we have always been responsible for the messes we make. Now we come to realize that we also own the wastes we create; until they are safely disposed of, they remain our responsibility.
3. We will subject all environmental claims, from friends and foes alike, to close scrutiny and a test of reasonableness. This rule ranges all the way from the safety of the disposal site that the contractor uses for our oil-based mud to the sometimes unsupported criticism of our activities leveled by environmental activists in the local newspaper. It is our obligation to look for the truth, whether positive or negative, in environmental matters.
4. We will educate our regulators. They run the

Table 4. Economically Recoverable Undiscovered Oil Equivalent

U.S.G.S. Region	U.S.G.S.[a] Assessment (BBOE)	Federal Lands: Fraction of Total Area	On Federal Lands (BBOE)
1 (AK)	8.1	0.81	6.6
2 (CA, OR, and WA)	5.3	0.44	2.3
3 (AZ, wCO, ID, NV, and wNM)	5.3	0.59	3.1
4 (eCO, MT, ND, SD, and WY)	6.3	0.21	1.3
		Total	13.3
Fraction of public lands off limits to drilling			x 0.43
		Total resource off limits	5.8

[a]U.S. Geological Survey and Minerals Management Service, 1989.

only game in town. Most of them are doing the best they can, often severely limited by lack of personnel, information, time, and money. Most regulatory problems arise from ignorance. The more a regulator knows about our business, the better his or her judgment and decisions will be concerning it. The less he or she knows, the more likely the response to a project will be negative. "No," if not the best answer, is almost always the safest answer.

5. We will educate our legislators. We must recognize that all issues are potentially political issues. The United States is perhaps the most vigorously representative and fiercely democratic society in the world. And we're still working on the rules. Just because things are one way today, does not ensure that they will be that way tomorrow. The only constant in law, regulation, and public perception is change. Whether we like it or not, our business also belongs to the government and the public. Those intent on stopping development altogether are "educating" with misconceptions and groundless fears about the oil industry. It is up to us to bring the facts—the reality of our operations—to the debate. We must tell them just how good we really are.
6. We will educate the public. We must never miss an opportunity to explain our science and our business to those around us, whether they're neighbors or reporters, church members or councilmen, onlookers or activists. As scientists and industry experts, our voices will carry more weight and credibility than even the largest industrial corporation. The debate over environmental protection is not "whether or not," but "how and how much." We have a responsibility to participate.

References Cited

Arnold, R., 1987, Ecology wars: Bellevue,WA, Free Enterprise Press, 182 pp.

Borlaug, N., 1972, In defense of DDT and other pesticides: UNESCO Courier, February.

Bowles, L. G., 1990, Marine geophysics and the environment are compatible, World Oil, v. 211, no. 3, p. 55

Bureau of Land Management, 1980, Buffalo resource area: Oil and gas environmental assessment: U.S. Dept. of the Interior, BLM, Casper District, Wyoming.

Carson, R., 1962, Silent spring: Boston, Houghton Mifflin Co., 368 pp.

Crow, P., 1990, Federal land access issues threaten activity: Oil and Gas Journal, v. 88, n. 3, p. 74.

Ehrlich, P., 1975, An ecologist's perspective on nuclear power: in Public Issue Report from the Federation of American Scientists, May/June.

Ethridge, M. A., 1984, Compatibility of oil and gas operations on federal onshore lands with rural and community values: American Petroleum Institute Discussion Paper, n. 033, p. 24.

Harper, W. P., 1989, Air quality issues presentation to President's Outer Continental Shelf Leasing and Development Task Force, Los Angeles (unpublished).

Kallman, R. E., and E. D. Wheeler, 1984, Coastal crude in a sea of conflict: San Luis Obispo, CA, Blake Printery and Publishing Co., 136 pp.

Meadows, D. H., D. L. Meadows, J. Randers, and W. W. Behrens III, 1972, The limits to growth: New York, Universe Books, 205 pp.

Minerals Management Service, 1989, Federal offshore statistics: 1988: Reston, VA, Minerals Management Service, MMS 89-0082, 102 pp.

National Academy of Science, 1983, Drilling discharges in the marine environment: Washington, D.C., National Academy Press, 180 pp.

President's Outer Continental Shelf Leasing and Development Task Force, 1990, A report to the President on lease sales 91, 95, and 116, Part II: Reston, VA, Minerals Management Service, draft report, 109 pp.

Richardson, W. J., C. R. Greene, and B. Wursig, 1985, Behavior, disturbance responses, and distribution of bowhead whales (*Balaena mysticetus*) in the Eastern Beaufort Sea, 1980–84: A summary: Reston, VA, Minerals Management Service, MMS 86-0034, 38 pp.

Schmertz, H., 1986, Goodbye to the low profile: Boston, Little, Brown and Co., 242 pp.

Tucker, W., 1982, Is nature too good for us?: Harpers, v. 264, n. 1582, p. 27-35

U.S. Geological Survey and Minerals Management Service, 1989, Estimates of undiscovered conventional oil and gas resources in the United States: A part of the nation's energy endowment: Washington, D.C., U.S. Government Printing Office, 44 pp.

Whelan, E. M., 1985, Toxic terror: Ottawa, IL, Jameson Books, 349 pp.

G. ROGGE MARSH has worked for Exxon for 38 years. He began his career as an exploration geologist in South Texas and subsequently worked in the areas of exploration economics, energy supply forecasting, computer applications, and econometric modeling. For the past 10 years, Mr. Marsh has been dealing with Congress, various Executive Agencies, and key state governments seeking reasonable treatment for both Exxon and industry exploration activities.

Glossary

James A. McCaleb[†]
Consultant
Batesville, Arkansas, U.S.A.

How is it that an NAC (New ACronym) roosts in so many phrases? Or, if an NAC must be brought to life, why not let the reader KWIM (Know What It Means)?
—Sven Treitel

INTRODUCTION

This Glossary contains a list of abbreviations and acronyms along with an extensive alphbetized list of terms and concepts relevant to the business of petroleum exploration. It also presents hypothetical examples that demonstrate the use of various leasehold, production, and financial terms.

The contents and definitions that follow come from many sources. First was the author's years of experience with Amoco in exploration operations, exploitation, and research and as manager of exploration training. The second source was the many contributors to this Handbook. Third, many standard books in the field were used, such as *An Introduction to Exploration Economics* by R. E. Megill and *Decision Analysis for Petroleum Exploration* by P. D. Newendorp. Finally, the author would especially like to thank Field Roebuck, Robert E. Megill, and Richard Steinmetz for their detailed reviews and editorial assistance.

ABBREVIATIONS AND ACRONYMS

ACRS—Accelerated Cost Recovery System

AFE—Authorization for Expenditure

AFIT—After Federal Income Taxes; usually used in connection with profit, as in "AFIT net income".

AMI—Area of Mutual Interest

AOFP—Absolute Open Flow Potential

ASTM—American Society for Testing Materials

Bbl—Barrel (of fluid)

BCFD—Billion Cubic Feet per Day (of gas)

BCPMM—Barrels of Condensed liquid Per Million (cubic feet of gas)

B/C—Benefits-to-Cost Ratio

BFIT—Before Federal Income Taxes

BFPD—Barrels of Fluid Per Day

B/I—Back-in Interest

BO—Barrels of Oil

BOE—Barrels of Oil Equivalent

BOPD, BPD, B/D—Barrels of Oil Per Day

BROR—Book Rate Of Return

BS&W—Basic Sediment and Water

Btu—British Thermal Unit

BW—Barrels of Water

CAP—Capital or Capitalized

CE—Capital Expenditure

CEA—Capital Employed Adjustment

CF—Compound Factor

CFG—Cubic Feet of Gas

CFGPD, CPD, CF/D—Cubic Feet of Gas per Day

COD—Cost Of Developing

COF—Cost Of Finding

COP—Cost Of Producing

COPAS—Council of Petroleum Accountants Societies

COST—Continental Offshore Stratigraphic Test (well)

CWI—Carried Working Interest

D&A—Dry and Abandoned (well)

DB—Declining Balance (depreciation method)

[†]Deceased March 27, 1991.

DCF—Discounted Cash Flow
DCFR—Discounted Cash Flow Rate
DDB—Double Declining Balance (depreciation method)
DD&A—Depletion, Depreciation, and Amortization
Depl.—Depletion
Depr.—Depreciation
DF—Discount Factor
DOE—Department of Energy
DPR—Discounted Profit-to-Investment Ratio
DROI—Discounted Return on Investment
EIS—Environmental Impact Statement
EOR—Enhanced Oil Recovery
EPA—Environmental Protection Agency
ERA—Energy Regulatory Agency
EV—Expected Value
Exp.—Expensed
FASB—Financial Accounting Standards Board
FCE—Full Cycle Economics
FERC—Federal Energy Regulatory Commission
FI, F/I—Farm-in
FIT—Federal Income Tax
FMV—Fair Market Value
FO, F/O—Farm-out
FPC—Federal Power Commission
FV—Future Value
FW—Future Worth
G & A—General and Accounting
G & G—Geological and Geophysical
GNP—Gross National Product
GOR—Gas-Oil Ratio
GROR, GRR—Growth Rate of Return
HBP—Held by Production
IDC—Intangible Drilling Costs
Inv.—Investment
IP—Initial Potential
IPF—Initial Potential Flowing
IPP—Initial Potential Pumping
IRR—Internal Rate of Return
ITC—Investment Tax Credit
IW—-Incremental Worth
JOA—Joint Operating Area
LNG—Liquified Natural Gas
LOE—Lease Operating Expenses
LORI—Landowner (mineral owner) Royalty Interest
MACRS—Modified Accelerated Cost Recovery System
MB/D—Thousand Barrels per Day (of fluid)
MCF—Thousand Cubic Feet (of gas)
MCF/D—Thousand Cubic Feet per Day (of gas)
MER—Maximum Efficient Rate (of production)
MI—Mineral Interest
MM—Million
MMB/D—Million Barrels per Day (of fluid)
MMCF—Million Cubic Feet (of gas)
MMCF/D—Million Cubic Feet per Day (of gas)
MMS—Minerals Management Service
MROR—Minimum Rate of Return
NCF—Net Cash Flow
NG—Natural Gas
NGL—Natural Gas Liquids
NPI—Net Profits Interest
NPPB—Net Profit Per Barrel
NPV—Net Present Value
NPW—Net Present Worth
NRI—Net Revenue Interest
NWI—Net Working Interest
OCS—Outer Continental Shelf
OFP—Open Flow Potential
OH—Overhead
OPEC—Organization of Petroleum Exporting Countries
Opcst.—Operating costs
ORI, ORRI—Overriding Royalty Interest
OSHA—Office of Safety and Health Administration
P & A—Plug and Abandon
PI—Profitability Index
P/I—Profit-to-Investment Ratio
PIA—Post-Installation Apprasial
PO, P/O—Payout
PPI—Production Payment Interest
Prod.—Production
Psia—Pounds per square inch absolute
Psig—Pounds per square inch gauge
PV—Present Value
PVF—Present Value Factor
P-V-T—Pressure-Volume-Temperature
PW—Present Worth

PWF—Present Worth Factor

RA—Reserve Additions

Res.—Reserves

Rev.—Revenue

RI—Royalty Interest

ROI—Return on Investment

ROR—Rate of Return

SC—Sunk Costs

SEC—Securities and Exchange Commission

S.F.—Shrinkage Factor

SL—Straight Line (depreciation method)

SPE—Society of Petroleum Engineers

SWOT—Strengths, Weaknesses, Opportunities, and Threats

S-Y-D—Sum of Year's Digits (depreciation method)

TCF—Trillion Cubic Feet (of gas)

TD—Total Depth

TRA 86—Tax Reform Act of 1986

UNIR—Ultimate Net Income Ratio

WC—Wildcat Well

WI—Working Interest

Y-T-S—Yet to Spend

TERMS

A

ABANDONMENT RATE—The producing rate at which well revenues equal direct operating costs, including field and administrative overhead. See ECONOMIC LIMIT.

ABANDONMENT VALUE—The amount that can be realized by liquidating a project. It constitutes the undepreciated investment balance at abandonment.

ABSOLUTE OPEN FLOW POTENTIAL (AOFP)—A theoretical value calculated upon initial completion of a gas well.

ABSTRACT OF TITLE—A collection of all the recorded instruments affecting the title to a tract of land. Some abstracts contain full and complete copies of the instruments on record while others summarize their effect.

ACCOUNTING METHODS:

A. FULL COST—Accounting system in which all exploration costs, (e.g., G & G and drilling) are charged against income on a units-of-production basis. This system is based on the view that, to find promising acreage, it is necessary to incur losses in unrelated areas. Full cost accounting involves capitalization of all costs. Costs of exploration for properties that are not acquired or are released are treated as additional indirect cost on acquired or retained acreage.

B. SUCCESSFUL EFFORTS—An accounting method that allows a firm to expense the exploration failures and capitalize its successes. FASB-19 states that

> under this method, the test-year exploration costs are allocated to basic types of oil or gas reservoirs on the basis of the intangible drilling costs of successful exploratory wells, as identified with those two types of reservoirs, or by other measures of the successful exploratory effort as identified with such reservoirs, such as footage drilled or the cost of acreage transferred from nonproducing to producing. Subsequent allocations of the cost thus allocated to reservoirs are made on a common-unit-of-measure (such as BTUs) or other basis, with the resultant cost thereby allocated to the lease liquids and gas then being attributed to test-year production volumes.

C. RESERVE RECOGNITION ACCOUNTING—An accounting system based upon the present worth of a firm's proved reserves.

ACCOUNTING RATE OF RETURN—See BOOK RATE OF RETURN

ACRES (OWNERSHIP):

A. FEE ACRES—Acreage on which the landowner has complete ownership, including both surface and mineral rights.

B. GROSS ACRES—Total surface acres on a property.

C. LEASEHOLD ACRES—Acreage on which a party (lessee) has acquired, by lease, the right to explore for and produce oil, gas, and/or other minerals in return for a stated royalty and possibly other considerations.

D. MINERAL ACRES—Acreage on which ownership of the minerals in place has not been divided into royalty and working interests.

E. NET ACRES—Gross acres times fractional interest owned or leased (working interest). Term can apply to leasehold, fee, mineral, or royalty.

F. ROYALTY ACRES—Acres on which a royalty interest is owned. A party owning such a royalty interest does not have the right to produce the minerals or reserves underlying such acreage.

ACRES (MEASUREMENT)—May be expressed on either a gross or net basis. Gross refers to total acres, while net refers to company-owned acres.

A. SURFACE ACRE—43,560 sq. ft of surface area.

B. PROVED ACRES—Relates to acres believed to contain economically producible reserves. May also be used in context of maximum limits of the proved portions of all producing formations (definition of *proved* may vary.)

ACREAGE—The leased area for which exclusive drilling rights are held.

ACREAGE-BASED ROYALTY—A term used to describe a royalty payable on a per acre basis, e.g., a royalty expressed as a fixed sum per acre on a shut-in well.

AD VALOREM TAX—See LOCAL TAXES.

ALLOCATION FORMULA—The formula employed in fixing the allowable production for a well, pool, or field.

ALLOWABLE—The amount of oil or gas that a well or leasehold is permitted to produce under proration orders of a state or federal regulatory body.

ALTERNATIVE FUNDING—A source of financing for a project other than internally generated funds from an individual company.

AMORTIZATION—A noncash deduction from current revenue on the financial books which provides for the recovery of a portion of the capitalized intangible investments due to use or passage of time. The term may also refer to the liquidation of a debt on an installment basis.

ANOMALY—A deviation in the geological structure or stratigraphy of a basin. Usually used for a seismic anomaly observed from seismic records.

AREA OF MUTUAL INTEREST (AMI)—Any area which, by prior agreement, is the subject of mutual sharing of ownership by two or more parties of any leasing rights acquired.

AREAL EXTENT—Area defined by the productive limits of a field.

ARTIFICIAL LIFT—Production of a well by some mechanical means other than natural flow. Not a recovery method.

ASSETS:

A. CAPITAL ASSETS—Assets intended for long, continued use or possession. May be further classified as TANGIBLE (e.g., land, buildings, producing wells, etc.) and INTANGIBLE (e.g., G & G expenditures incurred in connection with the acquisition or retention of acreage)

B. CURRENT ASSETS—Assets that are, or may be, converted to cash in a short period of time, generally one year or less.

ASSOCIATED GAS—Gas that occurs together with oil, either as free gas or in solution. Gas occurring alone in a reservoir is termed NONASSOCIATED GAS.

AUTHORIZATION FOR EXPENDITURE (AFE)—A request for approval of funds for a given project.

AVERAGE CAPITAL EMPLOYED—End of year capital employed plus beginning of year capital employed, divided by two. See CAPITAL EMPLOYED.

B

BACK-IN FARM-OUT—A farm-out agreement in which a retained nonoperating interest may be converted at a later date to a specified individual working interest.

BACK-IN INTEREST (B/I)—Working interest fraction available under a back-in option.

BARREL (bbl)—A standard U.S. oilfield barrel, which is 42 U.S. gallons or approximately 34.97 Imperial gallons. There are 6.2898 bbl in 1 m^3, and the weight of a barrel varies depending on the specific gravity of the crude oil.

BARRELS OF OIL EQUIVALENT (BOE)—The quantity of NGL or natural gas necessary to equate on a BTU or price basis with 1 bbl of crude oil. Federal agencies use 5.8×10^3 Btu as equivalent to 1 bbl of oil. For NGL, 1.455 bbl of NGL is equal to 1 BOE (U.S. Department of Energy definition).

BASE CASE—Usually refers to an economic scenario to which other cases are compared. It need not reflect the most likely scenario.

BASE ROYALTY—The royalty reserved by the landowner upon the creation of an oil and gas lease or the granting of an interest in oil and gas.

BASE–COST RATIO—Generic name for any of a number of ratios expressing value received or obtained per value invested.

BENCH MARK CRUDE—A produced crude oil considered to be typical or of average characteristics for an area or region and, hence, used as a basis for pricing of other crude oils produced within that area or region. Examples are Saudi Arabian Light, West Texas Intermediate, and North Sea Brent.

BLOCK—A group of leases, usually contiguous, held in one ownership.

BOND—A long-term debt instrument that obligates a borrower to repay a fixed amount at a given interest rate.

BONUS—The money paid by the lessee for the execution of an oil and gas lease by the mineral rights owner. Another form is called an *oil* or *royalty bonus*. This may be in the form of an overriding royalty reserved for the mineral rights owner in addition to the usual mineral owner royalty. Also, the cash payments offered at competitive bid sales of leases, such as offshore.

BOOK PROFIT—A financial book (accounting) assessment of earnings.

BOOK RATE OF RETURN (BROR)—A measure of corporate performance calculated as follows:

$$= \frac{\text{Financial book earnings} + \text{After tax interest expense}}{\text{Average capital employed}}$$

Note that some firms do not include interest in the numerator.

BOOK VALUE—Net amount at which an asset is valued on financial books. May vary from its market or intrinsic value. Book value is unrecovered cost, that is, original cost minus accumulated depreciation, depletion, amortization, and abandonments.

BOTTOM HOLE MONEY (or CONTRIBUTION)—Dollar contribution made to the drilling party for reaching a specified drilled depth or stratigraphic equivalent, irrespective of productivity. Other contributions, such as acreage, may be made to the drilling party.

BREAK-EVEN ANALYSIS—A quantitative technique to ascertain the point when total undiscounted revenues equal total undiscounted costs. Although break-even analysis may be before or after tax, the after tax calculation usually is of more use to the firm since tax is a very real portion of deductions from revenue. See PAYOUT.

BRITISH THERMAL UNIT (Btu)—The amount of heat required to raise the temperature of 1 lb of water 1 °F.

BUDGET—An estimation of income and expense over an annual cycle.

C

CALL—The right to purchase hydrocarbons from an outside party.

CAPITAL:

A. BORROWED CAPITAL—Amounts owed with an original maturity of more than one year. Also includes amounts owed on long-term debt maturing within the current year.

B. INVESTED CAPITAL—Total assets minus total liabilities. This equals shareholder's investment in a firm as determined on the financial books (shareholder net worth).

C. SOURCE OF CAPITAL—Capital may flow into the firm from long-term debt (repayment in more than one year), short-term debt (repayment in less than one year), equity, and retained earnings from operations.

D. TOTAL CAPITAL—Sum of borrowed and invested capital.

CAPITAL ASSETS—See ASSETS.

CAPITAL EMPLOYED—Capital expenditures minus noncash charges, that is, capital recovery, deferred tax, and amortization of nonproducing properties.

CAPITAL EMPLOYED ADJUSTMENT (CEA)—That portion of corporate expenses (excluding those charged directly) allocated to the company. These expenses are distributed to the subsidiaries based on average capital employed.

CAPITAL EXPENDITURES—Those expenditures that are recorded as assets on the financial books and are subtracted from revenues over an extended period of time through some form of capital recovery.

CAPITAL GAIN (OR LOSS)—The excess (or deficit) of proceeds realized from the sale or exchange of a capital asset over (or under) its book value. Capital gain may create an income tax liability.

CAPITAL RECOVERY—The method by which a capitalized investment is deducted over time from revenues in the determination of earnings or taxable income, that is, depreciation, depletion, and amortization.

CARRIED WORKING INTEREST (CWI)—An agreement under which one party (carrying party) agrees to pay a portion or all of the development and operating costs applicable to the share of the working interest owned by another party (carried party).

CASH BOOKS—Refers exclusively to the company's cash transactions irrespective of accounting treatment. Noncash charges are excluded. A company may not actually maintain separate "cash books."

CASH FLOW—A schedule or statement of all cash transactions. When both the outflow (expenses, capital expenditures, and taxes) and inflows are accounted for, the net summation of these flows is referred to as NET CASH FLOW (NCF), which equals the revenue minus investments, operating costs, overhead, and taxes.

CASH FROM OPERATIONS—Cash flow before capital expenditures or net revenue minus severance tax, operating cost, overhead, G & G expense, federal (tax book) tax, and exploratory dry holes.

CERTAINTY—The case of a single outcome of an event for which the probability of occurrence is 1.0 (100% certain).

COMMERCIAL WELL—A well that produces reserves (oil, gas, etc.) in sufficient quantities to be expected to be economical for commercial operation, that is, one in which the revenues are higher than the costs of operating and maintaining the well.

COMPLETION—Each separate physical arrangement within a well that provides for the production of hydrocarbons (generally from a single horizon, or a common source of supply as defined by a regulatory agency). Statistical summaries involving allowables, or allocations of overhead or operating costs, usually refer to completions. See WELL.

COMPOUND FACTOR (CF)—A multiplier to be applied to a current dollar amount to determine its subsequent future value at a given interest rate.

COMPOUND RATE—The rate at which an invested amount increases over time due to the accumulation of interest. See also COMPOUNDING, CONTINUOUS COMPOUNDING, DISCRETE COMPOUNDING, PERIODIC COMPOUNDING.

COMPOUNDING—The process of economic growth whereby the interest and principal from a previous period become the principal of the next period and grows at the original interest rate.

CONDENSATE—Liquid hydrocarbons produced with natural gas which are separated from the gas by cooling and various other means. Condensate generally has an API gravity of 50° to 120° and can be clear, straw, or bluish in color.

CONDENSATE RATIO—Barrels of condensate per million cubic feet of gas produced.

CONFIDENCE FACTOR—Discounting of reserves in relation to the degree of confidence their being produced. See various reserve classifications.

CONTINENTAL OFFSHORE STRATIGRAPHIC TEST (COST) WELL—A well funded by a number of firms for the purposes of obtaining geological information in frontier areas offshore. The well is drilled without the intention of striking reserves.

CONTINUOUS COMPOUNDING—The limit as the investment period approaches zero. Funds received as interest instantaneously begin to earn interest as well. Counterpart of PERIODIC COMPOUNDING.

CONVERSION CLAUSE—A clause in an assignment or other instrument providing for the conversion, either automatically or at the option of one of the parties, of one interest granted or reserved into another interest (e.g., conversion of an overriding royalty or net profits interest into a share of the working interest).

CONTRACT DEPTH—Depth of the wellbore at which the drilling contract is fulfilled.

CONVERTIBLE BOND—Bonds that may be exchanged at some future time for a specified number of shares of stock.

CONVERTIBLE INTEREST—An interest that is convertible to an interest of another type. By the terms of the instrument creating an overriding royalty, for example, it may be provided that at the owner's option, the interest may be converted after the completion of a well to a share of the working interest.

COUNCIL OF PETROLEUM ACCOUNTANTS SOCIETIES (COPAS)—Journal of accounting procedures for handling joint operations, particularly for allocating costs as either direct operating expenses or indirect overhead charges.

CORPORATE PROFITABILITY—See BOOK RATE OF RETURN.

CORPORATE GROWTH—Refers to increases over time in a firm's assets, net income, or dividends paid.

CORPORATE BOOKS—Method of accounting employed for reporting financial experience of a company to management and stockholders.

COST OF CAPITAL—The payment (expressed as a percentage) that must be made to sources of capital for the use of their funds. See WEIGHTED AVERAGE COST OF CAPITAL.

COST TO CONDEMN—The total cost of exploratory drilling and G & G necessary to decide that a lease or area is noncommercial.

COSTS—An expenditure or outlay of cash, other property, capital stock or services identified with goods or services purchased. Some costs may be expensed, others must be capitalized. See OPERATING COSTS.

COST OF FINDING—The amount of money spent per barrel of oil equivalent in locating reserves.

CUMULATIVE NET CASH FLOW—The running sum of all positive and/or negative annual net cash flows.

D

DD&A—Depletion, depreciation, and amortization; noncash charges; capital recovery; allowances for exhaustion of resources. See NONCASH CHARGES and CAPITAL RECOVERY.

DEBENTURE—A long-term debt instrument that is not secured by a mortgage on specific property.

DEBT RATIO—Total debt (including current liabilities) divided by total net worth (including total debt plus equity).

DECISION TREE—A device for illustrating diagrammatically the relationships between decisions and chance events and outcomes.

DECLINING BALANCE (DB)—An accelerated method of depreciation whereby the amount each year (or other period) is a multiple of the straight line rate and the remaining depreciable amount. See DOUBLE-DECLINING BALANCE.

DEDICATION OF RESERVES—The assurance of an adequate supply of natural gas to a pipeline company, usually accomplished by a gas purchase contract between a natural gas producer and the pipeline company. A typical contract will provide that "seller hereby dedicates to the performance of this contract all gas located in, under, or hereafter produced from the units, leases, and lands described in Exhibit 'A' (hereafter referred to as dedicated reserves)." Such contracts normally run for a period of a least 20 years.

DEFERRED BONUS—Landowner's bonus, the payment of which is made in installments spread over a number of years, as distinguished from the usual mode of payment, which is in a lump sum on execution and delivery of the lease.

DEFERRED PRODUCTION PAYMENT—A production payment that does not commence until after the operator has realized a specific sum from production on the lease or after a primary production payment.

DEFERRED TAX—A result of timing differences between recognition of tax liability and actual payment. Occurs when taxable income differs from financial book income, often because of different depreciation and amortization methods employed for tax and financial books.

DEFLATORS—Factors based on annual inflation rate and used to adjust all monetary amounts to a base year equivalent.

DELAY RENTAL—A payment, commonly made annually on a per-acre basis, to validate a lease in lieu of drilling.

DELINEATION WELL—A well drilled after discovery of a field to try to define or delineate the productive limits of the newly discovered field.

DEMURRAGE—A charge incurred by the shipper for detaining a vessel, freight car, or truck.

DEPLETION (Depl.)—Noncash deduction from current revenue which provides for the recovery of a portion of the capitalized costs of an oil or gas property due to exhaustion of resources.

A. COST DEPLETION—Acquisition costs are recovered in direct proportion to the amount of production in a given period.

B. PERCENTAGE DEPLETION—Allowed only on the first 1000 BOE per day. Currently defined as the lesser of 15% of gross revenue or 100% of net depletable profit from the company's economic interest in an oil or gas property. Has historically been subject to legislative changes.

DEPRECIATION (Depr.)—Noncash deduction from current revenue which provides for the recovery of a portion of the capitalized costs of tangible assets that are experiencing a decline in physical usefulness due to wear or the passing of time.

DEVELOPMENT COSTS—Expenses incurred developing a property for production.

DEVELOPMENT WELL—A well drilled within the proved area of an oil or gas reservoir to the depth of a stratigraphic horizon known to be productive. Sometimes called exploitation or field wells.

DIRECT OPERATING COSTS—See OPERATING COSTS.

DISCOUNT FACTOR (DF)—Inverse of COMPOUND FACTOR. Factor used to determine the present worth of a future sum of money. It is always related to an interest rate and is the reciprocal of the formula for FUTURE VALUE (FV). $DF = 1 \div (1.00 + i)^n$, where i is the interest rate and n is the number of years.

DISCOUNT RATE—The interest rate assumed in reducing future sums to their present worth.

DISCOUNTED CASH FLOW RATE OF RETURN—See INTERNAL RATE OF RETURN.

DISCOUNTED RETURN ON INVESTMENT (DROI)—The ratio of the project's present worth to the present worth of the project's major cash investments (after tax or before tax), discounted at some rate.

DISCOUNTING—The process of determining the present worth of future sums. Inverse of COMPOUNDING.

DISCOVERY WELL—An exploratory well that encounters a new and previously untapped petroleum deposit; a successful wildcat well. A discovery well may also open a new horizon in an established field.

DISCRETE COMPOUNDING—See PERIODIC COMPOUNDING.

DIVIDENDS—Payments to stockholders as compensation for the use of their funds.

DIVISION ORDER—A statement issued by the pipeline purchasing company setting out the names and fractional participations of the working interest and royalty ownership under a producing property.

DOUBLE-DECLINING BALANCE (DDB)—An accelerated method of depreciation whereby an asset is depreciated at a fraction, equal to double the straight line rate, of the remaining balance for each period.

DOWNSTREAM—The refining or marketing segment of the petroleum industry.

DRY HOLE—An unsuccessful well; a well drilled to a certain depth without finding commercial quantities of oil or gas.

DRY HOLE CONTRIBUTION (DRY HOLE MONEY)—Money or property given to an operator in payment for drilling of a well on property in which the contributor has no direct interest, but payable only in the event that the well is a dry hole.

E

EARNINGS—Generally, a company's financial book profit, minus unusual items, for an accounting or reporting period. See NET PROFIT.

ECONOMIC INTEREST—Any interest in minerals or reserves in place, in which the owner must look solely to the proceeds derived from the extraction of such minerals or reserves (if, as, and when produced) for a return of his capital. Types of economic interests are as follows:

A. CARRIED WORKING INTEREST (CWI)—An agreement under which one party (carrying party) agrees to pay a portion of all of the development and operating costs applicable to the share of the working interest owned by another party (carried party).

B. FEE INTEREST—An economic interest, percentage or fractional, in fee acreage.

C. MINERAL INTEREST (MI)—An economic interest, percentage or fractional, in the reserves in place without ownership of the surface.

D. NET PROFITS INTEREST (NPI)—An interest in a property which entitles the owner to receive a stated percentage of the net profit from that property, as defined in the instrument creating the interest.

E. NET WORKING INTEREST (NWI)—Working interest minus the royalty interest and any overriding royalty and production payment interests, if applicable.

F. OVERRIDING ROYALTY INTEREST (ORRI, ORI)—An economic interest created in addition to the mineral owner royalty named in the lease. It can be made payable out of all or a fraction of the revenue or production attributable to the net working interest, that is, working interest minus the royalty interest and production payment interest, if any. An ORRI is limited by the duration of the lease under which it is created.

G. ROYALTY INTEREST (RI)—The share of minerals reserved, free of expense, by a "mineral interest" or "fee" owner when leasing the property to another party.

H. WORKING INTEREST (WI)—The lessee's interest in a lease before deducting any overriding royalty and/or production payment interests. The working interest, therefore, reflects revenue due the lessee and represents the lessee's participation in exploration, development, and producing costs either on a cash, penalty, or carried basis.

ECONOMIC LIMIT—The point at which a project or well ceases to be profitable, or where the profit margin becomes zero. See ABANDONMENT.

EFFICIENT FRONTIER—The set of investments that provides the maximum expected return for a given risk.

EMPOWERMENT—How a company or individual develops people to become more independent.

ENHANCED OIL RECOVERY (EOR)—Oil recovered by a supplemental recovery method above and beyond that recoverable by conventional waterflooding.

EQUITY—Book value of a firm consisting of capital stock, paid-in surplus, and retained earnings.

EXCHANGE AGREEMENT—An agreement by one company to deliver oil or gas to another company at a specified location in exchange for oil or gas to be delivered by the latter to the former company at a different location. Also, a merger of ownerships in oil and gas on the basis of confidence factored reserves.

EXPENSE—Expenditure that is deducted from revenue in the current period. May include operating costs, non-cash charges, exploratory dryhole expense, and overhead. Usually, G&G is expensed only on the financial books while intangibles are expensed on the tax books.

EXPECTED VALUE (EV)—Mean or average probability weighted (risked) value.

EXPENDABLE WELL—A well drilled for information with no specific intention to produce the well if it encounters hydrocarbons. In offshore areas, such wells may be drilled to assess field size. The well may later be reentered if economically and mechanically feasible.

EXPLOITATION—Application of geological, geophysical, and engineering disciplines to the development of discovered oil and/or gas reserves for the purpose of maximizing efficient recovery and improving return on investment. It includes development of newly discovered fields, primary production, and improved and enhanced recovery.

EXPLORATION—The search for subsurface oil and/or gas accumulations that can be produced and sold at a profit. It represents the initial upstream phase of the petroleum industry and uses a wide variety of geological and geophysical disciplines. It also involves a series of economic investment decisions made under varying degrees of uncertainty.

EXPLORATORY WELL—The SEC defines this as a well drilled to find and produce oil or gas in an unproved area, to find a new reservoir in a field previously found to be productive of oil or gas in another reservoir, or to extend a known reservoir. Generally, an exploratory well is considered to be any well that is not a development well, a service well, or a stratigraphic test well. Compare WILDCAT WELL.

EXPONENTIAL DECLINE—A producing rate decline that is a logarithmic function of time. Also, constant percentage decline.

F

FARM-IN (FI, F/I)—A situation, formalized by an agreement, in which a second party agrees to extract reserves from a lease where the current lease holder is not himself intending to do so. See FARM-OUT.

FARM-OUT (FO, F/O)—A pact between oil operators whereby a working interest owner of a lease agrees to assign his interest, or a portion of it, to another operator who will drill on the acreage. The assignor may or may not retain an interest (royalty or carried interest) in the production.

FEE ROYALTY—The lessor's share of oil and gas production; landowner's royalty. Also, mineral owner royalty.

FEE ACREAGE, FEE LAND—See ACRES (OWNERSHIP).

FIELD—A producing area consisting of a single reservoir or multiple reservoirs all grouped on or related to the same individual geological structural feature and/or stratigraphic condition.

FIELD RULES—Set forth to ensure orderly development and exploitation in a specific field. May apply to one or more reservoirs in a field. Usually includes spacing rules, allocation of allowable production rates, and gas/oil ratio limits.

FINANCIAL ACCOUNTING STANDARDS BOARD (FASB)—A body recognized pursuant to Rule 203 of the Rules of Conduct of the American Institute of Certified Public Accountants (AICPA) to formulate and constitute accounting principles.

FINANCIAL BOOKS—Financial (accounting) records of a firm summarized by the balance sheet and income statement.

FINANCIAL BOOK EARNINGS—See NET PROFIT-CURRENT.

FINANCIAL BOOK TAXES—The tax liability that arises on a firm's financial books, as opposed to its tax books. A timing difference may lead to deferred taxes.

FULL CYCLE ECONOMICS (FCE)—Performance measure encompassing the entire span of the project regardless of the portion of the project already underway or completed.

FUTURE VALUE (FV)—FV of any investment (I) equals 1.00 plus the interest rate (i, expressed as a decimal fraction) taken to the power of the number of years (n) invested at that interest rate. The formula is FV of I = $(1.00 + i)^n$.

G-H

GAMBLER'S RUIN—(1) A situation in which a risk taker with limited funds goes broke through a continuous string of failures that exhausts his available funds. (2) A run of bad luck.

GAS–OIL RATIO (GOR)—The number of standard cubic feet of natural gas produced with a barrel of oil, expressed in cubic feet per barrel of oil (CF/B).

GEOLOGICAL AND GEOPHYSICAL (G & G)—Expenditures incurred by an operator in making geophysical surveys and geological studies of an area of interest.

GNP DEFLATOR—See IMPLICIT PRICE DEFLATOR.

GROSS—Adjective meaning exclusive of deductions.

GROSS INCOME—See GROSS REVENUE.

GROSS NATIONAL PRODUCT (GNP)—The sum of all goods and services produced by the economy, including net exports.

GROSS REVENUE—Total funds received or derived from the sale of products and services.

GROWTH RATE OF RETURN (GROR)—An earned interest rate on a stream of future cash flows reinvested at a fixed reinvestment rate and with future investments expressed as present worth equivalents.

HABENDUM CLAUSE—That which gives the lessee's ownership term in a mineral lease.

HYPERBOLIC DECLINE—A producing rate decline whereby the rate is a hyperbolic function of time. See EXPONENTIAL DECLINE.

I

IMPLICIT PRICE DEFLATOR—U.S. Department of Commerce index of factors for tracking the effect of inflation on GNP over time.

IMPROVED RECOVERY—Oil produced by means other than the natural energy mechanisms of a reservoir, invariably by some form of fluid injection. Includes enhanced recovery, but not artificial lift.

INCOME EARNINGS—See GROSS REVENUE.

INCREMENTAL ECONOMICS—Measures of project performance that compare one project to a mutually exclusive alternative.

INCREMENTAL WORTH (IW)—Measured by taking future value minus current year capital expenditure.

INDIRECT OPERATING COSTS—See OPERATING COSTS.

INFLATION—A rise in the price charged for a good or service not attributable to any real change in the product. Inflation is measured usually as percentage change over time.

$$\frac{\text{Cost}_{t2} - \text{Cost}_{t1}}{\text{Cost}_{t1}}$$

INJECTION WELL—General classification for a well used for introducing water or gas under pressure into an underground stratum. Injection wells may be employed for disposal of salt water produced with oil, for pressure maintenance, injecting water for secondary recovery, or other materials for improved or enhanced recovery (EOR) projects.

INTANGIBLE—Portion of an investment that is expensed for tax or financial book purposes.

INTANGIBLE DRILLING COSTS—(1) Expenditures incurred by an operator for labor, fuel, repairs, hauling, and supplies used in drilling and completing a well for production. (2) Portion of drilling costs that has no salvage value.

INTELLECTUAL CAPITAL—The net value provided an organization by its investment in nonoperating technical staff.

INTERNAL RATE OF RETURN (IRR)—The compound interest rate whose discount factors will make the present worth of a project's net cash flow equal zero. IRR is equal to the constant effective annual percentage earnings on the unrecovered capital in a project.

INVESTED CAPITAL—Shareholder's equity or net worth; total assets minus total liabilities.

INVESTMENT—An expenditure intended to acquire or improve property (real or personal, tangible or intangible), yielding revenue or services. Such expenditures are cash outlays when incurred and may or may not be capitalized on the financial or tax books.

INVESTMENT TAX CREDIT—A credit against taxes due in any given period on eligible investments. An incentive sometimes offered by the federal government to encourage such capitalized investments.

INVESTOR'S RATE OF RETURN—See INTERNAL RATE OF RETURN.

INVESTMENT EFFICIENCY—How much profit realized per dollar invested. For example, see GROWTH RATE OF RETURN.

L

LEASE—(1) The legal instrument by which a mineral leasehold is created in minerals. A contract that, for a stipulated sum, conveys to an operator the right to drill for oil and gas. The mineral lease is not to be confused with the usual lease of land or a building. (2) The location of production activity; oil installations and facilities.

LEASE ACQUISITION COSTS—Bonus payments to the mineral owner.

LEASE OPERATING EXPENSES—See OPERATING COSTS.

LEASEHOLD ACRES—See ACRES.

LEASEHOLD COSTS—Bonus or other consideration, legal fees, and capitalized (tax book) G & G costs.

LEASEHOLD INTEREST—See WORKING INTEREST.

LESSEE—The recipient of an oil and/or gas lease; the working interest.

LESSOR—The conveyor of an oil and/or gas lease; the mineral owner.

LIFTING COSTS—The cost of producing hydrocarbons (after exploration and development costs) from a well or a lease.

LIMITED PARTNERSHIP—An agreement between two or more parties providing that at least one of the parties has a limited liability.

LIQUIDITY—A relative measure of the ways with which assets may be converted into cash.

LOCAL TAXES—In general terms, usually composed of severance and ad valorem taxes imposed by counties, parishes, and cities; that is, below the state level. See SEVERANCE AND AD VALOREM TAX.

LONG-TERM DEBT—Amounts owed with maturities greater than one year.

M

MARGINAL REINVESTMENT RATE—The internal rate of return of the least desirable project of average risk which an operator is willing to accept.

MAXIMUM CASH OUTLAY (MAXIMUM CASH OUT-OF-POCKET)—The maximum after tax cumulative net cash deficit over a project's life.

MAXIMUM EFFICIENT RATE (MER)—The highest rate of production from an oil well consistent with proper pressure maintenance, gas–oil ratios, and other sound production practices.

MINERAL ACRES—Acreage on which ownership of the minerals in place has not been divided. See ACRES.

MINERAL INTEREST—Ownership of minerals in place without regard to ownership of the surface. See ECONOMIC INTEREST.

MINERAL RIGHTS—Perpetual ownership of oil, gas, and other minerals beneath the surface, conveyed by a deed. Often simply called "minerals."

MINIMUM RATE OF RETURN—The lowest discount rate that a company or investor will allow for any investment.

MINIMUM ROYALTY—A payment to be made regardless of production, such payment to be chargeable against future production, if any, accruing to the royalty interest.

MONTE CARLO SIMULATION—A mathematical model of a process in which variables cannot be fixed to single values but are represented by probability distributions. A sample value of each variable is randomly selected and answers are calculated. The process is repeated many times to derive a probability distribution of answers.

MUTUALLY EXCLUSIVE PROJECTS—Types of projects in which acceptance of one precludes the acceptance of its alternative (e.g., drilling wells with a 40-acre spacing versus 80-acre spacing.)

N

NATURAL GAS (NG)—Gaseous forms of petroleum consisting of mixtures of hydrocarbon gases and vapors, the more important of which are methane, ethane, propane, butane, pentane, and hexane.

NATURAL GAS LIQUIDS (NGL)—Hydrocarbons found in natural gas which may be extracted or isolated as ethane, propane, normal butane, isobutane, and debutanized natural gasoline (C_{5+}).

NET CASH FLOW—See NET PROFIT.

NET INCOME—See NET PROFIT.

NET INTEREST—A company's portion of the gross income after deducting the royalty interest.

NET PRODUCTION—Production of reserves, net of royalties, usage, and shrinkage.

NET PROFIT, CURRENT—The amount remaining from revenue after deducting all related cash expenses and noncash charges for the period.

NET PROFIT, TOTAL—Cumulative net cash position. Total revenue minus all cash expenses and all investments over the project life.

NET PROFITS INTEREST (NPI)—An interest in a property which entitles the owner to receive a stated percentage of the net operating profit as defined in the instrument creating the interest.

NET REVENUE INTEREST—A fraction of the gross working interest revenue minus an overriding royalty and production payment interests. See WORKING INTEREST.

NONCASH CHARGES—Charges against revenue, usually for the purpose of determining income taxes and/or financial book profit, which do not involve the actual expenditure of cash during the current period.

NONPARTICIPATING ROYALTY—See ROYALTY.

NONRECOURSE DEBT—A form of borrowing in which the general assets of the borrower's firm are not exposed to the claims of the lender. Lender's claims are limited to the revenue generated by the particular project for which he or she supplied proceeds.

O

OFFSET WELL—(1) A well drilled on an adjacent location to an existing well. The distance (spacing) from the first well to the offset well depends upon state or federal spacing regulations. (2) A well drilled on one tract of land to prevent the drainage of oil or gas to an adjoining tract where a well is being drilled or is already producing.

OIL AND/OR GAS PAYMENT—A fixed payment to a participant or lender in a contract area derived from a percentage of the gross income from production.

OPERATING AREA—A contract or lease granting exclusive drilling rights within a geographical area.

OPERATING COSTS (Opcst.)—Costs attributable to conducting operations designed to provide revenues (e.g., raw materials, fuel, operating labor, maintenance, and repairs.)

A. DIRECT OPERATING COSTS—Costs directly attributable to specific operations.
B. INDIRECT OPERATING COSTS—See OVERHEAD.

ORDINARY INCOME—Income from normal operations of a firm. Excludes income from sales of capital assets and other unusual activities.

OVERHEAD (OH)—All costs of materials and services which are not directly adding to, or readily identifiable with, specific operations, but which are necessary costs of doing business (e.g., accounting and personnel).

OVERRIDING ROYALTY INTEREST (ORI, ORRI)—See ECONOMIC INTEREST.

P

PARTICIPATING AREA—The part of a unit area to which production is allocated in the manner described as in a unit agreement.

PARTICIPATING ROYALTY—A royalty interest, independent of any existing lease, which shares in some lease benefits other than gross production, such as bonus, rental, or the right to join in the execution of leases. The term is ambiguous since it does not indicate in each particular case which other lease benefit is joined to the royalty interest.

PASS-THROUGH ROYALTY—A royalty paid on

production from a well drilled upon or through one tract (or horizon) and bottomed under another tract (or horizon).

PAY ZONE—That physical interval in a stratum in which oil or gas is found in commercial quantities.

PAYBACK PERIOD—See PAYOUT.

PAYOUT—An economic term referring to the length of time required for the cumulative cash position to reach zero (e.g., to break even on the total exploration, lease acquisition, drilling, production outlay, and other expenses). Payout is on an after tax basis unless otherwise designated and may be on an undiscounted or discounted cash flow basis.

PENALTY CLAUSE—That portion of a legal instrument (such as a unit agreement) providing for the assessment of a penalty under certain stated circumstances.

PERIODIC COMPOUNDING—Discrete compounding. Process in which funds received accumulate additional interest after a specified, finite period of time, such as yearly or monthly. See also COMPOUNDING.

PLAY—(1) An area of concentrated exploration activity and/or interest in a sedimentary basin. (2) A group of prospects and any related fields having common oil or gas sources, migration relations, reservoir formations, seals, and trap types.

PLAY BOUNDARIES—Coincide with any geological limits of adequacy or with any changes that affect group risks or potential field sizes. Boundaries may also be arbitrary, e.g., international, concession, water depth, or other economic ones chosen for convenience.

PLUG AND ABANDON (P&A)—To fill a well bore hole with cement or other impervious material and discontinue all further drilling or producing operations at that site.

POOLING—The bringing together of contiguous tracts of land sufficient for the granting of a well permit under applicable spacing rules and as provided for in the leases being pooled as distinguished from *unitization*, a term used to describe the joint operation of all or some portion of a producing reservoir. See UNITIZATION.

POINT-IN-TIME CASH FLOWS—Cash flows that are received or given out in a single lump sum at a given instant in time, such as a bonus payment or the sale of a major cash asset.

PRESENT WORTH, PRESENT VALUE (PW, PV)—The current value of future cash flows discounted at a specified discount rate.

PRESENT WORTH (PRESENT VALUE) PROFILE—A listing of present worths (present values) at a range of discount rates.

PRIMARY RECOVERY—Extraction of reserves by natural reservoir energy mechanisms. See SECONDARY RECOVERY, ENHANCED OIL RECOVERY.

PROBABILITY OF SUCCESS—The chance of a future event being a success. Given similar circumstances in the future, a past history of 20 out of 100 wells being productive would indicate a 20% probability of success. The probability of drilling a commercially productive well. See RISK.

PRODUCTION:

A. GROSS PRODUCTION—The total production from a property irrespective of how it is divided among the various economic interests.
B. WORKING INTEREST PRODUCTION—Gross production multiplied by the fractional interest owned. When the working interest is 100%, working interest production equals gross production minus the royalty fraction.
C. ROYALTY PRODUCTION—The share of total lease production that is reserved for the mineral interest or fee owner.
D. NET PRODUCTION—Production derived from ownership of (a) working interests minus any royalty or other outstanding interests, and/or (b) other economic interests such as royalty interest or overriding royalty interest.

PRODUCTION DECLINE—The natural decrease in the rate of production as reserves are depleted in a well. Can often be approximated using exponential or hyperbolic functions.

PRODUCTION PAYMENT INTEREST (PPI)—An economic interest limited to a specific portion of reserves in place. The limitation may be expressed as a stated amount of production, money (including interest and other charges), or time. A production payment cannot be liquidated out of revenue from any source other than reserves produced from the property against which it is applied, nor should it be guaranteed directly or indirectly.

A. CARVED OUT—A production payment in which the owner of the royalty or working interest sells a production payment while retaining the residual interest in the property.
B. PLEDGED RESERVES—A production payment in which the owner of the royalty or working interest sells the residual interest to a second party but reserves a production payment, which he or she may retain or sell simultaneously to a third party.
C. ABC PRODUCTION PAYMENT—Refers to the production payment where A is the seller of a given property, B is the party desiring ultimately to own A's property, and C is either a limited partnership or corporation that borrows from a commercial bank. The purpose of C is to reduce

B's outlay and risk. The bank has no recourse against B's general assets. The loan must be repaid solely from project proceeds.

PROFIT—See NET PROFIT.

PROFIT CONTRIBUTION—Funds passed through by a project to net profit.

PROFIT TO INVESTMENT RATIO—See RETURN ON INVESTMENT.

PROFITABILITY INDEX (PI)—See INTERNAL RATE OF RETURN.

PROJECT LIFE—The period of time over which a project can be expected to provide positive cash flows.

PROPERTY—As defined by the Department of Energy, the right that arises from a lease or from a fee interest to produce domestic oil or gas. Such leases typically describe the right to produce by surface boundaries, although they may also delineate particular underground strata to which rights are conferred.

PROPORTIONAL REDUCTION—A clause commonly included in a contemporary lease providing for the reduction of payments to a lessor if his interest is less than that which he purported to lease.

PRORATION—Restricted production, usually on basis of market demand.

PRORATION UNIT—An area attributed to a well by a regulatory agency in order to control development drilling in an oil or gas field.

PROSPECT—An area under which a geological trap with economically recoverable hydrocarbons is thought to exist. Also the specific area under consideration for locating and drilling an exploratory well commonly based on seismic information.

PROVED AREA—The part of a property to which proved reserves have been specifically attributed.

PURCHASE AGREEMENT—An agreement for the purchase and sale of oil and/or gas produced from designated leases, setting forth the terms and conditions of purchase and sale, and requirements as to quality and condition of the product and measurement or quantities.

R

RATE ACCELERATION PROJECT—Additional field development incremental to a field's existing depletion plan for the predominant purpose of shifting the existing depletion plan's expected production profile forward in time, rather than increasing the field's total production over time.

RECOMPLETION—The act of plugging a depleted or noncommercial zone in a well and then opening a new zone to production.

RECOVERABLE RESERVES—That portion of oil and/or gas in place which can be produced economically to the surface.

RECOVERY FACTOR—That fraction of the total hydrocarbons in a reservoir which it is believed can be produced.

REINVESTMENT—The purchase of a noncash asset with the proceeds from a prior investment.

RESERVES (Res.)—Recoverable or producible hydrocarbons or other mineral resources. See also Appendix to Chapter 12 (Wiggins, this volume) for definitions approved by the Society of Petroleum Engineers on Feb. 2, 1987.

A. GROSS RESERVES—Total reserves without regard to the owned interest breakdown.

B. NET RESERVES—Reserves derived from fractional ownership of (a) working interests and/or (b) other economic interests such as RI or ORI.

C. POSSIBLE RESERVES—Quantities of recoverable hydrocarbons estimated on the basis of geological and engineering data that are less complete and less conclusive than that necessary to support a higher classification. In some cases, economic and regulatory uncertainties may dictate a possible classification. Otherwise, *possible* reserves include (1) reserves that might be found if certain geological conditions exist that are indicated by structural extrapolation from developed areas; (2) reserves that might be found if reasonably definitive geophysical interpretations indicate a productive area larger than could be included within the proved and probable limits; (3) reserves that might be found in formations that have favorable log characteristics but leave a reasonable doubt as to their probability; (4) reserves that might exist in untested fault segments adjacent to proved reserves; and (5) reserves that might result from a planned improved recovery program that is not in operation and where fluid or formation characteristics are such that a reasonable doubt exists as to it success.

D. POTENTIAL RESERVES—Expected reserves at the outset of drilling.

E. PROBABLE RESERVES—Quantities of recoverable hydrocarbons estimated on the basis of engineering and geological data that are similar to those used for proved reserves but that lack, for various reasons, the certainty required to support a proved classification. Probable reserves include (1) reserves that appear to exist a reasonable distance beyond the proved limits of a productive reservoir, where a water contact has not been determined, and proved limits are established only by the lowest known structural occurrence of hydrocarbons; (2) reserves in formations that appear to be productive from log or core analyses but lack definitive tests; (3) reserves in a portion of a formation separated from the proved area by a sealing fault, provided that geological interpretation indicates the probable area to be productive; (4) reserves obtainable by improved recovery where an improved recovery program is planned but not yet in operation, a successful pilot test has been conducted, and formation characteristics appear favorable for its success; (5) reserves in the same reservoir as proved reserves that would be produced if a more efficient natural recovery mechanism develops than that assumed for the estimation of proved reserves; and (6) reserves that depend on successful workover, treatment, retreatment, change of equipment, or other mechanical procedures for recovery, unless such procedures have been proven successful in wells exhibiting similar behavior in the same reservoir.

F. PROVED RESERVES—Estimated volumes of crude oil, condensate, natural gas, or natural gas liquids, as of a specific date, which geological and engineering data demonstrate with a reasonable certainty to be recoverable in the future from known reservoirs under existing economic conditions. Reservoirs are considered to be proved if economic producibility is supported by actual production or formation tests. The proved area of a reservoir includes (1) that portion delineated by drilling and defined by fluid contacts, if any, and (2) the adjoining portions not yet drilled that reasonably can be judged economically productive on the basis of available geological, geophysical, and engineering data (frequently limited to offset locations). Reserves that can be produced economically through the application of established improved or enhanced recovery techniques are included in the proved classification when two qualifications are met: (a) testing by a pilot project or the operation of an installed program in the same reservoir or in one with similar rock and fluid properties is successful, and (b) it is reasonably certain the project will proceed.

G. PROVED DEVELOPED RESERVES—Those volumes expected to be recovered through existing wells (including behind-pipe reserves, sometimes referred to as *proved developed nonproducing*) with proved equipment and operating methods and without any additional major capital expenditures. Improved recovery reserves can be considered developed only after an improved recovery project has been installed.

H. PROVED UNDEVELOPED RESERVES—Those volumes expected to be recovered from (1) future drilling of wells, (2) deepening of existing wells to a different reservoir, or (3) the installation of an improved recovery project. These reserves require major capital investments to enable production.

I. WORKING INTEREST RESERVES—Gross reserves times the fractional net revenue interest owned by the working interest. (When the working interest is 100%, working interest reserves equal gross reserves minus royalty.)

RESERVES/PRODUCTION RATIO (R/P RATIO)—The ratio of booked reserves at year-end to production during the year. The ratio is expressed in years.

RESERVE STATUS CATEGORIES—Groupings that define the development and producing status of wells and/or reservoirs as approved by the Board of Directors, SPE, Feb. 27, 1987.

A. DEVELOPED—Developed reserves are expected to be recovered from existing wells (including reserves behind pipe). Improved recovery reserves are considered developed only after the necessary equipment has been installed, or when the costs to do so are relatively minor. Developed reserves may be subcategorized as producing or nonproducing.

(1) PRODUCING—Producing reserves are expected to be recovered from completion intervals open at the time of the estimate and producing. Improved recovery reserves are considered to be producing only after an improved recovery project is in operation.

(2) NONPRODUCING—Nonproducing reserves include shut-in and behind-pipe reserves. Shut-in reserves are expected to be recovered from completion intervals open at the time of the estimate, but which had not started producing, were shut in for market conditions or pipeline connection, or were not capable of production for mechanical reasons, and the time when sales will start is uncertain. Behind-pipe reserves are expected to be recovered from zones behind casing in existing wells which will require additional completion work or a future recompletion prior to the start of production.

B. UNDEVELOPED—Undeveloped reserves are expected to be recovered: (1) from new wells on undrilled acreage, (2) from deepening existing wells to a different reservoir, or (3) where a relatively large expenditure is required to recomplete an existing well or install production or transportation facilities for primary or improved recovery projects.

RESERVOIR—A porous and permeable underground formation containing a natural accumulation of producible oil and/or gas that is confined by impermeable rock or water barriers and is individual and separate from other reservoirs.

RESIDUAL INTEREST—An economic interest that is subjected to or diminished by an economic interest owned by another party.

RETURN ON INVESTMENT (ROI)—A measure of project performance, having the following formula:

$$ROI = \frac{\text{Total net cash flow}}{\text{Maximum cash outlay or total investment}}$$

REVENUE (Rev.)—See GROSS REVENUE.

RISK—(1) The relative dispersion of possible outcomes associated with an investment decision. The wider the range of possible outcomes, the riskier the project. Risk refers to a range of alternatives and does not connote that these alternatives are necessarily unfavorable. (2) The possibility of loss stated in terms of the probability of various outcomes. Expressed as a percentage chance of occurrence. (3) An opportunity for loss. See PROBABILITY OF SUCCESS.

RISK AVERSION—Proportion by which the expected value of a risk venture is discounted by an investor.

ROLL-OVER CONTRACTS—Contracts that expire and are renegotiated.

ROYALTY—The portion of oil and gas produced, or its economic equivalent, paid to the owner of the mineral rights under the terms of the lease or operating agreement.

ROYALTY INTEREST (RI)—The share of minerals reserved, free of expense, by a mineral interest or fee owner when leasing the property to another party. See ECONOMIC INTEREST.

RULE OF 72—Divide 72 by the percentage interest rate to obtain a good approximation of the number of years it takes to double the initial investment.

RUNS—Income or production from a lease, usually computed in a monthly statement, or run ticket, from the purchasing company.

S

SALVAGE VALUE—The value of a capital asset at the end of a specified period. It is the current market or "book" price of an asset being considered for replacement in a capital budgeting situation.

SCIENTIFIC RATE OF RETURN—See INTERNAL RATE OF RETURN.

SECONDARY RECOVERY—Methods designed to increase reservoir recovery at the end of the primary life. The most prevalent of these has been water flooding.

SECTION—A land area of 640 acres which measures 1 mile on each side.

SENSITIVITY ANALYSIS—A technique for testing, through multiple case comparisons, the degree to which a project is affected by the change in one or more components of the investment analysis (e.g., changes in price, production, royalty, and investment costs).

SERVICE WELL—A well drilled or completed for the purpose of supporting production in an existing field. Wells in this class are drilled for the following specific purposes: gas injection, water injection, steam injection, air injection, salt water disposal, water supply for injection, observation, or injection for *in situ* combustion.

SEVERANCE AND AD VALOREM TAX—A type of tax (usually state or local) based on some valuation of production revenue and/or remaining reserves.

SHUT-IN ROYALTY—Usually refers to payments made when a gas well, capable of producing in paying quantities, is shut in for lack of market or "reasonable" price for the gas. This type of royalty or some form of rental is usually required to prevent termination of the lease.

SLIDING SCALE ROYALTY—A royalty varying in accordance with the amount of production, e.g., a one-eighth royalty if the production is 100 bbl per day or less, and 3/16 royalty if the production is greater than 100 bbl per day.

SOUR CRUDE—Crude oil containing sulfur or sulfur compound contents sufficiently high to require their removal. The following guidelines aid in further classification of crude:

Crude	Sulfur Percent/Volume
Low sulfur	Up to and including 0.7%
Medium sulfur	>0.7% but ≤2.3%
High sulfur	>2.3%

SOUR GAS—Natural gas containing chemical impurities, notably hydrogen sulfide (H_2S) or other sulfur compounds. Gas is generally considered to be sour if it contains 10 or more grains of H_2S or 200 or more grains of total sulfur per MCF.

SPACING—The density with which development wells are drilled in a field. Spacing is usually regulated by governmental agencies. See WELL SPACING.

SPUD DATE—The date of the actual start of drilling a well.

STATE TAX—See SEVERANCE AND AD VALOREM TAX.

STEP-OUT WELL—A well drilled adjacent to a proven well but located in an unproven area; a well drilled as a "step out" from proven territory in an effort to determine the productive boundaries of a formation.

STOCKHOLDER (SHAREHOLDER) EQUITY—See EQUITY.

STRAIGHT-LINE (SL)—A method of depreciation in which the depreciable value is divided into equal annual (or periodic) installments.

STRATIGRAPHIC TEST WELL—A drilling effort, geologically directed, to obtain information pertaining to a specific geological condition. Such wells are customarily drilled without the intention of being completed for hydrocarbon production. This also includes wells identified as core tests and all types of expendable holes related to hydrocarbon exploration. Stratigraphic test wells are classified as (1) *exploratory type* if not drilled in a proved area, or (2) *development type* if drilled in a proved area.

STRIPPER GAS—Production from a well that produces gas at a rate not exceeding 60 MCF per day over 90 days.

STRIPPER OIL—Oil produced by a well efficiently producing less than 10 bbl per day for over 90 days.

SUCCESS:

A. COMMERCIAL SUCCESS—A discovery well that finds a field capable of paying for all subsequent costs of development drilling, completion, surface equipment, operation costs, and wellhead taxes, while returning a reasonable profit margin.
B. COMPLETION SUCCESS—A well that is completed for production.
C. GEOLOGICAL SUCCESS—A well that encounters a deposit of mobile and measurable hydrocarbons.
D. INCREMENTAL SUCCESS—A completed well that will return a profit only upon the investment of completing and operating it, ignoring (as sunk costs) the exploratory drilling costs required to find it.

SUM OF THE YEARS' DIGITS (S-Y-D)—An accelerated method of depreciation that applies a fraction times the original depreciable value each period. The fraction is constructed using a sum of the years' digits as the denominator, and the year numbers in reverse order as the numerator. For example, fractions for a 3-year project are 3/6, 2/6, and 1/6 and are multiplied by the depreciable value to determine the depreciation for each period.

SUNK COSTS (SC)—Those costs that have already been incurred and therefore do not affect yet-to-spend economics for a given project except for their effect on subsequent income taxes.

SWEET CRUDE—Crude not classified as sour. See SOUR CRUDE.

SWEET GAS—Natural gas containing no significant amount of hydrogen sulfide (H_2S). Gas is generally considered to be sweet if it contains fewer than 10 grains of H_2S or fewer than 200 grains of total sulfur per MCF.

SYMMETRIC DISTRIBUTION—A plot of possible outcomes versus probability of occurrence in which the expected value is also the mode, mean, and median value and the area of probability to the left of the mean is a mirror image of the right portion.

T

TANGIBLE—That portion of an investment that represents a lasting physical commodity (that is, production platform and equipment), otherwise that portion of an investment classified as tangible by the applicable tax law.

TANGIBLE COSTS—Items that have a salvage value. Must be capitalized.

TAX—An assessment by governmental agencies against profits. Federal, state, and local taxes are examples. See SEVERANCE AND AD VALOREM TAX, LOCAL TAXES, and DEFERRED TAX.

TAX BOOKS—A broad term referring to tax liability records maintained by an individual or firm. Differences develop between financial and tax books, based primarily upon timing differences in recognizing revenues and expenses.

TAX CREDITS—See INVESTMENT TAX CREDIT.

TAXABLE INCOME—An accounting figure used for determining federal income tax liability. For project evaluation purposes, taxable income is commonly calculated to equal revenue minus operating costs, overhead, expensed investment, capital recovery, and other taxes.

THIRD FOR A QUARTER—Seller receives a carried 25% working interest to some point (usually casing point) while the Buyer pays 33.33% of costs before casing point for right to pay 25% of costs after casing point. This is example of a CARRIED INTEREST.

TIGHT HOLE—A drilling well about which all information—depth, formations encountered, drilling rate, etc.—is kept secret by the operator.

TOP LEASE—A lease granted by land owner during the existence of a recorded lease which is to become effective if and when the existing lease expires or is terminated.

TOWNSHIP—A land area measurement which is equal to 36 sections. See SECTION.

TREND ANALYSIS—A technique to organize and use the available empirical facts to make strategic long-term exploration decisions in an objective manner.

TURNKEY—Relating to a project or installation built, supplied, or installed complete and ready to operate, e.g., a turnkey well.

U-V

ULTIMATE NEW INCOME RATIO (UNIR)—See RETURN ON INVESTMENT.

ULTIMATE NET CASH POSITION—See NET PROFIT.

ULTIMATE NET CASH FLOW—The sum of yearly net cash flows over the life of the project. See NET PROFIT.

UNCERTAINTY—(1) Refers to an event with unknown outcomes, or outcomes with unknown associated probabilities of occurrence. (2) The range of probabilities that some condition may exist or occur. See PROBABILITY OF SUCCESS and RISK.

UNIT AGREEMENT—An agreement for the recovery of oil and gas, where acreage is treated for operational purposes and for the allocation of costs and benefits as a single consolidated unit without regard to separate ownerships.

UNITS OF PRODUCTION METHOD—A method employed for the calculation of depletion, depreciation, or amortization based upon quantities produced in relation to reserves.

UNITIZATION—A term denoting the joint operation of all or some portion of a producing reservoir as distinguished from *pooling*, which is used to describe the bringing together of small tracts sufficient for the granting of a well permit under applicable spacing requirements.

VALUE—The dollar worth of an asset. In a general sense, the subjective intrinsic worth of that which is being evaluated.

VARIABLE ROYALTY—See SLIDING SCALE ROYALTY.

W-Z

WEIGHTED AVERAGE COST OF CAPITAL—A weighted average of the component costs of debt, preferred stock, and common equity. It represents the theoretically correct cost of capital.

WELL—A ground boring drilled to and completed in one or more subsurface horizons or producing formations, usually for the purpose of obtaining hydrocarbons. Also, may include service wells and stratigraphic tests.

A. GROSS WELLS—The total wells on a property regardless of the various economic interests involved.

B. NET WELLS—The total wells on a property multiplied by the fractional interest owned.

WELL SPACING—Regulation of number and location of wells producing from same reservoir. Normally spacing is stated as minimum permissible distances from lease lines and between wells. It essentially ensures that each well drains a specific number of productive acres within the operator's own lease.

WILDCAT WELL—A well drilled in an unproved area, a mile or more from existing production; an "exploratory well" in the truest sense of the term. May also refer to a well drilled to an unproved horizon.

WINNER'S CURSE—"If you win the bid, you paid too much."

WORKING CAPITAL—Refers to a firm's investment in short-term assets—cash, short-term securities, accounts receivable, and inventories.

A. GROSS WORKING CAPITAL—Defined as a firm's total current assets. If the term *working capital* is used without further qualification, it generally refers to gross working capital.

B. NET WORKING CAPITAL—Defined as current assets minus current liabilities.

WORKING INTEREST—The obligation to pay proportional costs and investments in return for some percentage of production and revenue. It represents the lessee's interest in a lease before deducting any overriding royalty and/or production payment.

YET-TO-SPEND ECONOMICS (Y-T-S)—Measures of project performance that consider only the project expenditures and benefits that have not yet occurred. Tax effects of past expenditures that impact future cash flows should be considered.

ZONE—An interval of a subsurface formation containing one or more reservoirs.

HYPOTHETICAL EXAMPLES DEMONSTRATING THE USE OF VARIOUS TERMS

1. COMPANY "A" OWNS A WORKING INTEREST (December 31, 1991)

ASSUMPTIONS	
Leasehold Acres	500
Gross Production	1000 bbl/day
Crude Price	$20.00/bbl
Operating Costs	$3.00/bbl
Royalty Interest	1/5
Company A's Working Interest Share	25%
Company A's Royalty Interest Share	None
Company A's Net Revenue Interest	20%

	TOTAL	COMPANY A
ACRES		
Gross Acres	500	N/A
Gross Leasehold Acres	500	N/A
Net Leasehold Acres	N/A	125
Royalty Acres	500	N/A
Net Royalty Acres	N/A	0
PRODUCTION—BOPD		
Gross Production	1000	N/A
Royalty Production	200	0
Working Interest Production	800	200
Net Production	800	200
REVENUE ($/year)		
Gross Revenue	7,300,000	N/A
Royalty Revenue	1,460,000	0
Working Interest Revenue	5,840,000	1,460,000
OPERATING COSTS ($/year)		
Operating Costs	1,095,000	273,750

N/A—Not applicable.

2. COMPANY "A" OWNS A ROYALTY INTEREST
(December 31, 1991)

ASSUMPTIONS	
Leasehold Acres	500
Gross Production	1000 bbl/day
Crude Price	$20.00/bbl
Operating Costs	$3.00/bbl
Royalty Interest	1/5
Company A's Working Interest Share	None
Company A's Royalty Interest Share	50%
Company A's Net Revenue Interest	none

	TOTAL	COMPANY A
ACRES		
Gross Acres	500	N/A
Gross Leasehold Acres	500	N/A
Net Leasehold Acres	N/A	0
Royalty Acres	500	N/A
Net Royalty Acres	N/A	250
PRODUCTION—BOPD		
Gross Production	1000	N/A
Royalty Production	200	100
Working Interest Production	800	0
Net Production	N/A	100
REVENUE ($/year)		
Gross Revenue	7,300,000	N/A
Royalty Revenue	1,460,000	730,000
Working Interest Revenue	5,840,000	0
OPERATING COSTS ($/year)		
Operating Costs	1,095,000	0

N/A—Not applicable.

3. ILLUSTRATION OF PETROLEUM FINANCIAL TERMS
XYZ OIL CORPORATION
(Year-ended December 31, 1991)

		AMOUNT (in $ thousands)
FINANCIAL BOOKS		
Gross Revenue (Includes XYZ's Share of Total Revenue from Leases, Plants, etc.)		549
Minus: Operating Costs		168
Depreciation, Depletion, and Amortization (Noncash Charges)		101
Exploration Expenses, including Dry Hole Costs		138
Taxes (including FIT)		48
Total Costs and Expenses		455
Net Profit (AFIT)		94
CASH FLOW		
Determined in one of two ways:		
1. Total Cash Transactions		
Gross Revenue		549
Minus Cash Costs:		
Operating Costs		168
Taxes		48
Investments:		
Expensed	138	
Capitalized	118	256
Total Cash Flow		77
2. Financial Book Cash Flow		
A. Net Earnings		94
Plus: Noncash Charges		101
Total Cash Flow		195
B. Gross Revenue		549
Minus Cash Costs:		
Operating Costs		168
Expensed Investments		138
Taxes		48
Total Cash Flow		195

JAMES A. McCALEB retired in 1989 after working in Amoco's Exploration Department for 25 years. He died March 27, 1991, while hospitalized in Little Rock, Arkansas. He held B.S. and M.S. degrees from the University of Arkansas and a Ph.D. from the University of Iowa. He was an operations geologist in exploitation and exploration in various supervisory positions in the Rocky Mountain and Gulf Coast Regions. Dr. McCaleb was also a research geologist at the Tulsa Research Center and served 10 years as Manager of Amoco's Exploration Training Center. He published widely on Carboniferous biostratigraphy in the U.S. Mid-Continent and on sandstone and carbonate reservoirs in Paleozoic fields in Wyoming.

Index

AAPG Code of Ethics, 334-335
AAPG Grievance Proceedings, 341-342
AAPG's Committee on Statistics of Drilling (CSD), 72, 74
AAPL Code of Ethics, 342
AAPL Standards of Practice, 342-343
Absolute ownership, 280
Accelerated Capital Recovery System (ACRS), 146-147
Accounting, methods of, 7-8
Acquisitions
 of inventory, 22
 of oil and gas property, 125, 157
Acreage
 control of, 268
 requirements for oil and gas activities, 348
Actuals, versus estimates, 31
Ad valorem tax, 141, 166
Adverse possession, of mineral estate, 284
Air quality, exploration effects on, 348, 350
Alaska, opportunity in, 190
Allowable depletion, 144
Alternative Minimum Tax (AMT), 167
American Petroleum Institute (API), 13
Amortization rate, profit oil share versus, 312
Annual budget, 237-253
API gravity, 140
Appraisal
 of equipment and facilities, 156-157
 of oil and gas properties, 126
 of undeveloped leasehold, 156
ARAMCO, 299
Argentina
 and multinational oil companies, 300
 service contracts in, 303
Assessment, of plays, 87-93
Assets, 7
Average error factor (AEF), 33
Average opportunity rate, 148
Average prospect chance of adequacy, 89
Back-in deal, 156
Bad faith trespasser, 283
Basin variance, versus prospect uncertainties, 40-42
Bell-shaped curve, 36-37, 60
Bias
 affecting risk decisions, 97
 definition of, 58
 in reserves estimates, 32-33
Binomial expansion, and gambler's ruin, 100
Binomial-lognormal distribution, 46-48, 58
"Blackball," 337
Blowouts, 346, 349
Bonus, definition of, 281
Budget
 annual, 237-253
 of drilling program, 238-240
 estimates for, 237-238
 format for, 250-251
 review of, 251-253
 variances in, 251-253
Business
 basic concepts of, 7-12
 budgeting for, 237-253
 creative environment in, 205-213
 deals in, 267-270
 economic aspects of, 105-183
 ethics in, 331-344
 exploration as misunderstood, 21-25
 legal aspects of, 279-318
 management of, 185-276
 nature of, 27-103
 salesmanship in, 263-265
 software for economic evaluation in, 163-173
 uniqueness of exploration, 23-25
Buyer, definition of, 257
California offshore
 environmental impact on, 350
 opportunity in, 190
Capen's 35/25 method, 66
Capital
 allocation of, 83-84
 intellectual, 202-203
 limited risk, 245-249
Capital expenditure, software calculation of, 166
Careers, planning of, 206, 207, 211
Career stages, definition of, 211
Carrying power, 241
Case study, of exploration, 21-23
Cash flow
 after-tax, 128, 147
 analysis of, 107-108
 before-tax, 127, 143
 calculation of, 238-239
 cost depletion calculation for, 145
 definition of, 8
 depreciation expense calculation for, 144
 discounting, 111
 investment yardsticks from, 111-113
 measures, 8-9
 NPV calculation of, 149
 operations, 142
 and present value, 108-111
 yardsticks from, 108
Cash flow stream, discounting, 110
CASOC, 299
Central limit theorem, 47
Chance factors, geological, 73-79
Chance of success
 definition of, 114-115
 estimating, 240
 and expected value, 72-73
 reluctance to estimate, 82
Clauses, in oil and gas lease, 286-296
Clifton v. Koontz, 286-287
Code of Ethics
 AAPG, 334-335
 AAPL, 342
 SEG, 344
Coin toss, 73, 95-96, 224
Commencement, legal definition of, 288
Commercial
 definition of, 49
 requirements of, 87
Commercial success
 definition of, 71, 74
 probability of, 80-81
Commitment, in workforce, 210-211
Company
 current status of, 190-192
 earnings of, 10-11
 goals of, 192-196
 income projections in, 237
 management of people in, 201-203
 mission of, 187
 operational plans of, 193-196
 organization of, 192, 202
 performance milestones in, 196-198
 personnel of, 192, 201-203
 strategies of, 192-196
Competitors, comparison with, 191
Completion chance, 80-81
Completion success, definition of, 71, 75
Compound interest factor, 147
Computer hardware, requirements of, 164-165
Computer programs, *see* Computer software
Computer software
 for cost calculations, 166
 design of, 164
 for economic evaluation, 163-173
 features of, 165
 new technology in, 168-169
 for tax calculations, 166
 in trend analysis, 217
 types of, 169
Concession agreement, 301
Consideration, legal definition of, 286
Continuous interest, 167-168
Contracts
 analysis example, 309-310
 comparisons of, 308
 ethics in making, 337
 international, 297-305, 307-318
 life of, 316-317
 structure of, 317
 types of, 302-304, 307-308
Control, of acreage, 268-269
Conversion, caused by negligence, 283-284
Core analysis, database of, 183
Cores, data from, 180-181
Correlative rights, 280
Cost depletion, 144-145
Cost of finding (COF) rule, 12
Cost oil, definition of, 308
Costs
 of development, 235
 environmental, 157
 of exploration, 234-235
 intangible drilling and development, 146
 of international ventures, 315
 leasehold, 142
 operating and overhead, 166
 of operation, 235
 of research, 202
 software calculation of, 166
 tangible/intangible, 142
 U.S. finding, 320, 321
Covenants, implied in contracts, 289-291
Creative leadership, definition of, 207
Creativity
 in deal making, 269
 leading people with, 210-211
 nature of, 207
 viewpoints in, 209-210
Crude oil prices, 140
Culture shock, 321
Cumulative distribution
 of field sizes, 64, 65
 graph of, 60
 of prospect sizes, 65
Cumulative frequency
 equation of, 38
 of net pay, 42-43
 of porosities, 39
 of productive area, 43
 of recovery, 44
Cumulative probability plot, definition of, 37-39
Cumulative production curve method, 136
Curiosity, in workforce, 210-211
Cuttings, environmental disturbance by, 349
Data
 conditioning of, 218-219
 gathering of, 56
 production, 219
 sources for trend analysis, 218
 types of commercial, 218
 well history, 175-183
Database
 conditioning of, 218-219
 and core analysis, 183
 definition of, 175
 ownership of, 176
 rock data in, 180-181
 for a trend analysis, 217-220
 and well history, 175-183
 well log data in, 181-182
Database management system (DBMS), 175
Dates, in a well history, 178
DCFROR, *see* Discounted cash flow rate of return
Deal
 ethics in making, 337
 oil and gas, 154-156
 philosophy of, 269
 presentation of, 268
 putting together a, 267-270
 sales of, 264
 structuring of, 270
Deal sheet, 273
Decision criteria, software applications to, 167-168
Decision frame, definition of, 97
Decisions, biases affecting, 97
Decline curve method, 133-136
Defeasable, definition of, 281
Defensive clauses, 288-289
Delay rental clause, 288, 293-294
Delay rentals, 281
Deliverability calculations, 138-139
Depletion, definition of, 144
Depletion allowance, 145
Depletion drive, 132
Depletion expense, 144
Depreciation, definition of, 143
Depreciation expense, 143-144
Depth, subsea, 180
Depth ranges, 178-179
Developed reserves, definition of, 161
Development
 of a deal, 267
 of people, 211-212
Development drilling funds, 259-260
Development economics, 117-123
Development project, example of, 118
Discounted cash flow, methods of, 9
Discounted cash flow rate (DCFR), 110
Discounted cash flow rate of return (DCFROR), 9-10, 110
Discounted profit to investment ratio (DPR), 152-153
Discount factor, equation of, 109
Discounting, of cash flow, 110-112
Discoveries, distribution of, 35
Discovery probability, 81-83
Discovery process modeling, 88
Discovery size distribution, 221-224
Dissolved gas drive, 132
Distribution
 of field sizes, 65, 221-224
 graphs of, 35-36
 lognormal, 239
 nature's versus discoveries, 35
 normal, 36-37
 of prospect sizes, 65, 90-93
 truncated, 52-53
Disturbance, definition of, 347

Divided ownership, 284-285
Division order, 287
Dollar accounting, 315-316
Domestic exploration, versus foreign, 319-320
Dominant mineral rights, 284
Double dipping rule, 12
Drilling
budget of program, 238-240
development of, 259-260
disturbance by, 347-348
exploratory, 258-259
funding of, 258-260
limited access to, 351-352
overseas operations, 329-330
Drilling decisions, Monte Carlo simulation for, 242
Drive mechanisms, in reservoirs, 131-133
Dry case, definition of, 111
Dry hole carrying capacity rule, 12
Dry hole clause, 288, 294
Due diligence process, 157
Earnings, of a company, 10-11
Earth, exploration effects on, 345-354
Economic evaluation
of hydrocarbon accumulations, 165
software for, 163-173
Economic limit calculation, 136
Economics
development, 117-123
drill versus farm-out, 120
of enhanced oil recovery, 121-122
of exploration, 107-115
outcomes and inventory, 22-23
rate acceleration and incremental, 120
uncertainty of, 80
Efficient frontier concept, 228
Eliff v. Texan Drilling Co., 280, 284
Employees, ethics of, 331-340
Endangered Species Act, 348
Enhanced oil recovery (EOR), economics of, 121-122
Enhancement techniques, for prospe
ct presentation, 271-276
Entrepreneur, definition of, 212
Environment
see also Working environment
exploration effects on, 345-350
impact on exploration, 350-353
Environmental costs, 157
Environmental Protection Agency (EPA), 349
Equipment, appraisal of, 156-157
Error factor, average, 33
Errors
surviving, 35
versus variance, 42
Estimates, versus actuals, 31
Ethical behavior
bases of, 332-335
guidelines for, 339-340
questions and issues in, 335-339
Ethics
in exploration business, 331-344
typical scenarios in, 338-339
EV, *see* Expected value
Executive right, 281
Expected monetary value (EMV), 111, 113
Expected reserves, definition of, 44
Expected value (EV)
calculation of, 119
and chance of success, 72-73
and coin toss example, 73
definition of, 44, 72
methods of, 240-242
of prospects, 72-73
and risk aversion, 95-96
risk-weighted, 242
unrisked, 241
Expenditures
for development, 14
history of exploration, 13-19
for production, 14
Experience, and ethics, 336
Exploration
budgeting for, 237-253
case study of, 21-23
cash flow in, 107-108
and chance of success, 71-85
development phase of, 117-118
domestic versus foreign, 319-320
economics of, 107-115
environmental issues of, 345-354
ethics in, 331-344
expenses in, 237-238, 315
funding of, 258-259
goals of, 215-216
history of, 298-301
international, 193, 196, 297-305, 319-330
inventory problems in, 22
managing a program of, 187-198
as misunderstood business, 21-25
portfolio analysis, 50-54
resources withdrawn from, 352, 354
risk behavior in, 95-103
success of, 201-202
surviving errors in, 35
tax on international, 313-315
uncertainties in, 29-60
Exploration business, *see* Business
Exploration drilling funding, 258-259
Exploration expenditures
for 1989, 14
categories of, 13-14
graph of, 15
history of, 13-19
and overhead, 17
tables of by year, 16, 17
and total net revenue, 19
Exploration wheel, 30
Exponential decline curve, 134, 135
Failure, probability of, 74, 77, 227
Failure case, definition of, 111
Fair market value, 153
definition of, 126
risk-adjusted value (RAV) and, 103
Fairway, definition of, 217
Farm-in, 268, 317, 323
Farm-out, 119-120, 155-156, 268
Federal income tax, 143
Federal lands, limited access to, 351-352, 354
Fee simple, 285
Field models, analysis of, 309-310
Field number, 90
Field size
cumulative frequency of, 67
distribution of, 63-64, 65, 221-224
Fill-up, 76
Financial risk aversion, 98, 100
Financing, problems for host country, 305
First well pays rule, 12
Flowability, 76
Force majeure clause, 289, 295-296
Foreign exploration, *see* International exploration
Foreign governments, independents and, 323
Foreign projects, outline of, 327-330
Formation name, 180
Fractile, 63
Framing effects, in risk decisions, 97
Freshwater, protection of, 347
Frontier basins, probability of success in, 81
F statistics, 31
Funding
repatriation of, 315
sources of, 260-262
types of, 257-260
of ventures, 257-265
Future value (FV), 198
equation of, 108-109
of one dollar, 110
Gambler's ruin, 100, 225-226, 227, 245, 246, 248
Gas cap drive, 131-132
Gas expansion drive, 132
Gas prospects, 275
Gas volume calculation, 129, 130
Geological age, 180
Geological chance factors
and probability of success, 73-79
subjective expression of, 82
systems using, 76-77
virtues of, 82-83
Geological play assessment, 89-93
Geological risks, 321
Geological shows, 179-180
Geological structure, as geological chance factor, 76
Geological success, definition of, 71, 75
Geological tops, 179-180
Geologists
ethics of, 334-335
necessity of, 202
Geophycists, necessity of, 202
Geophysical trespass, 284
Gillespie v. Ohio Oil Company, 287
Goals
career, 211-212
of a company, 192-196
of exploration, 215-216
Going nonconsent, 98
Goodness of fit, 39
Good shows, definition of, 74
Granting clause, 286, 292
Gravity drainage drive, 132
Grievance Proceedings, AAPG, 341-342
Gross production, determination of, 126-139
Groundwater, protection of, 347
Growth rate of return (GROR), 112, 123
Guide for Professional Conduct, SPE, 343-344
Gulf of Mexico, opportunity in, 190
Haberdum clause, 286-287, 292
Handbook
audience for, 3
authors of, 4-5
purposes of, 3
theme of, 3-4
utility of, 4
Hardware, requirements of, 164-165
Harmonic decline, 135
Heuristic bias, definition of, 97
Hidden hurdles, in risk aversion, 99
History
of exploration expenditures in U.S.A., 13-19
of industry, 188-189
of international oil and gas contracts, 297-305
Horizons
multiple, 46-47
single, 42-46
Host country
dealing with multinational oil companies, 297-305, 307-318
limited involvement in, 300-301
politics of, 320
problems in, 304-305
Humble Oil & Refining Co. v. Kishi, 283
Hurdle rate, 40/1, 11
Hybrid binomial-lognormal distribution, 46-48
Hydrocarbon accumulation
estimates of chance of, 114-115
requirements for, 73-74
Hydrocarbon charge, as geological chance factor, 76
Hyperbolic decline curves, 134
Hypothesis testing, 57-58
Ignorance, of inventory, 23-25
Implied covenant, definition of, 289
Income projection, 237
Incremental success, definition of, 71, 74
Incremental worth, 198
Independents
and foreign governments, 323
international exploration by, 319-330
Industry
environmental issues of, 345-354
as funding sources, 261-262
history of, 188-189
leadership in, 206
opportunity in, 189-192
Injected substances, as trespass, 283
Injection project
economics of, 121-122
timing analysis of, 122
Intangible drilling and development costs (IDCs), 142, 146
Intellectual capital, 202-203
Interest expense, 146
Interests, in oil and gas estate, 282
Internal rate of return (IRR), 110, 118, 120
International contracts
history of, 297-301
life of, 316-317
special considerations for, 315-318
types of, 305, 307-308
International exploration, 297-305
by independents, 319-330
problems of, 320-322
project outline, 327-330
International ventures, 315-318
Introduction, to Handbook, 3-5
Inventory
acquiring, 22
case study of, 23
and economic outcomes, 22-23
and expected value, 73
ignorance of, 23-25
prospect, 83
Investment efficiency index, 113
Investments
alternative, 118-121
costs, 142
definition of, 142
and expected value, 119
guidelines for decisions, 11-12
software for, 163-173
and trend analysis, 216
versus people, 201-202
Investment tax credit (ITC), 146-147
Investment yardsticks, 123
from cash flow concepts, 111-113
types of, 108, 118-121
Investor's interest rate (IIR), 110
Johnson plot, 310-312, 313, 314
Joint venture agreement, 301-302
Joint ventures, software applications to, 167
Keep It Simple, Stupid, 7, 238
KISS principle, 7, 238
Kolmogorov-Smirnov test, 41
Lahee Classification, 179
Land, unleased, 269
Land access, limited, 351-352
Land data, 218
Landowner's royalty, 281
Latitude and longitude, of wells, 177
Law
international, 297-305, 307-318
protecting environment, 350-352
in the United States, 279-296
Leadership, in industry, 206
Leading, creative people, 210-211
Lease
clauses in, 285-289
implied covenants in, 289-291
oil and gas, 285-289, 292-296
withdrawals of lands from, 351-352
Lease extension clause, 288
Leasehold, undeveloped, 156
Leasehold costs, 142
Leasing moratoriums, OCS, 353
Limited risk capital, 245-249
Linear decline, 135

Logarithms, 32
Lognormal distribution, 35-36
- cumulative, 37
- definition of, 239
- for reserves estimates, 239
- of reserve sizes, 63-64, 65

Logprobability paper, 63-64
Long-run breakeven, 246
Management
- of business, 185-276
- in estimating reserve sizes, 66
- of exploration people, 201-203, 210-213
- of exploration program, 187-198
- in international exploration, 323
- role in risk assessment, 30
- of values and vision, 212

Maps
- of plays, 88
- in prospect evaluation, 272-273
- scale of, 177-178
- and uncertainties, 47, 69
- in volumetric estimates, 129

Marine life, disturbance of, 349-350
Marketing, 263-265, 267
Markets, defining, 21
Material balance method, 137
Matrix, of data, 176
Maximum negative cash flow, 108-109
Mean
- definition of, 36
- derivation of, 58
- of prospect sizes, 91

Measures of success, 10
Median
- definition of, 36
- of prospect sizes, 91

Medical documents, needed overseas, 325-326
Menu-driven systems, 164
Mineral estate, 279-280, 280-282
- trespass of, 282-284

Minerals, as legal term, 279-280
Minimum royalties, 281
Mode, definition of, 36
Modern concession agreement, 301
Modified Accelerated Cost Recovery System (MACRS), 143-144, 167
Money-forward cost, 247
Monte Carlo simulation, 47, 54, 122-123, 228
- data development for, 231-236
- for drilling decisions, 242
- in trend analysis, 224-225, 226

Moratoriums, leasing, 353
Mud, environmental disturbance by, 349
Multinational oil companies, 297-305, 307-318
Multiple horizons
- prospects in, 77, 78, 79
- uncertainty in, 46-47

Multiple prospects, uncertainty in, 50-54
Multiple trend analysis, 227-229
Negative cash flow, 108-109
Negligence, conversion caused by, 283-284
Net cash flow
- estimated future, 122
- present value profile of, 121
- risk weighting of, 112

Net operating costs, 141-142
Net pay thickness
- cumulative frequency of, 42-43
- of prospect, 67

Net present value (NPV), 9, 103, 148
- risk weighted, 118

Net production, determination of, 139
Net profit per barrel, 198
Net profit interest, 156
Net revenue interest (NRI), 12, 139
Nominal dollars, software applications to, 167
Nomographic solution, of risk, 247-249
Nonliability, 280
Nonownership theory, 280
Nonparticipating royalty, 281
Nonproducing reserves, definition of, 161
Normal distribution, 36-37, 60
Notice before forfeiture clause, 289
Offshore lower 48
- environmental impact on, 348-350
- expenditures in, 14
- limited access to, 352

Oil concession, evolution of, 298-302
Oil and gas, as legal terms, 279-280
Oil and gas contracts, international, 297-305, 307-318
Oil and gas deals, 154-156
Oil and Gas Finance Sourcebook, 264
Oil and gas industry, *see* Industry
Oil and gas law
- international, 297-305, 307-318
- in the United States, 279-296

Oil and gas mineral estate, 280-282, *see also* Mineral estate
Oil and gas property, *see* Property evaluation
Oil and gas ventures, *see* Ventures
Oil prices, 140
Oil prospects, 275
Oil spills, 346, 349
Oil volume calculation, 129
Oil well blowout, 346, 349, 350
Onshore lower 48
- environmental impact on, 347-348
- estimates versus actuals, 31, 34
- expenditures in, 14
- limited access to, 351-352

OPEC, 188-189, 300-301
Operating costs, 141-142
Operational plans, of a company, 193-196
Operations cash flow, 142
Operations clause, 288-289, 294
Opportunity, in oil and gas industry, 189-192
Opportunity cost of capital, 148
Option pricing, 49
Or clause, 288
Organization, of a company, 192, 202
Organizational development, 207
Organizational risk aversion, 98-100, 102
Original oil in place (OOIP), estimates of, 34-35
Outer Continental Shelf (OCS) Lands Act, 349
Outer Continental Shelf leasing moratoriums, 353
Overhead, 316
- in exploration expenditure, 18

Override, 156
Overriding royalty, 156, 281
Overseas operations, 322, 324-325
Overseas travel, 325-327
Ownership
- of databases, 176
- divided, 284-285
- of ideas and skills, 336-337
- in place, 280

P/Z versus cumulative production curve, 138
Payout, 11-12
- calculation of, 120, 151-152
- period of, 108

PEMEX, 300
People
- creativity of, 205-213
- curiosity and commitment in, 210-211
- development of, 211-212
- ethics of, 331-344
- role in exploration success, 201-203

Percentage depletion, 145
Percentiles, 10th and 90th, 45, 66
Performance evaluation, 133-138
Performance milestones, of a company, 196-198
Personal computer software, *see* Computer software; Software
Personnel
- *see also* People
- of a company, 192, 201-203
- and overseas travel, 325-327
- requirements in international exploration, 322

Petroleum exploration, *see* Business; Exploration
Philosophy, of deal making, 269
Piecework digitizing, in databases, 181-182
Plan, from a vision, 209
Play chance, definition of, 93
Play chance of adequacy, 89
Plays
- *see also* Trend
- assessment methods for, 88-89
- definition of, 87, 217
- geological assessment procedures, 89-93
- mapping, 88
- probability of success of, 81
- selecting, 88

Plotting
- cumulative frequency, 43
- cumulative probability, 37-39
- of decline curves, 134
- field size distributions, 63-65
- of lognormal data, 37-39
- of profitability data, 310-312, 314, 315
- prospect sizes, 64-69

Political risks, in international ventures, 315, 320-321
Political uncertainty, 80
Pooling clause, 289, 293
Porosities, as lognormal data, 39
Portfolio
- analysis, 50-54, 228
- compositing of, 51

Possible, reserves, definition of, 161
Postdrilling analysis, 114-115
Postplay analysis, 114-115
Postprospect analysis, 114-115
Post-World War I, oil exploration, 298-300
Post-World War II, oil exploration, 300-301
Power to lease, 281
Predrilling analysis, reality checks in, 113-114
Preference theory, 96
Present value (PV), 9
- and cash flow analysis, 108-111
- definition of, 109, 110
- investment yardsticks from, 111-113
- profile of, 110-111

Present worth, 9
- definition of, 109

Preservation, 76
President's Outer Continental Shelf Leasing and Developmental Task Force, 349
Pressure decline method, 137
Pre-World War I, oil exploration, 298
Price, projection of, 235
Price caps, 316
Prices
- of crude oil, 139
- history of, 299-300
- posted, 317
- software calculation of, 166

Primary term, 286
Probability
- of all dry holes, 100
- of commercial success, 80-81
- of goal versus risk, 227
- short-run economic, 225-226
- of uncertainty, 29-60

Probability density
- definition of, 37
- functions of, 55
- graphs of, 36, 37

Probability of discovery
- estimates in, 81-83
- historical trends in, 81
- and prospect inventories, 83-84

Probability distributions, in trend analysis, 220
Probability of failure, 77, 227, 229
Probability of success, 77, 231-232, 241, 246
- geological chance factors and, 73-79
- in risk decisions, 98
- in trend analysis, 220-221

Probable, reserves, definition of, 160
Produce, legal definition of, 286
Producing reserves, definition of, 161
Production
- acquisition funding, 260
- data, 219
- decline curve method, 133-136
- determination of, 126-139
- payments deal, 156
- sharing contracts, 302-303, 307, 308, 311

Production schedules, software development of, 165-166
Productive area, cumulative frequency of, 43
Productivity, and ethics, 333
Products, defining, 21
Profession, definition of, 334
Professionalism, and ethics, 334
Profit
- definition of, 156
- equation of, 4, 7

Profitability
- data, 310-312, 313, 314
- field models of, 309-310
- measures of, 147-153

Profit to investment ratio, 152
Profit oil, definition of, 308
Profit oil share, versus amortization rate, 312
Profit to risk concept, 241
Profits, and ethics, 332-333
Profit split, 308, 309
Program forecasting, 83-84
Programs, *see* Software
Project portfolio, analysis, 50-54
Property
- acquiring, 125, 157-158
- divided ownership of, 285
- intellectual, 336-337

Property evaluation, 126
Proportionate reduction clause, 289
Prospect acquisition funding, 257-258
Prospect generators, 271-276, 321
Prospect inventories, probability of discovery and, 83-84
Prospects
- area, 67
- assessment of, 87-93
- and drilling budget, 238-240
- essential of, 272-273
- estimating sizes of, 63-69
- ethics in preparing, 335-340
- evaluating criteria of, 272
- expected value of, 72-73
- and field size, 64
- improving, 249
- inventory systems, 85
- multiple, 50-54
- net pay thickness, 67
- presentation of, 271-276
- questions about, 274-275
- ranking and selection of, 249-250
- recovery, 67
- rejections of, 275-276
- single horizon uncertainties, 42-46
- size distribution of, 90-93
- size estimates of, 114

Prospect uncertainties, versus basin variance, 40-42
Proved reserves, definition of, 160
Pseudoinvestment, 151
Pure service contract, 301, 303-304
PV, *see* Present value
Quality, accumulation, 71
Range approach, 168
Rapport
- with host country, 321
- with World Bank, 324

Rate of return (ROR), 11
- analysis of, 149-151
- definition of, 149
- graphical solution of, 151

Real dollars, software applications to, 167
Reality checks, in predrilling analysis, 113-114
Recoverable condensate, equation of, 131
Recoverable gas, equation for, 129, 132
Recoverable oil, equation for, 129, 131
Recovery, cumulative frequency of, 44
Recovery factors, calculation of, 131-133
Reference frame, in risk decisions, 97
Regression principles, 32
Regulation
 of oil and gas by state, 291
 protection of environment, 350-352
Reports
 improvement of, 271-276
 in oil and gas evaluation, 157
Research costs, justification of, 202
Reserves
 bias in estimates of, 32-33
 calculation of, 126-139, 222, 225
 components of, 44
 decline curve estimates of, 133-136
 definition of, 160
 equation of, 44
 estimates of, 32-35, 57-58, 114
 estimating, 219
 lognormal distribution of, 239
 risked versus unrisked, 226
 scheduling of, 138-139
 and size of prospect, 63-69
 types of, 24
Reservoir bulk volume, determination of, 129
Reservoir rock, as geological chance factor, 76
Reservoirs, drive mechanisms in, 131-133
Revenue, 140
Rights, in oil and gas law, 280-281
Risk
 definition of, 95
 in exploration versus development, 99
 graphical limits of, 101
 in multitrend analysis, 229
 nomographic solution of, 247-249
 in property evaluation, 154
 psychological reaction to, 96
 sharing the, 249
 sources of, 239
 versus achieving goal, 227
Risk-adjusted value (RAV), 246
 equation of, 101
 and fair market value, 103
Risk analysis
 methods of, 240-249
 and Monte Carlo simulation, 122-123
Risk assessment, example of, 29-30
Risk aversion
 cost of unwarranted, 99-100
 definition of, 95
 expected value (EV) and, 95-96
 financial, 98-100
 industry practices to deal with, 98-100
 organization, 98-102
 ubiquity of, 96
Risk behavior, in exploration, 95-103
Risk decisions
 biases affecting, 97-98
 consistency of, 102
 ways to improve, 100-102
Risked value, 241
Risk profiles, 229
Risk service contract, 301, 303, 304
Risk-weighted value, 241
Risk weighting, 72-73, 239
 of net cash flow, 112
Royalty
 definition of, 282
 overriding, 156, 281
 types of, 281-282
 versus working interest, 312-313
Royalty clause, 287, 292-293
Royalty interest, 139, 281
Rule of 72, 109
Rule of capture, 280
Rules of thumb, for investment, 11-12
Sales, 263-265
Salesmanship, 263-265
Salvage value, 142-143
Sample average, definition of, 38
Samples, data from, 180-181
Santa Barbara oil well blowout, 346, 349, 350
Scatter, acceptability of, 34
Scientific method, 57-58
Sealed trap, as geological chance factor, 76
Sealing capability, 76
Secondary term, 286
SEG Code of Ethics, 344
Seismic operations, overseas, 328-329
Seismic surveying, disturbance by, 347-348
Seller, definition of, 257
Sensitivity analysis, 242, 310, 311, 317
Service contracts, 303-304
Severance tax, 141, 166
Sharing contracts, 301, 302
Shrinkage factor, 130
Shut-in royalties, 281-282
Shut-in royalty clause, 289
Silent Spring, 346
Size distribution, *see* Distribution
Sizes
 of fields, 63-69
 of prospects, 63-69
 ranges of, 64-69
Slander of title, 284
Sliding scale terms, 316
Society, and ethics, 333-334
Software
 for cost calculations, 166
 design of, 164
 for economic evaluation, 163-173
 features of, 165
 new technology in, 168-169
 for tax calculations, 166
 in trend analysis, 217
 venders of, 169-183
Solution gas, 131
Solution gas drive, 132
Specific gravity
 of crude oil, 140
 of well stream gas, 130
SPE Guide for Professional Conduct, 343-344
Spinner wheel, 30
Stand alone system, 164
Standard deviation, 37
Standards of Practice, AAPL, 342-343
State income tax, 143
State regulation, of oil and gas, 291
Statistics
 in trend analysis, 220
 of uncertainty, 29-60
Strategy, of a company, 192-196, 202
Stratigraphic names, 182
Subsea depth, 180
Subsurface trespass, 283
Success
 chance of, 71-85, 240
 as defined by CSD, 72
 definition of, 7, 75
 equation of, 4
 of hydrocarbon accumulation, 73-79
 in managing exploration program, 187-198
 measures of, 10-12
 probability of, 73-79, 77
 rates versus estimates, 74-76
 systems of estimating, 76-77
 types of, 71, 74
Success ratio, 221, 222
 definition of, 89
 from prospect grading, 90
Surface disturbance, by exploration, 347
Surface rights, 279
Swanson's rule
 accuracy of, 45
 definition of, 44
"Taking," definition of, 348
Tangible costs, 142
Taxable income, 146
Taxes
 on income, 143, 146
 in international ventures, 313-315
 investment credit, 146-147
 payable, 147
 in property evaluation, 141
 software calculation of, 166
Tax Reform Act of 1986 (TRA 86), 143-144, 145, 146
Tax royalty contract, 307, 308, 312-313
Teamwork, 208
Technical efficiency, measures of, 253
Technology, transfer of, 304-305
Tenancy in common, 285
Term clause, 286-287, 292
Theme, of Handbook, 3-4
Three offsets rule, 12
Three-well portfolio, 52
Time value of money, 9, 147-148
Tool types, 182
Total net cash flow, 108
Tradeoffs, of contract terms, 312
Training, and ethics, 336
Transfer of technology, 304-305
Trap, 76
Travel, overseas, 325-327
Trend
 definition of, 217
 information database center for, 217-220
 simulation output of, 225
Trend analysis
 approximation methods in, 226-227
 computer software in, 217
 definition of, 215
 Monte Carlo simulation in, 224-225
 multiple, 227-229
 probability of success in, 220-221
 purpose of, 216-217
Trespass
 definition of, 282
 types of, 282-284
Triangular distributions, 231-232
Truncation point, 52
Truncations
 derivation of, 59-60
 double, 48
 of lognormal distributions, 48
t statistics, 31
Uncertainty
 calculation of, 56
 economic, 80
 in exploration, 29-60
 in multiple horizons, 46-47
 political, 80
 in property evaluation, 154
 in prospects, 40-42
 real examples of, 30-35
 in single horizon, 42-46
 sources of, 239
Undeveloped reserves, definition of, 161
Undiscovered oil, 354
Undivided interest, 285
Unethical explorers, 337
United States
 environmental effects of exploration in, 345-354
 history of exploration expenditures in, 13-19
 oil and gas industry in, 188-190
 oil and gas law in, 279-296
 tax on international ventures, 313-315
Unless clause, 288
Unproved reserves, definition of, 160
Unrisked mean, 93
Unrisked present worth, 241
Utility theory, 96, 246
Values
 in creative environments, 209, 211
 management of, 212
Variables, probabilistic, 231-236
Variance
 basin, 40
 definition of, 36
 derivation of, 58
 of truncated lognormals, 52-53
 versus errors, 42
Ventures
 analysis of international, 318
 costs of international, 315
 funding of, 257-265
 joint, 301-302
 overseas, 323
 U.S. taxation of, 313-315
Viewpoints, in creative environments, 209
Vision
 creating a, 206, 208
 management of, 212
 turning into a plan, 209
Volume of hydrocarbons, 4
Volumetric calculations, 129-131, 239
Warranty clause, 289, 295
Water drive, 131, 132
Water injection project, 123
Weaknesses, of a company, 191
Well classification, 179
Well depth, 178-179
Well head deliverability calculations, 138-139
Well head performance, multipoint test of, 139
Well history
 conditioning of data of, 218-219
 data for, 175-183, 218
 sequence of events in, 180
Well indentification, 176-177
Well location, 177-178, 179, 180
Well logs, and database development, 181-183
Well status, 179
West Wind basin play, 109
Wholesale prospect sales, 258
Wildcat wells, success of, 72
Wilderness
 disturbance of, 348
 limited access to, 351-352
Wilderness Act, 351-352
Wildlife habitat, disturbance of, 348
Williston basin, discoveries in, 40, 41
Windfall profit tax, 141
Window of opportunity, 189
Withdrawals, of lands from leasing, 351-352
Workforce, 206, 207
Working environment
 creativity in, 208-210
 individual contributors to, 206
Working interest, 139
 royalty versus, 312-313
World Bank, rapport with, 324
$(X - x)$ algebra, use of in exploration, 24
Yardsticks, *see* Investment yardsticks
YPF, 300, 303